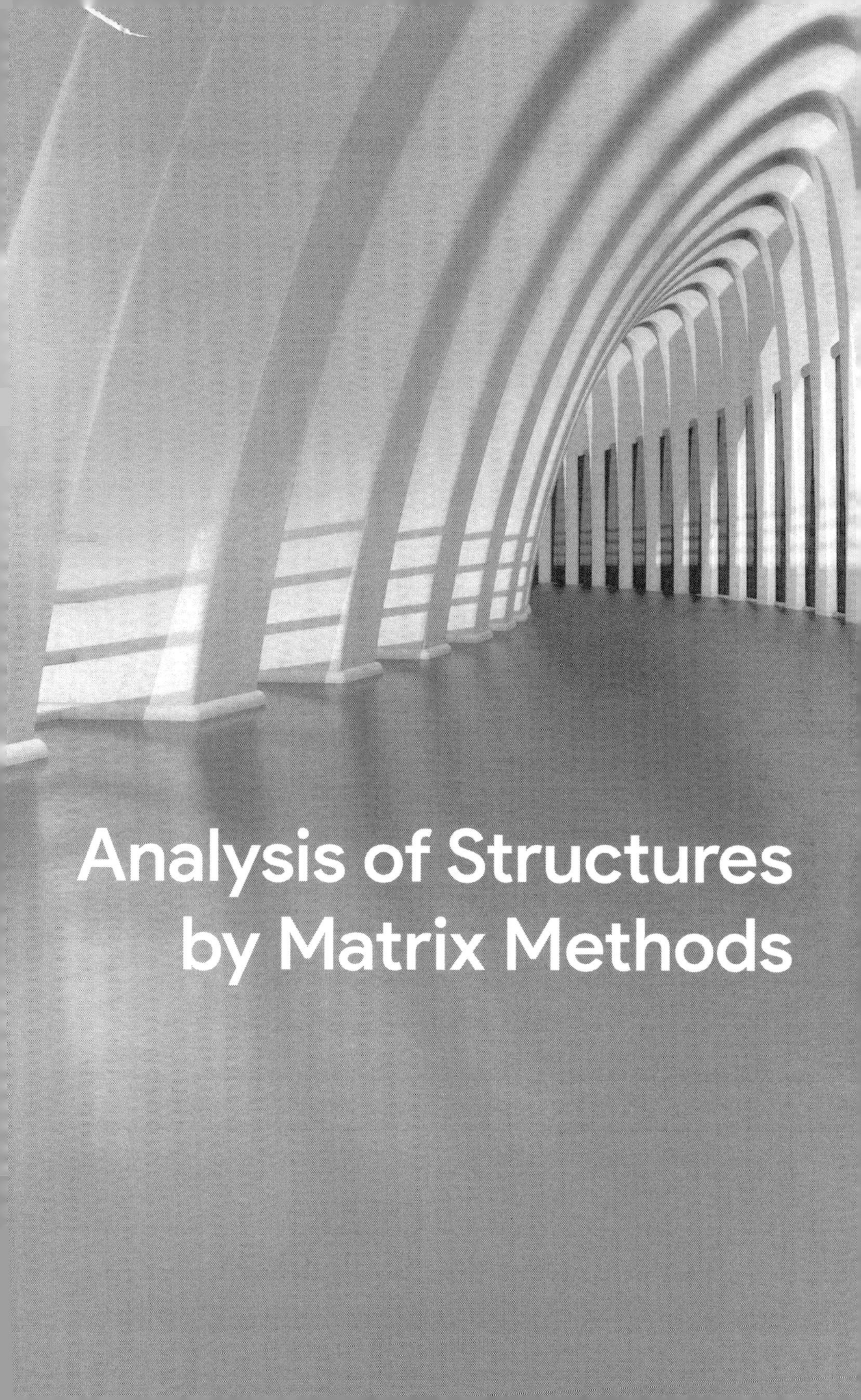

Analysis of Structures by Matrix Methods

Analysis of Structures by Matrix Methods

Fathi Al-Shawi

JENNY STANFORD
PUBLISHING

Published by

Jenny Stanford Publishing Pte. Ltd.
101 Thomson Road
#06-01, United Square
Singapore 307591

Email: editorial@jennystanford.com
Web: www.jennystanford.com

British Library Cataloguing-in-Publication Data
A catalogue record for this book is available from the British Library.

Analysis of Structures by Matrix Methods

ISBN 978-981-4968-19-5 (Hardcover)
ISBN 978-1-003-29130-5 (eBook)

Contents

Preface

In the past, the analysis of engineering structures has always been a challenge to engineers who used classical methods to quantify the response of a structure to the applied forces. These methods are suitable for the analysis of relatively simple structures that can be solved by hand calculations. When the structure gets more complicated, it is simplified to a model that can be solved by classical methods. The results obtained for the modified structure are approximations and their accuracy depends on what modifications were made to the original structure. The analysts' experience and judgment play an important role in the way the structure is simplified in order to get the best possible results. The absence of reasonably accurate methods for the analysis of large structures, which cannot be easily modified, limited the scope for engineers to invent complex structural forms.

In 1914, George Maney derived the slope-deflection equations for continuous beams. When these equations are applied at the various joints of the structure, a set of simultaneous equations with unknown displacements are obtained. The resulting set of simultaneous equations is solved for the unknown displacements and the results are further used to calculate the bending moments in members of the structure. For a relatively small structure where the number of unknown displacements is small, the set of equations can be solved by hand calculations. But for any structure of moderate size, the number of simultaneous equations is such that it is not practical to solve them by hand calculations. Hardy Cross in 1932 overcame this problem by devising a procedure for analysing continuous beams and rigidly jointed frames by what is called the moment distribution method. This is an iterative procedure where the joints of the structure are clamped and released alternately in cycles of calculations. The number of cycles in this iterative process depends on the desired degree of accuracy, i.e., the iteration is stopped when the difference in the calculated results between two successive cycles is within a set of prescribed small numbers. In essence, Hardy Cross was indirectly relaxing the slope-deflection equations one at a time rather than considering the full set of equations that would have resulted from the application of the slope-deflection equations to all the joints of the structure. It is a powerful method that was very popular in the past, but with the advent of electronic computers in the 1950s, engineers started developing systematic procedures for the analysis of structures.

When a given structure is subjected to loads its behaviour can be represented by a set of simultaneous equations, which are solved to give the response of the structure. For structures where the number of equations is large, hand calculations are not suitable, and a computer is used to obtain the required solution to the simultaneous equations.

The detailed work with simultaneous equations can be made in a general and compact form by using matrix notation leading to the development of the matrix methods of structural analysis.

There are two matrix methods that can be used: the flexibility method which was employed in the past but not commonly used at present and the stiffness method which is widely used and is followed in this book. It is worth mentioning that the stiffness method is regarded as the forerunner to and forms the basis of the finite element method of structural analysis.

The first chapter gives an introduction to matrix algebra, which explains the various operations of matrices. This is intended to help the reader gain an understanding of the basic principles and applications of matrix operations. Chapter 2 starts with setting out the general procedure of matrix formulation by considering the simple case of a bar to highlight the steps followed in the analysis. Some general notations and the treatment of other forms of structural members, by analogy with the bar problem, are also explained in this chapter.

Chapters 3 to 10 present the treatment of the linear static analysis of the various types of commonly used structures. Nonlinear analysis and dynamics of structures are dealt with in Chapters 11 and 12, respectively. A bibliography given at the end of the book provides a list of publications that readers can refer to, especially for the proofs of some of the statements made in the text.

Fathi Al-Shawi

Autumn 2022

List of Symbols

Symbols are defined appropriately where they occur in the text and the list below shows the general definition of the main symbols used. Symbols defining quantities relative to the local axis of the member will have a bar. For example, u is the displacement in the direction of the global x-axis while $\bar{u}$ is the displacement in the local $\bar{x}$-axis of the member.

A	Area, Amplitude
a	Acceleration
α, β	Constants
b	Width of cross section
E	Modulus of elasticity
E_S	Strain energy
E_P	Potential energy
e	Change in length
ε	Strain
δ	Displacement vector
F	Force vector
f	Subscript for the actions on the member due to the external forces
φ	Angle of rotation of the member
G	Modulus of rigidity
γ	Load factor
h	Height of cross section
I	Second moment of area
i, j	Subscripts for the two ends of member (element)
J	Polar second moment of area
K	Structure stiffness matrix
k	Member (element) stiffness matrix
L	Length
λ	Eigenvalue
M	Moment about the y-axis
m	Mass
N	Moment about the z-axis

n	Uniformly distributed load
P	Axial force
P_c	Critical (buckling) force
R	Reaction, radius of curvature
ρ	Density
S	Shear force in the y direction
s	Subscript for loads on joints of structure
σ	Stress
T	Moment about the x-axis
t	Time
u, v, w	Translational displacements in the x, y, and z directions
$\dot{u}$	Velocity in the x direction
$\ddot{u}$	Acceleration in the x direction
U	Work
V	Shear force in the z direction
W	Concentrated load
$\dot{w}$	Velocity in the z direction
$\ddot{w}$	Acceleration in the z direction
ω	Natural circular frequency of vibration
X, Y, Z	Forces in the x, y, and z directions
x, y, z	Global cartesian coordinates
Φ, θ, Ψ	Rotational displacements about the x, y, and z axes

Chapter 1

Introduction to Matrix Algebra

Throughout this book it will be seen that in the analysis of structural problems, sets of simultaneous equations will result. These sets are written in matrix form so that the computations are systematic and more manageable. In this chapter, the important aspects of matrix manipulations are presented for the benefit of the reader who has limited knowledge of matrix algebra. Although, there are computer programmes that deal with the various matrix operations, it is considered useful to learn the steps followed in these computations.

1.1 Matrix Operations

Consider the set of simultaneous equations

$$4x_1 - 5x_2 + 9x_3 = 8$$

$$3x_1 - 6x_2 - 4x_3 = -5$$

$$-8x_1 + 7x_2 + 2x_3 = 4$$

These equations can be written in matrix form as:

$$\begin{bmatrix} 4 & -5 & 9 \\ 3 & -6 & -4 \\ -8 & 7 & 2 \end{bmatrix} \begin{bmatrix} x_1 \\ x_2 \\ x_3 \end{bmatrix} = \begin{bmatrix} 8 \\ -5 \\ 4 \end{bmatrix} \quad \text{or} \quad Ax = b$$

Analysis of Structures by Matrix Methods
Fathi Al-Shawi

ISBN 978-981-4968-19-5 (Hardcover), 978-1-003-29130-5 (eBook)
www.jennystanford.com

where the matrices, $A = \begin{bmatrix} 4 & -5 & 9 \\ 3 & -6 & -4 \\ -8 & 7 & 2 \end{bmatrix}$, $x = \begin{bmatrix} 8 \\ -5 \\ 4 \end{bmatrix}$, and $b = \begin{bmatrix} 8 \\ -5 \\ 4 \end{bmatrix}$

Matrix A may be written in a general form as:

$$A = \begin{bmatrix} a_{11} & a_{12} & . & . & a_{1n} \\ a_{21} & a_{22} & . & . & a_{2n} \\ . & . & . & . & . \\ . & . & . & . & . \\ a_{m1} & a_{m2} & . & . & a_{mn} \end{bmatrix}$$

This is called an m×n matrix where m is the number of rows and n is the number of columns with coefficients a_{ij} where i is the number of the row and j is the number of the column at which a_{ij} occurs. A square matrix is when m = n, i.e. an n×n matrix.

A coefficient on the main diagonal of a matrix is defined by the a_{ii}.

A diagonal matrix is when there are coefficients only on the main diagonal and all other coefficients are zero, for example

$$\begin{bmatrix} a_{11} & 0 & 0 & 0 \\ 0 & a_{22} & 0 & 0 \\ 0 & 0 & a_{33} & 0 \\ 0 & 0 & 0 & a_{44} \end{bmatrix}$$

The unit matrix, I, is a diagonal matrix with coefficients on the main diagonal equal to 1, i.e. $a_{ii} = 1$, for example

$$I = \begin{bmatrix} 1 & 0 & 0 & 0 \\ 0 & 1 & 0 & 0 \\ 0 & 0 & 1 & 0 \\ 0 & 0 & 0 & 1 \end{bmatrix}$$

The zero (or null) matrix denoted by O is where all coefficients are equal to zero, for example

$$O = \begin{bmatrix} 0 & 0 & 0 & 0 \\ 0 & 0 & 0 & 0 \\ 0 & 0 & 0 & 0 \\ 0 & 0 & 0 & 0 \end{bmatrix}$$

Symmetric matrix is a matrix where $a_{ij} = a_{ji}$, for example

$$\begin{bmatrix} 2 & 5 & 0 & -1 \\ 5 & 4 & -7 & 3 \\ 0 & -7 & 9 & 0 \\ -1 & 3 & 0 & 6 \end{bmatrix}$$

An upper triangular matrix where there are coefficients along and above the main diagonal and the rest of the coefficients are zero, usually given the symbol U, for example

$$U = \begin{bmatrix} 4 & -1 & 0 & 8 \\ 0 & 5 & 7 & -3 \\ 0 & 0 & 6 & 2 \\ 0 & 0 & 0 & 9 \end{bmatrix}$$

A lower triangular matrix where there are coefficients along and below the main diagonal and the rest of the coefficients are zero, usually given the symbol L, for example

$$L = \begin{bmatrix} 7 & 0 & 0 & 0 \\ 2 & 3 & 0 & 0 \\ -9 & 0 & 5 & 0 \\ 8 & 1 & -4 & 6 \end{bmatrix}$$

Row vector is a matrix with only one row, $a = [a_1 \quad a_2 \quad . \quad . \quad a_n]$

Column vector is a matrix with only one column, $b = \begin{bmatrix} b_1 \\ b_2 \\ . \\ . \\ b_m \end{bmatrix}$

A zero vector O is where all coefficients are equal to zero, for example

$$O = \begin{bmatrix} 0 \\ 0 \\ 0 \\ 0 \\ 0 \end{bmatrix}$$

The transpose of matrix is where the coefficient in the pth row will become the coefficients in the pth column.

If matrix B is the transpose of matrix A written as $B = A^T$, then $b_{ij} = a_{ji}$.

Example 1.1

Find the transpose of the following matrices

$$\text{(i)}\quad A=\begin{bmatrix} 3 & 5 & -2 \\ 4 & -6 & 1 \\ -1 & 7 & -4 \end{bmatrix},\quad A^T=\begin{bmatrix} 3 & 4 & -1 \\ 5 & -6 & 7 \\ -2 & 1 & -4 \end{bmatrix}$$

$$\text{(ii)}\quad A=\begin{bmatrix} 5 \\ -2 \\ 8 \end{bmatrix},\quad A^T=\begin{bmatrix} 5 & -2 & 8 \end{bmatrix}$$

$$\text{(iii)}\quad A=\begin{bmatrix} -4 & 7 & 6 \end{bmatrix},\quad A^T=\begin{bmatrix} -4 \\ 7 \\ 6 \end{bmatrix}$$

$$\text{(iv)}\quad A=\begin{bmatrix} 5 & 7 & -4 \\ 6 & -3 & 2 \end{bmatrix},\quad A^T=\begin{bmatrix} 5 & 6 \\ 7 & -3 \\ -4 & 2 \end{bmatrix}$$

Matrix addition

The sum of two matrices A and B is matrix C, that is C = A + B, then the coefficients in matrix C are obtained by adding the coefficients in matrix B to the corresponding coefficients in matrix A. Thus $c_{ij} = a_{ij} + b_{ij}$.

Example 1.2

Given:

$$A=\begin{bmatrix} 5 & 2 \\ -4 & 3 \end{bmatrix},\quad B=\begin{bmatrix} 6 & -3 \\ 2 & 1 \end{bmatrix}$$

$$C=A+B=\begin{bmatrix} 5 & 2 \\ -4 & 3 \end{bmatrix}+\begin{bmatrix} 6 & -3 \\ 2 & 1 \end{bmatrix}=\begin{bmatrix} 5+6 & 2+(-3) \\ -4+2 & 3+1 \end{bmatrix}=\begin{bmatrix} 11 & -1 \\ -2 & 4 \end{bmatrix}$$

Matrix subtraction

Given matrices A and B then the difference between them C is C = A – B where the coefficients in matrix C are obtained by subtracting the coefficients in matrix B from the corresponding coefficients in matrix A. Thus $c_{ij} = a_{ij} - b_{ij}$.

Example 3

Given: $$A=\begin{bmatrix}7 & 4\\8 & -6\end{bmatrix},\qquad B=\begin{bmatrix}5 & -9\\2 & 3\end{bmatrix}$$

$$C=A-B=\begin{bmatrix}7 & 4\\8 & -6\end{bmatrix}-\begin{bmatrix}5 & -9\\2 & 3\end{bmatrix}=\begin{bmatrix}7-5 & 4-(-9)\\8-2 & -6-3\end{bmatrix}=\begin{bmatrix}2 & 13\\6 & -9\end{bmatrix}$$

Matrix multiplication

Given the m×n matrix $A = [a_{ij}]$ and the n×r matrix $B = [b_{ij}]$ the product $C = AB = [c_{ij}]$ where, c_{ij} is given by:

$$c_{ij}=\sum_{q=1}^{q=n} a_{iq}b_{qj}=a_{i1}b_{1j}+a_{i2}b_{2j}+a_{i3}b_{3j}+\ldots\ldots\ldots\ldots a_{in}b_{nj}$$

For the matrix product to be defined, the number of rows in matrix B must be equal to the number of columns in matrix A. So, if A is m×n matrix and B is n×r matrix then the resulting product C = AB is m×r matrix with m rows and r columns.

Note that in general AB ≠ BA except in special cases, for example when $B = A^{-1}$, i.e. $AA^{-1} = A^{-1}A = I$ (the unit matrix).

Example 4

Given: $$A=\begin{bmatrix}3 & 5 & -2\\4 & -6 & 1\\-1 & 7 & -4\end{bmatrix}\text{ and }B=\begin{bmatrix}8 & -2\\-3 & -7\\2 & 6\end{bmatrix}$$

Calculate the product AB.

Let the product $C = AB = [c_{ij}]$

$$c_{11} = 8\times3 + (-3)\times5 + 2\times(-2) = 5$$
$$c_{21} = 8\times4 + (-3)\times(-6) + 2\times1 = 52$$
$$c_{31} = 8\times(-1) + (-3)\times7 + 2\times(-4) = -37$$
$$c_{12} = (-2)\times3 + (-7)\times5 + 6\times(-2) = -53$$
$$c_{22} = (-2)\times4 + (-7)\times(-6) + 6\times1 = 40$$
$$c_{32} = (-2)\times(-1) + (-7)\times7 + 6\times(-4) = -71$$

$$\text{Therefore,}\quad \begin{bmatrix} 3 & 5 & -2 \\ 4 & -6 & 1 \\ -1 & 7 & -4 \end{bmatrix}\begin{bmatrix} 8 & -2 \\ -3 & -7 \\ 2 & 6 \end{bmatrix} = \begin{bmatrix} 5 & -53 \\ 52 & 40 \\ -37 & -71 \end{bmatrix}$$

Example 5

$$\text{Given: } A = \begin{bmatrix} 2 & -4 & 3 \\ 6 & 1 & -2 \\ -5 & 7 & 4 \end{bmatrix},\quad B\begin{bmatrix} 6 & -2 & 3 \\ -2 & 5 & -4 \\ 3 & -4 & 8 \end{bmatrix},\quad C = \begin{bmatrix} -3 & -1 & 4 \\ 2 & 6 & 8 \\ 7 & 3 & -5 \end{bmatrix}$$

Calculate the product ABC.
First calculate the product BC by premultiplying C by B

$$BC = \begin{bmatrix} 6 & -2 & 3 \\ -2 & 5 & -4 \\ 3 & -4 & 8 \end{bmatrix}\begin{bmatrix} -3 & -1 & 4 \\ 2 & 6 & 8 \\ 7 & 3 & -5 \end{bmatrix}$$

$$= \begin{bmatrix} (-3)\times 6 + 2\times(-2) + 7\times 3 & (-1)\times 6 + 6\times(-2) + 3\times 3 & 4\times 6 + 8\times(-2) + (-5)\times 3 \\ (-3)\times(-2) + 2\times 5 + 7\times(-4) & (-1)\times(-2) + 6\times 5 + 3\times(-4) & 4\times(-2) + 8\times 5 + (-5)\times(-4) \\ (-3)\times 3 + 2\times(-4) + 7\times 8 & (-1)\times 3 + 6\times(-4) + 3\times 8 & 4\times 3 + 8\times(-4) + (-5)\times 8) \end{bmatrix}$$

$$= \begin{bmatrix} -1 & -9 & -7 \\ -12 & 20 & 52 \\ 39 & -3 & -60 \end{bmatrix}$$

Now calculate A(BC) by premultiplying BC by A

$$A(BC) = \begin{bmatrix} 2 & -4 & 3 \\ 6 & 1 & -2 \\ -5 & 7 & 4 \end{bmatrix}\begin{bmatrix} -1 & -9 & -7 \\ -12 & 20 & 52 \\ 39 & -3 & -60 \end{bmatrix} = \begin{bmatrix} 163 & -107 & -402 \\ -96 & -28 & 130 \\ 77 & 173 & 159 \end{bmatrix}$$

$$= \begin{bmatrix} (-1)\times 2 + (-12)\times(-4) + 39\times 3 & (-9)\times 2 + 20\times(-4) + (-3)\times 3 & (-7)\times 2 + 52\times(-4) + (-60)\times 3 \\ (-1)\times 6 + (-12)\times 1 + 39\times(-2) & (-9)\times 6 + 20\times 1 + (-3)\times(-2) & (-7)\times 6 + 52\times 1 + (-60)\times(-2) \\ (-1)\times(-5) + (-12)\times 7 + 39\times 4 & (-9)\times(-5) + 20\times 7 + (-3)\times 4 & (-7)\times(-5) + 52\times 7 + (-60)\times 4 \end{bmatrix}$$

$$= \begin{bmatrix} 163 & -107 & -402 \\ -96 & -28 & 130 \\ 77 & 173 & 159 \end{bmatrix}$$

Alternatively, the product AB can be found first and this is post-multiplied by C to get the product ABC.

Product of a row vector times a column vector:

Example 6

Find the product of $\begin{bmatrix} 4 & 7 & -6 \end{bmatrix}$ and $\begin{bmatrix} 2 \\ -5 \\ 8 \end{bmatrix}$.

$$\begin{bmatrix} 4 & 7 & -6 \end{bmatrix} \begin{bmatrix} 2 \\ -5 \\ 8 \end{bmatrix} = \begin{bmatrix} 2 \times 4 + (-5) \times 7 + 8 \times (-6) \end{bmatrix} = \begin{bmatrix} -75 \end{bmatrix}$$

Product of a column vector by a row vector:

Example 7

Find the product of $\begin{bmatrix} 2 \\ -5 \\ 8 \end{bmatrix}$ and $\begin{bmatrix} 4 & 7 & -6 \end{bmatrix}$.

$$\begin{bmatrix} 2 \\ -5 \\ 8 \end{bmatrix} \begin{bmatrix} 4 & 7 & -6 \end{bmatrix} = \begin{bmatrix} 4 \times 2 & 7 \times 2 & -6 \times 2 \\ 4 \times (-5) & 7 \times (-5) & -6 \times (-5) \\ 4 \times 8 & 7 \times 8 & -6 \times 8 \end{bmatrix} = \begin{bmatrix} 8 & 14 & -12 \\ -20 & -35 & 30 \\ 32 & 56 & -48 \end{bmatrix}$$

Multiplication by a scalar:

When a matrix is multiplied by a scalar all the coefficients of the matrix are multiplied by that scalar.

Example 8

$$8 \begin{bmatrix} 3 & 5 & -2 \\ 4 & -6 & 1 \\ -1 & 7 & -4 \end{bmatrix} = \begin{bmatrix} 8 \times 3 & 8 \times 5 & 8 \times (-2) \\ 8 \times 4 & 8 \times (-6) & 8 \times 1 \\ 8 \times (-1) & 8 \times 7 & 8 \times (-4) \end{bmatrix} = \begin{bmatrix} 24 & 40 & -16 \\ 32 & -48 & 8 \\ -8 & 56 & -32 \end{bmatrix}$$

The transpose of the product of two matrices A and B is given by $(AB)^T = B^T A^T$.

Example 9

$$A=\begin{bmatrix}3 & -2\\ 1 & -4\end{bmatrix} \text{ and } B=\begin{bmatrix}-1 & 4\\ 2 & -3\end{bmatrix}.$$

$$AB=\begin{bmatrix}3 & -2\\ 1 & -4\end{bmatrix}\begin{bmatrix}-1 & 4\\ 2 & -3\end{bmatrix}=\begin{bmatrix}-7 & 18\\ -9 & 16\end{bmatrix},\ (AB)^T=\begin{bmatrix}-7 & 18\\ -9 & 16\end{bmatrix}^T=\begin{bmatrix}-7 & -9\\ 18 & 16\end{bmatrix}$$

$$A^T=\begin{bmatrix}3 & -2\\ 1 & -4\end{bmatrix}^T=\begin{bmatrix}3 & 1\\ -2 & -4\end{bmatrix},\ B^T=\begin{bmatrix}-1 & 4\\ 2 & -3\end{bmatrix}^T=\begin{bmatrix}-1 & 2\\ 4 & -3\end{bmatrix}$$

$$B^TA^T=\begin{bmatrix}-1 & 2\\ 4 & -3\end{bmatrix}\begin{bmatrix}3 & 1\\ -2 & -4\end{bmatrix}=\begin{bmatrix}-7 & -9\\ 18 & 16\end{bmatrix}. \text{ Thus } (AB)^T = B^TA^T.$$

Determinant of a matrix

Let A be a 2×2 matrix and is given by

$$A=\begin{bmatrix}a_{11} & a_{12}\\ a_{21} & a_{22}\end{bmatrix}$$

then the determinant D of matrix A is obtained from the difference of the cross multiplication of the coefficients:

$$D=\det A=\det\begin{bmatrix}a_{11} & a_{12}\\ a_{21} & a_{22}\end{bmatrix}=\begin{vmatrix}a_{11} & a_{12}\\ a_{21} & a_{22}\end{vmatrix}=a_{11}a_{22}-a_{12}a_{21}$$

$$\text{Thus if } A=\begin{bmatrix}5 & 3\\ 6 & 4\end{bmatrix} \text{ then } D=\begin{vmatrix}5 & 3\\ 6 & 4\end{vmatrix}=5\times4-3\times6=2$$

Now consider the 3×3 matrix

$$A=\begin{bmatrix}a_{11} & a_{12} & a_{13}\\ a_{21} & a_{22} & a_{23}\\ a_{31} & a_{32} & a_{33}\end{bmatrix}$$

$$\text{then } D=\det\begin{bmatrix}a_{11} & a_{12} & a_{13}\\ a_{21} & a_{22} & a_{23}\\ a_{31} & a_{32} & a_{33}\end{bmatrix}=\begin{vmatrix}a_{11} & a_{12} & a_{13}\\ a_{21} & a_{22} & a_{23}\\ a_{31} & a_{32} & a_{33}\end{vmatrix}$$

The entries in any row or column of D can be taken as the multipliers by the corresponding minors as shown in the example below where the multipliers are taken as the first column of matrix

A, i.e. a_{11}, a_{21}, and a_{31}. Their minors, which are the second order determinants obtained by deleting the row and column passing through that multiplier, and are given by:

Multiplier a_{11} with the minor $\begin{vmatrix} a_{22} & a_{23} \\ a_{32} & a_{33} \end{vmatrix}$

Multiplier a_{21} with the minor $\begin{vmatrix} a_{12} & a_{13} \\ a_{32} & a_{33} \end{vmatrix}$

Multiplier a_{31} with the minor $\begin{vmatrix} a_{12} & a_{13} \\ a_{22} & a_{23} \end{vmatrix}$

The general signs of the terms in the expansion of the determinant D alternate between + and – as: $\begin{bmatrix} + & - & + \\ - & + & - \\ + & - & + \end{bmatrix}$ thus

$$\begin{vmatrix} a_{11} & a_{12} & a_{13} \\ a_{21} & a_{22} & a_{23} \\ a_{31} & a_{32} & a_{33} \end{vmatrix} = +a_{11}\begin{vmatrix} a_{22} & a_{23} \\ a_{32} & a_{33} \end{vmatrix} - a_{21}\begin{vmatrix} a_{12} & a_{13} \\ a_{32} & a_{33} \end{vmatrix} + a_{31}\begin{vmatrix} a_{12} & a_{13} \\ a_{22} & a_{23} \end{vmatrix}$$

$$D = +a_{11}(a_{22}a_{33} - a_{23}a_{32}) - a_{21}(a_{12}a_{33} - a_{13}a_{32}) + a_{31}(a_{12}a_{23} - a_{13}a_{22})$$
$$= +a_{11}a_{22}a_{33} - a_{11}a_{23}a_{32} - a_{21}a_{12}a_{33} + a_{21}a_{13}a_{32} + a_{31}a_{12}a_{23} - a_{31}a_{13}a_{22}$$

Alternatively, if the coefficients in first row are taken as multipliers and noting that the signs are + – +, then:

$$\begin{vmatrix} a_{11} & a_{12} & a_{13} \\ a_{21} & a_{22} & a_{23} \\ a_{31} & a_{32} & a_{33} \end{vmatrix} = +a_{11}\begin{vmatrix} a_{22} & a_{23} \\ a_{32} & a_{33} \end{vmatrix} - a_{12}\begin{vmatrix} a_{21} & a_{23} \\ a_{31} & a_{33} \end{vmatrix} + a_{13}\begin{vmatrix} a_{21} & a_{22} \\ a_{31} & a_{32} \end{vmatrix}$$

$$D = +a_{11}(a_{22}a_{33} - a_{23}a_{32}) - a_{12}(a_{21}a_{33} - a_{23}a_{31}) + a_{13}(a_{21}a_{32} - a_{22}a_{31})$$

$$= +a_{11}a_{22}a_{33} - a_{11}a_{23}a_{32} - a_{12}a_{21}a_{33} + a_{12}a_{23}a_{31} + a_{13}a_{21}a_{32} - a_{13}a_{22}a_{31}$$

which is the same as the previous expansion.

Example 10

Calculate the determinant of matrix $A = \begin{bmatrix} 2 & -4 & 7 \\ 5 & 1 & -3 \\ -6 & 2 & 8 \end{bmatrix}$.

Taking the entries in the first column as multipliers

$$D=\begin{vmatrix}2 & -4 & 7\\ 5 & 1 & -3\\ -6 & 2 & 8\end{vmatrix}=+2\begin{vmatrix}1 & -3\\ 2 & 8\end{vmatrix}-5\begin{vmatrix}-4 & 7\\ 2 & 8\end{vmatrix}+(-6)\begin{vmatrix}-4 & 7\\ 1 & -3\end{vmatrix}$$

$$=+2[1\times8-(-3)\times2]-5[(-4)\times8-7\times2]-6[(-4)\times(-3)-7\times1]$$

$$=+228$$

Alternatively taking the entries in the first row as multipliers

$$D=\begin{vmatrix}2 & -4 & 7\\ 5 & 1 & -3\\ -6 & 2 & 8\end{vmatrix}=+2\begin{vmatrix}1 & -3\\ 2 & 8\end{vmatrix}-(-4)\begin{vmatrix}5 & -3\\ -6 & 8\end{vmatrix}+7\begin{vmatrix}5 & 1\\ -6 & 2\end{vmatrix}$$

$$=+2[1\times8-(-3)\times2]+4[5\times8-(-3)\times(-6)]+7[5\times2-1\times(-6)]$$

$$=+228$$

which is the same as the previous result.

A square matrix whose determinant is not equal to zero is called non-singular matrix and when its determinant is equal to zero it is singular, for example

$$A=\begin{bmatrix}5 & -5 & 0\\ -5 & 11 & -6\\ 0 & -6 & 6\end{bmatrix}$$

Taking the entries in the first column as multipliers

$$D=\begin{vmatrix}5 & -5 & 0\\ -5 & 11 & -6\\ 0 & -6 & 6\end{vmatrix}=+5\begin{vmatrix}11 & -6\\ -6 & 6\end{vmatrix}-(-5)\begin{vmatrix}-5 & 0\\ -6 & 6\end{vmatrix}+0\begin{vmatrix}-5 & 0\\ 11 & -6\end{vmatrix}$$

$$=5[11\times6-(-6)\times(-6)]+5[(-5)\times6-0\times(-6)]$$
$$+0[(-5)\times(-6)-0\times11)]=0$$

Therefore matrix A is singular.

When all the coefficients in any row are zeros, then the determinant of the matrix is equal to zero, for example

$$A=\begin{bmatrix}2 & -4 & 7\\ 0 & 0 & 0\\ -6 & 3 & 8\end{bmatrix}$$

Taking the entries in the first column as multipliers

$$D = 2[0 \times 8 - 0 \times 3] - 0[(-4) \times 8 - 7 \times 3] + (-6)[(-4) \times 0 - 7 \times 0] = 0$$

When all the coefficients in any column are zeros, then the determinant of the matrix is equal to zero, for example

$$A = \begin{bmatrix} 3 & 5 & 0 \\ 6 & -2 & 0 \\ -7 & 4 & 0 \end{bmatrix}$$

Taking the entries in the first column as multipliers

$$D = 3[(-2) \times 0 - 0 \times 4] - 6[5 \times 0 - 0 \times 4] + (-7)[5 \times 0 - 0 \times (-2)] = 0$$

1.2 Solution of Simultaneous Equations

In the analysis of structures by stiffness matrix methods a set of simultaneous linear equations with the displacements as the unknowns will result. An important step in the computations is the determination of these unknowns which will in turn lead to the calculation of external reactions at the supports of the structure and the forces developed in its members. Some of the methods that are commonly used in solving a set of simultaneous equations are presented in the subsequent sections.

If the set of equations is ill-conditioned the solution is sensitive to small changes in the coefficients of the matrix or in rounding off the numbers in the computations process. Also, such sets may converge slowly or may not converge to the correct solution when the iterative methods are used. One of the tests for ill conditioned set of equations is that the determinant of the matrix of coefficients is small compared with the absolute value of the largest coefficient. On the other hand a set of equations is well-conditioned if the coefficients on the main diagonal are large in absolute value in comparison with the off diagonal coefficients.

In general, the stiffness matrix method of structural analysis leads to well-conditioned sets of equations. These sets result in correct solutions when the direct methods are used and converge to the correct solution if the iterative methods are used.

1.2.1 Direct Methods

The most popular direct methods for the solution of simultaneous equations resulting from the application of matrix methods in structural analysis are the Gauss elimination and the Cholesky decomposition. An innovative method called the frontal solution is also used where the final solution is obtained without writing the full set of simultaneous equations. This method is outside the scope of this book and the reader is referred to specialised literature.

(i) Gauss elimination method

Consider the set of linear algebraic simultaneous equations

$$a_{11}x_1 + a_{12}x_2 + a_{13}x_3 = b_1$$

$$a_{21}x_1 + a_{22}x_2 + a_{23}x_3 = b_2$$

$$a_{31}x_1 + a_{32}x_2 + a_{33}x_3 = b_3$$

which can be written in matrix form as

$$\begin{bmatrix} a_{11} & a_{12} & a_{13} \\ a_{21} & a_{22} & a_{23} \\ a_{31} & a_{32} & a_{33} \end{bmatrix} \begin{bmatrix} x_1 \\ x_2 \\ x_3 \end{bmatrix} = \begin{bmatrix} b_1 \\ b_2 \\ b_3 \end{bmatrix}$$

The above matrix can be reduced to an upper triangular matrix as

$$\begin{bmatrix} c_{11} & c_{12} & c_{13} \\ 0 & c_{22} & c_{23} \\ 0 & 0 & c_{33} \end{bmatrix} \begin{bmatrix} x_1 \\ x_2 \\ x_3 \end{bmatrix} = \begin{bmatrix} r_1 \\ r_2 \\ r_3 \end{bmatrix}$$

The unknowns x_1, x_2, and x_3 are obtained by back-substitution in reverse order as follows:

From the third row of the above set of equations:

$$x_3 = \frac{r_3}{c_{33}}$$

Substitute this value of x_3 in second row to get

$$x_2 = \frac{r_2 - c_{23}x_3}{c_{22}}$$

Substitute the values of x_2 and x_3 in the first row to get

$$x_1 = \frac{r_1 - c_{12}x_2 - c_{13}x_3}{c_{11}}$$

Example 11

Find the unknowns x_1, x_2, and x_3 given by the matrix

$$\begin{bmatrix} 4 & 1 & -2 \\ -3 & 5 & 1 \\ 5 & -2 & -4 \end{bmatrix}\begin{bmatrix} x_1 \\ x_2 \\ x_3 \end{bmatrix} = \begin{bmatrix} 7 \\ -12 \\ 9 \end{bmatrix}$$ which can be written in the form

$$4x_1 + x_2 - 2x_3 = 7 \quad (1.1)$$

$$-3x_1 + 5x_2 + x_3 = -12 \quad (1.2)$$

$$5x_1 - 2x_2 - 4x_3 = 9 \quad (1.3)$$

Use x_1 in equation (1.1) as a pivot and eliminate x_1 from equations (1.2) and (1.3).

Multiply (1.1) by –(–3/4) and add to (1.2) to get

$$0 + 5.75x_2 - 0.50x_3 = -6.75 \quad (1.2')$$

Multiply (1.1) by –(5/4) and add to (1.3) to get

$$0 - 3.25x_2 - 1.50x_3 = 0.25 \quad (1.3')$$

The new set is

$$4x_1 + x_2 - 2x_3 = 7 \quad (1.1)$$

$$0 + 5.75x_2 - 0.50x_3 = -6.75 \quad (1.2')$$

$$0 - 3.25x_2 - 1.50x_3 = 0.25 \quad (1.3')$$

Now use x_2 in (1.2′) as a pivot and eliminate x_2 from (1.3′) by multiplying (1.2′) by –(–3.25/5.75) and adding to (1.3′) to get

$$0 + 0 - 1.78x_3 = -3.56 \quad (1.3'')$$

The new set is

$$4x_1 + x_2 - 2x_3 = 7 \quad (1.1)$$

$$0 + 5.75x_2 - 0.50x_3 = -6.75 \quad (1.2')$$

$$0 + 0 - 1.78x_3 = -3.56 \quad (1.3'')$$

The values of x_1, x_2, and x_3 are found by back substitution as follows:

From (1.3″), $x_3 = -3.56/(-1.78) = +2.00$

Substitute this value of x_3 in (1.2′) to get

$$0 + 5.75x_2 - (0.50)(2.00) = -6.75$$

$$x_2 = \frac{-6.75 + (0.50)(2.00)}{5.75} = -1.00$$

Substitute the values of x_2 and x_3 in (1.1) to get

$$4x_1 - 1.00 - (2)(2.00) = 7$$

$$x_1 = \frac{7.00 + 1.00 + (2)(2.00)}{4.00} = +3.00$$

(ii) Cholesky's method

Let $Ax = b$ where A is an n×n square matrix which can be decomposed into the product of two matrices L and U, i.e. A = LU, where L is a lower triangular matrix and U is an upper triangular matrix. For example if A is a 3×3 matrix then

$$L = \begin{bmatrix} l_{11} & 0 & 0 \\ l_{21} & l_{22} & 0 \\ l_{31} & l_{32} & l_{33} \end{bmatrix} \quad \text{and} \quad U = \begin{bmatrix} u_{11} & u_{12} & u_{13} \\ 0 & u_{22} & u_{23} \\ 0 & 0 & u_{33} \end{bmatrix}$$

If matrix A is symmetric, i.e. $A = A^T$, then $U = L^T$ leading to $A = LL^T$, hence $LL^Tx = b$.

Let $y = L^Tx$, then $Ly = b$ and the vector y is obtained by forward substitution in L. Then from the relation $y = L^Tx$ the required solution vector x is found by backward substitution in L^T.

$$A = \begin{bmatrix} a_{11} & a_{12} & a_{13} \\ a_{21} & a_{22} & a_{23} \\ a_{31} & a_{32} & a_{33} \end{bmatrix} = \begin{bmatrix} l_{11} & 0 & 0 \\ l_{21} & l_{22} & 0 \\ l_{31} & l_{32} & l_{33} \end{bmatrix} \begin{bmatrix} l_{11} & l_{21} & l_{31} \\ 0 & l_{22} & l_{32} \\ 0 & 0 & l_{33} \end{bmatrix}$$

Performing the multiplication of the two matrices at the right-hand side and equating the product to the corresponding coefficients of the matrix at the left-hand side results in relationships that will lead to the determination of the coefficients of matrix L as shown below.

$$l_{11}l_{11} = a_{11}, \quad l_{11} = \sqrt{a_{11}}$$

$$l_{11}l_{21} = a_{12}, \quad l_{21} = \frac{a_{12}}{l_{11}}$$

$$l_{11}l_{31} = a_{13}, \quad l_{31} = \frac{a_{13}}{l_{11}}$$

$$l_{21}^2 + l_{22}^2 = a_{22}, \quad l_{22} = \sqrt{a_{22} - l_{21}^2}$$

$$l_{21}l_{31} + l_{22}l_{32} = a_{23}, \quad l_{32} = \frac{a_{23} - l_{21}l_{31}}{l_{22}}$$

$$l_{31}^2 + l_{32}^2 + l_{33}^2 = a_{33}, \quad l_{33} = \sqrt{a_{33} - l_{31}^2 - l_{32}^2}$$

The above relationships can be written in general form as:

$$l_{11} = \sqrt{a_{11}}$$

$$l_{jj} = \sqrt{a_{jj} - \sum_{s=1}^{s=j-1} l_{js}^2} \qquad j = 2, 3, \ldots\ldots\ldots\ldots, n$$

$$l_{j1} = \frac{a_{j1}}{l_{11}} \qquad j = 2, 3, \ldots\ldots\ldots\ldots, n$$

$$l_{jk} = \frac{1}{l_{kk}}\left(a_{jk} - \sum_{s=1}^{s=k-1} l_{js}l_{ks}\right) \qquad j = k+1, k+2, \ldots\ldots\ldots\ldots, n.\ k \geq 2$$

Note that all the coefficients on the leading diagonal of matrix L, i.e. l_{11}, l_{22},...... l_{nn} have taken the positive value. This is one of the properties of positive definite matrices and the condition for the symmetric matrix A to be positive definite is that the quadratic form x^TAx is greater than zero for any non-zero vector x. It will be shown in Chapter 2 that the structure stiffness matrix is symmetric and positive definite.

Example 12

Solve the following set of simultaneous equations by Cholesky's method.

$$9x_1 - 3x_2 + 6x_3 = 27$$
$$-3x_2 + 5x_2 - 4x_3 = -3$$
$$6x_1 - 4x_2 + 6x_3 = 16$$

In matrix form

$$\begin{bmatrix} 9 & -3 & 6 \\ -3 & 5 & -4 \\ 6 & -4 & 6 \end{bmatrix}\begin{bmatrix} x_1 \\ x_2 \\ x_3 \end{bmatrix} = \begin{bmatrix} 27 \\ -3 \\ 16 \end{bmatrix}$$

$$A = LL^T$$

$$l_{11} = \sqrt{a_{11}} = \sqrt{9} = 3$$

$$l_{21} = \frac{a_{12}}{l_{11}} = \frac{-3}{3} = -1$$

$$l_{31} = \frac{a_{13}}{l_{11}} = \frac{6}{3} = 2$$

$$l_{22} = \sqrt{a_{22} - l_{21}^2} = \sqrt{5-(-1)^2} = 2$$

$$l_{32} = \frac{a_{23} - l_{21}l_{31}}{l_{22}} = \frac{-4-(-1)\times 2}{2} = -1$$

$$l_{33} = \sqrt{a_{33} - l_{31}^2 - l_{32}^2} = \sqrt{6-2^2-(-1)^2} = 1$$

We have, Ly = b

$$\begin{bmatrix} l_{11} & 0 & 0 \\ l_{21} & l_{22} & 0 \\ l_{31} & l_{32} & l_{33} \end{bmatrix}\begin{bmatrix} y_1 \\ y_2 \\ y_3 \end{bmatrix} = \begin{bmatrix} b_1 \\ b_2 \\ b_3 \end{bmatrix}$$

$$\begin{bmatrix} 3 & 0 & 0 \\ -1 & 2 & 0 \\ 2 & -1 & 1 \end{bmatrix}\begin{bmatrix} y_1 \\ y_2 \\ y_3 \end{bmatrix} = \begin{bmatrix} 27 \\ -3 \\ 16 \end{bmatrix}$$

By forward substitution, the first row gives,

$$y_1 = \frac{27}{3} = 9$$

From the second row and with the substitution of $y_1 = 9$

$$y_2 = \frac{-3+(1)(9)}{2} = 3$$

From the third row and with the substitution of $y_1 = 9$ and $y_2 = 3$

$$y_3 = \frac{16-(2)(9)+1\times 3}{1} = 1$$

And from, $y = L^T x$

$$\begin{bmatrix} l_{11} & l_{21} & l_{31} \\ 0 & l_{22} & l_{32} \\ 0 & 0 & l_{33} \end{bmatrix} \begin{bmatrix} x_1 \\ x_2 \\ x_3 \end{bmatrix} = \begin{bmatrix} y_1 \\ y_2 \\ y_3 \end{bmatrix}$$

$$\begin{bmatrix} 3 & -1 & 2 \\ 0 & 2 & -1 \\ 0 & 0 & 1 \end{bmatrix} \begin{bmatrix} x_1 \\ x_2 \\ x_3 \end{bmatrix} = \begin{bmatrix} 9 \\ 3 \\ 1 \end{bmatrix}$$

By backward substitution, the third row gives,

$$x_3 = \frac{1}{1} = 1$$

The second row and with the substitution of $x_3 = 1$, we get

$$x_2 = \frac{3+(1)(1)}{2} = 2$$

From the first row and with the substitution of $x_2 = 2$ and $x_3 = 1$ we get

$$x_1 = \frac{9+(1)(2)-(2)(1)}{3} = 3$$

(iii) Cramer's rule:

This method is suitable when the number of simultaneous equations is small. As will be seen later it involves the computation of determinants and for a large number of simultaneous equations the method is not efficient because of the large amount of computer time required for the calculation of large determinants. The introduction of Cramer's rule here is mainly to illustrate its use in the derivation of a condition for the existence of solutions in a certain class of cases known as eigenvalue problems as can be seen in Section 1.4.

Consider the following two simultaneous equations and find their solution, i.e. the unknown variables x_1 and x_2.

$$a_{11}x_1 + a_{12}x_2 = b_1 \qquad (1.4)$$

$$a_{21}x_1 + a_{22}x_2 = b_2 \qquad (1.5)$$

From equation (1.5)

$$x_2 = \frac{b_2 - a_{21}x_1}{a_{22}}$$

Substitute in equation (1.4)

$$a_{11}x_1 + a_{12}\left(\frac{b_2 - a_{21}x_1}{a_{22}}\right) = b_1$$

$$x_1 = \frac{b_1a_{22} - a_{12}b_2}{a_{11}a_{22} - a_{12}a_{21}} = \frac{D_1}{D}$$

where $D_1 = \begin{vmatrix} b_1 & a_{12} \\ b_2 & a_{22} \end{vmatrix}$ and $D = \begin{vmatrix} a_{11} & a_{12} \\ a_{21} & a_{22} \end{vmatrix}$

Back substitution in equation (1.5) gives

$$x_2 = \frac{1}{a_{22}}\left(b_2 - a_{21}\frac{b_1a_{22} - a_{12}b_2}{a_{11}a_{22} - a_{12}a_{21}}\right)$$

$$x_2 = \frac{a_{11}b_2 - b_1a_{21}}{a_{11}a_{22} - a_{12}a_{21}} = \frac{D_2}{D}$$

where $D_2 = \begin{vmatrix} a_{11} & b_1 \\ a_{21} & b_2 \end{vmatrix}$

The above is called Cramer's rule and the condition for the existence of a unique solution is that $D \neq 0$.

Example 13

Find the unknowns x_1 and x_2 given by the following two simultaneous equations

$$2x_1 - 3x_2 = 12$$

$$5x_1 + 4x_2 = 7$$

$$D = \begin{vmatrix} a_{11} & a_{12} \\ a_{21} & a_{22} \end{vmatrix} = \begin{vmatrix} 2 & -3 \\ 5 & 4 \end{vmatrix} = (2)(4) - (-3)(5) = 23$$

$$D_1 = \begin{vmatrix} b_1 & a_{12} \\ b_2 & a_{22} \end{vmatrix} = \begin{vmatrix} 12 & -3 \\ 7 & 4 \end{vmatrix} = (12)(4) - (-3)(7) = 69$$

$$D_2 = \begin{vmatrix} a_{11} & b_1 \\ a_{21} & b_2 \end{vmatrix} = \begin{vmatrix} 2 & 12 \\ 5 & 7 \end{vmatrix} = (2)(7) - (12)(5) = -46$$

$$x_1 = \frac{D_1}{D} = \frac{69}{23} = 3$$

$$x_2 = \frac{D_2}{D} = \frac{-46}{23} = -2$$

1.2.2 Iterative Methods

These methods are particularly useful when dealing with sparsely populated matrices, i.e. matrices containing a relatively small number of non-zero coefficients. Since only non-zero coefficients are involved, this will greatly reduce storage requirements but convergence to the final solution might be slow in some cases.

The procedure is to start the iteration process with assumed values of the variables and find new (corrected) values. The iteration is continued until a convergence criterion is reached and this is defined by the desired degree of accuracy, i.e. $x_i^{(r+1)} - x_i^{(r)} < \in$ where $x_i^{(r+1)}$ and $x_i^{(r)}$ are the values of the unknown variable x_i obtained from two successive iteration cycles r^{th} and $(r + 1)^{th}$ and $\in$ is a small prescribed number.

(i) Jacobi method

Example 14

Calculate the unknowns x_1, x_2, and x_3 in the set of simultaneous equations below

$$4x_1 - 2x_2 + x_3 = 30 \qquad (1.6)$$

$$-2x_1 + 5x_2 - x_3 = -29 \qquad (1.7)$$

$$x_1 - x_2 + 3x_3 = 20 \qquad (1.8)$$

From equation (1.6), find x_1 in the $(r + 1)^{th}$ cycle from the values of x_2 and x_3 obtained from the rth cycle as

$$x_1^{(r+1)} = \frac{1}{4}(2x_2^{(r)} - x_3^{(r)} + 30)$$

Similarly find x_2 and x_3 from equations (1.7) and (1.8) respectively as

$$x_2^{(r+1)} = \frac{1}{5}(2x_1^{(r)} + x_3^{(r)} - 29)$$

$$x_3^{(r+1)} = \frac{1}{3}(-x_1^{(r)} + x_2^{(r)} + 20)$$

where r is the cycle number in the iteration process

First cycle: r = 0

Start the iteration with assumed initial values such as

$$x_1^{(0)} = 0, \quad x_2^{(0)} = 0, \quad \text{and} \quad x_3^{(0)} = 0$$

$$x_1^{(1)} = \frac{1}{4}\left(2x_2^{(0)} - x_3^{(0)} + 30\right) = \frac{1}{4}\left[2(0) - 0 + 30\right] = 7.500$$

$$x_2^{(1)} = \frac{1}{5}\left(2x_1^{(0)} + x_3^{(0)} - 29\right) = \frac{1}{5}\left[2(0) + 0 - 29\right] = -5.800$$

$$x_3^{(1)} = \frac{1}{3}\left(-x_1^{(0)} + x_2^{(0)} + 20\right) = \frac{1}{3}\left[-0 + 0 + 20\right] = 6.667$$

Second cycle: r = 1, with $x_1 = 7.500$, $x_2 = -5.800$, and $x_3 = 6.667$

$$x_1^{(2)} = \frac{1}{4}\left(2x_2^{(1)} - x_3^{(1)} + 30\right) = \frac{1}{4}\left[2(-5.800) - 6.667 + 30\right] = 2.933$$

$$x_2^{(2)} = \frac{1}{5}\left(2x_1^{(1)} + x_3^{(1)} - 29\right) = \frac{1}{5}\left[2(7.500) + 6.667 - 29\right] = -1.467$$

$$x_3^{(2)} = \frac{1}{3}\left(-x_1^{(1)} + x_2^{(1)} + 20\right) = \frac{1}{3}\left[-7.500 + (-5.800) + 20\right] = 2.233$$

and so on. After twenty-six cycles the values of the unknown variables converged to $x_1 = 5.000$, $x_2 = -3.000$, and $x_3 = 4.000$ correct to three decimal places.

(The exact solution is: $x_1 = 5$, $x_2 = -3$, and $x_3 = 4$.)

(ii) Gauss–Seidel method

In this method the new values of the variables are used as soon as they are calculated (within the same cycle of iteration). Generally, this will lead to faster convergence compared with the Jacobi method in which *all* the variables are calculated first and then they are used in the next cycle as explained in the preceding section. In order to compare this method with Jacobi method, the same set of simultaneous equations is used.

The new value of x_1 is calculated from the old values of x_2 and x_3

$$x_1^{(r+1)} = \frac{1}{4}(2x_2^{(r)} - x_3^{(r)} + 30)$$

The new value of x_2 is calculated from new value of x_1, as obtained above, but the old value of x_3.

$$x_2^{(r+1)} = \frac{1}{5}(2x_1^{(r+1)} + x_3^{(r)} - 29)$$

The new value of x_3 is calculated from new values of x_1 and x_2.

$$x_3^{(r+1)} = \frac{1}{3}(-x_1^{(r+1)} + x_2^{(r+1)} + 20)$$

First cycle: r = 0

Start the iteration with assumed initial values such as

$$x_1^{(0)} = 0, \quad x_2^{(0)} = 0, \quad \text{and} \quad x_3^{(0)} = 0$$

$$x_1^{(1)} = \frac{1}{4}\left(2x_2^{(0)} - x_3^{(0)} + 30\right) = \frac{1}{4}\left[2(0) - 0 + 30\right] = 7.500$$

$$x_2^{(1)} = \frac{1}{5}\left(2x_1^{(1)} + x_3^{(0)} - 29\right) = \frac{1}{5}\left[2(7.500) + 0 - 29\right] = -2.800$$

$$x_3^{(1)} = \frac{1}{3}\left(-x_1^{(1)} + x_2^{(1)} + 20\right) = \frac{1}{3}\left[-7.500 + (-2.800) + 20\right] = 3.233$$

Second cycle: r = 1

$$x_1^{(1)} = 7.500, \quad x_2^{(1)} = -2.800, \quad \text{and} \quad x_3^{(1)} = 3.233$$

$$x_1^{(2)} = \frac{1}{4}\left(2x_2^{(1)} - x_3^{(1)} + 30\right) = \frac{1}{4}\left[2(-2.800) - 3.233 + 30\right] = 5.292$$

$$x_2^{(2)} = \frac{1}{5}\left(2x_1^{(2)} + x_3^{(1)} - 29\right) = \frac{1}{5}\left[2(5.292) + 3.233 - 29\right] = -3.037$$

$$x_3^{(2)} = \frac{1}{3}\left(-x_1^{(2)} + x_2^{(2)} + 20\right) = \frac{1}{3}\left[-5.292 + (-3.037) + 20\right] = 3.890$$

and so on. After seven cycles the values of the unknown variables converged to $x_1 = 5.000$, $x_2 = -3.000$, and $x_3 = 4.000$ correct to three decimal places.

It is seen in this example that in the Gauss–Seidel method the number of cycles to reach convergence to the final solution correct to three decimal places is reduced from twenty six to seven cycles as compared with the Jacobi method. The number of cycles to convergence can be reduced even further by using the so called successive over-relaxation technique which involves calculating a new modified (weighted) value for the variables from the following relationship:

$[x_i^{(r+1)}]_{new} = \omega x_i^{(r+1)} + (1-\omega)x_i^{(r)}$, where ω is called the over-relaxation factor.

So, the procedure is to calculate $x_i^{(r+1)}$ as shown in the previous example then modify it to $[x_i^{(r+1)}]_{new}$ and use the modified value in place of $x_i^{(r+1)}$.

In order to get the fastest convergence we use $\omega_{optimum}$ whose value is in the range 1 to 2. In practice we often deal with the same problem many times and in that case it may be worth exploring the optimum value of ω since this will have a repeated use. A simple and straight forward way of finding $\omega_{optimum}$ is by experimenting with different values and finding the one that gives the fastest convergence.

1.3 Matrix Inverse

Consider a set of linear simultaneous equations in matrix form as Ax = b then the required solution vector $x = A^{-1}b$ where A^{-1} is called the inverse of matrix A, i.e. $AA^{-1} = I$ where I is the unit matrix. Although this is not the best method for determining, x particularly for large sets of equations, it is sometimes used to achieve economy when dealing with the same matrix but with many values of the right-hand vector b.

A non-singular matrix is a matrix that has an inverse while a matrix that does not have an inverse is called a singular matrix.

Let $A^{-1} = C$ and premultiply both sides by A, i.e. $AA^{-1} = AC$ hence AC = I.

So to find A^{-1} the coefficients of matrix C have to be computed as shown below.

Given matrix $A = \begin{bmatrix} a_{11} & a_{12} & a_{13} \\ a_{21} & a_{22} & a_{23} \\ a_{31} & a_{32} & a_{33} \end{bmatrix}$ and it is required to find its inverse which is given by matrix $C = \begin{bmatrix} c_{11} & c_{12} & c_{13} \\ c_{21} & c_{22} & c_{23} \\ c_{31} & c_{32} & c_{33} \end{bmatrix}$ that is AC = I

$$\begin{bmatrix} a_{11} & a_{12} & a_{13} \\ a_{21} & a_{22} & a_{23} \\ a_{31} & a_{32} & a_{33} \end{bmatrix} \begin{bmatrix} c_{11} & c_{12} & c_{13} \\ c_{21} & c_{22} & c_{23} \\ c_{31} & c_{32} & c_{33} \end{bmatrix} = \begin{bmatrix} 1 & 0 & 0 \\ 0 & 1 & 0 \\ 0 & 0 & 1 \end{bmatrix}$$

The expanded form of the above is

$$\begin{bmatrix} a_{11} & a_{12} & a_{13} \\ a_{21} & a_{22} & a_{23} \\ a_{31} & a_{32} & a_{33} \end{bmatrix} \begin{bmatrix} c_{11} \\ c_{21} \\ c_{31} \end{bmatrix} = \begin{bmatrix} 1 \\ 0 \\ 0 \end{bmatrix}$$ which is solved to yield the vector

$$\begin{bmatrix} c_{11} \\ c_{21} \\ c_{31} \end{bmatrix}$$

$$\begin{bmatrix} a_{11} & a_{12} & a_{13} \\ a_{21} & a_{22} & a_{23} \\ a_{31} & a_{32} & a_{33} \end{bmatrix} \begin{bmatrix} c_{12} \\ c_{22} \\ c_{32} \end{bmatrix} = \begin{bmatrix} 0 \\ 1 \\ 0 \end{bmatrix}$$ to give the vector $\begin{bmatrix} c_{12} \\ c_{22} \\ c_{32} \end{bmatrix}$

$$\begin{bmatrix} a_{11} & a_{12} & a_{13} \\ a_{21} & a_{22} & a_{23} \\ a_{31} & a_{32} & a_{33} \end{bmatrix} \begin{bmatrix} c_{13} \\ c_{23} \\ c_{33} \end{bmatrix} = \begin{bmatrix} 0 \\ 0 \\ 1 \end{bmatrix}$$ leading to the vector $\begin{bmatrix} c_{13} \\ c_{23} \\ c_{33} \end{bmatrix}$

So, in order to find the matrix C, the original set of equations is solved n times for n right-hand sides and each time the result represents one column of the matrix C. This is not an efficient method and a more practical approach to find the inverse of a matrix is explained below.

Gauss–Jordan method

This is one of the methods that can be used to compute the inverse of a given matrix and is basically an extended form of Gauss elimination method. The original matrix is first reduced to an upper triangular matrix which in turn is reduced to a unit matrix, I, i.e. with each of the coefficients on the main diagonal equal to 1 as explained below.

Given a matrix A and it is required to find its inverse A^{-1}.

Consider the augmented matrix [A:I] and premultiply by A^{-1} then

$$A^{-1}[A:I] = [A^{-1}A:A^{-1}] = [I:A^{-1}]$$

The process followed in this method is to perform operations similar to the Gauss elimination method on A and I simultaneously to transform matrix A to a unit matrix I and the unit matrix I to A^{-1}.

Example 15

Use the Gauss–Jordan method to compute the inverse of matrix

$$A = \begin{bmatrix} 5 & -2 & 4 \\ -2 & 8 & -3 \\ 4 & -3 & 6 \end{bmatrix}$$

$$\begin{bmatrix} \begin{bmatrix} 5 & -2 & 4 \\ -2 & 8 & -3 \\ 4 & -3 & 6 \end{bmatrix} & \begin{matrix} : \\ : \\ : \end{matrix} & \begin{bmatrix} 1 & 0 & 0 \\ 0 & 1 & 0 \\ 0 & 0 & 1 \end{bmatrix} \end{bmatrix} \begin{matrix} \text{Row1} \\ \text{Row2} \\ \text{Row3} \end{matrix}$$

$$\begin{bmatrix} \begin{bmatrix} 5 & -2 & 4 \\ 0 & 7.2 & -1.4 \\ 0 & -1.4 & 2.8 \end{bmatrix} & \begin{matrix} : \\ : \\ : \end{matrix} & \begin{bmatrix} 1 & 0 & 0 \\ 0.4 & 1 & 0 \\ -0.8 & 0 & 1 \end{bmatrix} \end{bmatrix} \begin{matrix} \\ -(-2/5)\times \text{Row1} + \text{Row2} \\ -(4/5)\times \text{Row1} + \text{Row3} \end{matrix}$$

$$\begin{bmatrix} \begin{bmatrix} 5 & -2 & 4 \\ 0 & 7.2 & -1.4 \\ 0 & 0 & 2.528 \end{bmatrix} & \begin{matrix} : \\ : \\ : \end{matrix} & \begin{bmatrix} 1 & 0 & 0 \\ 0.4 & 1 & 0 \\ -0.722 & 0.194 & 1 \end{bmatrix} \end{bmatrix} \begin{matrix} \\ \\ -(-1.4/7.2)\times \text{Row2} + \text{Row3} \end{matrix}$$

Now make the coefficients of the main diagonal equal to 1 by multiplying rows 1, 2 and 3 by 1/5, 1/7.2 and 1/2.528 respectively to get:

$$\begin{bmatrix} \begin{bmatrix} 1 & -0.4 & 0.8 \\ 0 & 1 & -0.194 \\ 0 & 0 & 1 \end{bmatrix} & \begin{matrix} : \\ : \\ : \end{matrix} & \begin{bmatrix} 0.2 & 0 & 0 \\ 0.056 & 0.139 & 0 \\ -0.286 & 0.077 & 0.396 \end{bmatrix} \end{bmatrix}$$

The left upper triangular matrix is reduced to a unit matrix as follows

$$\begin{bmatrix} \begin{bmatrix} 1 & -0.4 & 0 \\ 0 & 1 & 0 \\ 0 & 0 & 1 \end{bmatrix} & \begin{matrix} : \\ : \\ : \end{matrix} & \begin{bmatrix} 0.429 & -0.062 & -0.317 \\ 0 & 0.154 & 0.077 \\ -0.286 & 0.077 & 0.396 \end{bmatrix} \end{bmatrix} \begin{matrix} -(0.8)\times \text{Row3} + \text{Row1} \\ -(-0.194)\times \text{Row3} + \text{Row2} \\ \end{matrix}$$

$$\begin{bmatrix} \begin{bmatrix} 1 & 0 & 0 \\ 0 & 1 & 0 \\ 0 & 0 & 1 \end{bmatrix} & \begin{matrix} : \\ : \\ : \end{matrix} & \begin{bmatrix} 0.429 & 0 & -0.286 \\ 0 & 0.154 & 0.077 \\ -0.286 & 0.077 & 0.396 \end{bmatrix} \end{bmatrix} \begin{matrix} -(-0.4)\times \text{Row2} + \text{Row1} \\ \\ \end{matrix}$$

The above augmented matrix is equivalent to [I: C], i.e. [I: A^{-1}], therefore

$$A^{-1} = \begin{bmatrix} 0.429 & 0 & -0.286 \\ 0 & 0.154 & 0.077 \\ -0.286 & 0.077 & 0.396 \end{bmatrix}$$

Notice that the original matrix is symmetric and its inverse is also symmetric.

Check the result

$$A^{-1}A = \begin{bmatrix} 0.429 & 0 & -0.286 \\ 0 & 0.154 & 0.077 \\ -0.286 & 0.077 & 0.396 \end{bmatrix} \begin{bmatrix} 5 & -2 & 4 \\ -2 & 8 & -3 \\ 4 & -3 & 6 \end{bmatrix} = \begin{bmatrix} 1.001 & 0 & 0 \\ 0 & 1.001 & 0 \\ 0 & 0 & 1.001 \end{bmatrix}$$

The matrix on the right should be a unit matrix and the small differences are due to rounding off the computations to three decimal places.

The inverse of a diagonal matrix is a diagonal matrix with coefficients equal to the reciprocals of the corresponding coefficients in the original matrix.

$$A = \begin{bmatrix} a_{11} & 0 & 0 \\ 0 & a_{22} & 0 \\ 0 & 0 & a_{33} \end{bmatrix}, \text{ then } A^{-1} = \begin{bmatrix} 1/a_{11} & 0 & 0 \\ 0 & 1/a_{22} & 0 \\ 0 & 0 & 1/a_{33} \end{bmatrix}$$

Example 16

Find the inverse of the diagonal matrix, A.

$$A = 10\begin{bmatrix} 2 & 0 & 0 \\ 0 & 5 & 0 \\ 0 & 0 & 4 \end{bmatrix}$$

$$A^{-1} = (1/10)\begin{bmatrix} 1/2 & 0 & 0 \\ 0 & 1/5 & 0 \\ 0 & 0 & 1/4 \end{bmatrix} = \begin{bmatrix} 0.050 & 0 & 0 \\ 0 & 0.020 & 0 \\ 0 & 0 & 0.025 \end{bmatrix}$$

or, $$= 10\begin{bmatrix} 2 & 0 & 0 \\ 0 & 5 & 0 \\ 0 & 0 & 4 \end{bmatrix} = \begin{bmatrix} 20 & 0 & 0 \\ 0 & 50 & 0 \\ 0 & 0 & 40 \end{bmatrix}, \quad A^{-1} = \begin{bmatrix} 0.050 & 0 & 0 \\ 0 & 0.020 & 0 \\ 0 & 0 & 0.025 \end{bmatrix}$$

1.4 Eigenvalues and Eigenvectors

In the application of matrix methods for the solution of stability and vibration of structures, as will be seen in Chapters 11 and 12, an eigenvalue problem arises. In such cases the right-hand-side vector, b, of the set of simultaneous equations is zero and a trivial solution for the unknown vector x is zero. A non-trivial solution can be obtained by the so called eigenvalue procedure which is explained below.

1.4.1 The Algebraic Method

Consider the following set of equations

$$A_{11}x_1 + A_{12}x_2 + A_{13}x_3 = \lambda x_1$$

$$A_{21}x_1 + A_{22}x_2 + A_{23}x_3 = \lambda x_2$$

$$A_{31}x_1 + A_{32}x_2 + A_{33}x_3 = \lambda x_3$$

Which can be written as $Ax = \lambda x$ or $(A - \lambda I)x = 0$, where $A = \begin{bmatrix} a_{11} & a_{12} & a_{13} \\ a_{21} & a_{22} & a_{23} \\ a_{31} & a_{32} & a_{33} \end{bmatrix}$, $x = \begin{bmatrix} x_1 \\ x_2 \\ x_3 \end{bmatrix}$, $\begin{bmatrix} 1 & 0 & 0 \\ 0 & 1 & 0 \\ 0 & 0 & 1 \end{bmatrix}$ and λ is a constant called the eigenvalue of matrix A, thus

$$(a_{11} - \lambda)\, x_1 + a_{12}x_2 + a_{13}x_3 = 0$$

$$(a_{21}x_1 + (a_{22} - \lambda)x_2 + a_{23}x_3 = 0$$

$$a_{31}x_1 + a_{32}x_2 + (a_{33} - \lambda)x_3 = 0$$

A set of equations with the right-hand-side vector b, equal to zero is called a homogeneous system of simultaneous equations. The solution for the unknown vector, x, is given by Cramer's rule as

$$x_1 = \frac{D_1}{D}, \quad x_2 = \frac{D_2}{D}, \quad \text{and} \quad x_3 = \frac{D_3}{D}$$

where, $D_1 = \begin{vmatrix} b_1 & a_{12} & a_{13} \\ b_2 & a_{22} - \lambda & a_{23} \\ b_3 & a_{32} & a_{33} - \lambda \end{vmatrix}$, $D_2 = \begin{vmatrix} a_{11} - \lambda & b_1 & a_{13} \\ a_{21} & b_2 & a_{23} \\ a_{31} & b_3 & a_{33} - \lambda \end{vmatrix}$, and

$$D_3 = \begin{vmatrix} a_{11}-\lambda & a_{12} & b_1 \\ a_{21} & a_{22}-\lambda & b_2 \\ a_{31} & a_{32} & b_3 \end{vmatrix}.$$

A determinant with zero coefficients in any column is equal to zero and since

$$b = \begin{bmatrix} b_1 \\ b_2 \\ b_3 \end{bmatrix} = \begin{bmatrix} 0 \\ 0 \\ 0 \end{bmatrix} \text{ then } D_1 = 0,\ D_2 = 0, \text{ and } D_3 = 0.$$

$$x_1 = \frac{D_1}{D} = \frac{0}{D} = 0,\quad x_2 = \frac{D_2}{D} = \frac{0}{D} = 0,\quad \text{and}\quad x_3 = \frac{D_3}{D} = \frac{0}{D} = 0.$$

The above solution vector $x = 0$ is called the trivial solution of the set of simultaneous equations.

A non-trivial solution exists, i.e. the vector x does not equal to zero if the determinant D of the matrix $(A - \lambda I)$ is zero.

In this case, $x_1 = D_1/D = 0/0$ which is not defined, i.e. indeterminate and can have any value. Similarly, $x_2 = D_2/D = 0/0$ and $x_3 = D_2/D = 0/0$. The condition of $D = 0$ will give the required values of λ.

Example 17

Find the eigenvalues and eigenvectors of the following matrix given by the following two homogeneous equations:

$$3x_1 - x_2 = \lambda x_1$$

$$-6x_1 - 4x_2 = \lambda x_2$$

which can be written as:

$$(3 - \lambda)x_1 - x_2 = 0 \tag{1.9}$$

$$-6x_1 + (4 - \lambda)x_2 = 0 \tag{1.10}$$

or, $$(A - \lambda I)x = 0, \text{ where } x = \begin{bmatrix} x_1 \\ x_2 \end{bmatrix}.$$

The trivial solution of the above equations is $x_1 = 0$ and $x_2 = 0$, but a non-trivial solution exists if the determinant of matrix $(A - \lambda I)$ is equal to zero.

$$A-\lambda I=\begin{bmatrix}3 & -1\\ -6 & 4\end{bmatrix}-\lambda\begin{bmatrix}1 & 0\\ 0 & 1\end{bmatrix}=\begin{bmatrix}3 & -1\\ -6 & 4\end{bmatrix}-\begin{bmatrix}\lambda & 0\\ 0 & \lambda\end{bmatrix}=\begin{bmatrix}3-\lambda & -1\\ -6 & 4-\lambda\end{bmatrix}$$

$$D=\det(A-\lambda I)=\begin{vmatrix}3-\lambda & -1\\ -6 & 4-\lambda\end{vmatrix}=(3-\lambda)(4-\lambda)-(-1)(-6)$$

$$D = \lambda^2 - 7\lambda + 6$$

The eigenvalues are given by D = 0, i.e. the roots of $\lambda^2 - 7\lambda + 6 = 0$ which is called the characteristic equation of matrix A and is a polynomial of degree 2 since A in this case is a 2×2 matrix, hence there are two roots.

(For an n×n matrix, the characteristic equation is a polynomial of degree n with n roots, i.e. n eigenvalues)

The solution of the above quadratic equation is given by:

$$\lambda=\frac{7\pm\sqrt{(-7)^2-(4)(6)}}{2}=1 \text{ and } 6$$

Therefore, the eigenvalues of matrix A are: $\lambda_1 = 1$ and $\lambda_2 = 6$

For each eigenvalue there is a corresponding eigenvector, x, i.e. values of the unknowns, in this case, x_1 and x_2 as shown here.

Substitute $\lambda_1 = 1$ in equations (1.9) and (1.10)

$$(3 - 1)x_1 - x_2 = 0 \qquad (1.9a)$$

$$-6x_1 + (4 - 1)x_2 = 0 \qquad (1.10a)$$

or

$$2x_1 - x_2 = 0$$

$$-6x_1 + 3x_2 = 0$$

The second equation gives $x_2 = 2x_1$

Putting an arbitrary value of $x_1 = 1$ will give $x_2 = 2$ and the eigenvector is

$$x=\begin{bmatrix}x_1\\ x_2\end{bmatrix}=\begin{bmatrix}1\\ 2\end{bmatrix}$$

The values of the unknowns are not absolute but rather relative to each other and for this reason the vector of unknowns is sometimes normalised by making the largest coefficient in absolute value equal to unity. Thus dividing the above vector by 2 to give:

$$x = \begin{bmatrix} x_1 \\ x_2 \end{bmatrix} = \begin{bmatrix} 1 \\ 2 \end{bmatrix} = \alpha \begin{bmatrix} 0.500 \\ 1.000 \end{bmatrix} \text{ where } \alpha \text{ is any scalar multiplier.}$$

Substitute $\lambda_2 = 6$ in equations (1.9) and (1.10)

$$(3 - 6)x_1 - x_2 = 0 \tag{1.9b}$$

$$-6x_1 + (4 - 6)x_2 = 0 \tag{1.10b}$$

or

$$-3x_1 - x_2 = 0$$

$$-6x_1 + 2x_2 = 0$$

The second equation gives $x_2 = -3x_1$.

Putting an arbitrary value of $x_1 = 1$ will give $x_2 = -3$ and the eigenvector is

$$x = \begin{bmatrix} x_1 \\ x_2 \end{bmatrix} = \begin{bmatrix} 1 \\ -3 \end{bmatrix}.$$

The values of the unknowns are not absolute but rather relative to each other and for this reason the vector of unknowns is sometimes normalised by making the largest coefficient in absolute value equal to unity. Thus dividing the above vector by 3 to give the normalised eigenvector

$$x = \beta \begin{bmatrix} 0.333 \\ -1.000 \end{bmatrix} \text{ where } \beta \text{ is any scalar multiplier.}$$

To summarise:

Eigenvalue $\lambda_1 = 1$ and the corresponding normalised eigenvector, $x = \begin{bmatrix} 0.500 \\ 1.000 \end{bmatrix}$.

Eigenvalue $\lambda_2 = 6$ and the corresponding normalised eigenvector, $x = \begin{bmatrix} 0.333 \\ -1.000 \end{bmatrix}$.

Note that the trace of a matrix, which is defined as the sum of the coefficients of the main diagonal, is equal to the sum of the eigenvalues of the matrix.

Trace 3 + 4 = 7 and the sum of eigenvalues = $\lambda_1 + \lambda_2 = 1 + 6 = 7$ and this will provide a check on the previous computations.

In the above example, we had two equations and the determinant resulted in a quadratic characteristic equation, i.e. a polynomial of the second degree whose two roots were calculated algebraically.

In the analysis of structures by matrix methods we often deal with large sets of equations and the degree of the resulting polynomials will be high. The algebraic method of finding the roots of these polynomials is not practical and other more efficient methods are used such as the one explained in the next section.

1.4.2 The Direct Evaluation of Determinant

In the preceding section, a polynomial was obtained as the characteristic equation. The eigenvalues are the roots of the polynomial which were found algebraically. The degree of the resulting polynomial is equal to the number of equations and for large sets of equations the algebraic method of solution to calculate the roots is not practical. One of the alternative methods is the direct evaluation of the determinant for different values of λ and finding the eigenvalues by either tabular or graphical methods as shown in the following example.

Example 18

Use the direct evaluation of determinant method to calculate the eigenvalues and eigenvectors of the matrix:

$$A = \begin{bmatrix} 5 & -2 & 4 \\ -2 & 9 & -3 \\ 4 & -3 & 8 \end{bmatrix}$$

$$A - \lambda I = \begin{bmatrix} 5-\lambda & -2 & 4 \\ -2 & 9-\lambda & -3 \\ 4 & -3 & 8-\lambda \end{bmatrix}$$

$$\det(A - \lambda I) = D = \begin{vmatrix} 5-\lambda & -2 & 4 \\ -2 & 9-\lambda & -3 \\ 4 & -3 & 8-\lambda \end{vmatrix}$$

The determinant D is calculated for different values of λ and the eigenvalues of matrix A are those which give D = 0. This occurs between points where D changes sign and the values of λ are calculated by interpolation as shown below.

λ	D	λ	D	λ	D
1	80	6	−5	11	110
2	11	7	26	12	91
3	−26	8	59	13	44
4	−37	9	88	14	−37
5	−28	10	107	15	−158

By linear interpolation: D = 0 when

$$\lambda_1 = 2 + \frac{11}{11+26} = 2.297$$

$$\lambda_2 = 6 + \frac{5}{5+26} = 6.161$$

$$\lambda_3 = 13 + \frac{44}{44+37} = 13.543$$

Sum of eigenvalues = $\lambda_1 + \lambda_2 + \lambda_3$ = 2.297 + 6.161 + 13.543 = 22.001, and as a check, this should be equal to the trace of the matrix which is defined as the sum of the coefficients on the leading diagonal of the matrix.

Trace = 5 + 9 + 8 + = 22 , which agrees with the sum of eigenvalues.

Alternatively, the eigenvalues can be found by plotting D against λ and the points where the curve intersects with the λ axis give the required eigenvalues as shown below.

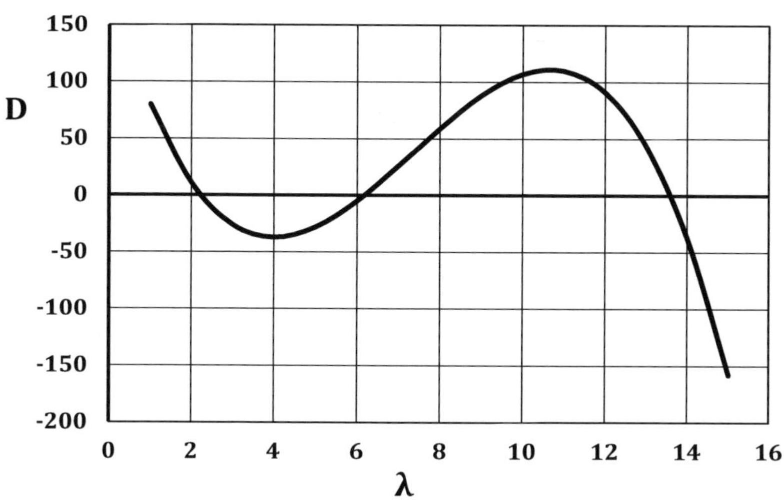

The eigenvectors of a matrix represent solutions of the unknown variables for the different eigenvalues of the matrix.

$$\begin{bmatrix} 5-\lambda & -2 & 4 \\ -2 & 9-\lambda & -3 \\ 4 & -3 & 8-\lambda \end{bmatrix} \begin{bmatrix} x_1 \\ x_2 \\ x_3 \end{bmatrix} = \begin{bmatrix} 0 \\ 0 \\ 0 \end{bmatrix}$$

Let $x_1 = 1$ and delete the first row.
The second and third rows become

$$(9 - \lambda)x_2 - 3x_3 = 2$$

$$-3x_2 + (8 - \lambda)x_3 = -4$$

Substitute the first eigenvalue, i.e. $\lambda = \lambda_1 = 2.297$ to get

$$6.703x_2 - 3x_3 = 2$$

$$-3x_2 + 5.703x_3 = -4$$

The solution of the above simultaneous equations is:
$x_2 = -0.020$ and $x_3 = -0.712$. So the full solution vector is:

$$x = \begin{bmatrix} x_1 \\ x_2 \\ x_3 \end{bmatrix} = \begin{bmatrix} 1.000 \\ -0.020 \\ -0.712 \end{bmatrix}.$$

Similarly, for $\lambda = \lambda_2 = 6.161$, we get $x = \begin{bmatrix} x_1 \\ x_2 \\ x_3 \end{bmatrix} = \begin{bmatrix} 1.000 \\ 2.202 \\ 1.417 \end{bmatrix}$

Divide by 2.202 to get the normalised eigenvector $x = \begin{bmatrix} 0.454 \\ 1.000 \\ 0.644 \end{bmatrix}$

Similarly, for $\lambda = \lambda_3 = 13.543$, we get $\begin{bmatrix} x_1 \\ x_2 \\ x_3 \end{bmatrix} = \begin{bmatrix} 1.000 \\ -1.427 \\ 1.494 \end{bmatrix}$ and the normalised eigenvector is $x = \begin{bmatrix} 0.669 \\ -0.955 \\ 1.000 \end{bmatrix}$.

To summarise:

$\lambda_1 = 2.297$, the corresponding eigenvector is $x = \begin{bmatrix} 1.000 \\ -0.020 \\ -0.712 \end{bmatrix}$,

$\lambda_2 = 6.161,\ x = \begin{bmatrix} 0.454 \\ 1.000 \\ 0.644 \end{bmatrix}$, and for $\lambda_3 = 13.543,\ x = \begin{bmatrix} 0.669 \\ -0.955 \\ 1.000 \end{bmatrix}$.

There are other more efficient methods for the determination of matrix eigenvalues particularly when dealing with large sets of simultaneous equations. These methods are beyond the scope of this book and the reader can refer to specialised literature on the subject.

Example 19

Given matrix $A = \begin{bmatrix} -8 & -3 \\ 1 & 7 \end{bmatrix}$ and matrix $B = \begin{bmatrix} 2 & 4 \\ 3 & 5 \end{bmatrix}$, find the eigenvalues λ and the corresponding eigenvectors of the following relationship:

$$(A - \lambda B) = 0$$

In order to reduce the above equation to a standard eigenvalue problem premultiply by the inverse of matrix B, i.e.

$$(B^{-1}A - \lambda B^{-1}B) = 0$$

$$(B^{-1}A - \lambda I) = 0 \text{ or } (C - \lambda I) = 0, \text{ where } C = B^{-1}A$$

$$B^{-1} = \begin{bmatrix} 2 & 4 \\ 3 & 5 \end{bmatrix}^{-1} = \begin{bmatrix} -2.5 & 2.0 \\ 1.5 & -1 \end{bmatrix},$$

$$(\text{check } B^{-1}B = \begin{bmatrix} -2.5 & 2.0 \\ 1.5 & -1 \end{bmatrix}\begin{bmatrix} 2 & 4 \\ 3 & 5 \end{bmatrix} = \begin{bmatrix} 1 & 0 \\ 0 & 1 \end{bmatrix} = I)$$

$$C = B^{-1}A = \begin{bmatrix} -2.5 & 2.0 \\ 1.5 & -1 \end{bmatrix}\begin{bmatrix} -8 & -3 \\ 1 & 7 \end{bmatrix} = \begin{bmatrix} 22.0 & 21.5 \\ -13.0 & -11.5 \end{bmatrix}$$

$$(C - \lambda I) = \begin{bmatrix} 22.0 & 21.5 \\ -13.0 & -11.5 \end{bmatrix} - \lambda\begin{bmatrix} 1 & 0 \\ 0 & 1 \end{bmatrix} = \begin{bmatrix} 22.0 - \lambda & 21.5 \\ -13.0 & -11.5 - \lambda \end{bmatrix}$$

$$\text{Det}\begin{bmatrix} 22.0-\lambda & 21.5 \\ -13.0 & -11.5-\lambda \end{bmatrix} = (22.0-\lambda)(-11.5-\lambda) - 21.5(-13.0) = 0$$

$$\lambda^2 - 10.5\lambda + 26.5 = 0$$

$$\lambda_1, \lambda_2 = \frac{10.5 \pm \sqrt{(-10.5)^2 - (4)(26.5)}}{2} = 4.22 \text{ and } 6.28.$$

Calculation of eigenvectors

$$(C - \lambda I)x = 0$$

$$\begin{bmatrix} 22.0-\lambda & 21.5 \\ -13.0 & -11.5-\lambda \end{bmatrix}\begin{bmatrix} x_1 \\ x_2 \end{bmatrix} = \begin{bmatrix} 0 \\ 0 \end{bmatrix}$$

$$\lambda_1 = 4.22$$

$$\begin{bmatrix} 22.0-4.22 & 21.5 \\ -13.0 & -11.5-4.22 \end{bmatrix}\begin{bmatrix} x_1 \\ x_2 \end{bmatrix} = \begin{bmatrix} 0 \\ 0 \end{bmatrix}$$

$$17.78x_1 + 21.50x_2 = 0$$

$$-13.00x_1 - 15.72x_2 = 0$$

Let, $x_1 = +1.00$ and from the second of the above two equations, we get

$$x_2 = -\frac{13.00}{15.72} = -0.83$$

The first eigenvector is: $\begin{bmatrix} x_1 \\ x_2 \end{bmatrix} = \begin{bmatrix} +1.00 \\ -0.83 \end{bmatrix}$.

Similarly, for $\lambda_2 = 6.28$, the second eigenvector is: $\begin{bmatrix} x_1 \\ x_2 \end{bmatrix} = \begin{bmatrix} +1.00 \\ -0.73 \end{bmatrix}$.

Alternatively, particularly when matrix inversion is to be avoided

$$(A - \lambda B) = \begin{bmatrix} -8 & -3 \\ 1 & 7 \end{bmatrix} - \lambda\begin{bmatrix} 2 & 4 \\ 3 & 5 \end{bmatrix} = \begin{bmatrix} -8-2\lambda & -3-4\lambda \\ 1-3\lambda & 7-5\lambda \end{bmatrix}$$

$$\text{Det}\begin{bmatrix} -8-2\lambda & -3-4\lambda \\ 1-3\lambda & 7-5\lambda \end{bmatrix} = -2\lambda^2 + 21\lambda - 53 = 0$$

$\lambda^2 - 10.521\lambda + 26.5 = 0$ which is the same characteristic equation obtained previously and its roots are

$$\lambda_1 = 4.22 \text{ and } \lambda_2 = 6.28.$$

The eigenvectors are given by

$$\begin{bmatrix} -8-2\lambda & -3-4\lambda \\ 1-3\lambda & 7-5\lambda \end{bmatrix}\begin{bmatrix} x_1 \\ x_2 \end{bmatrix} = \begin{bmatrix} 0 \\ 0 \end{bmatrix}.$$

For $\lambda_1 = 4.22$

$$\begin{bmatrix} -8-2\times 4.22 & -3-4\times 4.22 \\ 1-3\times 4.22 & 7-5\times 4.22 \end{bmatrix}\begin{bmatrix} x_1 \\ x_2 \end{bmatrix} = \begin{bmatrix} 0 \\ 0 \end{bmatrix}$$

$$\begin{bmatrix} -16.44 & -19.88 \\ -11.66 & -14.10 \end{bmatrix}\begin{bmatrix} x_1 \\ x_2 \end{bmatrix} = \begin{bmatrix} 0 \\ 0 \end{bmatrix}$$

Let, $x_1 = +1.00$ and from the second of the above two equations, we get

$$x_2 = -\frac{11.66}{14.10} = -0.83.$$

The first eigenvector is: $\begin{bmatrix} x_1 \\ x_2 \end{bmatrix} = \begin{bmatrix} +1.00 \\ -0.83 \end{bmatrix}$.

Similarly, for $\lambda_2 = 6.28$, the second eigenvector is: $\begin{bmatrix} x_1 \\ x_2 \end{bmatrix} = \begin{bmatrix} +1.00 \\ -0.73 \end{bmatrix}$.

Problems

P1.1. Given, $A = \begin{bmatrix} 3 & 1 & -2 \\ 4 & 6 & -4 \\ 5 & -3 & 4 \end{bmatrix}$ and $B = \begin{bmatrix} 2 & -5 & -1 \\ 4 & 1 & 4 \\ 2 & 3 & -5 \end{bmatrix}$,

find A + B and A – B.

Answer:

$$A+B = \begin{bmatrix} 5 & -4 & -3 \\ 8 & 7 & 0 \\ 7 & 0 & -1 \end{bmatrix},\ A-B = \begin{bmatrix} 1 & 6 & -1 \\ 0 & 5 & -8 \\ 3 & -6 & 9 \end{bmatrix}$$

P1.2. Given, A = [5   –7   4] and $B = \begin{bmatrix} 3 \\ 6 \\ -2 \end{bmatrix}$, find AB.

Answer:

AB = [–35]

P1.3. Given, $A = \begin{bmatrix} -4 \\ 2 \\ 5 \end{bmatrix}$ and B = [6   4   –3], find AB.

Answer:

$$AB = \begin{bmatrix} -24 & -16 & 12 \\ 12 & 8 & -6 \\ 30 & 20 & -15 \end{bmatrix}$$

P1.4. Given, $A = \begin{bmatrix} 2 & 1 & -4 \\ 4 & -5 & 2 \\ 3 & 6 & -1 \end{bmatrix}$ and $B = \begin{bmatrix} 5 \\ 6 \\ -3 \end{bmatrix}$, find AB.

Answer:

$$AB = \begin{bmatrix} 28 \\ -16 \\ 54 \end{bmatrix}$$

P1.5. $A = \begin{bmatrix} 5 & -2 \\ -4 & 6 \end{bmatrix}$ and $B = \begin{bmatrix} -3 & 6 \\ -5 & 2 \end{bmatrix}$, find 3AB.

Answer:

$$3AB = \begin{bmatrix} -15 & 78 \\ -54 & -36 \end{bmatrix}$$

P1.6. Given, $A = \begin{bmatrix} 2 & -1 \\ 5 & 3 \end{bmatrix}$, $B = \begin{bmatrix} -4 & 3 \\ 2 & -5 \end{bmatrix}$, and $C = \begin{bmatrix} -3 & 5 \\ 4 & -2 \end{bmatrix}$, find ABC.

Answer:

$$ABC = \begin{bmatrix} 74 & -72 \\ 42 & -70 \end{bmatrix}$$

P1.7. Given, $A = \begin{bmatrix} 4 & 1 & -2 \\ 0 & -3 & 4 \\ 5 & 2 & 6 \end{bmatrix}$, $B = \begin{bmatrix} 2 & 5 & 0 \\ 3 & 6 & -7 \\ -4 & 1 & 5 \end{bmatrix}$, and

$C = \begin{bmatrix} 3 & 5 & 4 \\ 2 & 1 & 2 \\ -6 & 3 & 4 \end{bmatrix}$, find ABC.

Answer:

$$ABC = \begin{bmatrix} 207 & 68 & 56 \\ -349 & -16 & 36 \\ -34 & 51 & 118 \end{bmatrix}$$

P1.8. Given, $A = \begin{bmatrix} 3 & 6 \\ 5 & -2 \end{bmatrix}$ and $B = \begin{bmatrix} 2 & 4 \\ -3 & 5 \end{bmatrix}$. Show that $(AB)^T = B^T A^T$.

Answer:

$$(AB)^T = \begin{bmatrix} -12 & 16 \\ 42 & 10 \end{bmatrix} = B^T A^T$$

P1.9. Find the determinant of the following matrices:

$$\text{(i)} \begin{bmatrix} 6 & 5 \\ 3 & 4 \end{bmatrix}, \text{(ii)} \begin{bmatrix} 5 & 4 \\ 3 & -7 \end{bmatrix}, \text{(iii)} \begin{bmatrix} 4 & 6 \\ -2 & 5 \end{bmatrix},$$

$$\text{(iv)} \begin{bmatrix} 3 & 4 \\ 6 & 8 \end{bmatrix}, \text{(v)} \begin{bmatrix} 7 & -2 & 4 \\ 2 & 5 & 3 \\ -6 & 4 & 1 \end{bmatrix}.$$

Answer:

(i) 9, (ii) –47, (iii) 32, (iv) 0, (v) 143

P1.10. Use Gauss elimination to calculate the unknowns x_1, x_2, and x_3 that satisfy the following three simultaneous equations:

$4x_1 - x_2 - 5x_3 = 3$

$-2x_1 + 5x_2 + 3x_3 = -9$

$-3x_1 + 4x_2 + 8x_3 = 4$

Answer:

$x_1 = 4$, $x_2 = -2$, $x_3 = 3$

P1.11. Use Cholesky's method to calculate the unknowns x_1, x_2, and x_3 in the following simultaneous equations:

$6x_1 - x_2 + 3x_3 = -8$

$-x_1 + 8x_2 - 2x_3 = 11$

$3x_1 - 2x_2 + 5x_3 = 7$

Answer:

$x_1 = -3$, $x_2 = 2$, $x_3 = 4$

P1.12. Use Jacobi iteration to calculate the unknowns x_1, x_2, and x_3 in the following equations:

$5x_1 - x_2 + 2x_3 = 20$

$$-x_1 + 7x_2 - 4x_3 = -42$$
$$2x_1 - 4x_2 + 6x_3 = 38$$

Answer:

$$x_1 = 2, x_2 = -4, x_3 = 3$$

P1.12. Repeat P.1.10 to find the unknowns by using Gauss–Seidel iteration and compare the required number of iterations with that obtained by the Jacobi iteration method to achieve the same accuracy.

P1.13. Use Gauss–Jordan method to compute the inverse of the matrix

$$\begin{bmatrix} 7 & -2 & 3 \\ -2 & 8 & 4 \\ 3 & 4 & 5 \end{bmatrix}$$

Answer:

$$\begin{bmatrix} 0.857 & 0.786 & -1.143 \\ 0.786 & 0.929 & -1.214 \\ -1.143 & -1.214 & 1.857 \end{bmatrix}$$

P1.14. Calculate the eigenvalues and normalised eigenvectors of the matrix

$$\begin{bmatrix} 6 & 8 \\ 1 & 4 \end{bmatrix}$$

Answer:

$$\lambda_1 = 2, \begin{bmatrix} x_1 \\ x_2 \end{bmatrix} = \begin{bmatrix} 1 \\ -0.50 \end{bmatrix}, \lambda_2 = 8, \begin{bmatrix} x_1 \\ x_2 \end{bmatrix} = \begin{bmatrix} 1 \\ 0.25 \end{bmatrix}$$

P1.15. Calculate the eigenvalues and normalised eigenvectors of the matrix

$$\begin{bmatrix} 4 & -1 & 2 \\ -1 & 7 & 3 \\ 2 & 3 & 6 \end{bmatrix}$$

Answer:

$$\lambda_1 = 1.525, \begin{bmatrix} x_1 \\ x_2 \\ x_3 \end{bmatrix} = \begin{bmatrix} 1.000 \\ 0.676 \\ -0.901 \end{bmatrix}, \lambda_2 = 5.876, \begin{bmatrix} x_1 \\ x_2 \\ x_3 \end{bmatrix} = \begin{bmatrix} 1.000 \\ -0.691 \\ 0.592 \end{bmatrix},$$

$$\lambda_3 = 9.600, \begin{bmatrix} x_1 \\ x_2 \\ x_3 \end{bmatrix} = \begin{bmatrix} 0.143 \\ 1.000 \\ 0.911 \end{bmatrix}$$

P1.16. Given matrix $A = \begin{bmatrix} 8 & 1 & -3 \\ 1 & 4 & 0 \\ -3 & 0 & 7 \end{bmatrix}$ and matrix $B = \begin{bmatrix} 4 & -2 & 0 \\ -2 & 6 & 3 \\ 0 & 3 & 5 \end{bmatrix}$,

find the eigenvalues λ and the corresponding normalised eigenvectors of the following relationship:
$(A - \lambda B) = 0$

Answer:

$$\lambda_1 = 0.530, \begin{bmatrix} x_1 \\ x_2 \\ x_3 \end{bmatrix} = \begin{bmatrix} +0.253 \\ -1.000 \\ -0.191 \end{bmatrix}, \ \lambda_2 = 0.985, \begin{bmatrix} x_1 \\ x_2 \\ x_3 \end{bmatrix} = \begin{bmatrix} +0.875 \\ -0.187 \\ +1.000 \end{bmatrix},$$

$$\lambda_3 = 5.423, \begin{bmatrix} x_1 \\ x_2 \\ x_3 \end{bmatrix} = \begin{bmatrix} +1.000 \\ +0.928 \\ -0.900 \end{bmatrix}$$

Chapter 2

General Principles

2.1 Bar Element Subjected to an Axial Force

In order to illustrate the general principles of matrix methods for the analysis of structures, the simplest type of element will be considered in this chapter and treated in detail first. This will make it easy for the reader to follow and grasp the basic principles involved in the derivation of a stiffness matrix and the process of computing the resulting displacements and forces developed in the members of the structure. The element of this type is subjected to an axial force, causing either uniform tensile or compressive stress across the whole cross section. Consequently, the element will deform by increasing or decreasing in length depending on whether it is in tension or compression. The complete analysis of an isolated individual bar by classical methods is quite straight forward and quick. But its treatment here by matrix methods is to show the process and general principles followed in the derivation of relationships between the various variables involved. These principles can be applied for the treatment of more complicated structures as will be seen later.

2.1.1 Derivation of Stiffness Matrix

The equation relating the change in length of a bar subjected to an axial force can be derived from elementary mechanics of materials as follows:

Analysis of Structures by Matrix Methods
Fathi Al-Shawi

ISBN 978-981-4968-19-5 (Hardcover), 978-1-003-29130-5 (eBook)
www.jennystanford.com

Axial force = stress × cross-sectional area

$$X = \sigma A$$

Stress = strain × modulus of elasticity

$$\sigma = \varepsilon E$$

$$\text{Strain} = \frac{\text{change in length}}{\text{initial length}}$$

$$\varepsilon = \frac{u}{L}$$

Therefore,

$$X = \frac{EA}{L}u \tag{2.1}$$

where X is the applied axial force,
u is the change in length,
L is the initial length of the bar,
A is the cross-sectional area, and
E is the modulus of elasticity of the material of the bar.

The stiffness matrix of a bar is the relationship between the forces and displacements at the ends of the bar.

The derivation of the stiffness matrix is based on the local coordinates system $\bar{x},\bar{y},\bar{z}$ with the $\bar{x}$-axis running along the axis of the bar. The displacements and forces are relative to the local $\bar{x}$-axis thus they are written with a bar.

Consider a bar subjected to axial forces $\bar{X}_i$ and $\bar{X}_j$ acting at the nodes i and j respectively. The corresponding axial displacements at the ends of the bar are $\bar{u}_i$ and $\bar{u}_j$ as shown in Fig. 2.1 and we want to derive a relationship between these displacements and forces acting at the nodes.

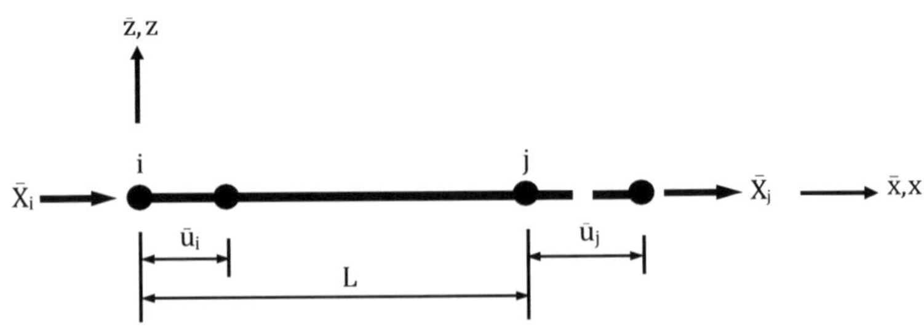

Figure 2.1 Bar element.

The derivation is carried out in two stages using a single prime for the first stage and double primes for the second. In the first stage assume that node i undergoes a displacement of $\bar{u}_i$ along the $\bar{x}$ -axis and node j is fixed as shown in Fig. 2.2, then

$$X = \bar{X}_i' \text{ and } u = \bar{u}_i$$

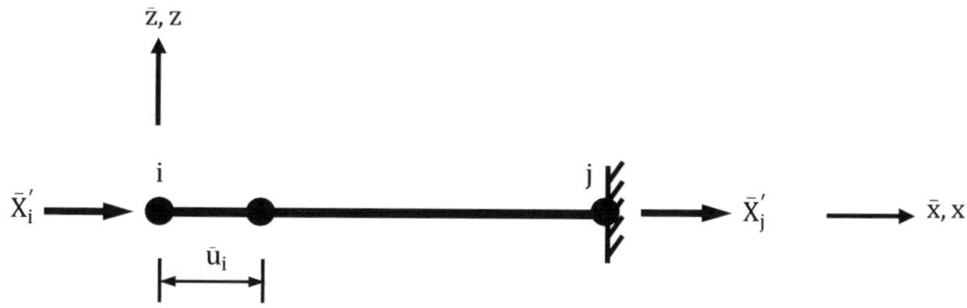

Figure 2.2

Substituting these values in equation (2.1) to get:

$$\bar{X}_i' = \frac{EA}{L}\bar{u}_i \qquad \text{(single prime is used for this case)}$$

From equilibrium of forces in the $\bar{x}$ -direction

$$\bar{X}_i' + \bar{X}_j' = 0 \quad \text{giving} \quad \bar{X}_j' = -\bar{X}_i'$$

$$\bar{X}_j' = -\frac{EA}{L}\bar{u}_i$$

For the second stage, assume that node j undergoes a displacement of $\bar{u}_j$ along the $\bar{x}$ -axis and node i is fixed as shown in Fig. 2.3, then

$$X = \bar{X}_j'' \text{ and } u = \bar{u}_j$$

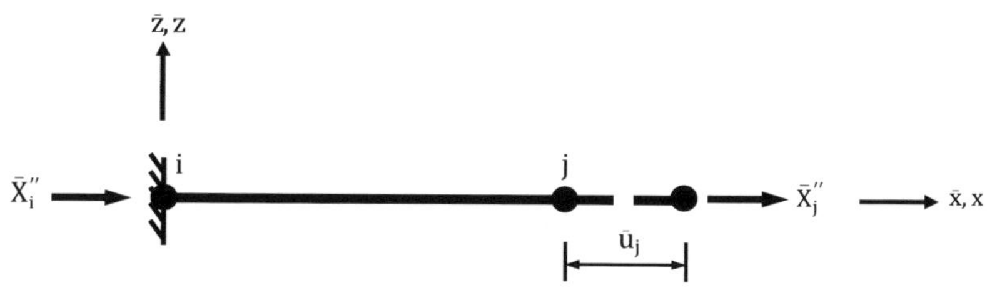

Figure 2.3

Substituting these values in equation (2.1) to get:

$$\bar{X}_j'' = \frac{EA}{L}\bar{u}_j \qquad \text{(double prime is used for this case)}$$

From equilibrium of forces in the $\bar{x}$-direction

$$\bar{X}_i'' + \bar{X}_j'' = 0 \quad \text{giving} \quad \bar{X}_i'' = -\bar{X}_j''$$

$$\bar{X}_i'' = -\frac{EA}{L}\bar{u}_j$$

The final result is obtained by combining the above two cases, thus

$$\bar{X}_i = \bar{X}_i' + \bar{X}_i''$$

$$\bar{X}_i = \frac{EA}{L}\bar{u}_i - \frac{EA}{L}\bar{u}_j \tag{2.2}$$

$$\bar{X}_j = \bar{X}_j' + \bar{X}_j''$$

$$\bar{X}_j = -\frac{EA}{L}\bar{u}_i + \frac{EA}{L}\bar{u}_j \tag{2.3}$$

Equations (2.2) and (2.3) are written in matrix form as:

$$\begin{bmatrix} \bar{X}_i \\ \bar{X}_j \end{bmatrix} = \begin{bmatrix} \frac{EA}{L} & -\frac{EA}{L} \\ -\frac{EA}{L} & \frac{EA}{L} \end{bmatrix} \begin{bmatrix} \bar{u}_i \\ \bar{u}_j \end{bmatrix} \tag{2.4}$$

The above relationship can be written as:

$$\bar{F} = \bar{k}\bar{\delta} \tag{2.5}$$

where $\bar{k} = \begin{bmatrix} \frac{EA}{L} & -\frac{EA}{L} \\ -\frac{EA}{L} & \frac{EA}{L} \end{bmatrix}$ is the stiffness matrix of a bar element subjected to axial forces at its ends and $\bar{\delta} = \begin{bmatrix} \bar{\delta}_i \\ \bar{\delta}_j \end{bmatrix}$ is the displacement vector which is composed of the displacements at nodes i and j.

Since there is only one degree of freedom, namely the translational displacement $\bar{u}$ in the $\bar{x}$-direction, it follows that, $\bar{\delta}_i = \bar{u}_i$ and $\bar{\delta}_j = \bar{u}_j$, thus

$$\bar{\delta} = \begin{bmatrix} \bar{u}_i \\ \bar{u}_j \end{bmatrix}$$

The right-hand side of (2.4) is the load vector which is composed of the forces at nodes i and j and these correspond to the relevant displacements, i.e.

$$\bar{F} = \begin{bmatrix} \bar{F}_i \\ \bar{F}_j \end{bmatrix} = \begin{bmatrix} \bar{X}_i \\ \bar{X}_j \end{bmatrix}$$

The above relationships can alternatively be derived by a finite element approach using the so-called interpolation polynomial which defines the displacement along the element as explained in Appendix 1.

2.1.2 The Overall Structure Matrix

The overall stiffness matrix is assembled relative to global coordinate system. So, the first step is to find the stiffness matrices of the bar elements relative to the global coordinate system. Since the local $\bar{x}$-axis coincides with the global x-axis then the stiffness matrices and displacements derived relative to local coordinates will have the same values relative to global coordinates. Thus $\bar{u}_i = u_i$, $\bar{u}_j = u_j$, $\bar{X}_i = X_i$ and $\bar{X}_j = X_j$, and the relations in (2.4) become

$$\begin{bmatrix} \frac{EA}{L} & -\frac{EA}{L} \\ -\frac{EA}{L} & \frac{EA}{L} \end{bmatrix} \begin{bmatrix} u_i \\ u_j \end{bmatrix} = \begin{bmatrix} X_i \\ X_j \end{bmatrix} \tag{2.6}$$

where $k = \begin{bmatrix} \frac{EA}{L} & -\frac{EA}{L} \\ -\frac{EA}{L} & \frac{EA}{L} \end{bmatrix}$, $\delta = \begin{bmatrix} u_i \\ u_j \end{bmatrix}$, and $F = \begin{bmatrix} X_i \\ X_j \end{bmatrix}$

Thus (2.6) in terms of global coordinates is written as:

$$k\delta = F \tag{2.7}$$

The stiffness matrix is written in the following general form

$$k = \begin{bmatrix} k_{ii} & k_{ij} \\ k_{ji} & k_{jj} \end{bmatrix} \text{where } k_{ii} = \frac{EA}{L}, k_{ij} = -\frac{EA}{L}, k_{ji} = -\frac{EA}{L}, \text{and } k_{jj} = \frac{EA}{L}$$

Note that the element stiffness matrix is symmetric since $k_{ij} = k_{ji}$.

One of the steps in the analysis of a structure by matrix methods is the assembly of the overall structure stiffness matrix which is built up from the stiffness matrices of its constituent (individual) elements. Thus, for the overall structure the general relationship is

$$K\delta = F \tag{2.8}$$

where K is overall stiffness matrix of the structure, δ is the vector of displacements at the nodes, and F is the load vector of external forces acting at the nodes of the structure with all these quantities written relative to global coordinates.

When the local $\bar{x}$ -axis of any member of the structure does not coincide with the global x-axis transformation from local coordinates to global coordinates is required as will be explained in subsequent chapters.

Example 1

Calculate the displacements and the forces developed at nodes 1, 2, and 3 of the stepped aluminium bar shown in Fig. 2.4 which is free at node 1 and fixed at node 3 for the following data:

Element 1, $L_1 = 0.42$ m, $A_1 = 150 \times 10^{-6}$ m^2,
Element 2, $L_2 = 0.56$ m, $A_2 = 240 \times 10^{-6}$ m^2,
The modulus of elasticity $E = 70 \times 10^6$ kN/m^2.

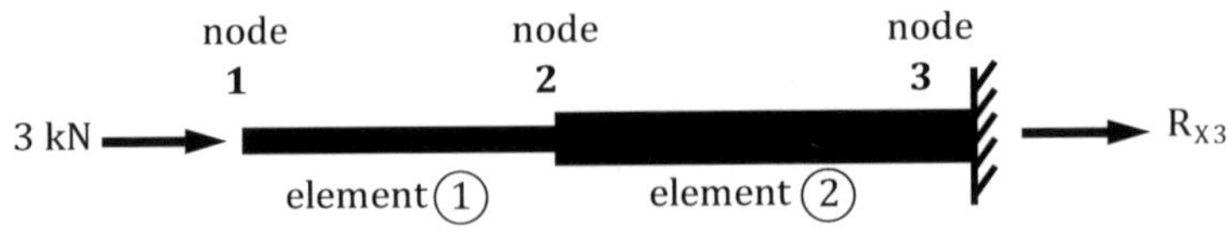

Figure 2.4

This example will be treated in detail showing the calculations step by step to highlight the general procedure followed in a typical computer program.

Element stiffness matrices

These are obtained from (2.6) as shown below.

Element 1

$$\frac{EA}{L} = \frac{70 \times 10^6 \times 150 \times 10^{-6}}{0.42} = 25000 \text{ kN/m}$$

From (2.6a)

$$k^1 = \begin{bmatrix} 25000 & -25000 \\ -25000 & 25000 \end{bmatrix} = \begin{bmatrix} k_{ii}^1 & k_{ij}^1 \\ k_{ji}^1 & k_{jj}^1 \end{bmatrix}$$

(superscript 1 indicates element 1)

i.e., $k_{ii}^1 = 25000, k_{ij}^1 = -25000, k_{ji}^1 = -25000, k_{jj}^1 = 25000$

Element 2

$$\frac{EA}{L} = \frac{70 \times 10^6 \times 240 \times 10^{-6}}{0.56} = 30000 \text{ kN/m}$$

$$k^2 = \begin{bmatrix} 30000 & -30000 \\ -30000 & 30000 \end{bmatrix} = \begin{bmatrix} k_{ii}^2 & k_{ij}^2 \\ k_{ji}^2 & k_{jj}^2 \end{bmatrix}$$

(superscript 2 indicates element 2)

i.e., $k_{ii}^2 = 30000, k_{ij}^2 = -30000, k_{ji}^2 = -30000, k_{jj}^2 = 30000$

Assembly of the overall structure stiffness matrix

The overall structure stiffness matrix K in (2.8), is assembled by starting with a square matrix of order n where n is the total number of degrees of freedom. Since there is only one degree of freedom at each of the three nodes, namely the displacement u in the x direction, it follows that the overall structure stiffness matrix is a 3×3 matrix whose coefficients are denoted by K_{ij} where the subscripts i and j refer to the number of row and number of column, respectively. Thus

$$\begin{array}{cc} & \begin{array}{ccc} u_1 & u_2 & u_3 \end{array} \\ K = & \begin{bmatrix} K_{11} & K_{12} & K_{13} \\ K_{21} & K_{22} & K_{23} \\ K_{31} & K_{32} & K_{33} \end{bmatrix} \begin{array}{c} u_1 \\ u_2 \\ u_3 \end{array} \end{array}$$

In order to find the coefficients of the above matrix, it is useful to relate the element address to the structure address as explained below.

Element 1 has its i node at node1 of the structure and its j node at node 2 of the structure. This leads to the following relationship between element address and structure address.

Element 1 address	i	j
The corresponding structure address	1	2

From the above correspondence between nodes 1 and 2 of the structure and the element nodes, i and j, it follows that element 1 will contribute to the coefficients of the overall structure stiffness matrix K as shown below:

subscript of coefficient in K	11	12	21	22
subscript of contributing coefficient from k^1	ii	ij	ji	jj

So the contribution of element 1 to the overall structure matrix is

$$K_{11}^1 = k_{ii}^1 = 25000, K_{12}^1 = k_{ij}^1 = -25000, K_{21}^1 = k_{ji}^1 = -25000,$$

$$K_{22}^1 = k_{jj}^1 = 25000$$

Element 2 has its i node at node 2 of the structure and its j node at node 3 of the structure. This leads to the following relationship between element address and structure address.

Element 2 address	i	j
The corresponding structure address	2	3

From the above correspondence between nodes 2 and 3 of the structure and the element nodes i and j, it follows that element 2 will contribute to the coefficients of the overall structure stiffness matrix K as shown below:

subscript of coefficient in K	22	23	32	33
subscript of contributing coefficient from k^2	ii	ij	ji	jj

So the contribution of element 2 to the overall structure matrix is

$$K_{22}^2 = k_{ii}^2 = 30000, K_{23}^2 = k_{ij}^2 = -30000, K_{32}^2 = k_{ji}^2 = -30000,$$

$$K_{33}^2 = k_{jj}^2 = 30000$$

Note that the superscript in k indicates the number of the element and superscript in K indicates the number of the contributing element to the overall stiffness matrix.

Steps in the construction of the overall structure stiffness matrix K

(1) Write the value of zero for all coefficients and call it the zero matrix K°. Parts of this matrix will be filled and the rest will remain to have zeros as can be seen later.

$K^{\circ} =$

	u_1	u_2	u_3
u_1	0	0	0
u_2	0	0	0
y_3	0	0	0

Strictly speaking, the above table should be written in the usual matrix notation but the use of tabular form, here and in other parts of the book, makes the presentation clearer particularly when one matrix is superimposed on another.

(2) Enter the contribution of element 1 and call it K^1

$K^1 =$

	u_1	u_2	u_3
u_1	k^1_{ii} (=25000)	k^1_{ij} (=−25000)	0
u_2	k^1_{ji} (=−25000)	k^1_{jj} (=25000)	0
u_3	0	0	0

(The superscript in K indicates the number of the contributing element, in this case it is element 1.)

(3) Enter the contribution of element 2 and call it K^2

$K^2 =$

	u_1	u_2	u_3
u_1	0	0	0
u_2	0	k^2_{ii} (=30000)	k^2_{ij} (=−30000)
u_3	0	k^2_{ji} (=−30000)	k^2_{jj} (=30000)

(The superscript in K indicates the number of the contributing element, in this case it is member 2.)

(4) The final overall structure matrix K is obtained by adding the contributions of all elements of the structure, i.e. $K = K^{\circ} + K^1 + K^2$ simply by superimposing them one on top of the other to get

	u_1	u_2	u_3	
	k_{ii}^1 (=25000)	k_{ij}^1 (=−25000)	0	u_1
K =	k_{ji}^1 (=−25000)	$k_{jj}^1 + k_{ii}^2$ (=25000+30000)	k_{ij}^2 (=−30000)	u_2
	0	k_{ji}^2(=−30000)	k_{jj}^2 (=30000)	u_3

$$K = \begin{bmatrix} 25000 & -25000 & 0 \\ -25000 & 55000 & -30000 \\ 0 & -30000 & 30000 \end{bmatrix} \tag{2.9}$$

To summarise the procedure of assembly of the overall structure matrix K is to start with all the coefficients equal to zero and then dumping k_{ii}, k_{ij}, k_{ji}, and k_{jj} of the element stiffness sub-matrices in the appropriate location in the K matrix.

The displacement vector, $\delta = \begin{bmatrix} \delta_1 \\ \delta_2 \\ \delta_3 \end{bmatrix}$ consists of the displacements at the three nodes.

Since there is only one degree of freedom at each node which is the translation u in the global x-direction it follows that, $\delta_1 = u_1$, $\delta_2 = u_2$, and $\delta_3 = u_3$, therefore

$$\delta = \begin{bmatrix} u_1 \\ u_2 \\ u_3 \end{bmatrix} \tag{2.10}$$

Load vector

The external load vector $F = \begin{bmatrix} F_1 \\ F_2 \\ F_3 \end{bmatrix}$ consists of the external forces (loads) acting at the nodes and since there is only one degree of freedom at each node these forces will be acting in the direction of that degree of freedom. Thus $F_1 = X_1$, $F_2 = X_2$, and $F_3 = X_3$, therefore

$$F = \begin{bmatrix} X_1 \\ X_2 \\ X_3 \end{bmatrix}$$

The forces X_1, X_2, and X_3 are the external forces applied to the structure at the nodes 1, 2, and 3, respectively, and are written relative to the global coordinates. Therefore, at node 1, $X_1 = +3$ kN and at node 2, $X_2 = 0$. At node 3 where the bar (structure) is fixed the displacement is known (i.e., $u_3 = 0$) but the force, which is the reaction of the support on the structure, is unknown. If we denote the reaction by R_{X3} then the force at node 3, X_3, will take the value of R_{X3}. Hence the load vector F is:

$$F = \begin{bmatrix} +3 \\ 0 \\ R_{X3} \end{bmatrix} \tag{2.11}$$

Setting up the full set of equations

Substituting (2.9), (2.10), and (2.11) in (2.8) to get

$$\begin{bmatrix} 25000 & -25000 & 0 \\ -25000 & 55000 & -30000 \\ 0 & -30000 & 30000 \end{bmatrix} \begin{bmatrix} u_1 \\ u_2 \\ u_3 \end{bmatrix} = \begin{bmatrix} 3 \\ 0 \\ R_{X3} \end{bmatrix} \tag{2.12}$$

The structure matrix in (2.9) and (2.12) is singular, i.e. its determinant is equal to zero. Therefore it has no inverse and consequently no unique solution can be obtained and the structure will move as a rigid body. In order to obtain a solution, some constraints must be imposed on the structure and these are called the boundary conditions as shown in the following section.

The relationship in (2.12) can be written as a set of simultaneous equations

$$25000u_1 - 25000u_2 + 0u_3 = 3 \tag{2.12a}$$

$$-25000u_1 + 55000u_2 - 30000u_3 = 0 \tag{2.12b}$$

$$-0u_1 - 30000u_2 + 30000u_3 = R_{X3} \tag{2.12c}$$

Applying the boundary conditions

The next step is to introduce the boundary conditions (constraints) which in this case is the fixed end, i.e. node 3, thus $u_3 = 0$ resulting in the so-called reduced matrix. This can be enforced by deleting the corresponding row (number 3) and the corresponding column (number 3) to give

$$25000u_1 - 25000u_2 = 3$$

$$-25000u_1 + 55000u_2 = 0$$

Notice that the determinant of the matrix $\begin{bmatrix} 25000 & -25000 \\ -25000 & 55000 \end{bmatrix}$ is not zero, therefore it is non-singular and there exists a unique solution of the above simultaneous equations.

Solution of the reduced set of equations

The above two simultaneous equations are solved by any of the methods explained in Chapter 1 to give the displacements as u_1 = 0.00022 m and u_2 = 0.00010 m. With the boundary condition at the fixed end, i.e. $u_3 = 0$, the full displacement vector is:

$$\delta = \begin{bmatrix} \delta_1 \\ \delta_2 \\ \delta_3 \end{bmatrix} = \begin{bmatrix} u_1 \\ u_2 \\ u_3 \end{bmatrix} = \begin{bmatrix} 0.00022 \\ 0.00010 \\ 0 \end{bmatrix}$$

Calculation of reactions at the constrained nodes

These are usually calculated from the appropriate equations of the original (unreduced) matrix by using the values of the displacements obtained from the previous step.

The reactions at the nodes where there are constraints are calculated relative to global coordinates. In this example there is only one reaction R_{X3} at the support (node 3) and is obtained from (2.12c)

$$-30000u_2 + 30000u_3 = R_{X3}$$

$$R_{X3} = -30000 \times 0.00010 + 30000 \times 0 = -3.00 \text{ kN}$$

The negative sign means that it is actually in the opposite direction to that shown, i.e. in the negative x-direction.

Calculation of actions (forces) developed in the elements

The internal forces (also called actions) induced in each of the individual elements are calculated relative to the local coordinates from the relation $\bar{F} = \bar{k}\bar{\delta}$ whose expanded form is given in (2.4).

Element 1

$$\bar{F}^1 = \bar{k}^1\bar{\delta}^1$$

$$\bar{F}^1 = \begin{bmatrix} \bar{X}_1^1 \\ \bar{X}_2^1 \end{bmatrix}, \bar{k}^1 = \begin{bmatrix} 25000 & -25000 \\ -25000 & 25000 \end{bmatrix}, \bar{\delta}^1 = \begin{bmatrix} \bar{u}_i^1 \\ \bar{u}_j^1 \end{bmatrix} = \begin{bmatrix} u_1 \\ u_2 \end{bmatrix} = \begin{bmatrix} 0.00022 \\ 0.00010 \end{bmatrix}$$

$$\begin{bmatrix} \bar{X}_1^1 \\ \bar{X}_2^1 \end{bmatrix} = \begin{bmatrix} 25000 & -25000 \\ -25000 & 25000 \end{bmatrix} \begin{bmatrix} 0.00022 \\ 0.00010 \end{bmatrix} = \begin{bmatrix} +3.00 \\ -3.00 \end{bmatrix}$$

(The subscript indicates the node number and the superscript represents the number of the element.)

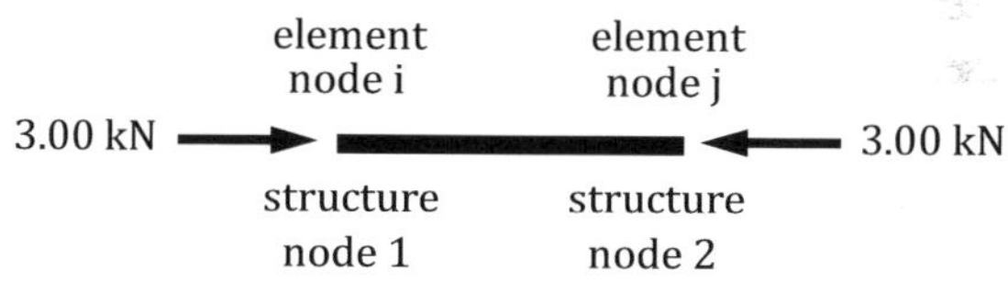

Figure 2.5

Notice that the force at node 1 (or i for the element) is positive and that at node 2 (or j for the element) is negative and this means that the element is in compression as shown in Fig. 2.5.

Element 2

$$\bar{F}^2 = \bar{k}^2\bar{\delta}^2$$

$$\bar{F}^2 = \begin{bmatrix} \bar{X}_2^2 \\ \bar{X}_3^2 \end{bmatrix}, \bar{k}^2 = \begin{bmatrix} 30000 & -30000 \\ -30000 & 30000 \end{bmatrix}, \bar{\delta}^2 = \begin{bmatrix} \bar{u}_i^2 \\ \bar{u}_j^2 \end{bmatrix} = \begin{bmatrix} u_2 \\ u_3 \end{bmatrix} = \begin{bmatrix} 0.00010 \\ 0 \end{bmatrix}$$

$$\begin{bmatrix} \bar{X}_2^2 \\ \bar{X}_3^2 \end{bmatrix} = \begin{bmatrix} 30000 & -30000 \\ -30000 & 30000 \end{bmatrix} \begin{bmatrix} 0.00010 \\ 0 \end{bmatrix} = \begin{bmatrix} +3.00 \\ -3.00 \end{bmatrix},$$

i.e. the element is in compression as shown in Fig. 2.6.

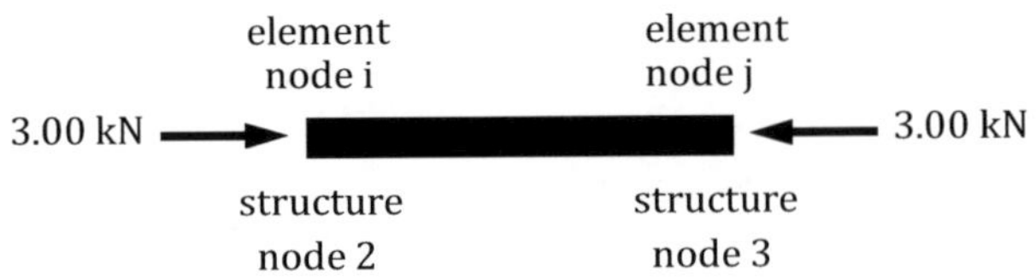

Figure 2.6

This is a very simple example that can be solved quite easily (and quickly) by elementary methods. However, it illustrates the procedure followed in the analysis of structures by matrix methods.

2.1.3 Bar Elements with Variable Cross Section

Sometimes the bar has a continuously variable cross section and one of the methods to deal with such a case is to derive the stiffness matrix from first principles for the specific type of cross section variation. Alternatively, the bar is divided into elements each of which is assumed to have a constant average cross section resulting in a stepped bar as shown in the example below. The solution is of course approximate and the accuracy can be improved by increasing the number of elements.

Example 2

A steel bar with its ends fixed at nodes 1 and 4 has uniform thickness 0.006 m and its width is given by the equation $b = 0.12 + 2x^2$ as shown in Fig. 2.7. Calculate the displacements and forces induced in the bar at nodes 1, 2, 3, and 4. The modulus of elasticity of steel $E = 210 \times 10^6\ \text{kN/m}^2$.

The bar is divided into three elements and the width of each element is assumed to be equal to the width at the middle of the element as shown in Fig. 2.7. The resulting analysis model is a stepped bar and is treated in the same way as in Example 1.

$$\text{Element 1: } A_1 = b_1 t = 0.14 \times 0.006 = 840 \times 10^{-6}\ \text{m}^2$$

$$\text{Element 2: } A_2 = b_2 t = 0.30 \times 0.006 = 1800 \times 10^{-6}\ \text{m}^2$$

$$\text{Element 3: } A_3 = b_3 t = 0.62 \times 0.006 = 3720 \times 10^{-6}\ \text{m}^2$$

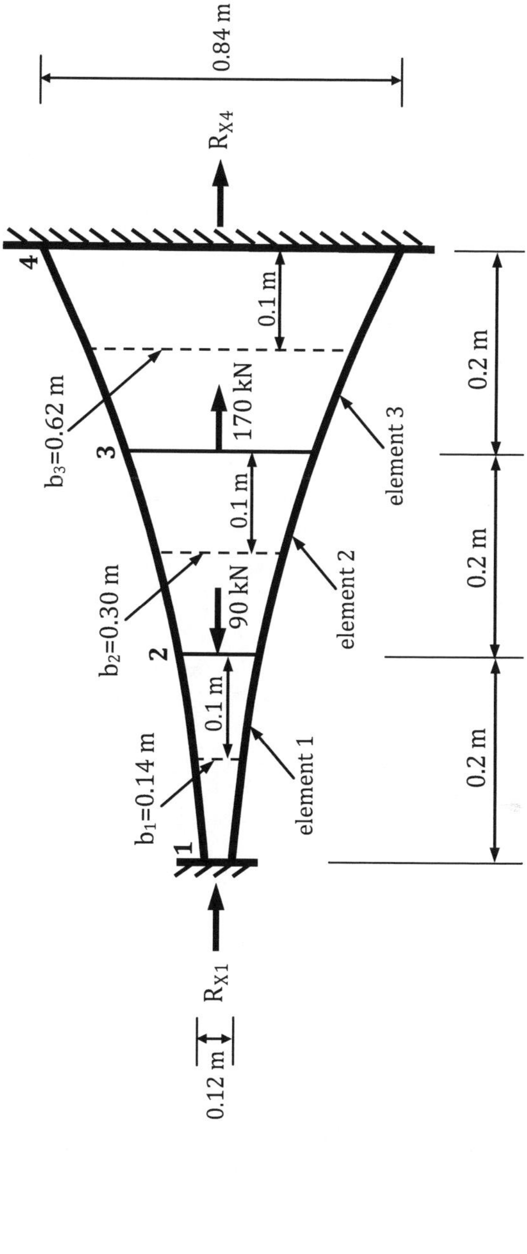
0.84 m
R_{X4}
4
0.1 m
b_3=0.62 m
170 kN
3
element 3
0.2 m
0.1 m
b_2=0.30 m
90 kN
2
element 2
0.2 m
0.1 m
b_1=0.14 m
element 1
0.2 m
1
R_{X1}
0.12 m

Figure 2.7

Element stiffness matrices

These are obtained from (2.6) as shown below.

Element 1

$$\frac{EA}{L} = \frac{210 \times 10^6 \times 840 \times 10^{-6}}{0.2} = 882 \times 10^3 \text{ kN/m}$$

$$k^1 = 10^3 \begin{bmatrix} 882 & -882 \\ -882 & 882 \end{bmatrix}$$

Element 2

$$\frac{EA}{L} = \frac{210 \times 10^6 \times 1800 \times 10^{-6}}{0.2} = 1890 \times 10^3 \text{ kN/m}$$

$$k^2 = 10^3 \begin{bmatrix} 1890 & -1890 \\ -1890 & 1890 \end{bmatrix}$$

Element 3

$$\frac{EA}{L} = \frac{210 \times 10^6 \times 3720 \times 10^{-6}}{0.2} = 3906 \times 10^3 \text{ kN/m}$$

$$k^3 = 10^3 \begin{bmatrix} 3906 & -3906 \\ -3906 & 3906 \end{bmatrix}$$

Following the same procedure of example 1 the following matrix is obtained by inspection.

$K = 10^3$

u_1	u_2	u_3	u_4	
k^1_{ii} (= 882)	k^1_{ij} (= −882)	0	0	u_1
k^1_{ji} (= −882)	$k^1_{jj} + k^2_{ii}$ (=882 + 1890)	k^2_{ij} (= −1890)	0	u_2
0	k^2_{ji} (= −1890)	$k^2_{jj} + k^3_{ii}$ (= 1890 + 3906)	k^3_{ij} (= −3906)	u_3
0	0	k^3_{ji} (= −3906)	k^3_{jj} (= 3906)	u_4

$$K = 10^3 \begin{bmatrix} 882 & -882 & 0 & 0 \\ -882 & 2772 & -1890 & 0 \\ 0 & -1890 & 5796 & -3906 \\ 0 & 0 & -3906 & 3906 \end{bmatrix} \tag{2.13}$$

The displacement vector for the structure is

$$\delta = \begin{bmatrix} \delta_1 \\ \delta_2 \\ \delta_3 \\ \delta_4 \end{bmatrix} = \begin{bmatrix} u_1 \\ u_2 \\ u_3 \\ u_4 \end{bmatrix} \tag{2.14}$$

Load vector

The force vector, F, is composed of the external forces acting at the nodes, i.e. at node 1, $X_1 = R_{X1}$; at node 2, $X_2 = -90$ kN; at node 3, $X_3 = +170$ kN; and $X_4 = R_{X4}$ at node 4. Hence the force vector F is given by:

$$F = \begin{bmatrix} F_1 \\ F_2 \\ F_3 \\ F_4 \end{bmatrix} = \begin{bmatrix} X_1 \\ X_2 \\ X_3 \\ X_4 \end{bmatrix} = \begin{bmatrix} R_{X1} \\ -90 \\ +170 \\ R_{X4} \end{bmatrix} \tag{2.15}$$

Setting up the full set of equations

Substitute (2.13), (2.14), and (2.15) in (2.8) to get:

$$10^3 \begin{bmatrix} 882 & -882 & 0 & 0 \\ -882 & 2772 & -1890 & 0 \\ 0 & -1890 & 5796 & -3906 \\ 0 & 0 & -3906 & 3906 \end{bmatrix} \begin{bmatrix} u_1 \\ u_2 \\ u_3 \\ u_4 \end{bmatrix} = \begin{bmatrix} R_{X1} \\ -90 \\ +170 \\ R_{X4} \end{bmatrix} \tag{2.16}$$

The boundary conditions are the fixed ends of the bar, i.e. $u_1 = 0$ and $u_4 = 0$, hence delete rows and columns 1 and 4 respectively to get

$$2772 \times 10^3 u_2 - 1890 \times 10^3 u_3 = -90$$

$$-1890 \times 10^3 u_2 - 5796 \times 10^3 u_3 = +170$$

The solution of the above set of simultaneous equations is:

$u_2 = -0.0000160$ m and $u_3 = +0.0000241$ m

The full displacement vector is:

$$\delta = \begin{bmatrix} u_1 \\ u_2 \\ u_3 \\ u_4 \end{bmatrix} = \begin{bmatrix} 0 \\ -0.0000160 \\ +0.0000241 \\ 0 \end{bmatrix}$$

The reaction R_{X1} at the left support can be found from the first row of (2.16)

$$882 \times 10^3 u_1 - 882 \times 10^3 u_2 = R_{X1}$$

$$R_{X1} = 882 \times 10^3 \times 0 - 882 \times 10^3 \times (-0.0000160) = +14.11 \text{ kN}$$

The reaction R_{X4} at the right support can be found from the fourth row of (2.16)

$$-3906 \times 10^3 u_3 + 3906 \times 10^3 u_4 = R_{X4}$$

$$R_{X4} = -3906 \times 10^3 \times 0.0000241 + 3906 \times 10^3 \times 0 = -94.14 \text{ kN}$$

The exact values are: $u_2 = -0.0000163$ m, $u_3 = +0.0000248$ m, $R_{X1} = +14.66$ kN and $R_{X4} = -94.66$ kN. The largest difference between the values obtained from dividing the member into only three elements and the exact values is –3.75%. Higher accuracy can be achieved if the member is divided into a larger number of elements.

Calculation of actions (forces) developed in the elements

The final step is to calculate the internal forces induced in the elements relative to local coordinates using (2.4).

Element 1

$$\begin{bmatrix} \bar{X}_1^1 \\ \bar{X}_2^1 \end{bmatrix} = 10^3 \begin{bmatrix} 882 & -882 \\ -882 & 882 \end{bmatrix} \begin{bmatrix} 0 \\ -0.0000160 \end{bmatrix} = \begin{bmatrix} +14.11 \\ -14.11 \end{bmatrix}$$

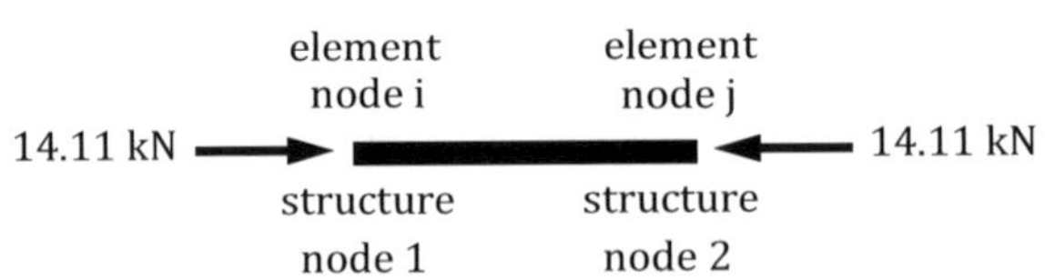

Notice that the force at node 1 (or i for the element) is positive and that at node 2 (or j for the element) is negative and this means that the element is in compression.

Element 2

$$\begin{bmatrix} \bar{X}_2^2 \\ \bar{X}_3^2 \end{bmatrix} = 10^3 \begin{bmatrix} 1890 & -1890 \\ -1890 & 1890 \end{bmatrix} \begin{bmatrix} -0.0000160 \\ +0.0000241 \end{bmatrix} = \begin{bmatrix} -75.79 \\ +75.79 \end{bmatrix}$$

(i.e. the element is in tension)

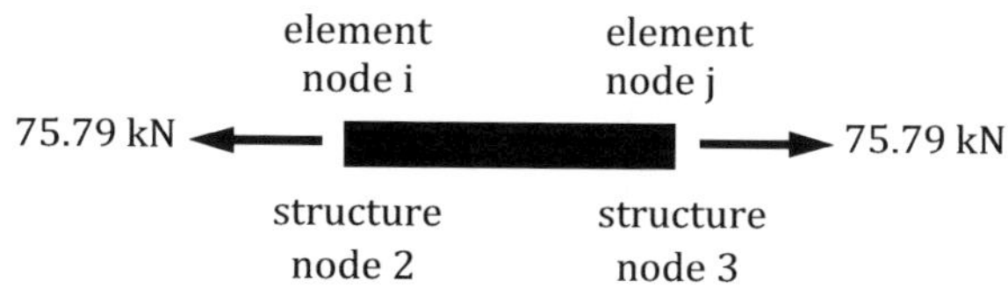

Element 3

$$\begin{bmatrix} \bar{X}_3^3 \\ \bar{X}_4^3 \end{bmatrix} = 10^3 \begin{bmatrix} 3906 & -3906 \\ -3906 & 3906 \end{bmatrix} \begin{bmatrix} +0.0000241 \\ 0 \end{bmatrix} = \begin{bmatrix} +94.14 \\ -94.14 \end{bmatrix}$$

(i.e. the element is in compression)

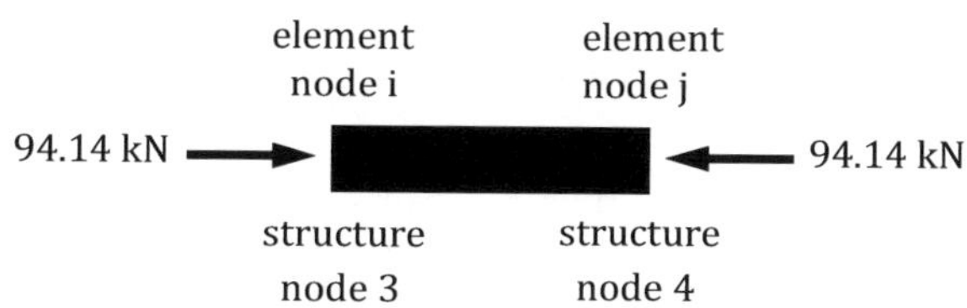

2.1.4 Some Important Properties of the Stiffness Matrix

It can be seen in the above examples that the structure stiffness matrix is symmetric since $K_{ij} = K_{ji}$. This important property is a consequence of Maxwell reciprocal theorem which states that the displacement at node j produced by a unit load applied at node i is equal to the displacement at node i produced by a unit load applied at node j. The displacement may be translational or rotational and the load may be a force or a moment.

Another useful property of the structure stiffness matrix is that it is positive definite, i.e. the quadratic form $\delta^T K \delta$ is always positive as explained below.

Consider a set of forces F_1, F_2, ... F_n acting on a structure and the resulting displacements are δ_1, δ_2, ... δ_n, then the work done U is given by

$$U = \frac{1}{2}F_1\delta_1 + \frac{1}{2}F_2\delta_2 \ldots\ldots + \frac{1}{2}F_n\delta_n = \frac{1}{2}\begin{bmatrix} F_1 & F_2 & . & . & F_n \end{bmatrix}\begin{bmatrix} \delta_1 \\ \delta_2 \\ . \\ . \\ \delta_n \end{bmatrix} = \frac{1}{2}F^T\delta$$

$$\text{But } F = K\delta \text{ then } U = \frac{1}{2}(K\delta)^T\delta$$

Since $(K\delta)^T = \delta^T K^T$ and $K^T = K$ because K is symmetric, therefore

$$U = \frac{1}{2}\delta^T K\delta$$

The work done U is always positive, hence $\delta^T K\delta$ is positive for any non-zero displacement vector δ thus K is positive definite.

2.2 Coordinate Systems

The standard right-handed xyz cartesian coordinates system is formed by the right hand where the thumb represents the x-axis, the index finger the y-axis and the middle finger the z-axis.

The three systems shown in Fig. 2.8 are all right-handed xyz coordinate systems and they are the same except that they are viewed from different points. The orientation of the coordinate system shown in Fig. 2.8(i) is the one used in this book where the xy plane is the horizontal plane and the z-axis is perpendicular to it.

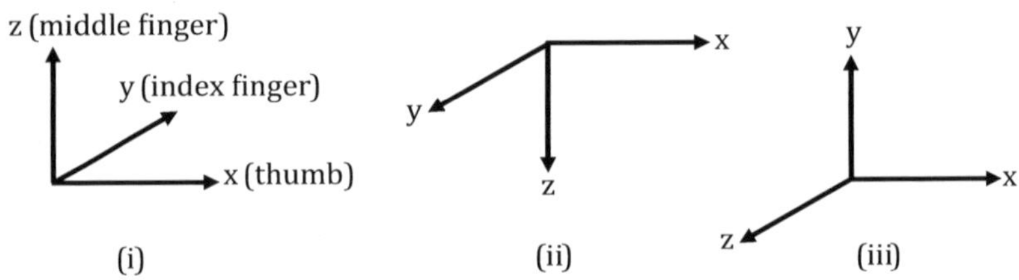

Figure 2.8 Right-handed xyz coordinate systems.

Definitions and sign convention

u, v, and w: Translational displacements in the x, y, and z directions.
Φ, θ, and Ψ: Rotational displacements about the positive x, y, and z axes.
X, Y, and Z: Forces in the x, y, and z directions.
T, M, and N: Moments about the x, y, and z axes.

Translational displacements and forces are positive when they are in the positive direction of the relevant axis. Rotational displacements and moments are positive when they are clockwise about the positive direction of the relevant axis as shown in Fig. 2.9.

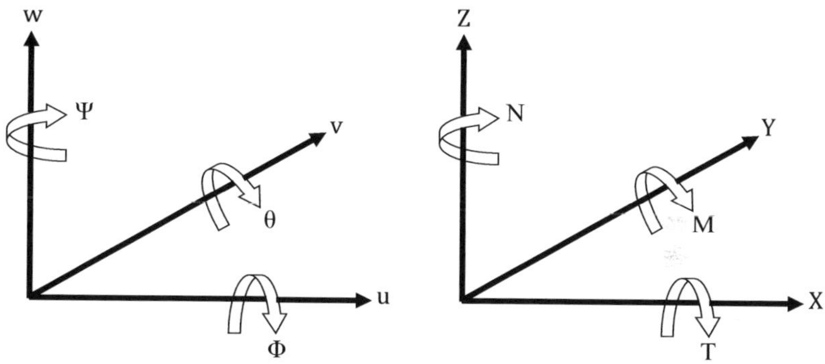

Figure 2.9 Sign convention.

2.3 Extension of Bar Stiffness Matrix to Other Types of Structural Elements

Due to the systematic nature of the subject of matrix methods in the analysis of structures, the detailed steps followed in this chapter can be applied to more complex problems. The principles are the same but some modifications might be necessary where appropriate as will be seen in the chapters that follow. Therefore, a thorough understanding of the contents of this chapter is essential since this will make the reader familiar with the procedures which will help in making good progress through the book and leave more time for understanding any additional development that may occur as the material gets more involved.

In general, for any individual element with two ends the element stiffness matrix, k, is made of 4 sub-matrices, k_{ii}, k_{ij}, k_{ji}, and k_{jj}. For a bar element it was seen that k_{ii}, for example, is a 1×1 sub-matrix (i.e. just one number). This is a consequence of the fact that the displacement at each node is defined by only one degree of freedom, namely, the translation u along the x-axis. The load is defined by only one force acting in the x-direction at each node. For skeletal structures the stiffness matrix of any element is analogous to that of an individual bar except that the number of degrees of freedom is generally more than one. For example in a pin-connected plane frame lying in the xz plane there are two degrees of freedom at each joint (node), namely, the translations, u and w, in the x and z directions respectively. As a consequence, k_{ii}, for example, is a 2×2 sub-matrix and the displacement at each node is a 2×1 vector made of the translation u along the x-axis and the translation w along the z-axis. The load at each node is a 2×1 vector made of a force in the x direction and a force in the z direction. It follows that for an element with n degrees of freedom at each end sub-matrices such as k_{ii} will be of size n×n. Similarly the displacement and load vectors at each end will be vectors of size n×1. The above statement is illustrated in the cases considered below.

The general relationship for an element with two ends i and j is:

$$k\delta = F \text{ or } \begin{bmatrix} k_{ii} & k_{ij} \\ k_{ji} & k_{jj} \end{bmatrix} \begin{bmatrix} \delta_i \\ \delta_j \end{bmatrix} = \begin{bmatrix} F_i \\ F_j \end{bmatrix} \tag{2.17}$$

k_{ii}, k_{ij}, k_{ji}, and k_{jj} are the four sub-matrices of the stiffness matrix, k.

δ_i and δ_j are the two sub-matrices of the displacement vector, δ.
F_i and F_j are the two sub-matrices of the load vector, F.

(1) Bar element (Fig. 2.10)

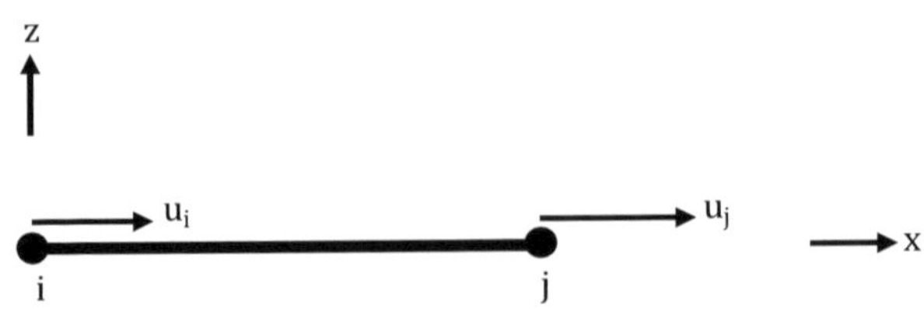

Figure 2.10

k: each of the four sub-matrices is 1×1, for example $k_{ii} = [EA/L]$

δ: each of the two sub-vectors is 1 coefficient, for example $\delta_i = [u_i]$

F: each of the two sub-vectors is 1 coefficient, for example $F_i = [X_i]$

(2) Element in a pin-connected plane frame (Fig. 2.11)

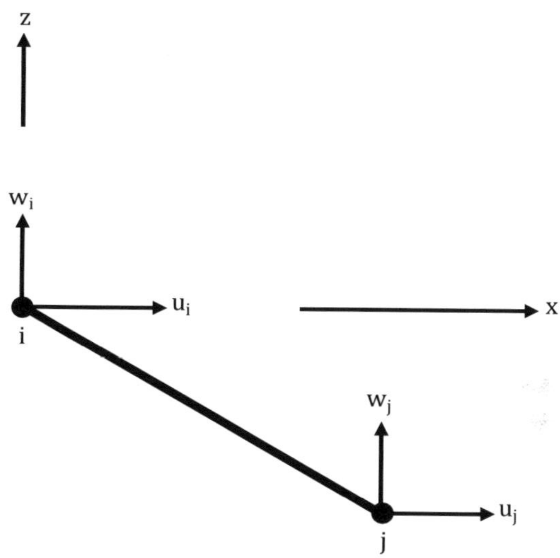

Figure 2.11

k: each of the four sub-matrices is 2×2; $k_{ii} = \begin{bmatrix} * & * \\ * & * \end{bmatrix}$

(The asterisk means one number.)

δ: each of the two sub-vectors is 2 coefficients; $\delta_i = \begin{bmatrix} u_i \\ w_i \end{bmatrix}$

F: each of the two sub-vectors is 2 coefficients; $F_i = \begin{bmatrix} X_i \\ Z_i \end{bmatrix}$

(3) Element in a beam (Fig. 2.12)

k: each of the four sub-matrices is 2×2; $k_{ii} = \begin{bmatrix} * & * \\ * & * \end{bmatrix}$

δ: each of the two sub-vectors is 2 coefficients; $\delta_i = \begin{bmatrix} w_i \\ \theta_i \end{bmatrix}$

F: each of the two sub-vectors is 2 coefficients; $F_i = \begin{bmatrix} Z_i \\ M_i \end{bmatrix}$

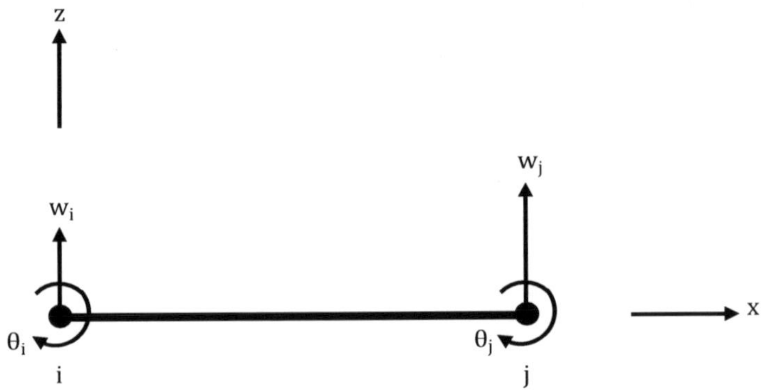

Figure 2.12

(4) Element in a rigidly connected frame (Fig. 2.13)

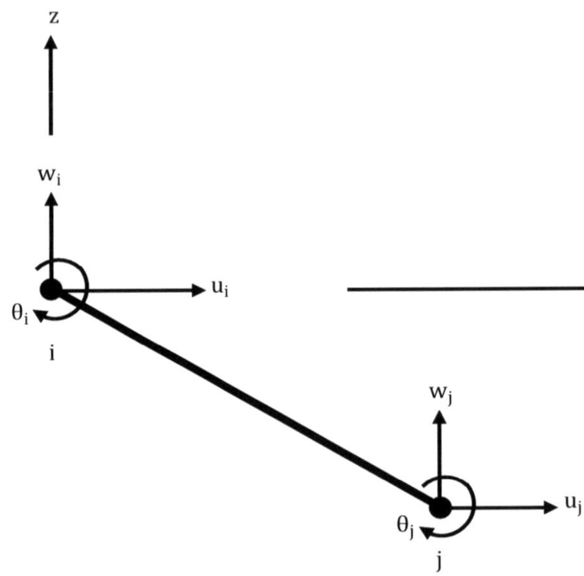

Figure 2.13

k: each of the four sub-matrices is 3×3, $k_{ii} = \begin{bmatrix} * & * & * \\ * & * & * \\ * & * & * \end{bmatrix}$

δ: each of the two sub-vectors is 3 coefficients; $\delta_i = \begin{bmatrix} u_i \\ w_i \\ \theta_i \end{bmatrix}$

F: each of the two sub-vectors is 3 coefficients, $F_i = \begin{bmatrix} X_i \\ Z_i \\ M_i \end{bmatrix}$

(5) Element in an arch (Fig. 2.14)

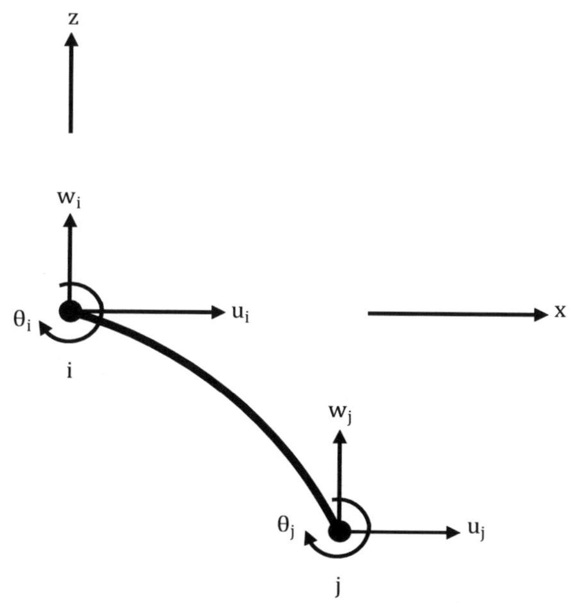

Figure 2.14

k: each of the four sub-matrices is 3×3; $k_{ii} = \begin{bmatrix} * & * & * \\ * & * & * \\ * & * & * \end{bmatrix}$

δ: each of the two sub-vectors is 3 coefficients; $\delta_i = \begin{bmatrix} u_i \\ w_i \\ \theta_i \end{bmatrix}$

F: each of the two sub-vectors is 3 coefficients; $F_i = \begin{bmatrix} X_i \\ Z_i \\ M_i \end{bmatrix}$

(6) Element in a grillage (Fig. 2.15)

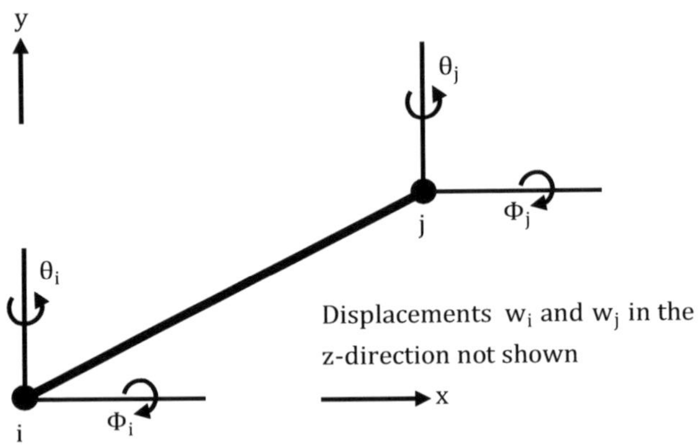

Figure 2.15

k: each of the four sub-matrices is 3×3; $k_{ii} = \begin{bmatrix} * & * & * \\ * & * & * \\ * & * & * \end{bmatrix}$

δ: each of the two sub-vectors is 3 coefficients; $\delta_i = \begin{bmatrix} w_i \\ \Phi_i \\ \theta_i \end{bmatrix}$

F: each of the two sub-vectors is 3 coefficients; $F_i = \begin{bmatrix} Z_i \\ T_i \\ M_i \end{bmatrix}$

(7) Element in a beam curved in plan (Fig. 2.16)

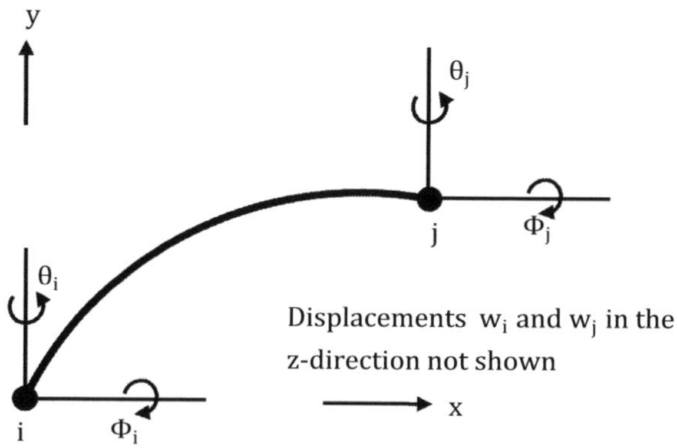

Figure 2.16

k: each of the four sub-matrices is 3×3; $k_{ii} = \begin{bmatrix} * & * & * \\ * & * & * \\ * & * & * \end{bmatrix}$

δ: each of the two sub-vectors is 3 coefficients; $\delta_i = \begin{bmatrix} w_i \\ \Phi_i \\ \theta_i \end{bmatrix}$

F: each of the two sub-vectors is 3 coefficients; $F_i = \begin{bmatrix} Z_i \\ T_i \\ M_i \end{bmatrix}$

(8) Element in a pin-connected space frame (Fig. 2.17)

k: each of the four sub-matrices is 3×3; $k_{ii} = \begin{bmatrix} * & * & * \\ * & * & * \\ * & * & * \end{bmatrix}$

δ: each of the two sub-vectors is 3 coefficients; $\delta_i = \begin{bmatrix} u_i \\ v_i \\ w_i \end{bmatrix}$

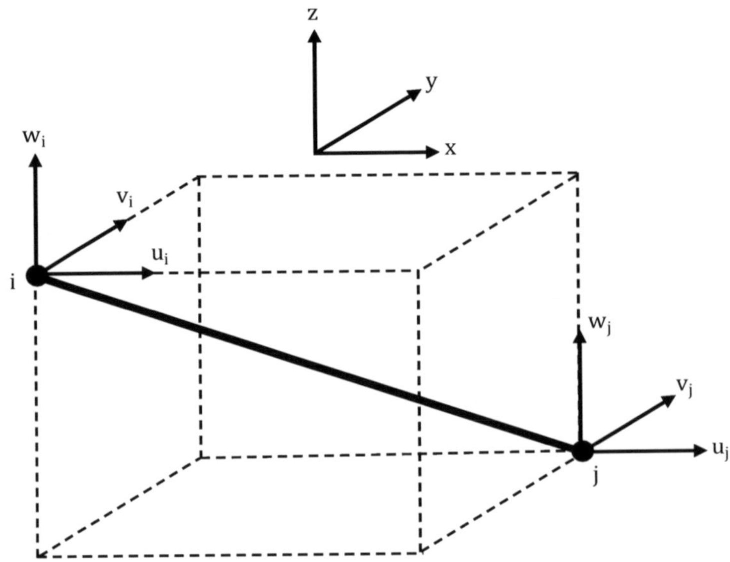

Figure 2.17

F: each of the two sub-vectors is 3 coefficients; $F_i = \begin{bmatrix} X_i \\ Y_i \\ Z_i \end{bmatrix}$

(9) Element in a rigidly connected space frame (Fig. 2.18)

k: each of the four sub-matrices is 6×6; $k_{ii} = \begin{bmatrix} * & * & * & * & * & * \\ * & * & * & * & * & * \\ * & * & * & * & * & * \\ * & * & * & * & * & * \\ * & * & * & * & * & * \\ * & * & * & * & * & * \end{bmatrix}$

δ: each of the two sub-vectors is 6 coefficients; $\delta_i = \begin{bmatrix} u_i \\ v_i \\ w_i \\ \Phi_i \\ \theta_i \\ \Psi_i \end{bmatrix}$

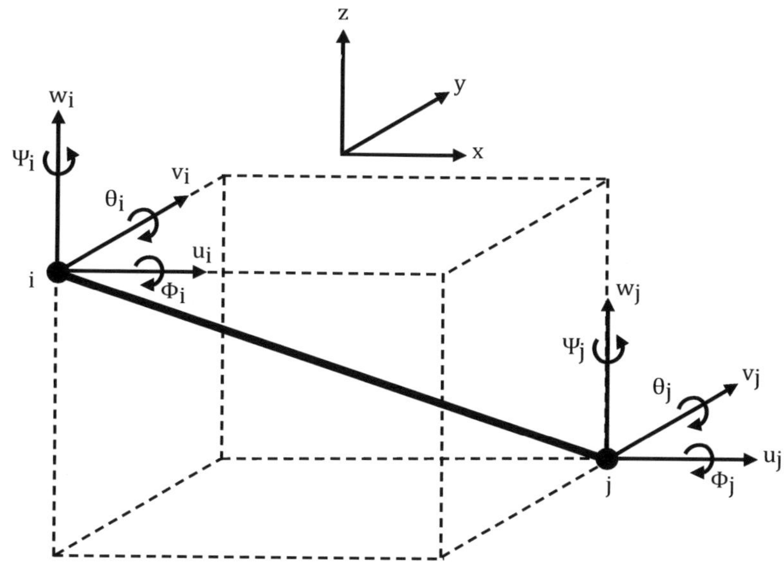

Figure 2.18

F: each of the two sub-vectors is 6 coefficients; $F_i = \begin{bmatrix} X_i \\ Y_i \\ Z_i \\ T_i \\ M_i \\ N_i \end{bmatrix}$

2.4 Banded Matrix

In most applications of matrix methods of structural analysis the matrix is not fully populated but rather banded with zero coefficients outside the band. In order to reduce the computer storage requirements only those coefficients within the band are stored and subsequently used in the computations. So, the smaller the band width the more efficient it is and this depends on the way the nodes of the structure are numbered.

Table 2.1 Summary of the relation between the number of degrees of freedom per node and the size of sub-matrices in the standard form (2.17) for various types of structures

	Degrees of freedom per node						Size of any sub-matrix	Size of any sub-vector
Type of structure	**u**	**v**	**w**	**Φ**	**θ**	**Ψ**	$\mathbf{k_{ii}, k_{ij}, k_{ji}, k_{jj}}$	$\delta_i, \delta_j, \mathbf{F_i, F_j}$
Bar	√						1×1	1×1
Pin-connected plane frame	√		√				2×2	2×1
Continuous beam			√		√		2×2	2×1
Rigidly connected plane frame	√		√		√		3×3	3×1
Arch	√		√		√		3×3	3×1
Grillage			√	√	√		3×3	3×1
Beam curved in plan			√	√	√		3×3	3×1
Pin-connected space frame	√	√	√				3×3	3×1
Rigidly connected space frame	√	√	√	√	√	√	6×6	6×1

Since the stiffness matrix is symmetrical, only the semi-band width is considered which is defined by the coefficient on the main diagonal and the non-zero coefficients to its right side. To illustrate this consider the frame shown in Fig. 2.19a, where nodes are numbered in the short direction. The semi-band width is 3 and since the number of degrees of freedom per node is 3, consisting of u, w, and θ, i.e. each asterisk is a 3×3 sub-matrix, so the number of coefficients in the semi-band width is 3×3=9. The total number of rows is 3×14=42 giving the number of coefficients to be stored as 9×42=378 compared with the total number of coefficients in the stiffness matrix which is, 42×42=1746. Thus there is significant saving in computer storage in this small problem but for large matrices the saving can be very large.

If the nodes are numbered in the long direction as shown in Fig. 2.19b, then the semi-band width will be 7 and the total number of coefficients to be stored is 7×3×42=882. Thus for this case and in general, numbering in the short direction requires less storage than numbering in the long direction.

δ_3	δ_4	δ_5	δ_6	δ_7	δ_8	δ_9	δ_{10}	δ_{11}	δ_{12}	δ_{13}	δ_{14}	
*	*	*										δ_3
*	*		*									δ_4
*		*	*	*								δ_5
	*	*	*		*							δ_6
		*		*	*	*						δ_7
			*	*	*		*					δ_8
				*		*	*	*				δ_9
					*	*	*		*			δ_{10}
						*		*	*	*		δ_{11}
							*	*	*		*	δ_{12}
								*		*	*	δ_{13}
									*	*	*	δ_{14}

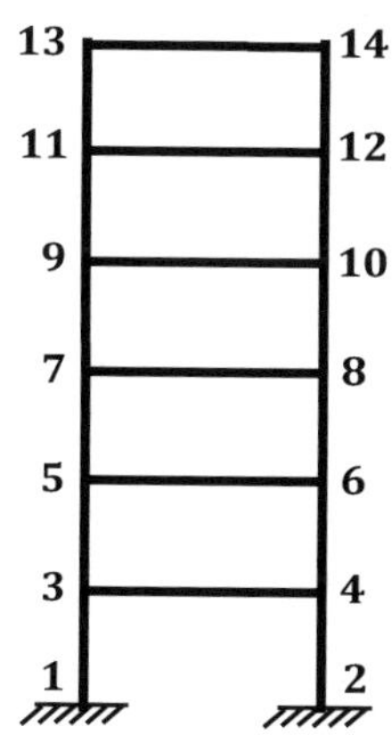

Figure 2.19a

	δ_2	δ_3	δ_4	δ_5	δ_6	δ_7	δ_9	δ_{10}	δ_{11}	δ_{12}	δ_{13}	δ_{14}
δ_2	*	*					*					
δ_3	*	*	*					*				
δ_4		*	*	*					*			
δ_5			*	*	*					*		
δ_6				*	*	*					*	
δ_7					*	*						*
δ_9	*						*	*				
δ_{10}		*					*	*	*			
δ_{11}			*					*	*	*		
δ_{12}				*					*	*	*	
δ_{13}					*					*	*	*
δ_{14}						*					*	*

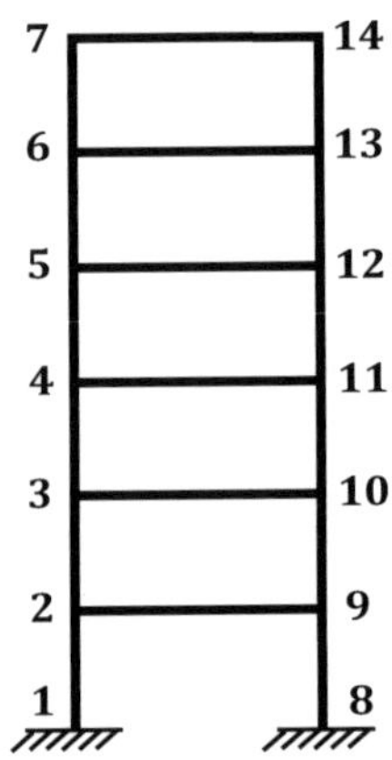

Figure 2.19b

(Each asterisk is an r×r sub-matrix where r is the number of degrees of freedom per node.)

Problems

Calculate the displacements at the nodes and forces developed in the elements of the structures shown in Problems P2.1 to P2.4.

P2.1. Element 1, $L_1 = 0.56$ m, $A_1 = 720 \times 10^{-6}$ m^2, element 2, $L_2 = 0.35$ m, $A_2 = 240 \times 10^{-6}$ m^2. The material is aluminium with a modulus of elasticity $E = 70 \times 10^6$ kN/m^2.

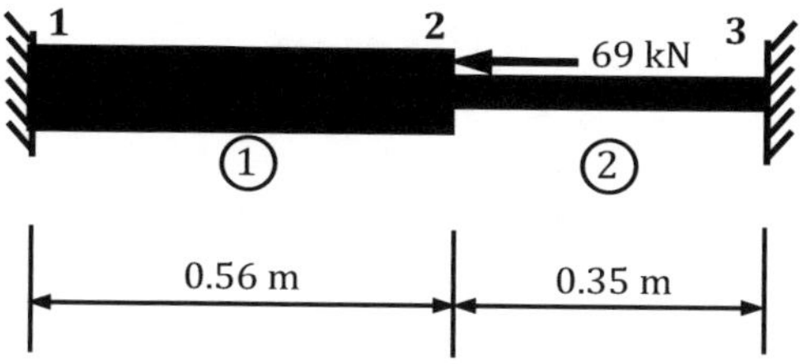

Figure P2.1

Answer:

$u_1 = 0$, $u_2 = -0.0005$ m, $u_3 = 0$, $R_{X1} = 45.00$ kN, $R_{X3} = 24.00$ kN

$$\text{Element 1: } \begin{bmatrix} \bar{X}_1^1 \\ \bar{X}_2^1 \end{bmatrix} = \begin{bmatrix} +45.00 \\ -45.00 \end{bmatrix} \text{kN (compression)}$$

$$\text{Element 2: } \begin{bmatrix} \bar{X}_2^2 \\ \bar{X}_3^2 \end{bmatrix} = \begin{bmatrix} -24.00 \\ +24.00 \end{bmatrix} \text{kN (tension)}$$

P2.2. Element 1, $L_1 = 0.45$ m, $A_1 = 5000 \times 10^{-6}$ m^2, element 2, $L_2 = 0.36$ m, $A_2 = 8000 \times 10^{-6}$ m^2. The material is timber with a modulus of elasticity $E = 9 \times 10^6$ kN/m^2.

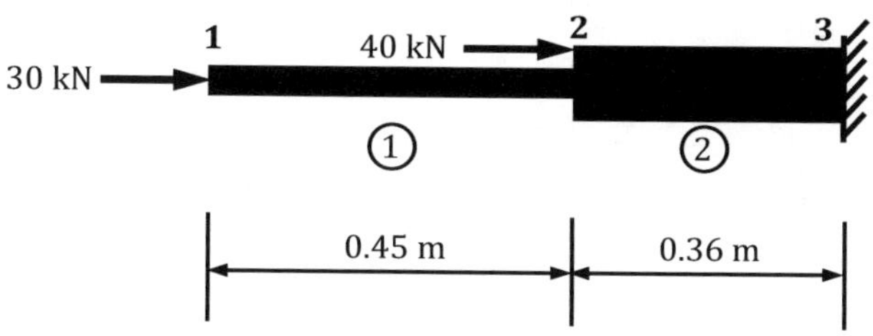

Figure P2.2

Answer:

$u_1 = 0.00065$ m, $u_2 = 0.00035$ m, $u_3 = 0$, $R_{X3} = -70.00$ kN

$$\text{Element 1: } \begin{bmatrix} \bar{X}_1^1 \\ \bar{X}_2^1 \end{bmatrix} = \begin{bmatrix} +30.00 \\ -30.00 \end{bmatrix} \text{kN (compression)}$$

$$\text{Element 2: } \begin{bmatrix} \bar{X}_2^2 \\ \bar{X}_3^2 \end{bmatrix} = \begin{bmatrix} +70.00 \\ -70.00 \end{bmatrix} \text{kN (compression)}$$

P2.3. Element 1, $L_1 = 0.18$ m, $A_1 = 300 \times 10^{-6}$ m^2, element 2, $L_2 = 0.42$ m, $A_2 = 800 \times 10^{-6}$ m^2, element 3, $L_3 = 0.25$ m, $A_3 = 500 \times 10^{-6}$ m^2. The material is steel with a modulus of elasticity $E = 210 \times 10^6$ kN/m^2.

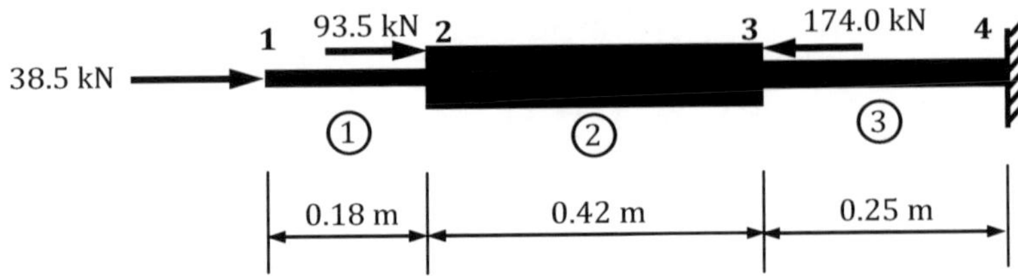

Figure P2.3

Answer:

$u_1 = 0.00034$ m, $u_2 = 0.00023$ m, $u_3 = -0.00010$ m, $u_4 = 0$, $R_{X4} = 42.00$ kN

$$\text{Element 1: } \begin{bmatrix} \bar{X}_1^1 \\ \bar{X}_2^1 \end{bmatrix} = \begin{bmatrix} +38.50 \\ -38.50 \end{bmatrix} \text{kN (compression)}$$

$$\text{Element 2: } \begin{bmatrix} \bar{X}_2^2 \\ \bar{X}_3^2 \end{bmatrix} = \begin{bmatrix} +132.00 \\ -132.00 \end{bmatrix} \text{kN (compression)}$$

$$\text{Element 3: } \begin{bmatrix} \bar{X}_3^3 \\ \bar{X}_4^3 \end{bmatrix} = \begin{bmatrix} -42.00 \\ +42.00 \end{bmatrix} \text{kN (tension)}$$

P2.4. A block of concrete of thickness 0.120 m and its other dimensions are as shown in Fig. P2.4 is fixed at nodes 1 and 4. Calculate the displacements and the forces developed at the nodes of the block. The modulus of elasticity of concrete $E = 20 \times 10^6$ kN/m^2.

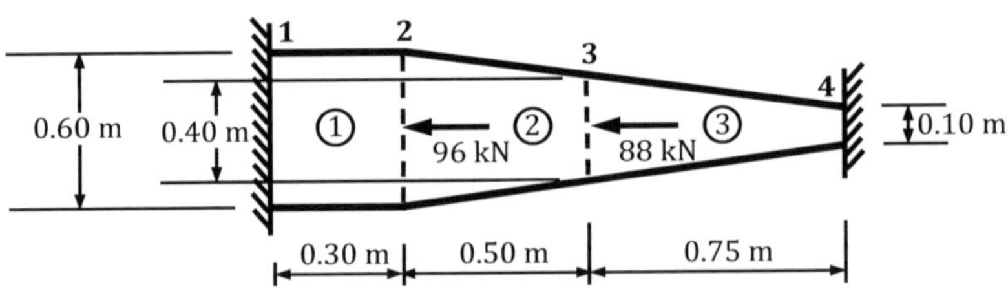

Figure P2.4

Answer:

$u_1 = 0$, $u_2 = -0.00003$ m, $u_3 = -0.00005$ m, $u_4 = 0$

$R_{X1} = 144.00$ kN, $R_{X4} = 40.00$ kN

Element 1: $\begin{bmatrix} \bar{X}_1^1 \\ \bar{X}_2^1 \end{bmatrix} = \begin{bmatrix} +144.00 \\ -144.00 \end{bmatrix}$ kN (compression)

Element 2: $\begin{bmatrix} \bar{X}_2^2 \\ \bar{X}_3^2 \end{bmatrix} = \begin{bmatrix} +48.00 \\ -48.00 \end{bmatrix}$ kN (compression)

Element 3: $\begin{bmatrix} \bar{X}_3^3 \\ \bar{X}_4^3 \end{bmatrix} = \begin{bmatrix} -40.00 \\ +40.00 \end{bmatrix}$ kN (tension)

Chapter 3

Pin-Connected Plane Frames

Structures are usually three dimensional but the type of connections between the members and the framing can be so arranged that the main support structure can be analysed as a plane frame. This is a simplifying assumption that gives results not too far from the actual behaviour of the overall structure particularly when the secondary members are pin-connected to the main supporting structure. Secondary members can be in the form of purlins supporting roof decking which are in turn supported by the main frame. There are many types of main frames used in practice and one of these is the truss which is treated in this chapter.

Roof trusses are used when a single member will not be an efficient structural design option in certain situations such as large span column free spaces as shown in Fig. 3.1. Also, one of the options of bridge design is the use of trusses when these are considered as a suitable choice for a particular span and applied loads as shown in Fig. 3.2. The members in such frames are usually assumed to be pin-connected to each other although in practice they might not have physical pins at their ends. This assumption means that the joints of the frame are not capable of transferring moments. Of course, the actual construction of such frames must be consistent with this assumption in that the connections are detailed in such a way that they can transfer forces but are not capable of resisting moments.

Analysis of Structures by Matrix Methods
Fathi Al-Shawi

ISBN 978-981-4968-19-5 (Hardcover), 978-1-003-29130-5 (eBook)
www.jennystanford.com

The applied loads on the frame are usually applied at the joints and as a consequence, the members of a pin-connected frame will develop axial forces only.

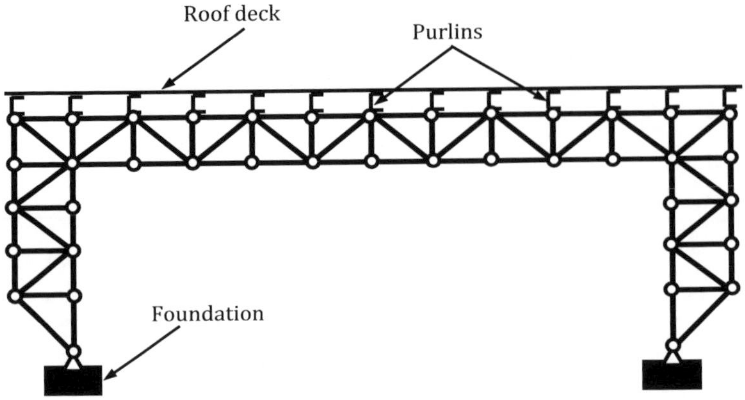

Figure 3.1 Large span pin-connected frame (side rails and cladding not shown).

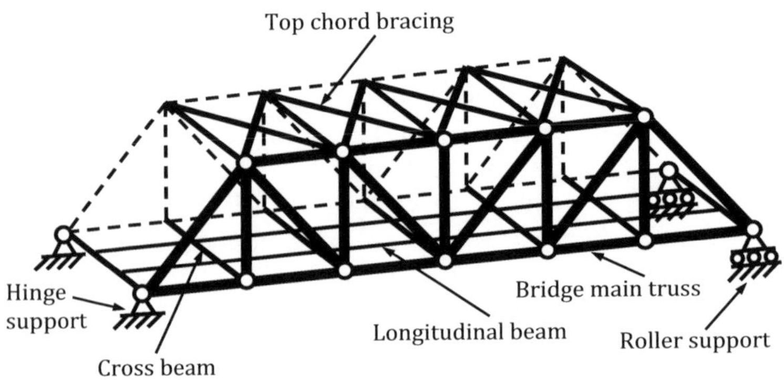

Figure 3.2 Pin-connected truss as the main structure of a bridge (only main truss pins are shown and deck not shown for clarity).

3.1 Derivation of Stiffness Matrix

From Chapter 2, of bars subjected to axial forces where each end has one degree of freedom, we had a relationship between the stiffness, displacements, and forces at the ends of a bar as:

$$\begin{bmatrix} \frac{EA}{L} & -\frac{EA}{L} \\ -\frac{EA}{L} & \frac{EA}{L} \end{bmatrix} \begin{bmatrix} \bar{u}_i \\ \bar{u}_j \end{bmatrix} = \begin{bmatrix} \bar{X}_i \\ \bar{X}_j \end{bmatrix} \tag{3.1}$$

The joints in a pin-connected plane frame have two degrees of freedom defined by displacements in the x- and z-directions and in order to make relation (3.1) applicable to such cases we introduce displacements, $\bar{w}_i$ and $\bar{w}_j$ in the $\bar{z}$ -direction as shown in Fig. 3.3.

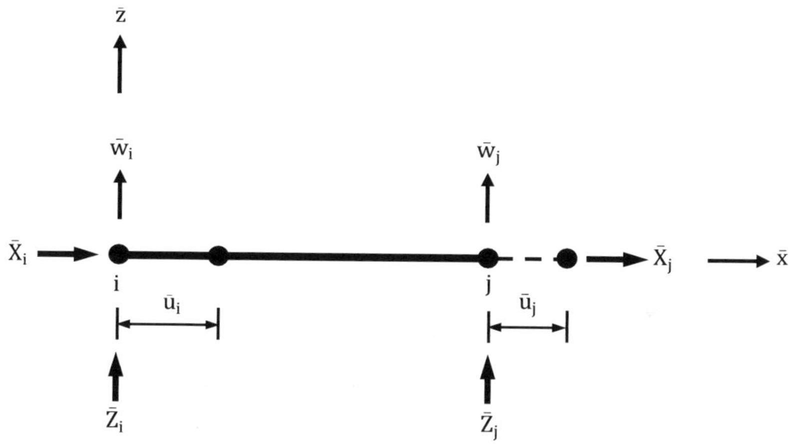

Figure 3.3 Bar in local coordinates system, $\bar{x}$ and $\bar{z}$.

The matrix in (3.1) is extended to incorporate the displacements in the $\bar{z}$ -direction to give

$$\begin{bmatrix} \frac{EA}{L} & 0 & -\frac{EA}{L} & 0 \\ 0 & 0 & 0 & 0 \\ -\frac{EA}{L} & 0 & \frac{EA}{L} & 0 \\ 0 & 0 & 0 & 0 \end{bmatrix} \begin{bmatrix} \bar{u}_i \\ \bar{w}_i \\ \bar{u}_j \\ \bar{w}_j \end{bmatrix} = \begin{bmatrix} \bar{X}_i \\ \bar{Z}_i \\ \bar{X}_j \\ \bar{Z}_j \end{bmatrix} \tag{3.2}$$

Notice that the expansion of the above matrix gives the forces $\bar{Z}_i$ and $\bar{Z}_j$ in the $\bar{z}$ -direction at the ends of the member equal to zero which means that the state of the applied forces is not changed and the bar is still subjected to only axial forces.

Equation (3.2) can be written in the general form

$$\bar{k}\bar{\delta} = \bar{F} \tag{3.3}$$

where the stiffness matrix relative to local coordinates is:

$$\bar{k} = \begin{bmatrix} \frac{EA}{L} & 0 & -\frac{EA}{L} & 0 \\ 0 & 0 & 0 & 0 \\ -\frac{EA}{L} & 0 & \frac{EA}{L} & 0 \\ 0 & 0 & 0 & 0 \end{bmatrix} \tag{3.4}$$

the displacement vector, $\bar{\delta} = \begin{bmatrix} \bar{\delta}_i \\ \bar{\delta}_j \end{bmatrix} = \begin{bmatrix} \bar{u}_i \\ \bar{w}_i \\ \bar{u}_j \\ \bar{w}_j \end{bmatrix}$ and the action vector,

$$\bar{F} = \begin{bmatrix} \bar{F}_i \\ \bar{F}_j \end{bmatrix} = \begin{bmatrix} \bar{X}_i \\ \bar{Z}_i \\ \bar{X}_j \\ \bar{Z}_j \end{bmatrix}.$$

3.2 Transformation from Local to Global Coordinates

Quantities in (3.2) are relative to the local $\bar{x}$- and $\bar{z}$-axes where $\bar{x}$ represents the direction of the longitudinal axis of the member and when this does not coincide with the global x-axis these quantities have to be transformed to become relative to the global x- and z-axes. Such transformation is necessary because the overall structure stiffness matrix is written relative to global coordinates.

Suppose that a member lies initially along the global x-axis and then it is moved to take the final position shown in Fig. 3.4. The new position of the member is achieved by a rotation about the $\bar{y}$-axis by a clockwise (i.e. positive) angle of $\varphi_{\bar{y}}$. Notice that the $\bar{y}$- and y-axes are still coincident but the local $\bar{x}$-axis is now making an angle $\varphi_{\bar{y}}$ with the global x-axis.

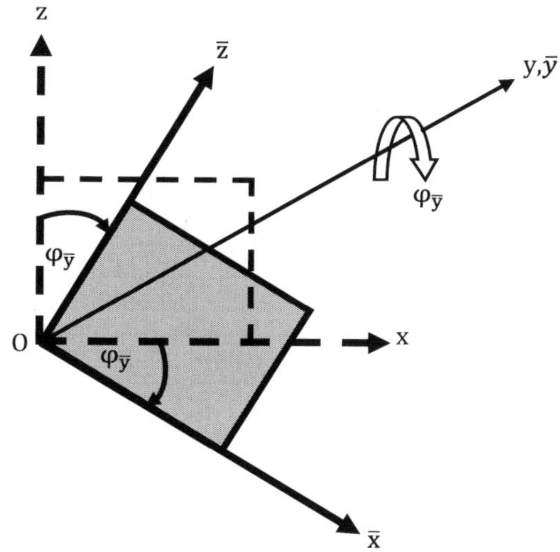

Figure 3.4

The displacements and forces relative to local coordinates are transformed to be relative to the global coordinates as follows:

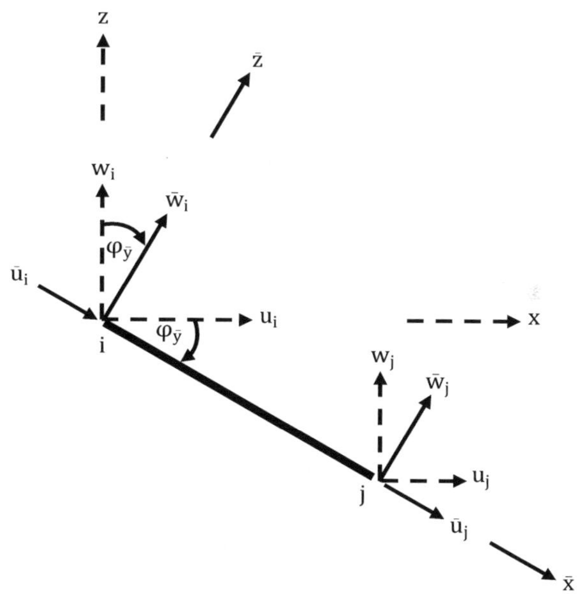

Figure 3.5

3.2.1 Transformation of Displacements

With reference to Fig. 3.5, the displacement $\bar{u}_i$ along the $\bar{x}$-axis is equal to the algebraic sum of the components of the displacements u_i and w_i and is given by:

$$\bar{u}_i = u_i \cos\varphi_{\bar{y}} - w_i \sin\varphi_{\bar{y}}$$

Similarly, the displacement $\bar{w}_i$ along the $\bar{z}$-axis is equal to the algebraic sum of the components of the displacements u_i and w_i and is given by:

$$\bar{w}_i = u_i \sin\varphi_{\bar{y}} + w_i \cos\varphi_{\bar{y}}$$

and in matrix form

$$\begin{bmatrix} \bar{u}_i \\ \bar{w}_i \end{bmatrix} = \begin{bmatrix} \cos\varphi_{\bar{y}} & -\sin\varphi_{\bar{y}} \\ \sin\varphi_{\bar{y}} & \cos\varphi_{\bar{y}} \end{bmatrix} \begin{bmatrix} u_i \\ w_i \end{bmatrix}$$

$$\begin{bmatrix} \bar{u}_i \\ \bar{w}_i \end{bmatrix} = \rho_{\bar{y}} \begin{bmatrix} u_i \\ w_i \end{bmatrix}, \text{ where } \rho_{\bar{y}} = \begin{bmatrix} \cos\varphi_{\bar{y}} & -\sin\varphi_{\bar{y}} \\ \sin\varphi_{\bar{y}} & \cos\varphi_{\bar{y}} \end{bmatrix}$$

Sometimes it is more convenient to write the above transformation matrix in a general form as

$$\rho_{\bar{y}} = \begin{bmatrix} \lambda_{\bar{x}x} & \lambda_{\bar{x}z} \\ \lambda_{\bar{z}x} & \lambda_{\bar{z}z} \end{bmatrix}$$

where the λ's are called the direction cosines in vector analysis and are defined as follows:

$\lambda_{\bar{x}x}$ is the cosine of the angle made by the local $\bar{x}$-axis with the global x-axis $= \cos\varphi_{\bar{y}}$.

$\lambda_{\bar{x}z}$ is the cosine of the angle made by the local $\bar{x}$-axis with the global z-axis $= \cos\left(\varphi_{\bar{y}} + 90\right) = -\sin\varphi_{\bar{y}}$.

$\lambda_{\bar{z}x}$ is the cosine of the angle made by the local $\bar{z}$-axis with the global x-axis $= \cos\left(90 - \varphi_{\bar{y}}\right) = \sin\varphi_{\bar{y}}$.

$\lambda_{\bar{z}z}$ is the cosine of the angle made by the local $\bar{z}$-axis with the global z-axis $= \cos\varphi_{\bar{y}}$.

Similarly for node j

$$\begin{bmatrix} \bar{u}_j \\ \bar{w}_j \end{bmatrix} = \rho_{\bar{y}} \begin{bmatrix} u_j \\ w_j \end{bmatrix}$$

For nodes i and j, $\begin{bmatrix} \bar{u} \\ \bar{w}_i \\ \bar{u}_j \\ \bar{w}_j \end{bmatrix} = \begin{bmatrix} \rho_{\bar{y}} & 0 \\ 0 & \rho_{\bar{y}} \end{bmatrix} \begin{bmatrix} u_i \\ w_i \\ u_j \\ w_j \end{bmatrix}$, where $0 = \begin{bmatrix} 0 & 0 \\ 0 & 0 \end{bmatrix}$

The full transformation matrix is

$$\begin{bmatrix} \bar{u} \\ \bar{w}_i \\ \bar{u}_j \\ \bar{w}_j \end{bmatrix} = \begin{bmatrix} \cos\varphi_{\bar{y}} & -\sin\varphi_{\bar{y}} & 0 & 0 \\ \sin\varphi_{\bar{y}} & \cos\varphi_{\bar{y}} & 0 & 0 \\ 0 & 0 & \cos\varphi_{\bar{y}} & -\sin\varphi_{\bar{y}} \\ 0 & 0 & \sin\varphi_{\bar{y}} & \cos\varphi_{\bar{y}} \end{bmatrix} \begin{bmatrix} u_i \\ w_i \\ u_j \\ w_j \end{bmatrix}$$

or

$$\bar{\delta} = r\delta \tag{3.5}$$

where r is the transformation matrix which is given by:

$$r = \begin{bmatrix} \cos\varphi_{\bar{y}} & -\sin\varphi_{\bar{y}} & 0 & 0 \\ \sin\varphi_{\bar{y}} & \cos\varphi_{\bar{y}} & 0 & 0 \\ 0 & 0 & \cos\varphi_{\bar{y}} & -\sin\varphi_{\bar{y}} \\ 0 & 0 & \sin\varphi_{\bar{y}} & \cos\varphi_{\bar{y}} \end{bmatrix} \tag{3.6}$$

3.2.2 Transformation of Forces

With reference to Fig. 3.6, the force $\bar{X}_i$ is equal to the algebraic sum of the components of the forces X_i and Z_i along the $\bar{x}$-axis and is given by:

$$\bar{X}_i = X_i\cos\varphi_{\bar{y}} - Z_i\sin\varphi_{\bar{y}}$$

Similarly, the force $\bar{Z}_i$ is equal to the algebraic sum of the components of the forces X_i and Z_i along the $\bar{z}$-axis and is given by:

$$\bar{Z}_i = X_i\sin\varphi_{\bar{y}} + Z_i\cos\varphi_{\bar{y}}$$

and in matrix form

$$\begin{bmatrix} \bar{X}_i \\ \bar{Z}_i \end{bmatrix} = \begin{bmatrix} \cos\varphi_{\bar{y}} & -\sin\varphi_{\bar{y}} \\ \sin\varphi_{\bar{y}} & \cos\varphi_{\bar{y}} \end{bmatrix} \begin{bmatrix} X_i \\ Z_i \end{bmatrix}$$

$$\begin{bmatrix} \bar{X}_i \\ \bar{Z}_i \end{bmatrix} = \rho_{\bar{y}} \begin{bmatrix} X_i \\ Z_i \end{bmatrix}$$

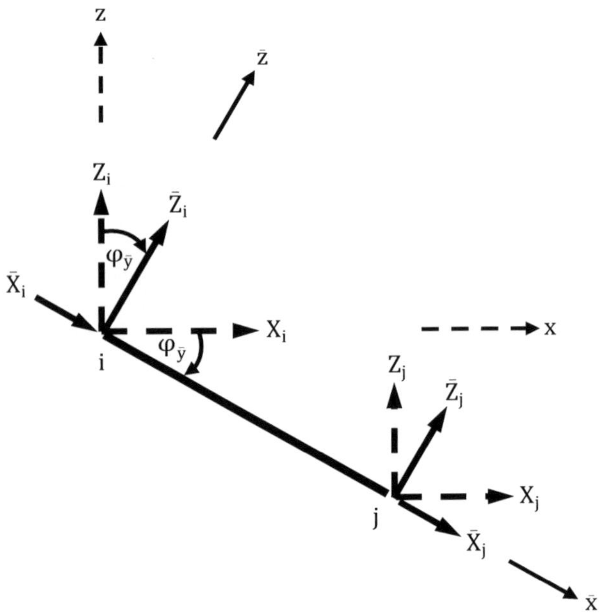

Figure 3.6

Similarly for node j

$$\begin{bmatrix} \bar{X}_j \\ \bar{Z}_j \end{bmatrix} = \rho_{\bar{y}} \begin{bmatrix} X_j \\ Z_j \end{bmatrix}$$

For nodes i and j, $\begin{bmatrix} \bar{X}_i \\ \bar{Z}_i \\ \bar{X}_j \\ \bar{Z}_j \end{bmatrix} = \begin{bmatrix} \rho_{\bar{y}} & O \\ O & \rho_{\bar{y}} \end{bmatrix} \begin{bmatrix} X_i \\ Z_i \\ X_j \\ Z_j \end{bmatrix}$, where $O = \begin{bmatrix} 0 & 0 \\ 0 & 0 \end{bmatrix}$

The full transformation matrix is

$$\begin{bmatrix} \overline{X}_i \\ \overline{Z}_i \\ \overline{X}_j \\ \overline{Z}_j \end{bmatrix} = \begin{bmatrix} \cos\varphi_{\bar{y}} & -\sin\varphi_{\bar{y}} & 0 & 0 \\ \sin\varphi_{\bar{y}} & \cos\varphi_{\bar{y}} & 0 & 0 \\ 0 & 0 & \cos\varphi_{\bar{y}} & -\sin\varphi_{\bar{y}} \\ 0 & 0 & \sin\varphi_{\bar{y}} & \cos\varphi_{\bar{y}} \end{bmatrix} \begin{bmatrix} X_i \\ Z_i \\ X_j \\ Z_j \end{bmatrix}$$

or

$$\overline{F} = rF \tag{3.7}$$

Notice that matrix r for the transformation of forces from local coordinates to global coordinates is the same as that for the transformation of displacements as given by (3.6) because both of them are vectors having the same respective directions relative to the relevant coordinate axes.

The transformation matrix, r given by (3.6) can be written in a more convenient form by expressing $\sin\varphi_{\bar{y}}$ and $\cos\varphi_{\bar{y}}$ in terms of the coordinates at the ends of the member as shown in Fig. 3.7.

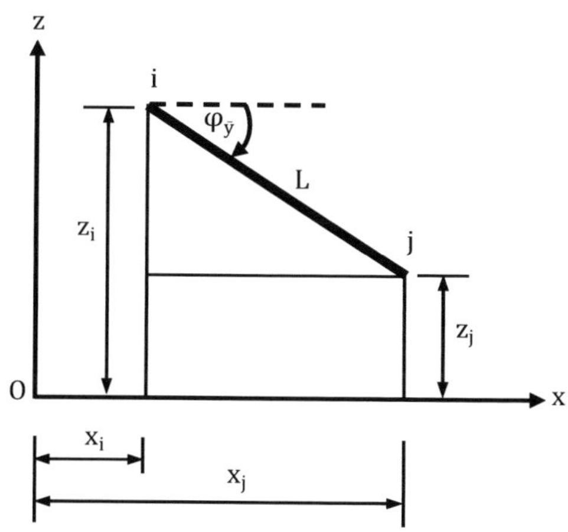

Figure 3.7

$$\cos\varphi_{\bar{y}} = \frac{x_j - x_i}{L} = \frac{x_{ij}}{L}$$

$$\sin\varphi_{\bar{y}} = \frac{z_i - z_j}{L} = -\frac{z_j - z_i}{L} = -\frac{z_{ij}}{L}$$

Notice that for positive rotation $\varphi_{\bar{y}}$, $z_j < z_i$ and hence z_{ij} is negative.

$$L = \sqrt{(x_j - x_i)^2 + (z_j - z_i)^2} = \sqrt{x_{ij}^2 + z_{ij}^2}$$

Thus the transformation matrix r becomes:

$$r = \begin{bmatrix} x_{ij}/L & z_{ij}/L & 0 & 0 \\ -z_{ij}/L & x_{ij}/L & 0 & 0 \\ 0 & 0 & x_{ij}/L & z_{ij}/L \\ 0 & 0 & -z_{ij}/L & x_{ij}/L \end{bmatrix} \tag{3.8}$$

3.3 Stiffness Matrix Relative to Global Coordinates

The overall structure matrix is written relative to global coordinates, therefore the stiffness matrices of the members of the structure have to be transformed and written relative to global coordinates as explained below.

We have from (3.3), $\bar{k}\bar{\delta} = \bar{F}$ (relative to local coordinates).

Substitute $\bar{\delta} = r\delta$ and $\bar{F} = rF$ from (3.5) and (3.7) in the above equation to get

$$\bar{k}(r\delta) = rF$$

Premultiply both sides by r^{-1}

$r^{-1}\bar{k}(r\delta) = r^{-1}rF$ and since $r^{-1}r = I$ (the unit matrix)

$$r^{-1}\bar{k}r\delta = F$$

One of the properties of the transformation matrix is that its inverse is equal to its transpose, i.e. $r^{-1} = r^T$

$(r^T\bar{k}r)\delta = F$, which can be written as

$$K\delta = F \tag{3.9}$$

where k is the stiffness matrix of the member relative to global coordinates and is given by:

$k = r^T \bar{k} r$ with $\bar{k}$ and r as given by (3.4) and (3.8) respectively. Thus

$$k = \begin{bmatrix} x_{ij}/L & -z_{ij}/L & 0 & 0 \\ z_{ij}/L & x_{ij}/L & 0 & 0 \\ 0 & 0 & x_{ij}/L & -z_{ij}/L \\ 0 & 0 & z_{ij}/L & x_{ij}/L \end{bmatrix} \begin{bmatrix} \frac{EA}{L} & 0 & -\frac{EA}{L} & 0 \\ 0 & 0 & 0 & 0 \\ -\frac{EA}{L} & 0 & \frac{EA}{L} & 0 \\ 0 & 0 & 0 & 0 \end{bmatrix}$$

$$\begin{bmatrix} x_{ij}/L & z_{ij}/L & 0 & 0 \\ -z_{ij}/L & x_{ij}/L & 0 & 0 \\ 0 & 0 & x_{ij}/L & z_{ij}/L \\ 0 & 0 & -z_{ij}/L & x_{ij}/L \end{bmatrix}$$

$$k = \begin{bmatrix} \frac{EAx_{ij}^2}{L^3} & \frac{EAx_{ij}z_{ij}}{L^3} & -\frac{EAx_{ij}^2}{L^3} & -\frac{EAx_{ij}z_{ij}}{L^3} \\ \frac{EAx_{ij}z_{ij}}{L^3} & \frac{EAz_{ij}^2}{L^3} & -\frac{EAx_{ij}z_{ij}}{L^3} & -\frac{EAz_{ij}^2}{L^3} \\ -\frac{EAx_{ij}^2}{L^3} & -\frac{EAx_{ij}z_{ij}}{L^3} & \frac{EAx_{ij}^2}{L^3} & \frac{EAx_{ij}z_{ij}}{L^3} \\ -\frac{EAx_{ij}z_{ij}}{L^3} & -\frac{EAz_{ij}^2}{L^3} & \frac{EAx_{ij}z_{ij}}{L^3} & \frac{EAz_{ij}^2}{L^3} \end{bmatrix} \quad (3.10)$$

Equation (3.9) can be written as:

$$\begin{bmatrix} k_{ii} & k_{ij} \\ k_{ji} & k_{jj} \end{bmatrix} \begin{bmatrix} \delta_i \\ \delta_j \end{bmatrix} = \begin{bmatrix} F_i \\ F_j \end{bmatrix}$$

$$k_{ii} = \begin{bmatrix} \frac{EAx_{ij}^2}{L^3} & \frac{EAx_{ij}z_{ij}}{L^3} \\ \frac{EAx_{ij}z_{ij}}{L^3} & \frac{EAz_{ij}^2}{L^3} \end{bmatrix}, \quad k_{ij} = \begin{bmatrix} -\frac{EAx_{ij}^2}{L^3} & -\frac{EAx_{ij}z_{ij}}{L^3} \\ -\frac{EAx_{ij}z_{ij}}{L^3} & -\frac{EAz_{ij}^2}{L^3} \end{bmatrix},$$

$$k_{ji} = \begin{bmatrix} -\dfrac{EAx_{ij}^2}{L^3} & -\dfrac{EAx_{ij}z_{ij}}{L^3} \\ -\dfrac{EAx_{ij}z_{ij}}{L^3} & -\dfrac{EAz_{ij}^2}{L^3} \end{bmatrix}, \quad \text{and} \quad k_{jj} = \begin{bmatrix} \dfrac{EAx_{ij}^2}{L^3} & \dfrac{EAx_{ij}z_{ij}}{L^3} \\ \dfrac{EAx_{ij}z_{ij}}{L^3} & \dfrac{EAz_{ij}^2}{L^3} \end{bmatrix}$$

Note: Use the global stiffness matrix k in (3.10) to calculate the displacements δ and hence the external reactions. Then use the local stiffness matrix $\bar{k}$ in (3.4) and $\bar{\delta}$ as calculated from (3.5) to find the actions on the member as explained in the example below.

Example 1

Calculate the displacements in the x- and z-directions at nodes 1, 2, 3, and 4 of the pin-connected plane frame shown in Fig. 3.8. Hence find the reactions at the supports 1 and 4 and the internal forces developed in the members of the frame. The modulus of elasticity of all the members is $E = 210 \times 10^6$ kN/m^2 and the cross-sectional area of the members is as follows:

$A_1 = 350 \times 10^{-6}$ m^2, $A_2 = 320 \times 10^{-6}$ m^2,
$A_3 = 440 \times 10^{-6}$ m^2, $A_4 = 350 \times 10^{-6}$ m^2,
$A_5 = 680 \times 10^{-6}$ m^2.

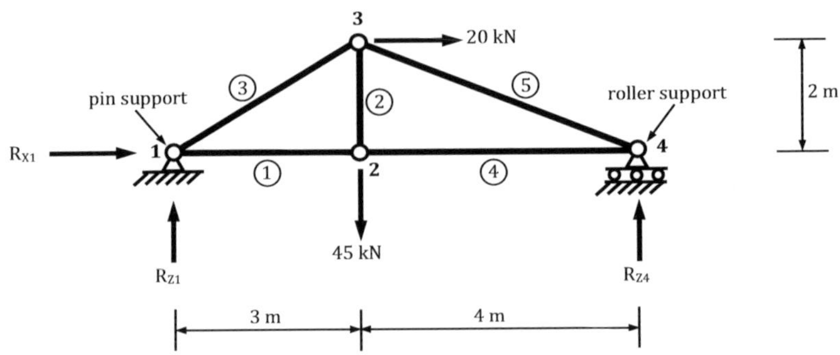

(Node 1 is taken as the origin of the global coordinates system.)

Figure 3.8

Calculation of stiffness matrices of the members of the structure

Member 1

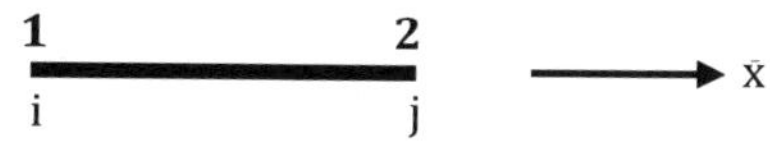

Member address in the member stiffness matrix k: i j

Structure address in the overall structure matrix K: 1 2

The above correspondence means that member 1 contributes to only nodes 1 and 2 of the overall structure.

$E = 210 \times 10^6\ kN/m^2$, $A = 350 \times 10^{-6}\ m^2$

$x_i = 0$, $x_j = 3$ m, $x_{ij} = x_j - x_i = 3 - 0 = 3$ m

$z_i = 0$, $z_j = 0$, $z_{ij} = z_j - z_i = 0 - 0 = 0$

$$L = \sqrt{x_{ij}^2 + z_{ij}^2} = \sqrt{3^2 + 0^2} = 3\ \text{m}$$

Substitute the above values in (3.10) to get the stiffness matrix of member 1 relative to global coordinates as:

$$\overbrace{\overbrace{\delta_i = \delta_1}^{}\quad \overbrace{\delta_j = \delta_2}^{}}^{\delta^1}$$

$$\begin{matrix} u_i & w_i & u_j & w_j \\ u_1 & w_1 & u_2 & w_2 \end{matrix} \tag{3.11}$$

$$k^1 = 10^3 \begin{bmatrix} 24.50 & 0 & -24.50 & 0 \\ 0 & 0 & 0 & 0 \\ -24.50 & 0 & 24.50 & 0 \\ 0 & 0 & 0 & 0 \end{bmatrix} \begin{matrix} u_1 \\ w_1 \\ u_2 \\ w_2 \end{matrix}$$

$$k_{ii}^1 = 10^3 \begin{bmatrix} 24.5 & 0 \\ 0 & 0 \end{bmatrix}, \quad k_{ij}^1 = 10^3 \begin{bmatrix} -24.50 & 0 \\ 0 & 0 \end{bmatrix},$$

$$k_{ji}^1 = 10^3 \begin{bmatrix} -24.5 & 0 \\ 0 & 0 \end{bmatrix}, \quad k_{jj}^1 = 10^3 \begin{bmatrix} 24.50 & 0 \\ 0 & 0 \end{bmatrix}.$$

Member 2

Member address: i j

Structure address: 2 3

(i.e. member 2 contributes to only nodes 2 and 3 of the structure.)

$E = 210 \times 10^6\ \text{kN/m}^2$, $A = 320 \times 10^{-6}\ \text{m}^2$

$x_i = 3$ m, $x_j = 3$ m, $x_{ij} = x_j - x_i = 3 - 3 = 0$

$z_i = 0$, $z_j = 2$ m, $z_{ij} = z_j - z_i = 2 - 0 = 2$ m

$$L = \sqrt{x_{ij}^2 + z_{ij}^2} = \sqrt{0^2 + 2^2} = 2\ \text{m}$$

Substitute in (3.10) to get

$$\overbrace{\overbrace{\begin{matrix} u_i & w_i \\ u_2 & w_2 \end{matrix}}^{\delta_i = \delta_2} \quad \overbrace{\begin{matrix} u_j & w_j \\ u_3 & w_3 \end{matrix}}^{\delta_j = \delta_3}}^{\delta^2} \tag{3.12}$$

$$k^2 = 10^3 \begin{bmatrix} 0 & 0 & 0 & 0 \\ 0 & 33.60 & 0 & -33.60 \\ 0 & 0 & 0 & 0 \\ 0 & -33.60 & 0 & 33.60 \end{bmatrix} \begin{matrix} u_2 \\ w_2 \\ u_3 \\ w_3 \end{matrix}$$

$$k_{ii}^2 = 10^3 \begin{bmatrix} 0 & 0 \\ 0 & 33.60 \end{bmatrix}, \quad k_{ij}^2 = 10^3 \begin{bmatrix} 0 & 0 \\ 0 & -33.60 \end{bmatrix},$$

$$k_{ji}^2 = 10^3 \begin{bmatrix} 0 & 0 \\ 0 & -33.60 \end{bmatrix}, \quad k_{jj}^2 = 10^3 \begin{bmatrix} 0 & 0 \\ 0 & 33.60 \end{bmatrix}.$$

Member 3

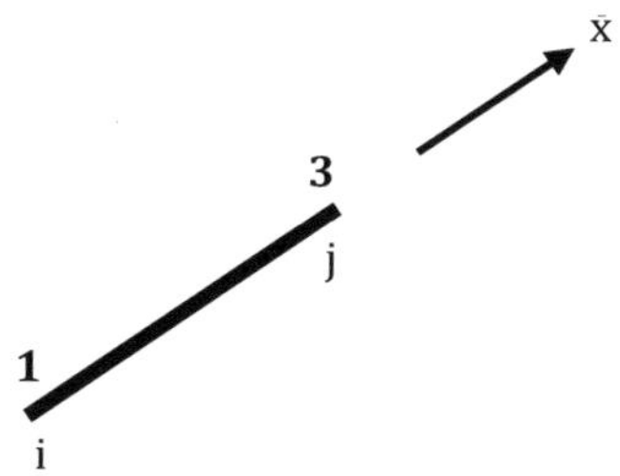

Member address: i j
Structure address: 1 3
(i.e. member 3 contributes to only nodes 1 and 3 of the structure.)
$E = 210 \times 10^6$ kN/m^2, $A = 440 \times 10^{-6}$ m^2
$x_i = 0$, $x_j = 3$ m, $x_{ij} = x_j - x_i = 3 - 0 = 3$ m
$z_i = 0$, $z_j = 2$ m, $z_{ij} = z_j - z_i = 2 - 0 = 2$ m

$$L = \sqrt{x_{ij}^2 + z_{ij}^2} = \sqrt{3^2 + 2^2} = 3.606 \text{ m}$$

Substitute in (3.10) to get

$$\overbrace{\overbrace{\begin{matrix} u_i & w_i \\ u_1 & w_1 \end{matrix}}^{\delta_i = \delta_1} \quad \overbrace{\begin{matrix} u_j & w_j \\ u_3 & w_3 \end{matrix}}^{\delta_j = \delta_3}}^{\delta^3} \tag{3.13}$$

$$k^3 = 10^3 \begin{bmatrix} 17.740 & 11.83 & -17.74 & -11.83 \\ 11.83 & 7.89 & -11.83 & -7.89 \\ -17.74 & -11.83 & 17.74 & 11.83 \\ -11.83 & -7.89 & 11.83 & 7.89 \end{bmatrix} \begin{matrix} u_1 \\ w_1 \\ u_3 \\ w_3 \end{matrix}$$

$$k_{ii}^3 = 10^3 \begin{bmatrix} 17.74 & 11.83 \\ 11.83 & 7.89 \end{bmatrix}, \quad k_{ij}^3 = 10^3 \begin{bmatrix} -17.74 & -11.83 \\ -11.83 & -7.89 \end{bmatrix},$$

$$k_{ji}^3 = 10^3 \begin{bmatrix} -17.74 & -11.83 \\ -11.83 & -7.89 \end{bmatrix}, \quad k_{jj}^3 = 10^3 \begin{bmatrix} 17.74 & 11.83 \\ 11.83 & 7.89 \end{bmatrix}.$$

Member 4

Member address: i j

Structure address: 2 4

(i.e. member 4 contributes to only nodes 2 and 4 of the structure.)

$E = 210 \times 10^6$ kN/m^2, $A = 350 \times 10^{-6}$ m^2

$x_i = 3$ m, $x_j = 7$ m, $x_{ij} = x_j - x_i = 7 - 3 = 4$ m

$z_i = 0$, $z_j = 0$, $z_{ij} = z_j - z_i = 0 - 0 = 0$

$$L = \sqrt{x_{ij}^2 + z_{ij}^2} = \sqrt{4^2 + 0^2} = 4\text{ m}$$

Substitute in (3.10) to get

$$\overbrace{\overbrace{\delta_i = \delta_2}^{}\quad \overbrace{\delta_j = \delta_4}^{}}^{\delta^4}$$

$$\begin{matrix} u_i & w_i & u_j & w_j \\ u_2 & w_2 & u_4 & w_4 \end{matrix} \tag{3.14}$$

$$k^3 = 10^3 \begin{bmatrix} 18.38 & 0 & -18.38 & 0 \\ 0 & 0 & 0 & 0 \\ -18.38 & 0 & 18.38 & 0 \\ 0 & 0 & 0 & 0 \end{bmatrix} \begin{matrix} u_2 \\ w_2 \\ u_4 \\ w_4 \end{matrix}$$

$$k_{ii}^4 = 10^3 \begin{bmatrix} 18.38 & 0 \\ 0 & 0 \end{bmatrix}, \quad k_{ij}^4 = 10^3 \begin{bmatrix} -18.38 & 0 \\ 0 & 0 \end{bmatrix},$$

$$k_{ji}^4 = 10^3 \begin{bmatrix} -18.38 & 0 \\ 0 & 0 \end{bmatrix}, \quad k_{jj}^4 = 10^3 \begin{bmatrix} 18.38 & 0 \\ 0 & 0 \end{bmatrix}.$$

Member 5

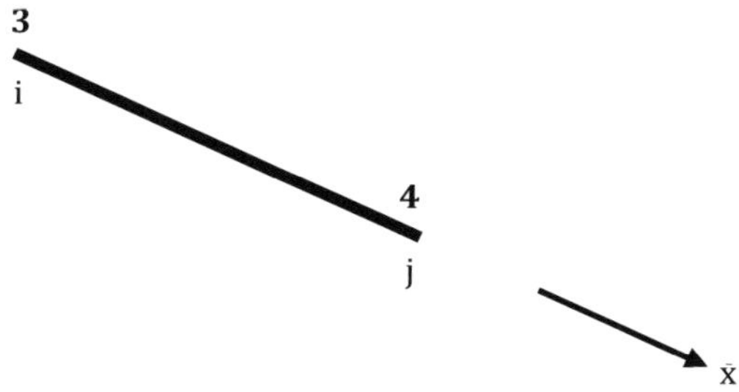

Member address: i j
Structure address: 3 4
(i.e. member 5 contributes to only nodes 3 and 4 of the structure.)
$E = 210 \times 10^6\ \text{kN/m}^2$, $A = 680 \times 10^{-6}\ \text{m}^2$
$x_i = 3$ m, $x_j = 7$ m, $x_{ij} = x_j - x_i = 7 - 3 = 4$ m
$z_i = 2$ m, $z_j = 0$, $z_{ij} = z_j - z_i = 0 - 2 = -2$ m

$$L = \sqrt{x_{ij}^2 + z_{ij}^2} = \sqrt{4^2 + (-2)^2} = 4.472\ \text{m}$$

Substitute in (3.10) to get

$$\overbrace{\overbrace{\begin{matrix} u_i & w_i \\ u_3 & w_3 \end{matrix}}^{\delta_i = \delta_3} \quad \overbrace{\begin{matrix} u_j & w_j \\ u_4 & w_4 \end{matrix}}^{\delta_j = \delta_4}}^{\delta^3} \tag{3.15}$$

$$k^5 = 10^3 \begin{bmatrix} 25.54 & -12.77 & -25.54 & 12.77 \\ -12.77 & 6.39 & 12.77 & -6.39 \\ -25.54 & 12.77 & 25.54 & -12.77 \\ 12.77 & -6.39 & -12.77 & 6.39 \end{bmatrix} \begin{matrix} u_3 \\ w_2 \\ u_4 \\ w_4 \end{matrix}$$

$$k_{ii}^5 = 10^3 \begin{bmatrix} 25.54 & -12.77 \\ -12.77 & 6.39 \end{bmatrix}, \quad k_{ij}^5 = 10^3 \begin{bmatrix} -25.54 & 12.77 \\ 12.77 & -6.39 \end{bmatrix},$$

$$k_{ji}^5 = 10^3 \begin{bmatrix} -25.54 & 12.77 \\ 12.77 & -6.39 \end{bmatrix}, \quad k_{jj}^5 = 10^3 \begin{bmatrix} 25.54 & -12.77 \\ -12.77 & 6.39 \end{bmatrix}.$$

Assembly of the overall stiffness matrix relative to global coordinates

The general relationship for the overall structure is $K\delta = F$ which can be written as

$$\begin{array}{cccc} \overbrace{u_1 \qquad w_1}^{\delta_1} & \overbrace{u_2 \qquad w_2}^{\delta_2} & \overbrace{u_3 \qquad w_3}^{\delta_3} & \overbrace{u_4 \qquad w_4}^{\delta_4} \end{array}$$

$$\begin{bmatrix} K_{11} & K_{12} & K_{13} & K_{14} \\ K_{21} & K_{22} & K_{23} & K_{24} \\ K_{31} & K_{32} & K_{33} & K_{34} \\ K_{41} & K_{42} & K_{43} & K_{44} \end{bmatrix} \begin{bmatrix} u_1 \\ w_1 \\ u_2 \\ w_2 \\ u_3 \\ w_3 \\ u_4 \\ w_4 \end{bmatrix} \begin{matrix} \delta_1 \\ \\ \delta_2 \\ \\ \delta_3 \\ \\ \delta_4 \end{matrix} = \begin{bmatrix} X_1 \\ Z_1 \\ X_2 \\ Z_2 \\ X_3 \\ Z_3 \\ X_4 \\ Z_4 \end{bmatrix} \begin{matrix} F_1 \\ \\ F_2 \\ \\ F_3 \\ \\ F_4 \end{matrix}$$

Since there are two degrees of freedom (u and w) at each node, the coefficients in the overall structure stiffness matrix, K, are 2×2 sub-matrices rather than single numbers. Any coefficient in K is derived from the summation of the contributions of the members in the structure to that coefficient, i.e.

$$K_{ij} = \sum_{g=1}^{g=m} K_{ij}^{g}$$

where K_{ij}^{g} is a 2×2 sub-matrix representing the contribution of the gth member to the ij^{th} coefficient in K, and m is the number of members in the structure (=5 in this example).

Each of the members of the frame will contribute to the overall structure stiffness matrix, K, according to the relationship between the member address and the structure address.

Contribution of member 1 to the overall stiffness matrix, K, will be denoted by K^1 and found as follows:

Member address	i	j
Structure address	1	2

From the above correspondence between nodes 1 and 2 of the structure and the member nodes, i and j, it follows that member 1 will contribute to the coefficients of the overall structure stiffness matrix, K, as shown in the table below:

subscript of coefficient in K	11	12	21	22
subscript of contributing coefficient from k^1	ii	ij	ji	jj

From (3.11)

$$K_{11}^1 = k_{ii}^1 = 10^3\begin{bmatrix} 24.5 & 0 \\ 0 & 0 \end{bmatrix}, \quad K_{12}^1 = k_{ij}^1 = 10^3\begin{bmatrix} -24.50 & 0 \\ 0 & 0 \end{bmatrix},$$

$$K_{21}^1 = k_{ji}^1 = 10^3\begin{bmatrix} -24.5 & 0 \\ 0 & 0 \end{bmatrix}, \quad \text{and} \quad K_{22}^1 = k_{jj}^1 = 10^3\begin{bmatrix} 24.50 & 0 \\ 0 & 0 \end{bmatrix}.$$

Contribution of member 2

Member address	i	j
Structure address	2	3

subscript of coefficient in K	22	23	32	33
subscript of contributing coefficient from k^2	ii	ij	ji	jj

From (3.12)

$$K_{22}^2 = k_{ii}^2 = 10^3\begin{bmatrix} 0 & 0 \\ 0 & 33.60 \end{bmatrix}, \quad K_{23}^2 = k_{ij}^2 = 10^3\begin{bmatrix} 0 & 0 \\ 0 & -33.60 \end{bmatrix},$$

$$K_{32}^2 = k_{ji}^2 = 10^3\begin{bmatrix} 0 & 0 \\ 0 & -33.60 \end{bmatrix}, \quad \text{and} \quad K_{33}^2 = k_{jj}^2 = 10^3\begin{bmatrix} 0 & 0 \\ 0 & 33.60 \end{bmatrix}.$$

Contribution of member 3

Member address	i	j
Structure address	1	3

subscript of coefficient in K	11	13	31	33
subscript of contributing coefficient from k^3	ii	ij	ji	jj

From (3.13)

$$K_{11}^3 = k_{ii}^3 = 10^3\begin{bmatrix} 17.74 & 11.83 \\ 11.83 & 7.89 \end{bmatrix}, \quad K_{13}^3 = k_{ij}^3 = 10^3\begin{bmatrix} -17.74 & -11.83 \\ -11.83 & -7.89 \end{bmatrix},$$

$$K_{31}^3 = k_{ji}^3 = 10^3\begin{bmatrix} -17.74 & -11.83 \\ -11.83 & -7.89 \end{bmatrix}, \quad \text{and} \quad K_{33}^3 = k_{jj}^3 = 10^3\begin{bmatrix} 17.74 & 11.83 \\ 11.83 & 7.89 \end{bmatrix}.$$

Contribution of member 4

Member address	i	j
Structure address	2	4

subscript of coefficient in K	22	24	42	44
subscript of contributing coefficient from k^4	ii	ij	ji	jj

From (3.14)

$$K_{22}^4 = k_{ii}^4 = 10^3\begin{bmatrix} 18.38 & 0 \\ 0 & 0 \end{bmatrix}, \quad K_{24}^4 = k_{ij}^4 = 10^3\begin{bmatrix} -18.38 & 0 \\ 0 & 0 \end{bmatrix},$$

$$K_{42}^4 = k_{ji}^4 = 10^3\begin{bmatrix} -18.38 & 0 \\ 0 & 0 \end{bmatrix}, \quad \text{and} \quad K_{44}^4 = k_{jj}^4 = 10^3\begin{bmatrix} 18.38 & 0 \\ 0 & 0 \end{bmatrix}.$$

Contribution of member 5

Member address	i	j
Structure address	3	4

subscript of coefficient in K	33	34	43	44
subscript of contributing coefficient from k^5	ii	ij	ji	jj

From (3.15)

$$K_{33}^5 = k_{ii}^5 = 10^3\begin{bmatrix} 25.54 & -12.77 \\ -12.77 & 6.39 \end{bmatrix}, \quad K_{34}^5 = k_{ij}^5 = 10^3\begin{bmatrix} -25.54 & 12.77 \\ 12.77 & -6.39 \end{bmatrix},$$

$$K_{43}^5 = k_{ji}^5 = 10^3\begin{bmatrix} -25.54 & 12.77 \\ 12.77 & -6.39 \end{bmatrix}, \quad \text{and} \quad K_{44}^5 = k_{jj}^5 = 10^3\begin{bmatrix} 25.54 & -12.77 \\ -12.77 & 6.39 \end{bmatrix}.$$

Note that in general, not all the members contribute to a particular value of K_{ij}.

$$K_{11} = K_{11}^1 + K_{11}^2 + K_{11}^3 + K_{11}^4 + K_{11}^5 = k_{ii}^1 + 0 + k_{ii}^3 + 0 + 0$$

$$= 10^3\begin{bmatrix} 24.50 & 0 \\ 0 & 0 \end{bmatrix} + 10^3\begin{bmatrix} 17.74 & 11.83 \\ 11.83 & 7.89 \end{bmatrix} = 10^3\begin{bmatrix} 42.24 & 11.83 \\ 11.83 & 7.89 \end{bmatrix}$$

(i.e. only members 1 and 3 contribute to K_{11})

$$K_{12} = K_{12}^1 + K_{12}^2 + K_{12}^3 + K_{12}^4 + K_{12}^5 = k_{ij}^1 + 0 + 0 + 0 + 0$$

$$= 10^3 \begin{bmatrix} -24.50 & 0 \\ 0 & 0 \end{bmatrix}$$

(i.e. only member 1 contributes to K_{12})

$$K_{13} = K_{13}^1 + K_{13}^2 + K_{13}^3 + K_{13}^4 + K_{13}^5 = 0 + 0 + k_{ij}^3 + 0 + 0$$

$$= 10^3 \begin{bmatrix} -17.74 & -11.83 \\ -11.83 & -7.89 \end{bmatrix}$$

(i.e. only member 3 contributes to K_{13})

$$K_{14} = K_{14}^1 + K_{14}^2 + K_{14}^3 + K_{14}^4 + K_{14}^5 = 0 + 0 + 0 + 0 + 0 = \begin{bmatrix} 0 & 0 \\ 0 & 0 \end{bmatrix}$$

(i.e. no member contributes to K_{14})

$$K_{21} = K_{21}^1 + K_{21}^2 + K_{21}^3 + K_{21}^4 + K_{21}^5 = k_{ji}^1 + 0 + 0 + 0 + 0$$

$$= 10^3 \begin{bmatrix} -24.50 & 0 \\ 0 & 0 \end{bmatrix}$$

$$K_{22} = K_{22}^1 + K_{22}^2 + K_{22}^3 + K_{22}^4 + K_{22}^5 = k_{jj}^1 + k_{ii}^2 + 0 + k_{ii}^4 + 0$$

$$= 10^3 \begin{bmatrix} 24.50 & 0 \\ 0 & 0 \end{bmatrix} + 10^3 \begin{bmatrix} 0 & 0 \\ 0 & 33.60 \end{bmatrix} + 10^3 \begin{bmatrix} 18.38 & 0 \\ 0 & 0 \end{bmatrix} = 10^3 \begin{bmatrix} 42.88 & 0 \\ 0 & 33.60 \end{bmatrix}$$

$$K_{23} = K_{23}^1 + K_{23}^2 + K_{23}^3 + K_{23}^4 + K_{23}^5 = 0 + k_{ij}^2 + 0 + 0 + 0 = 10^3 \begin{bmatrix} 0 & 0 \\ 0 & -33.60 \end{bmatrix}$$

$$K_{24} = K_{24}^1 + K_{24}^2 + K_{24}^3 + K_{24}^4 + K_{24}^5 = 0 + 0 + 0 + k_{ij}^4 + 0 = 10^3 \begin{bmatrix} -18.38 & 0 \\ 0 & 0 \end{bmatrix}$$

$$K_{31} = K_{31}^1 + K_{31}^2 + K_{31}^3 + K_{31}^4 + K_{31}^5 = 0 + 0 + k_{ji}^3 + 0 + 0$$

$$= 10^3 \begin{bmatrix} -17.74 & -11.83 \\ -11.83 & -7.89 \end{bmatrix}$$

$$K_{32} = K_{32}^1 + K_{32}^2 + K_{32}^3 + K_{32}^4 + K_{32}^5 = 0 + k_{ji}^2 + 0 + 0 + 0 = 10^3 \begin{bmatrix} 0 & 0 \\ 0 & -33.60 \end{bmatrix}$$

$$K_{33} = K_{33}^1 + K_{33}^2 + K_{33}^3 + K_{33}^4 + K_{33}^5 = 0 + k_{jj}^2 + k_{jj}^3 + 0 + k_{ii}^5$$

$$= 10^3 \begin{bmatrix} 0 & 0 \\ 0 & 33.60 \end{bmatrix} + 10^3 \begin{bmatrix} 17.74 & 11.83 \\ 11.83 & 7.89 \end{bmatrix} + 10^3 \begin{bmatrix} 25.54 & -12.77 \\ -12.77 & 6.39 \end{bmatrix}$$

$$= 10^3 \begin{bmatrix} 43.28 & -0.94 \\ -0.94 & 47.88 \end{bmatrix}$$

$$K_{34} = K_{34}^1 + K_{34}^2 + K_{34}^3 + K_{34}^4 + K_{34}^5 = 0 + 0 + 0 + 0 + k_{ij}^5$$

$$= 10^3 \begin{bmatrix} -25.54 & 12.77 \\ 12.77 & -6.39 \end{bmatrix}$$

$$K_{41} = K_{41}^1 + K_{41}^2 + K_{41}^3 + K_{41}^4 + K_{41}^5 = 0 + 0 + 0 + 0 + 0 = \begin{bmatrix} 0 & 0 \\ 0 & 0 \end{bmatrix}$$

$$K_{42} = K_{42}^1 + K_{42}^2 + K_{42}^3 + K_{42}^4 + K_{42}^5 = 0 + 0 + 0 + k_{ji}^4 + 0 = 10^3 \begin{bmatrix} -18.38 & 0 \\ 0 & 0 \end{bmatrix}$$

$$K_{43} = K_{43}^1 + K_{43}^2 + K_{43}^3 + K_{43}^4 + K_{43}^5 = 0 + 0 + 0 + 0 + k_{ji}^5$$

$$= 10^3 \begin{bmatrix} -25.54 & 12.77 \\ 12.77 & -6.39 \end{bmatrix}$$

$$K_{44} = K_{44}^1 + K_{44}^2 + K_{44}^3 + K_{44}^4 + K_{44}^5 = 0 + 0 + 0 + k_{jj}^4 + k_{jj}^5$$

$$= 10^3 \begin{bmatrix} 18.38 & 0 \\ 0 & 0 \end{bmatrix} + 10^3 \begin{bmatrix} 25.54 & -12.77 \\ -12.77 & 6.39 \end{bmatrix} = 10^3 \begin{bmatrix} 43.92 & -12.77 \\ -12.77 & 6.39 \end{bmatrix}$$

Or by inspection (based on the correspondence of the member and structure addresses) as:

K= (3.17)

$k_{ii}^1 + k_{ii}^3$	k_{ij}^1	k_{ij}^3	0
k_{ji}^1	$k_{jj}^1 + k_{ii}^2 + k_{ii}^4$	k_{ij}^2	k_{ij}^4
k_{ji}^3	k_{ji}^2	$k_{jj}^2 + k_{jj}^3 + k_{ii}^5$	k_{ij}^5
0	k_{ji}^4	k_{ji}^5	$k_{jj}^4 + k_{jj}^5$

Load vector

The load vector, which is written in terms of global coordinates, will be composed of the forces acting at the nodes and is given by:

$$F = \begin{bmatrix} F_1 \\ F_2 \\ F_3 \\ F_4 \end{bmatrix}$$

Since there are two degrees of freedom at each node which are defined by the displacements u and w in the x- and z-directions respectively, the corresponding forces will be X and Z. Thus:

$F_1 = \begin{bmatrix} X_1 \\ Z_1 \end{bmatrix} = \begin{bmatrix} R_{X1} \\ R_{Z1} \end{bmatrix}$, where R_{x1} and R_{z1} are the reactions in the x- and z-directions respectively at the hinged support 1.

$$F_2 = \begin{bmatrix} X_2 \\ Z_2 \end{bmatrix} = \begin{bmatrix} 0 \\ -45 \end{bmatrix}, \quad F_3 = \begin{bmatrix} X_3 \\ Z_3 \end{bmatrix} = \begin{bmatrix} 20 \\ 0 \end{bmatrix}, \quad \text{and} \quad F_4 = \begin{bmatrix} X_4 \\ Z_4 \end{bmatrix} = \begin{bmatrix} 0 \\ R_{Z4} \end{bmatrix},$$

where R_{z4} is the reaction in the z-direction at the roller support 4.

$$F = \begin{bmatrix} R_{X1} \\ R_{Z1} \\ 0 \\ -45 \\ 20 \\ 0 \\ 0 \\ R_{Z4} \end{bmatrix} \qquad (3.18)$$

Substitute the values of K in (3.17) and F in (3.18) to get:

$$10^3 \begin{bmatrix} 42.24 & 11.83 & -24.50 & 0 & -17.74 & -11.83 & 0 & 0 \\ 11.83 & 7.89 & 0 & 0 & -11.83 & -7.89 & 0 & 0 \\ -24.50 & 0 & 42.88 & 0 & 0 & 0 & -18.38 & 0 \\ 0 & 0 & 0 & 33.60 & 0 & -33.60 & 0 & 0 \\ -17.74 & -11.83 & 0 & 0 & 43.28 & -0.94 & -25.54 & 12.77 \\ -11.83 & -7.89 & 0 & -33.60 & -0.94 & 47.88 & 12.77 & -6.39 \\ 0 & 0 & -18.38 & 0 & -25.54 & 12.77 & 43.92 & -12.77 \\ 0 & 0 & 0 & 0 & 12.77 & -6.39 & -12.77 & 6.39 \end{bmatrix} \begin{bmatrix} u_1 \\ w_1 \\ u_2 \\ w_2 \\ u_3 \\ w_3 \\ u_4 \\ w_4 \end{bmatrix} = \begin{bmatrix} R_{X1} \\ R_{Z1} \\ 0 \\ -45 \\ 20 \\ 0 \\ 0 \\ R_{Z4} \end{bmatrix} \qquad (3.19)$$

Boundary conditions

The next step is to apply the boundary conditions as follows:

At node 1 where there is a pinned support allowing no displacement, then $u_1 = 0$; and in order to enforce this boundary condition, delete row 1 and column 1 of the above matrix.

Similarly, $w_1 = 0$; and in order to enforce this boundary condition, delete row 2 and column 2 of the above matrix.

At node 4 where there is a roller support allowing displacement in the x-direction only but not in the z-direction, then, $w_4 = 0$; and in order to enforce this boundary condition, delete row 8 and column 8 of the above matrix.

In some cases the matrix needs to be compacted after deletion of some of the rows and columns. However, matrix compaction is not necessary in this particular example.

The resulting 'reduced' matrix is:

$$10^3 \begin{bmatrix} 42.88 & 0 & 0 & 0 & -18.38 \\ 0 & 33.60 & 0 & -33.60 & 0 \\ 0 & 0 & 43.28 & -0.94 & -25.54 \\ 0 & -33.60 & -0.94 & 47.88 & 12.77 \\ -18.38 & 0 & -25.54 & 12.77 & 43.92 \end{bmatrix} \begin{bmatrix} u_2 \\ w_2 \\ u_3 \\ w_3 \\ u_4 \end{bmatrix} = \begin{bmatrix} 0 \\ -45 \\ 20 \\ 0 \\ 0 \end{bmatrix}$$

Calculation of displacements

The above set is written in the form of simultaneous equations as:

$$42880u_2 + 0w_2 + 0u_3 + 0w_3 - 18380u_4 = 0$$
$$0u_2 + 33600w_2 + 0u_3 - 336000w_3 + 0u_4 = -45$$
$$0u_2 + 0w_2 + 43280u_3 - 940w_3 - 25540u_4 = 20$$
$$0u_2 - 33600w_2 - 9480u_3 + 47880w_3 + 12770u_4 = 0$$
$$-18380u_2 + 0w_2 - 25540u_3 + 12770w_3 + 43920u_4 = 0$$

The solution of the above set of simultaneous equations is:

$u_2 = 0.002039$ m, $w_2 = -0.008540$ m, $u_3 = 0.003113$ m, $w_2 = -0.007200$ m, and $u_4 = 0.004757$ m.

We also have, from the boundary conditions, $u_1 = 0$, $w_1 = 0$, and $w_4 = 0$.

So the full column vector of displacements for the whole structure is:

$$\delta = \begin{bmatrix} \delta_1 \\ \delta_2 \\ \delta_3 \\ \delta_4 \end{bmatrix} = \begin{bmatrix} u_1 \\ w_1 \\ u_2 \\ w_2 \\ u_3 \\ w_3 \\ u_4 \\ w_4 \end{bmatrix} = \begin{bmatrix} 0 \\ 0 \\ 0.002039 \\ -0.008540 \\ 0.003113 \\ -0.007200 \\ 0.004757 \\ 0 \end{bmatrix} \quad (3.20)$$

And the deformed shape of the frame is shown in Fig. 3.9.

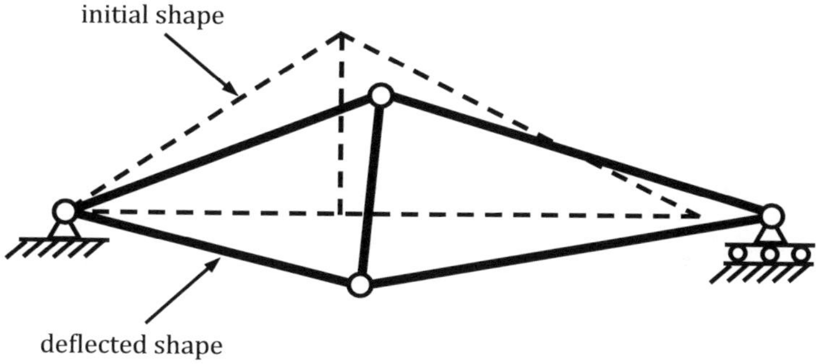

Figure 3.9 Deformed structure.

Calculation of the reactions at the supports

The reactions can be found from the full matrix in (3.19) as follows:

The reaction at support 1 in the x-direction is calculated from the first row

$42240u_1 + 11830w_1 - 24500u_2 - 17740u_3 - 11830w_3 = Ru_{X1}$

$R_{X1} = 42240 \times 0 + 11830 \times 0 - 24500 \times 0.002039 - 17740 \times 0.003113 - 11830 \times (-0.007200) = -20.00$ kN

The reaction at support 1 in the z-direction is calculated from the second row

$11830u_1 + 7890w_1 - 11830u_3 - 7890w_3 = R_{z1}$

$R_{z1} = 11830 \times 0 + 7890 \times 0 - 11830 \times 0.003113 - 7890 \times (-0.007200) = 19.98$ kN

The reaction at support 4 in the z-direction is calculated from the eighth row

$12770u_3 - 6390w_3 - 12770u_4 + 6390w_4 = R_{z4}$

$R_{z4} = 12770 \times 0.003113 - 6390 \times (-0.007200) - 12770 \times 0.004757 + 6390 \times 0 = 25.01$ kN

Calculation of actions (forces) developed in the members

These are usually calculated relative to the local coordinates of the member from (3.3) as $\bar{F} = \bar{k}\bar{\delta}$, where $\bar{k}$ is given by (3.4), the displacement vector $\bar{\delta}$ is obtained from (3.5) as $\bar{\delta} = r\delta$, r is the transformation matrix which is given in (3.8) and δ is the vector of displacements at the ends of the member relative to global coordinates obtained from (3.20).

Member 1

$$\frac{EA}{L} = \frac{210 \times 10^6 \times 350 \times 10^{-6}}{3} = 24500 \text{ kN/m}$$

Substitute in (3.4) to get

$$\bar{k}^1 = \begin{bmatrix} 24500 & 0 & -24500 & 0 \\ 0 & 0 & 0 & 0 \\ -24500 & 0 & 24500 & 0 \\ 0 & 0 & 0 & 0 \end{bmatrix}$$

$x_{ij} = 3$ m, $z_{ij} = 0$, $L = 3$ m,

Substitute in (3.8) to get

$$r^1 = \begin{bmatrix} 1 & 0 & 0 & 0 \\ 0 & 1 & 0 & 0 \\ 0 & 0 & 1 & 0 \\ 0 & 0 & 0 & 1 \end{bmatrix} = I \qquad \text{(the unit matrix)}$$

The above result could have been found by inspection since the local $\bar{x}$-axis of member 1 coincides with and pointing in the positive direction (i.e. node j is to the right of node i) of the global x-axis.

$$\delta^1 = \begin{bmatrix} \delta_i^1 \\ \delta_j^1 \end{bmatrix} = \begin{bmatrix} \delta_1 \\ \delta_2 \end{bmatrix} \begin{bmatrix} u_1 \\ w_1 \\ u_2 \\ w_2 \end{bmatrix} = \begin{bmatrix} 0 \\ 0 \\ 0.002039 \\ -0.008540 \end{bmatrix},$$

$$\bar{\delta}^1 = r^1\delta^1 = I\delta^1 = \delta^1 = \begin{bmatrix} 0 \\ 0 \\ 0.002039 \\ -0.008540 \end{bmatrix}$$

$$\begin{bmatrix} \bar{X}_1^1 \\ \bar{Z}_1^1 \\ \bar{X}_2^1 \\ \bar{Z}_2^1 \end{bmatrix} = \bar{F}^1 = \bar{k}^1\bar{\delta}^1 = \begin{bmatrix} 24500 & 0 & -24500 & 0 \\ 0 & 0 & 0 & 0 \\ -24500 & 0 & 24500 & 0 \\ 0 & 0 & 0 & 0 \end{bmatrix} \begin{bmatrix} 0 \\ 0 \\ 0.002039 \\ -0.008540 \end{bmatrix} = \begin{bmatrix} -49.96 \\ 0 \\ +49.96 \\ 0 \end{bmatrix}$$

The negative force at node i indicates tension.

Member 2

$$\frac{EA}{L} = \frac{210\times10^6\times320\times10^{-6}}{2} = 33600\ \text{kN/m}$$

$$\bar{k}^2 = \begin{bmatrix} 33600 & 0 & -33600 & 0 \\ 0 & 0 & 0 & 0 \\ -33600 & 0 & 33600 & 0 \\ 0 & 0 & 0 & 0 \end{bmatrix}$$

$x_{ij} = 0$, $z_{ij} = 2$ m, $L = 2$ m,

$$r^2 = \begin{bmatrix} 0 & 1 & 0 & 0 \\ -1 & 0 & 0 & 0 \\ 0 & 0 & 0 & 1 \\ 0 & 0 & -1 & 0 \end{bmatrix}$$

$$\delta^2 = \begin{bmatrix} \delta_i^2 \\ \delta_j^2 \end{bmatrix} = \begin{bmatrix} \delta_2 \\ \delta_3 \end{bmatrix} \begin{bmatrix} u_2 \\ w_2 \\ u_3 \\ w_3 \end{bmatrix} = \begin{bmatrix} 0.002039 \\ -0.008540 \\ 0.003113 \\ -0.007200 \end{bmatrix}$$

$$\bar{\delta}^2 = r^2\delta^2 = \begin{bmatrix} 0 & 1 & 0 & 0 \\ -1 & 0 & 0 & 0 \\ 0 & 0 & 0 & 1 \\ 0 & 0 & -1 & 0 \end{bmatrix} \begin{bmatrix} 0.002039 \\ -0.008540 \\ 0.003113 \\ -0.007200 \end{bmatrix} = \begin{bmatrix} -0.008540 \\ -0.002039 \\ -0.007200 \\ -0.003113 \end{bmatrix}$$

$$\begin{bmatrix} \bar{X}_2^2 \\ \bar{Z}_2^2 \\ \bar{X}_3^2 \\ \bar{Z}_3^2 \end{bmatrix} = \bar{F}^2 = \bar{k}^2\bar{\delta}^2 = \begin{bmatrix} 33600 & 0 & -33600 & 0 \\ 0 & 0 & 0 & 0 \\ -33600 & 0 & 33600 & 0 \\ 0 & 0 & 0 & 0 \end{bmatrix} \begin{bmatrix} -0.008540 \\ -0.002039 \\ -0.007200 \\ -0.003113 \end{bmatrix} = \begin{bmatrix} -45.02 \\ 0 \\ +45.02 \\ 0 \end{bmatrix}$$

The negative force at node i indicates tension.

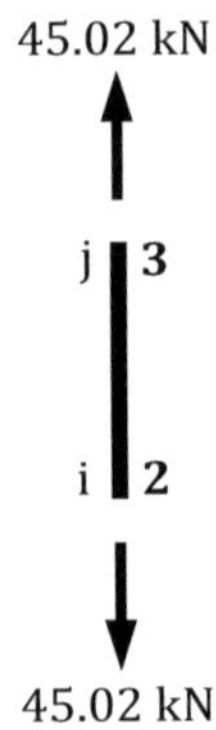

Member 3

$$\frac{EA}{L} = \frac{210 \times 10^6 \times 440 \times 10^{-6}}{3.606} = 25624 \text{ kN/m}$$

$$\bar{k}^3 = \begin{bmatrix} 25624 & 0 & -25624 & 0 \\ 0 & 0 & 0 & 0 \\ -25624 & 0 & 25624 & 0 \\ 0 & 0 & 0 & 0 \end{bmatrix}, \; x_{ij} = 3\text{ m}, \; z_{ij} = 2\text{ m}, \; L = 3.606\text{ m}$$

$$r^3 = \begin{bmatrix} 0.832 & 0.555 & 0 & 0 \\ -0.555 & 0.832 & 0 & 0 \\ 0 & 0 & 0.832 & 0.555 \\ 0 & 0 & -0.555 & 0.832 \end{bmatrix}$$

$$\delta^3 = \begin{bmatrix} \delta_i^3 \\ \delta_j^3 \end{bmatrix} = \begin{bmatrix} \delta_1 \\ \delta_3 \end{bmatrix} \begin{bmatrix} u_1 \\ w_1 \\ u_3 \\ w_3 \end{bmatrix} = \begin{bmatrix} 0 \\ 0 \\ 0.003113 \\ -0.007200 \end{bmatrix}$$

$$\bar{\delta}^3 = r^3\delta^3 = \begin{bmatrix} 0.832 & 0.555 & 0 & 0 \\ -0.555 & 0.832 & 0 & 0 \\ 0 & 0 & 0.832 & 0.555 \\ 0 & 0 & -0.555 & 0.832 \end{bmatrix} \begin{bmatrix} 0 \\ 0 \\ 0.003113 \\ -0.007200 \end{bmatrix} = \begin{bmatrix} 0 \\ 0 \\ -0.001406 \\ -0.007718 \end{bmatrix}$$

$$\begin{bmatrix} \bar{X}_1^3 \\ \bar{Z}_1^3 \\ \bar{X}_3^3 \\ \bar{Z}_3^3 \end{bmatrix} = \bar{F}^3 = \bar{k}^3\bar{\delta}^3 = \begin{bmatrix} 25624 & 0 & -25624 & 0 \\ 0 & 0 & 0 & 0 \\ -25624 & 0 & 25624 & 0 \\ 0 & 0 & 0 & 0 \end{bmatrix} \begin{bmatrix} 0 \\ 0 \\ -0.001406 \\ -0.007718 \end{bmatrix} = \begin{bmatrix} +36.03 \\ 0 \\ -36.03 \\ 0 \end{bmatrix}$$

The positive force at node i indicates compression.

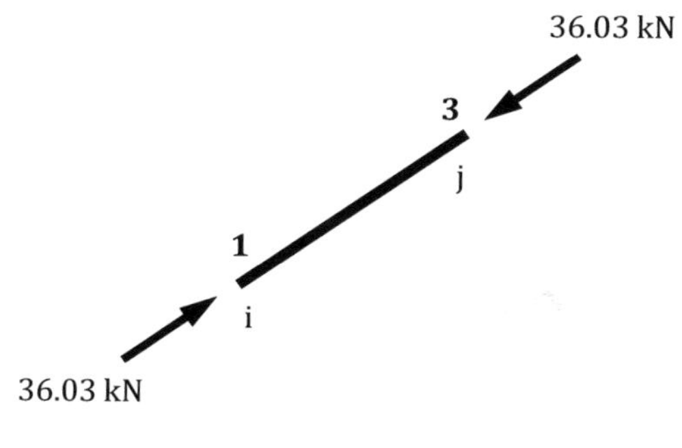

Member 4

$$\frac{EA}{L} = \frac{210 \times 10^6 \times 350 \times 10^{-6}}{4} = 18375\ \text{kN/m}$$

$$\bar{k}^4 = \begin{bmatrix} 18375 & 0 & -18375 & 0 \\ 0 & 0 & 0 & 0 \\ -18375 & 0 & 18375 & 0 \\ 0 & 0 & 0 & 0 \end{bmatrix}$$

Since the member local axis lies along the global x-axis, $r^4 = I$.

$$\delta^4 = \begin{bmatrix} \delta_i^4 \\ \delta_j^4 \end{bmatrix} = \begin{bmatrix} \delta_2 \\ \delta_4 \end{bmatrix} \begin{bmatrix} u_2 \\ w_2 \\ u_4 \\ w_4 \end{bmatrix} = \begin{bmatrix} 0.002039 \\ -0.008540 \\ 0.004757 \\ 0 \end{bmatrix},$$

$$\bar{\delta}^4 = r^4\delta^4 = I\delta^4 = \delta^4 = \begin{bmatrix} 0.002039 \\ -0.008540 \\ 0.004757 \\ 0 \end{bmatrix}$$

$$\begin{bmatrix} \bar{X}_2^4 \\ \bar{Z}_2^4 \\ \bar{X}_4^4 \\ \bar{Z}_4^4 \end{bmatrix} = \bar{F}^4 = \bar{k}^4\bar{\delta}^4 = \begin{bmatrix} 18375 & 0 & -18375 & 0 \\ 0 & 0 & 0 & 0 \\ -18375 & 0 & 18375 & 0 \\ 0 & 0 & 0 & 0 \end{bmatrix} \begin{bmatrix} 0.002039 \\ -0.008540 \\ 0.004757 \\ 0 \end{bmatrix} = \begin{bmatrix} -49.94 \\ 0 \\ +49.94 \\ 0 \end{bmatrix}$$

The member is in tension.

Member 5

$$\frac{EA}{L} = \frac{210 \times 10^6 \times 680 \times 10^{-6}}{4.472} = 31932\,\text{kN/m}$$

$$\bar{k}^5 = \begin{bmatrix} 31932 & 0 & -31932 & 0 \\ 0 & 0 & 0 & 0 \\ -31932 & 0 & 31932 & 0 \\ 0 & 0 & 0 & 0 \end{bmatrix},$$

$x_{ij} = 4$ m, $z_{ij} = -2$ m, L = 4.472 m

$$r^5 = \begin{bmatrix} 0.895 & -0.447 & 0 & 0 \\ 0.447 & 0.895 & 0 & 0 \\ 0 & 0 & 0.895 & -0.447 \\ 0 & 0 & 0.447 & 0.895 \end{bmatrix}$$

$$\delta^5 = \begin{bmatrix} \delta_i^5 \\ \delta_j^5 \end{bmatrix} = \begin{bmatrix} \delta_3 \\ \delta_4 \end{bmatrix} \begin{bmatrix} u_3 \\ w_3 \\ u_4 \\ w_4 \end{bmatrix} = \begin{bmatrix} 0.003113 \\ -0.007200 \\ 0.004757 \\ 0 \end{bmatrix}$$

$$\bar{\delta}^5 = r^5\delta^5 = \begin{bmatrix} 0.895 & -0.447 & 0 & 0 \\ 0.447 & 0.895 & 0 & 0 \\ 0 & 0 & 0.895 & -0.447 \\ 0 & 0 & 0.447 & 0.895 \end{bmatrix} \begin{bmatrix} 0.003113 \\ -0.007200 \\ 0.004757 \\ 0 \end{bmatrix} = \begin{bmatrix} 0.006005 \\ -0.005053 \\ 0.004258 \\ 0.002126 \end{bmatrix}$$

$$\begin{bmatrix} \bar{X}_3^5 \\ \bar{Z}_3^5 \\ \bar{X}_4^5 \\ \bar{Z}_4^5 \end{bmatrix} = \bar{F}^5 = \bar{k}^5\bar{\delta}^5 = \begin{bmatrix} 31932 & 0 & -31932 & 0 \\ 0 & 0 & 0 & 0 \\ -31932 & 0 & 31932 & 0 \\ 0 & 0 & 0 & 0 \end{bmatrix} \begin{bmatrix} 0.006005 \\ -0.005053 \\ 0.004258 \\ 0.002126 \end{bmatrix} = \begin{bmatrix} +55.79 \\ 0 \\ -55.79 \\ 0 \end{bmatrix}$$

The member is in compression.

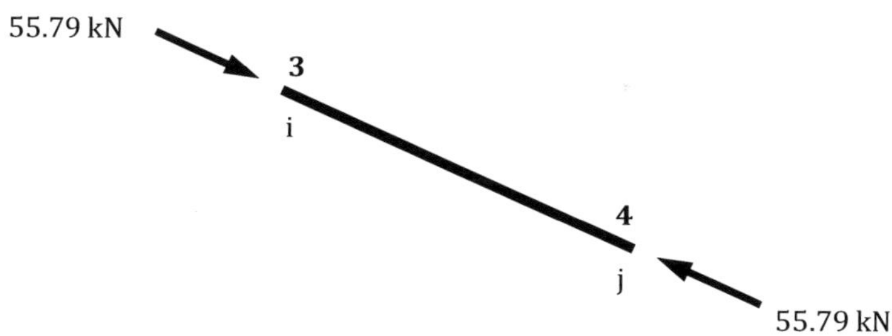

Problems

P3.1. Calculate the displacements at the nodes, the reactions at the supports, and the forces developed in the members of the pin-connected plane frame shown in Fig. P3.1. $A_1 = 500 \times 10^{-6}\,m^2$, $A_2 = 800 \times 10^{-6}\,m^2$ and both members are made of steel with modulus of elasticity, $E = 210 \times 10^6\,kN/m^2$.

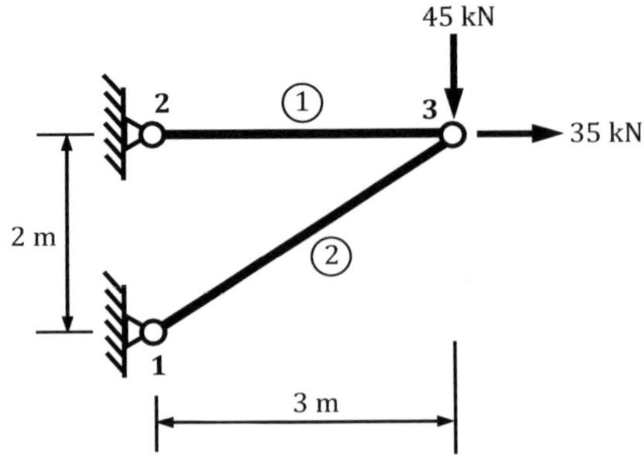

Figure P3.1

Answer:

$u_1 = 0$, $w_1 = 0$, $u_2 = 0$, $w_2 = 0$, $u_3 = 0.002929$ m, $w_3 = -0.007532$ m,
$R_{X1} = 67.50$ kN, $R_{Z1} = 45.00$ kN, $R_{X2} = -102.50$ kN, $R_{Z2} = 0$.

$$\text{Member 1:}\ \begin{bmatrix} \bar{X}_2^1 \\ \bar{Z}_2^1 \\ \bar{X}_3^1 \\ \bar{Z}_3^1 \end{bmatrix} = \begin{bmatrix} -102.50 \\ 0 \\ +102.50 \\ 0 \end{bmatrix} \text{kN (Tension)},$$

$$\text{Member 2:}\ \begin{bmatrix} \bar{X}_1^2 \\ \bar{Z}_1^2 \\ \bar{X}_3^2 \\ \bar{Z}_3^2 \end{bmatrix} = \begin{bmatrix} +81.13 \\ 0 \\ -81.13 \\ 0 \end{bmatrix} \text{kN (Compression)}.$$

P3.2. Calculate the displacements at the nodes, the reactions at the supports, and the forces developed in the members of the pin-connected plane frame shown in Fig. P3.2. $A_1 = A_5 = 900 \times 10^{-6}\,m^2$, $A_2 = 300 \times 10^{-6}\,m^2$, $A_3 = A_4 = 600 \times 10^{-6}\,m^2$ and all members are made of steel with modulus of elasticity, $E = 210 \times 10^6\,kN/m^2$.

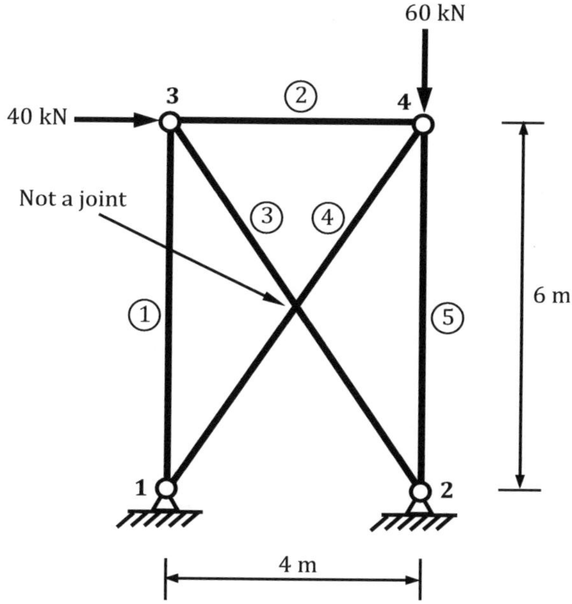

Figure P3.2

Answer:

$u_1 = 0$, $w_1 = 0$, $u_2 = 0$, $w_2 = 0$, $u_3 = 0.006986$ m, $w_3 = 0.001292$ m,
$u_4 = 0.006169$ m, $w_4 = -0.002517$ m,
$R_{X1} = -12.86$ kN, $R_{Z1} = -60.00$ kN, $R_{X2} = -27.14$ kN, $R_{Z2} = 120.00$ kN.

$$\text{Member 1}: \begin{bmatrix} \bar{X}_1^1 \\ \bar{Z}_1^1 \\ \bar{X}_3^1 \\ \bar{Z}_3^1 \end{bmatrix} = \begin{bmatrix} -40.70 \\ 0 \\ +40.70 \\ 0 \end{bmatrix} \text{kN} \quad (\text{Tension}),$$

$$\text{Member 2}: \begin{bmatrix} \bar{X}_3^2 \\ \bar{Z}_3^2 \\ \bar{X}_4^2 \\ \bar{Z}_4^2 \end{bmatrix} = \begin{bmatrix} +12.86 \\ 0 \\ -12.86 \\ 0 \end{bmatrix} \text{kN} \quad (\text{Compression}),$$

$$\text{Member 3:}\begin{bmatrix}\bar{X}_3^3\\ \bar{Z}_3^3\\ \bar{X}_2^3\\ \bar{Z}_2^3\end{bmatrix}=\begin{bmatrix}+48.92\\ 0\\ -48.92\\ 0\end{bmatrix}\text{ kN (Compression),}$$

$$\text{Member 4:}\begin{bmatrix}\bar{X}_1^4\\ \bar{Z}_1^4\\ \bar{X}_4^4\\ \bar{Z}_4^4\end{bmatrix}=\begin{bmatrix}-23.19\\ 0\\ +23.19\\ 0\end{bmatrix}\text{ kN (Tension),}$$

$$\text{Member 5:}\begin{bmatrix}\bar{X}_2^5\\ \bar{Z}_2^5\\ \bar{X}_4^5\\ \bar{Z}_4^5\end{bmatrix}=\begin{bmatrix}+79.30\\ 0\\ -79.30\\ 0\end{bmatrix}\text{ kN (Compression).}$$

P3.3. Calculate the displacements at the nodes, the reactions at the supports, and the forces developed in the members of the pin-connected plane frame shown in Fig. P3.3. $A_1 = 7000 \times 10^{-6}\ m^2$, $A_2 = A_3 = A_6 = 3000 \times 10^{-6}\ m^2$, $A_4 = A_5 = 4000 \times 10^{-6}\ m^2$ and all members are made of timber with modulus of elasticity, $E = 8 \times 10^6\ kN/m^2$.

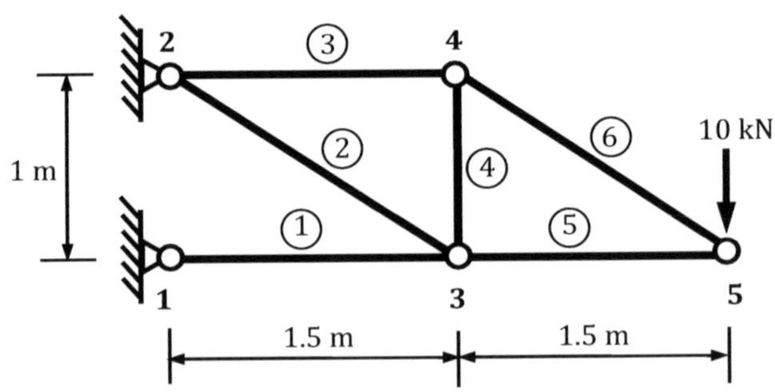

Figure P3.3

Answer:

$u_1 = 0$, $w_1 = 0$, $u_2 = 0$, $w_2 = 0$, $u_3 = -0.000804$ m, $w_3 = 0.003647$ m,
$u_4 = 0.000937$ m, $w_4 = -0.003959$ m, $u_5 = -0.001507$ m,
$w_5 = -0.010067$ m,
$R_{X1} = 30.00$ kN, $R_{Z1} = 0$, $R_{X2} = -30.00$ kN, $R_{Z2} = 10.00$ kN.

$$\text{Member 1}: \begin{bmatrix} \overline{X}_1^1 \\ \overline{Z}_1^1 \\ \overline{X}_3^1 \\ \overline{Z}_3^1 \end{bmatrix} = \begin{bmatrix} +30.00 \\ 0 \\ -30.00 \\ 0 \end{bmatrix} \text{kN} \quad (\text{Compression}),$$

$$\text{Member 2}: \begin{bmatrix} \overline{X}_2^2 \\ \overline{Z}_2^2 \\ \overline{X}_3^2 \\ \overline{Z}_3^2 \end{bmatrix} = \begin{bmatrix} -18.03 \\ 0 \\ +18.03 \\ 0 \end{bmatrix} \text{kN} \quad (\text{Tension}),$$

$$\text{Member 3}: \begin{bmatrix} \overline{X}_2^3 \\ \overline{Z}_2^3 \\ \overline{X}_4^3 \\ \overline{Z}_4^3 \end{bmatrix} = \begin{bmatrix} -15.00 \\ 0 \\ +15.00 \\ 0 \end{bmatrix} \text{kN} \quad (\text{Tension}),$$

$$\text{Member 4}: \begin{bmatrix} \overline{X}_3^4 \\ \overline{Z}_3^4 \\ \overline{X}_4^4 \\ \overline{Z}_4^4 \end{bmatrix} = \begin{bmatrix} +10.00 \\ 0 \\ -10.00 \\ 0 \end{bmatrix} \text{kN} \quad (\text{Compression}),$$

$$\text{Member 5}: \begin{bmatrix} \overline{X}_3^5 \\ \overline{Z}_3^5 \\ \overline{X}_5^5 \\ \overline{Z}_5^5 \end{bmatrix} = \begin{bmatrix} +15.00 \\ 0 \\ -15.00 \\ 0 \end{bmatrix} \text{kN} \quad (\text{Compression}),$$

$$\text{Member 6}: \begin{bmatrix} \overline{X}_4^6 \\ \overline{Z}_4^6 \\ \overline{X}_5^6 \\ \overline{Z}_5^6 \end{bmatrix} = \begin{bmatrix} -18.03 \\ 0 \\ +18.03 \\ 0 \end{bmatrix} \text{kN} \quad (\text{Tension}).$$

P3.4. Calculate the displacements at the nodes, the reactions at the supports, and the forces developed in the members of the pin-connected plane frame shown in Fig. P3.4. All members have the same cross-sectional area, $A = 120 \times 10^{-6}$ m^2 and are made of aluminium with modulus of elasticity, $E = 70 \times 10^6$ kN/m^2.

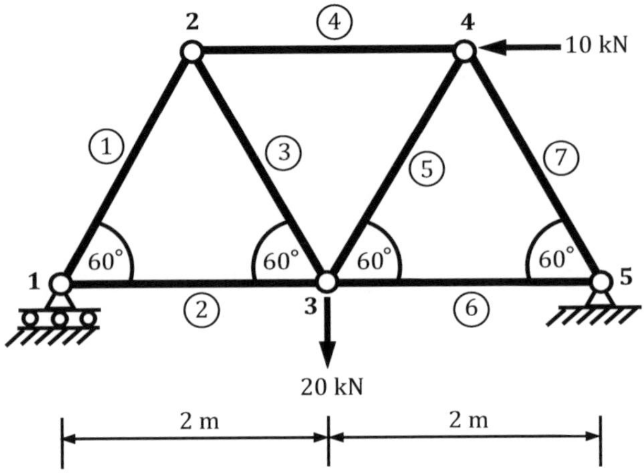

Figure P3.4

Answer:

$u_1 = -0.005130$ m, $w_1 = 0$, $u_2 = -0.003274$, $w_2 = -0.005621$ m, $u_3 = -0.003160$ m, $w_3 = -0.010105$ m, $u_4 = -0.007213$ m, $w_4 = -0.005965$ m, $u_5 = 0$, $w_5 = 0$,
$R_{X1} = 0$, $R_{Z1} = 14.33$ kN, $R_{X5} = 10.00$ kN, $R_{Z5} = 5.67$ kN.

$$\text{Member 1:} \begin{bmatrix} \overline{X}_1^1 \\ \overline{Z}_1^1 \\ \overline{X}_2^1 \\ \overline{Z}_2^1 \end{bmatrix} = \begin{bmatrix} +16.55 \\ 0 \\ -16.55 \\ 0 \end{bmatrix} \text{ kN (Compression),}$$

$$\text{Member 2:} \begin{bmatrix} \overline{X}_1^2 \\ \overline{Z}_1^2 \\ \overline{X}_3^2 \\ \overline{Z}_3^2 \end{bmatrix} = \begin{bmatrix} -8.27 \\ 0 \\ +8.27 \\ 0 \end{bmatrix} \text{ kN (Tension),}$$

$$\text{Member 3}: \begin{bmatrix} \bar{X}_2^3 \\ \bar{Z}_2^3 \\ \bar{X}_3^3 \\ \bar{Z}_3^3 \end{bmatrix} = \begin{bmatrix} -16.55 \\ 0 \\ +16.55 \\ 0 \end{bmatrix} \text{kN} \quad (\text{Tension}),$$

$$\text{Member 4}: \begin{bmatrix} \bar{X}_2^4 \\ \bar{Z}_2^4 \\ \bar{X}_4^4 \\ \bar{Z}_4^4 \end{bmatrix} = \begin{bmatrix} +16.55 \\ 0 \\ -16.55 \\ 0 \end{bmatrix} \text{kN} \quad (\text{Compression}),$$

$$\text{Member 5}: \begin{bmatrix} \bar{X}_4^5 \\ \bar{Z}_4^5 \\ \bar{X}_3^5 \\ \bar{Z}_3^5 \end{bmatrix} = \begin{bmatrix} -6.55 \\ 0 \\ +6.55 \\ 0 \end{bmatrix} \text{kN} \quad (\text{Tension}),$$

$$\text{Member 6}: \begin{bmatrix} \bar{X}_3^6 \\ \bar{Z}_3^6 \\ \bar{X}_5^6 \\ \bar{Z}_5^6 \end{bmatrix} = \begin{bmatrix} -13.27 \\ 0 \\ +13.27 \\ 0 \end{bmatrix} \text{kN} \quad (\text{Tension}),$$

$$\text{Member 7}: \begin{bmatrix} \bar{X}_4^7 \\ \bar{Z}_4^7 \\ \bar{X}_5^7 \\ \bar{Z}_5^7 \end{bmatrix} = \begin{bmatrix} +6.55 \\ 0 \\ -6.55 \\ 0 \end{bmatrix} \text{kN} \quad (\text{Compression}).$$

Chapter 4

Bending of Beams

Beams are straight members subjected to loads acting between the supports and usually in the direction of gravity, i.e. along the z-axis. The boundary conditions may include fixed, hinged, roller, elastically restrained, or free types of support. The beam may be prismatic, i.e. with a constant cross section otherwise it is non-prismatic. In this book only prismatic beams are treated in detail. When a case of non-prismatic beam is encountered then the beam may be divided into elements, each of which is assumed to have constant cross section leading to an approximate solution and the accuracy can be improved if the number of elements is increased.

The beam under load will deform undergoing deflections and rotations due to the curvature of the deflection curve and the actions developed in the beam are bending moments and shear forces.

4.1 Derivation of Beam Stiffness Matrix

The stiffness matrix of a beam is the relationship between the actions (forces and moments) and displacements (translational and rotational) at the ends of the beam.

The derivation of the stiffness matrix is based on the local coordinates system $\bar{x}, \bar{y}, \bar{z}$ with the $\bar{x}$-axis running along the axis of the beam. The displacements and forces are relative to the local $\bar{x}$-axis thus they are written with a bar.

Analysis of Structures by Matrix Methods
Fathi Al-Shawi

ISBN 978-981-4968-19-5 (Hardcover), 978-1-003-29130-5 (eBook)
www.jennystanford.com

Consider a beam subjected to forces and moments $\bar{Z}_i$ and $\bar{M}_i$ at node i and $\bar{Z}_j$ and $\bar{M}_j$ at node j as shown in Fig. 4.1. The beam will deform from its initial straight horizontal position into the shape shown with the translational and rotational displacements $\bar{w}_i$ and $\bar{\theta}_i$ at node i and $\bar{w}_j$ and $\bar{\theta}_j$ at node j. The moment $\bar{M}$ and the rotation $\bar{\theta}$ are both about the local $\bar{y}$-axis.

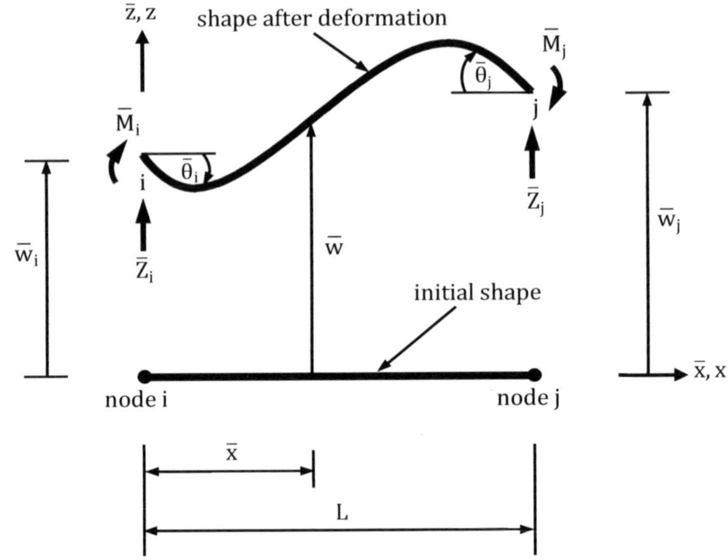

Figure 4.1 A beam element.

Castigliano's theorem is used to find a relationship between the actions (forces and moments) applied at the ends of the beam and the corresponding displacements (translational and rotational).

The derivation is divided into two parts: the first part assumes that the beam is fixed at nod j and the second part assumes that the beam is fixed at node i and then these two parts are superimposed linearly to get the final result. For the first part where the beam is fixed at node j, assume that the beam undergoes a translational displacement, $\bar{w}_i$ and a rotational displacement, $\bar{\theta}_i$ at node i. The forces and moments (actions) $\bar{Z}_i^{'}$, $\bar{M}_i^{'}$, $\bar{Z}_j^{'}$, and $\bar{M}_j^{'}$ at the ends i and j of the beam are as shown in Fig. 4.2 and these will be found in terms of $\bar{w}_i$ and $\bar{\theta}_i$.

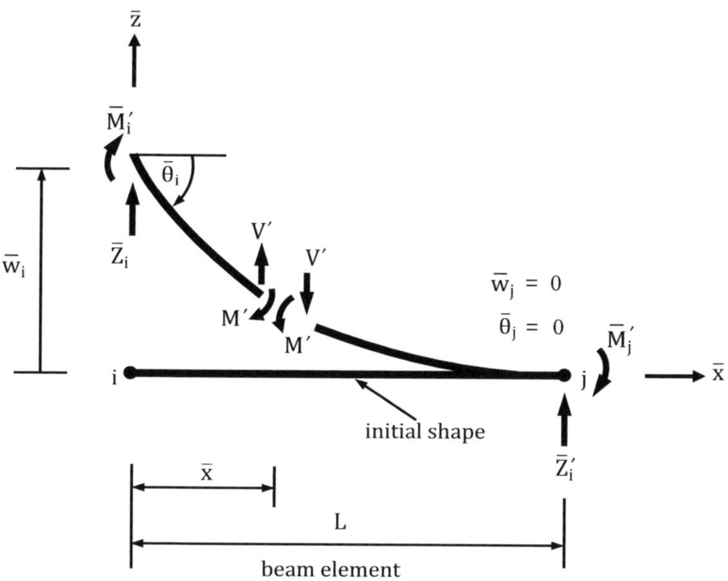

Figure 4.2

Cut a section at a distance $\overline{x}$ from node i and consider the equilibrium of the left potion of the beam. At the right end of the left portion of the beam the clockwise bending moment $M^{'}$ and the shear force $V^{'}$ in the positive z-direction, shown in Fig. 4.2, are both positive.

From equilibrium of the left part of the beam and by taking moments about the cut section, we get

$$\overline{M}_i^{'} + \overline{Z}_i^{'}\overline{x} + M^{'} = 0$$

$$M^{'} = -\overline{M}_i^{'} - \overline{Z}_i^{'}\overline{x} \tag{4.1}$$

The strain energy in bending is:

$$E_S^{'} = \int_0^L \frac{M^{'2} d\overline{x}}{2EI} \tag{4.2}$$

where E is the modulus of elasticity and I is the second moment of area of the cross section about its centroidal $\overline{y}$ -axis. Strictly, it should be written as $I_{\overline{y}}$ but because bending is about the $\overline{y}$ -axis only and for simplicity it is written as I. This notation is used in subsequent chapters except in Chapter 10 where there is bending about the $\bar{y}$-and $\bar{z}$-axes and the appropriate subscripts will be used.

Castigliano's theorem states that the deflection at a point in the structure is equal to the partial derivative of the strain energy in with respect to the force acting at that point. Thus:

$$\bar{w}_i = \frac{\partial E_S'}{\partial \bar{Z}_i'} = \frac{\partial E_S'}{\partial M'} \frac{\partial M'}{\partial \bar{Z}_i'} \tag{4.3}$$

From equation (4.2), $\dfrac{\partial E_S'}{\partial \bar{M}'} = \displaystyle\int_0^L \frac{M' d\bar{x}}{EI}$

From equation (4.1), $\dfrac{\partial M'}{\partial \bar{Z}_i'} = -\bar{x}$

Equation (4.3) becomes, $\bar{w}_i = \displaystyle\int_0^L \frac{M'}{EI}(-\bar{x})d\bar{x} = \int_0^L \frac{(-\bar{M}_i' - \bar{Z}_i'\bar{x})(-\bar{x})}{EI} d\bar{x}$

$$\bar{w}_i = \frac{1}{EI}\left(\bar{M}_i' \frac{L^2}{2} + \bar{Z}_i' \frac{L^3}{3} \right) \tag{4.4}$$

Similarly, Castigliano's theorem states that the rotation at a point in the structure is equal to the partial derivative of the strain energy with respect to the moment acting at that point. Thus:

$$\bar{\theta}_i = \frac{\partial E_S'}{\partial \bar{M}_i'} = \frac{\partial E_S'}{\partial M'} \frac{\partial M'}{\partial \bar{M}_i'} \tag{4.5}$$

From equation (4.1), $\dfrac{\partial M'}{\partial \bar{M}_i'} = -1$

Equation (4.5) becomes, $\bar{\theta}_i = \displaystyle\int_0^L \frac{M'}{EI}(-1)d\bar{x} = \int_0^L \frac{(-\bar{M}_i' - \bar{Z}_i'\bar{x})(-1)}{EI} d\bar{x}$

$$\bar{\theta}_i = \frac{1}{EI}\left(\bar{M}_i' L + \bar{Z}_i' \frac{L^2}{2} \right) \tag{4.6}$$

Solving equations (4.4) and (4.6) simultaneously for the unknowns $\bar{Z}_i'$ and $\bar{M}_i'$ to get:

$$\bar{Z}_i' = \frac{12EI}{L^3}\bar{w}_i - \frac{6EI}{L^2}\bar{\theta}_i \tag{4.7}$$

$$\bar{M}_i' = -\frac{6EI}{L^2}\bar{w}_i + \frac{4EI}{L}\bar{\theta}_i \tag{4.8}$$

Equations (4.7) and (4.8) represent $\bar{k}_{ii}$ of the stiffness matrix.

The force and moment at end j are calculated from consideration of equilibrium of the whole beam.

Summation of the forces in the $\bar{z}$-direction is zero

$\bar{Z}_i' + \bar{Z}_j' = 0$, $\bar{Z}_j' = -\bar{Z}_i'$ and substitute equation (4.7) we get

$$\bar{Z}_j' = -\frac{12EI}{L^3}\bar{w}_i + \frac{6EI}{L^2}\bar{\theta}_i \tag{4.9}$$

Summation of the moments about node j is zero

$\bar{M}_i' + \bar{Z}_i' L + \bar{M}_j' = 0$, $\bar{M}_j' = -\bar{M}_i' - \bar{Z}_i' L$ and substitute equations (4.7) and (4.8) to get

$$\bar{M}_j' = -\frac{6EI}{L^2}\bar{w}_i + \frac{2EI}{L}\bar{\theta}_i \tag{4.10}$$

Equations (4.9) and (4.10) represent $\bar{k}_{ji}$ of the stiffness matrix.

For the second part of the derivation, assume that the beam is fixed at node i and has a translational displacement, $\bar{w}_j$ and a rotational displacement, $\bar{\theta}_i$ at node j. The forces and moments (actions) $\bar{Z}_i''$, $\bar{M}_i''$, $\bar{Z}_j''$, and $\bar{M}_j''$ at the ends i and j of the beam are shown in Fig. 4.3 and these will be found in terms of $\bar{w}_j$ and $\bar{\theta}_j$.

The bending moment M'' at a distance $\bar{x}$ from end i is now found in terms of the moment $\bar{M}_j''$ and force $\bar{Z}_j''$ at node j as follows:

From the equilibrium of the whole beam, $\bar{Z}_i'' = -\bar{Z}_j''$ and the summation of the moments about node i is zero, $\bar{M}_i'' = -\bar{M}_j'' + \bar{Z}_j'' L$.

From the equilibrium of the left part of the beam

$$M'' = -\bar{M}_i'' - \bar{Z}_i'' \bar{x}$$

Substitute for $\bar{Z}_i''$ and $\bar{M}_i''$ in term of $\bar{Z}_j''$ and $\bar{M}_j''$, respectively to get

$$M'' = \bar{M}_j'' - \bar{Z}_j''(L - \bar{x}) \tag{4.11}$$

(The above equation could have been derived directly by considering the equilibrium of the right part of the beam, but that might cause some confusion to the reader with the signs of the forces and moments.)

$$E_S'' = \int_0^L \frac{M''^2 d\bar{x}}{2EI} \tag{4.12}$$

$$\bar{w}_j = \frac{\partial E_S''}{\partial \bar{Z}_j''} = \frac{\partial E_S''}{\partial M''}\frac{\partial M''}{\partial \bar{Z}_j''} \tag{4.13}$$

From equation (4.12), $\dfrac{\partial E_S''}{\partial M''} = \displaystyle\int_0^L \frac{\bar{M}'' d\bar{x}}{EI}$

From equation (4.11), $\dfrac{\partial M''}{\partial \bar{Z}_i''} = -(L-\bar{x})$

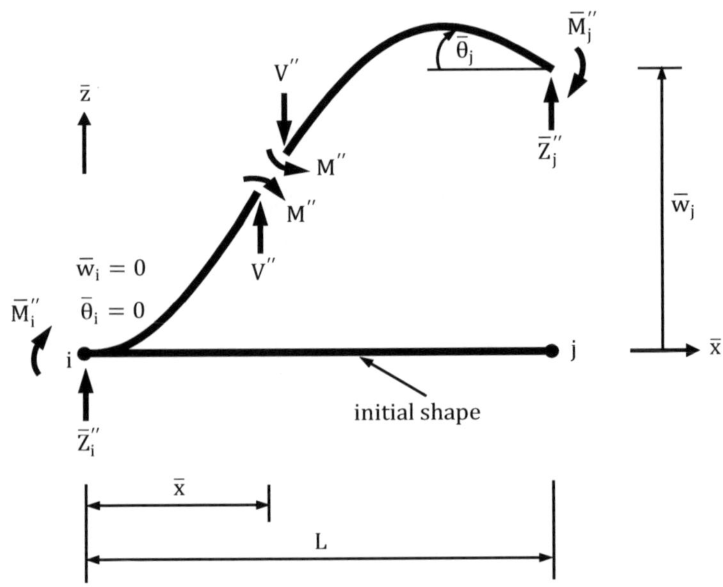

Figure 4.3

Equation (4.13) becomes

$$\bar{w}_j = \int_0^L -\frac{M''}{EI}(L-\bar{x})d\bar{x} = \int_0^L -\frac{[\bar{M}_j'' - \bar{Z}_j''(L-\bar{x})](L-\bar{x})}{EI}d\bar{x}$$

$$\bar{w}_j = \frac{1}{EI}\left(-\bar{M}_j''\frac{L^2}{2} + \bar{Z}_j''\frac{L^3}{3}\right) \tag{4.14}$$

$$\bar{\theta}_i = \frac{\partial E_S''}{\partial \bar{M}_j''} = \frac{\partial E_S''}{\partial M''} \frac{\partial M''}{\partial \bar{M}_j''} \tag{4.15}$$

From equation (4.11), $\dfrac{\partial M''}{\partial \bar{M}_j''} = +1$

Equation (4.15) becomes $\bar{\theta}_j = \int_0^L \dfrac{M''}{EI}(+1)d\bar{x} = \int_0^L \dfrac{[\bar{M}_j'' - \bar{Z}_j''(L-\bar{x})]}{EI} d\bar{x}$

$$\bar{\theta}_j = \frac{1}{EI}\left(\bar{M}_j'' L - \bar{Z}_j'' \frac{L^2}{2}\right) \tag{4.16}$$

Solving (4.14) and (4.16) simultaneously to get

$$\bar{Z}_j'' = \frac{12EI}{L^3}\bar{w}_j + \frac{6EI}{L^2}\bar{\theta}_j \tag{4.17}$$

$$\bar{M}_j'' = \frac{6EI}{L^2}\bar{w}_j + \frac{4EI}{L}\bar{\theta}_j \tag{4.18}$$

Equations (4.17) and (4.18) represent $\bar{k}_{jj}$ of the stiffness matrix.
From equilibrium of the whole beam
$\bar{Z}_i'' + \bar{Z}_j'' = 0$, $\bar{Z}_i'' = -\bar{Z}_j''$, substitute (4.17) to get

$$\bar{Z}_i'' = -\frac{12EI}{L^3}\bar{w}_j - \frac{6EI}{L^2}\bar{\theta}_j \tag{4.19}$$

Summation of the moments about node i is zero
$\bar{M}_i'' - \bar{Z}_j'' L + \bar{M}_j'' = 0$, $\bar{M}_i'' = -\bar{M}_j'' + \bar{Z}_j'' L$, substitute (4.17) and (4.18) to get

$$\bar{M}_i'' = \frac{6EI}{L^2}\bar{w}_j + \frac{2EI}{L}\bar{\theta}_j \tag{4.20}$$

Equations (4.19) and (4.20) represent $\bar{k}_{ij}$ of the stiffness matrix.

The final end forces and moments are obtained by adding the appropriate equations obtained from cases one and two as follows:

$\bar{Z}_i = \bar{Z}_i' + \bar{Z}_i''$, and from (4.7) and (4.19) we get

$$\bar{Z}_i = \frac{12EI}{L^3}\bar{w}_i - \frac{6EI}{L^2}\bar{\theta}_i - \frac{12EI}{L^3}\bar{w}_j - \frac{6EI}{L^2}\bar{\theta}_j \tag{4.21}$$

$\bar{M}_i = \bar{M}_i' + \bar{M}_i''$, and from (4.8) and 4.20) we get

$$\bar{M}_i = -\frac{6EI}{L^2}\bar{w}_i + \frac{4EI}{L}\bar{\theta}_i + \frac{6EI}{L^2}\bar{w}_j + \frac{2EI}{L}\bar{\theta}_j \tag{4.22}$$

$\bar{Z}_j = \bar{Z}_j^{'} + \bar{Z}_j^{''}$, and from (4.9) and (4.17) we get

$$\bar{Z}_j = -\frac{12EI}{L^3}\bar{w}_i + \frac{6EI}{L^2}\bar{\theta}_i + \frac{12EI}{L^3}\bar{w}_j + \frac{6EI}{L^2}\bar{\theta}_j \tag{4.23}$$

$\bar{M}_j = \bar{M}_j^{'} + \bar{M}_j^{''}$, and from (4.10) and (4.18) we get

$$\bar{M}_j = -\frac{6EI}{L^2}\bar{w}_i + \frac{2EI}{L}\bar{\theta}_i + \frac{6EI}{L^2}\bar{w}_j + \frac{4EI}{L}\bar{\theta}_j \tag{4.24}$$

Writing equations (4.21) to (4.24) in matrix form leads to:

$$\begin{bmatrix} \bar{Z}_i \\ \bar{M}_i \\ \bar{Z}_j \\ \bar{M}_j \end{bmatrix} = \begin{bmatrix} \frac{12EI}{L^3} & -\frac{6EI}{L^2} & -\frac{12EI}{L^3} & -\frac{6EI}{L^2} \\ -\frac{6EI}{L^2} & \frac{4EI}{L} & \frac{6EI}{L^2} & \frac{2EI}{L} \\ -\frac{12EI}{L^3} & \frac{6EI}{L^2} & \frac{12EI}{L^3} & \frac{6EI}{L^2} \\ -\frac{6EI}{L^2} & \frac{2EI}{L} & \frac{6EI}{L^2} & \frac{4EI}{L} \end{bmatrix} \begin{bmatrix} \bar{w}_i \\ \bar{\theta}_i \\ \bar{w}_j \\ \bar{\theta}_j \end{bmatrix} \tag{4.25}$$

or

$$\bar{F} = \bar{k}\bar{\delta} \tag{4.26}$$

where $\bar{F} = \begin{bmatrix} \bar{Z}_i \\ \bar{M}_i \\ \bar{Z}_j \\ \bar{M}_j \end{bmatrix}$ is the action vector, $\bar{\delta} = \begin{bmatrix} \bar{w}_i \\ \bar{\theta}_i \\ \bar{w}_j \\ \bar{\theta}_j \end{bmatrix}$ is the displacement vector, and

$$\bar{k} = \begin{bmatrix} \frac{12EI}{L^3} & -\frac{6EI}{L^2} & -\frac{12EI}{L^3} & -\frac{6EI}{L^2} \\ -\frac{6EI}{L^2} & \frac{4EI}{L} & \frac{6EI}{L^2} & \frac{2EI}{L} \\ -\frac{12EI}{L^3} & \frac{6EI}{L^2} & \frac{12EI}{L^3} & \frac{6EI}{L^2} \\ -\frac{6EI}{L^2} & \frac{2EI}{L} & \frac{6EI}{L^2} & \frac{4EI}{L} \end{bmatrix} \text{(is the stiffness matrix)}. \tag{4.27}$$

The above relationships can alternatively be derived by a finite element approach using the so-called interpolation polynomial which defines the displacement along the element as explained in Appendix 2.

The overall stiffness matrix is assembled relative to global coordinates. So, the first step is to find the stiffness matrices of the members relative to the global coordinates. For a beam element where its $\bar{x}$-axis coincides with the global x-axis the displacements and stiffness matrices derived relative to local coordinates will have the same values relative to global coordinates, i.e. they do not need to be transformed. This follows from the fact that in this case the transformation matrix r will be equal to the unit matrix I and by noting that $k = r^T\bar{k}r = I^T\bar{k}I = \bar{k}$. Thus, the stiffness matrix of the beam relative to local coordinates is used as the stiffness matrix relative to global coordinates, thus:

$$k = \begin{bmatrix} \frac{12EI}{L^3} & -\frac{6EI}{L^2} & -\frac{12EI}{L^3} & -\frac{6EI}{L^2} \\ -\frac{6EI}{L^2} & \frac{4EI}{L} & \frac{6EI}{L^2} & \frac{2EI}{L} \\ -\frac{12EI}{L^3} & \frac{6EI}{L^2} & \frac{12EI}{L^3} & \frac{6EI}{L^2} \\ -\frac{6EI}{L^2} & \frac{2EI}{L} & \frac{6EI}{L^2} & \frac{4EI}{L} \end{bmatrix} \quad (4.28)$$

The stiffness matrix derived above is for an individual member and structures are normally composed of more than one member that are connected together to form the structure. The required overall structure stiffness matrix is assembled by adding the contributions of the individual members' stiffness matrices to any joint that is common to these members.

Example 1

Calculate and draw the shear force and bending moment diagrams and the deflected shape for the continuous beam shown in Fig. 4.4.

The beam is made of concrete with modulus of elasticity $E = 35 \times 10^6$ kN/m^2 and has the following properties:

Member 1: $L = 7$ m, $I = 490 \times 10^{-6}$ m^4

Member 2: $L = 6$ m, $I = 660 \times 10^{-6}$ m^4

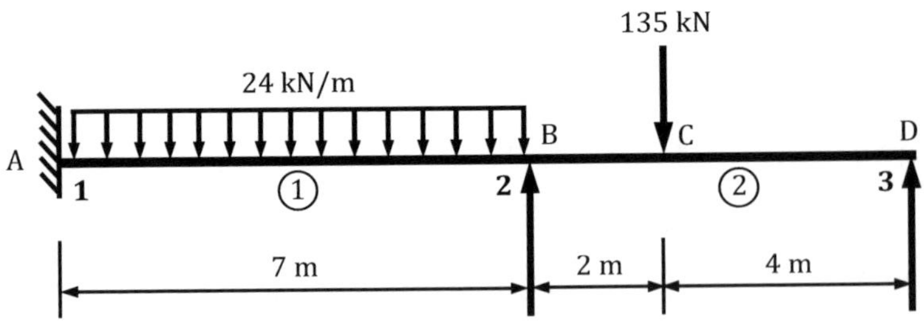

Figure 4.4

Stiffness matrices of the members

Member 1

Member address: i j
Structure address: 1 2
L = 7 m, I = 490 × 10^{-6} m^4, E = 35 × 10^6 kN/m^2.
Substitute the above values in (4.28) to get

$$\overbrace{\overbrace{\begin{matrix} w_i & \theta_i \\ w_1 & \theta_1 \end{matrix}}^{\delta_i = \delta_1} \quad \overbrace{\begin{matrix} w_j & \theta_j \\ w_2 & \theta_2 \end{matrix}}^{\delta_j = \delta_2}}^{\delta^1}$$

$$k^1 = \begin{bmatrix} 600 & -2100 & -600 & -2100 \\ -2100 & 9800 & 2100 & 4900 \\ -600 & 2100 & 600 & 2100 \\ -2100 & 4900 & 2100 & 9800 \end{bmatrix} \begin{matrix} w_1 \\ \theta_1 \\ w_2 \\ \theta_2 \end{matrix}$$

where $k_{ii}^1 = \begin{bmatrix} 600 & -2100 \\ -2100 & 9800 \end{bmatrix}$, $k_{ij}^1 = \begin{bmatrix} -600 & -2100 \\ 2100 & 4900 \end{bmatrix}$,

$k_{ji}^1 = \begin{bmatrix} -600 & 2100 \\ -2100 & 4900 \end{bmatrix}$, and $k_{jj}^1 = \begin{bmatrix} 600 & 2100 \\ 2100 & 9800 \end{bmatrix}$.

Member 2

Member address: i j
Structure address: 2 3
$L = 6$ m, $I = 660 \times 10^{-6}$ m^4, $E = 35 \times 10^6$ kN/m^2.
Substitute the above values in (4.28)

$$\overbrace{\overbrace{\begin{matrix} w_i & \theta_i \\ w_2 & \theta_2 \end{matrix}}^{\delta_i = \delta_2} \quad \overbrace{\begin{matrix} w_j & \theta_j \\ w_3 & \theta_3 \end{matrix}}^{\delta_j = \delta_3}}^{\delta^2}$$

$$k^2 = \begin{bmatrix} 1283 & -3850 & -1283 & -3850 \\ -3850 & 15400 & 3850 & 7700 \\ -1283 & 3850 & 1283 & 3850 \\ -3850 & 7700 & 3850 & 15400 \end{bmatrix} \begin{matrix} w_2 \\ \theta_2 \\ w_3 \\ \theta_3 \end{matrix}$$

$$\text{where } k_{ii}^2 = \begin{bmatrix} 1283 & -3850 \\ -3850 & 15400 \end{bmatrix}, \; k_{ij}^2 = \begin{bmatrix} -1283 & -3850 \\ 3850 & 7700 \end{bmatrix},$$

$$k_{ji}^2 = \begin{bmatrix} -1283 & 3850 \\ -3850 & 7700 \end{bmatrix}, \text{ and } k_{jj}^2 = \begin{bmatrix} 1283 & 3850 \\ 3850 & 15400 \end{bmatrix}.$$

The general relationship for the whole structure is

$$K\delta = F \tag{4.29}$$

where K is the structure overall stiffness matrix relative to global coordinates, δ is the vector of displacements at the nodes of the structure, and F is the load vector.

Assembly of the overall stiffness matrix relative to global coordinates

There are three nodes in the structure therefore the overall stiffness matrix K will consist of 3×3 submatrices. And since each node has two degrees of freedom, w and θ, then each submatrix will be 2×2 as shown below.

Each of the members of the frame will contribute to the structure stiffness matrix K according to the relationship between the member address and the structure address. In the following, the translation w and rotation θ at the nodes will be treated as one displacement δ. This will make the assembly process more manageable and easily handled. Thus, the overall stiffness matrix of the structure is

K=

	δ_1	δ_2	δ_3	
	k^1_{ii}	k^1_{ij}	0	δ_1
	k^1_{ji}	$k^1_{jj}+k^2_{ii}$	k^2_{ij}	δ_2
	0	k^2_{ji}	k^2_{jj}	δ_3

K= (4.30)

	w_1	θ_1	w_2	θ_2	w_3	θ_3	
	600	-2100	-600	-2100	0	0	w_1
	-2100	9800	2100	4900	0	0	θ_1
	-600	2100	1883	-1750	-1283	-3850	w_2
	-2100	4900	-1750	25200	3850	7700	θ_2
	0	0	-1283	3850	1283	3850	w_3
	0	0	-3850	7700	3850	15400	θ_3

4.2 Load Vector

In the previous chapter we had pin-connected frames where the forces on the structure were considered as point loads acting on the joints. For continuous beams most of the loads are acting on the members rather than directly on the joints and they have to be transferred to the joints as equivalent forces (as well as moments). For this purpose the members are first assumed fixed at their ends and the forces and moments acting at these ends are calculated as explained in Appendix 5. The forces and moments acting on any joint are obtained by adding up the contributions from the members meeting at that joint.

Consider beam AB with fixed ends at nodes i and j and is subjected to an arbitrary load as shown in Fig. 4.5.

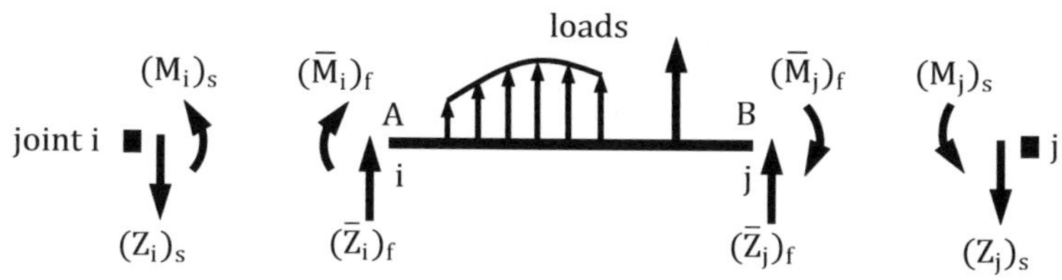

Figure 4.5

The forces $(\bar{Z}_i)_f$ and $(\bar{Z}_j)_f$ and moments $(\bar{M}_i)_f$ and $(\bar{M}_j)_f$ are called actions on the beam due to the applied forces on the span of the beam and are calculated assuming that the beam is fixed at its ends i and j. The forces $(Z_i)_S$ and $(Z_j)_S$ and moments $(M_i)_S$ and $(M_j)_S$ are called loads on joints i and j of the structure and from the beam-joint section equilibrium they have the same magnitude of forces and moments as those acting on the beam but in the opposite direction.

Note that all forces and moments acting on the beam are drawn in the positive directions as shown in Fig. 4.5 and if any of them turns out to be negative then its actual direction is opposite to that assumed.

The subscript (f) is for the vector of *actions* (forces and moments) on the ends of the member due to the forces acting on the span of the member. These actions are calculated relative to the local coordinates of the member. The subscript (s) is for the *load* vector on the joints of the structure. For equilibrium at the section between the member end and the joint the quantities in the load vector on the joint have opposite signs to those in the vector of actions on the member. The load vector on the joints of the structure is calculated relative to global coordinates, but because the beam local $\bar{x}$-axis coincides with the global x-axis transformation is not required.

When two beams meet at a continuous rigid joint then the force and moment applied on that joint are given by the algebraic sum of forces and moments at the meeting ends of the two beams.

The total force and moment acting on any joint of the structure are calculated from the combined effect of the applied load on the beams and the reactions exerted by their support.

The load vector for the overall structure is obtained from the magnitudes and directions of the forces and moments acting at all joints of the structure.

To build up the load vector for the whole structure a similar procedure as that in building up the structure stiffness matrix is followed.

The total load vector F acting on the joints of the structure is composed of forces and moments F_S due to the forces acting on the span of the members in addition to the forces and moments F_C exerted by the restraints, for example the reactions of the supports on the structure at the nodes, hence

$$F = F_S + F_C$$

Member 1

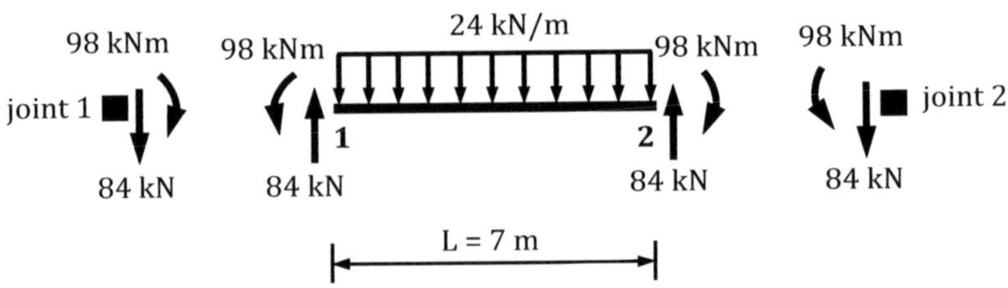

Figure 4.6

With reference to Fig. 4.5 for the notation and n = –24 kN/m

$$(\bar{Z}_1^1)_f = -\frac{nL}{2} = -\frac{(-24)\times 7}{2} = +84 \text{ kN}$$

$$(\bar{Z}_2^1)_f = -\frac{nL}{2} = -\frac{(-24)\times 7}{2} = +84 \text{ kN}$$

$$(\bar{M}_1^1)_f = +\frac{nL^2}{12} = +\frac{(-24)\times 7^2}{12} = -98 \text{ kNm}$$

$$(\bar{M}_2^1)_f = -\frac{nL^2}{12} = -\frac{(-24)\times 7^2}{12} = +98 \text{ kNm}$$

Hence the action vector for member 1 is:

$$\bar{F}_f^1 = \begin{bmatrix} (\bar{Z}_1^1)_f \\ (\bar{M}_1^1)_f \\ (\bar{Z}_2^1)_f \\ (\bar{M}_2^1)_f \end{bmatrix} = \begin{bmatrix} +84 \\ -98 \\ +84 \\ +98 \end{bmatrix} \tag{4.31}$$

From equilibrium of the beam-joint section at node 1

$$(Z_1^1)_S + (\bar{Z}_1^1)_f = 0,\ (Z_1^1)_S = -\left(\bar{Z}_1^1\right)_f = -84\ \text{kN}$$

$$(M_1^1)_S + (\bar{M}_1^1)_f = 0,\ (M_1^1)_S = -\left(\bar{M}_1^1\right)_f = +98\ \text{kNm}$$

From equilibrium of the beam-joint section at node 2

$$(Z_2^1)_S + (\bar{Z}_2^1)_f = 0,\ (Z_2^1)_S = -\left(\bar{Z}_2^1\right)_f = -84\ \text{kN}$$

$$(M_2^1)_S + (\bar{M}_2^1)_f = 0,\ (M_2^1)_S = -\left(\bar{M}_2^1\right)_f = -98\ \text{kNm}$$

or, simply, $F_S^1 = -\bar{F}_f^1$

$$F_S^1 = \begin{bmatrix} (Z_1^1)_S \\ (M_1^1)_S \\ (Z_2^1)_S \\ (M_2^1)_S \end{bmatrix} = \begin{bmatrix} -84 \\ +98 \\ -84 \\ -98 \end{bmatrix} \tag{4.32}$$

Member 2

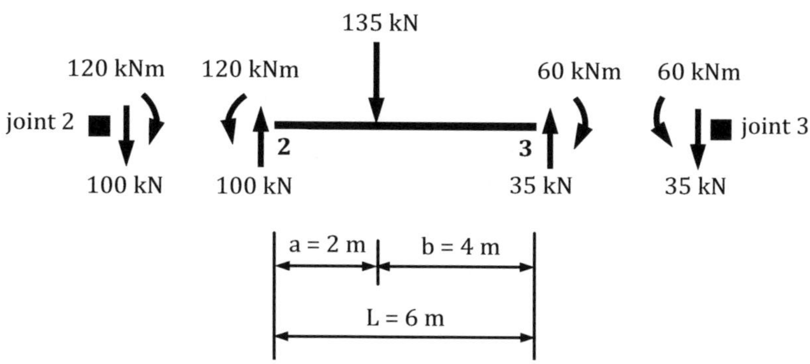

Figure 4.7

With reference to Fig. 4.5 for the notation and W = –135 kN

$$(\bar{Z}_2^2)_f = -\frac{Wb}{L^3}\left(L^2 + ab - a^2\right) = -\frac{(-135)\times 4}{6^3}\left(6^2 + 2\times 4 - 2^2\right) = +100\text{ kN}$$

$$(\bar{Z}_3^2)_f = -\frac{Wa}{L^3}\left(L^2 + ab - b^2\right) = -\frac{(-135)\times 2}{6^3}\left(6^2 + 2\times 4 - 4^2\right) = +35\text{ kN}$$

$$(\bar{M}_2^2)_f = +\frac{Wab^2}{L^2} = +\frac{(-135)\times 2\times 4^2}{6^2} = -120\text{ kNm}$$

$$(\bar{M}_3^2)_f = -\frac{Wa^2b}{L^2} = -\frac{(-135)\times 2^2\times 4}{6^2} = +60\text{ kNm}$$

The action vector for member 2 is:

$$\bar{F}_f^2 = \begin{bmatrix} (\bar{Z}_2^2)_f \\ (\bar{M}_2^2)_f \\ (\bar{Z}_3^2)_f \\ (\bar{M}_3^2)_f \end{bmatrix} = \begin{bmatrix} +100 \\ -120 \\ +35 \\ +60 \end{bmatrix} \qquad (4.33)$$

From equilibrium, the joint load vector is given by:

$$F_S^2 = -\bar{F}_f^2$$

$$F_S^2 = \begin{bmatrix} (Z_2^2)_S \\ (M_2^2)_S \\ (Z_3^2)_S \\ (M_3^2)_S \end{bmatrix} = \begin{bmatrix} -100 \\ +120 \\ -35 \\ -60 \end{bmatrix} \qquad (4.34)$$

Now calculate the resultant forces and moments acting on the joints of the structure.

Joint 1: This joint is acted upon by a downward force of 84 kN and a clockwise moment of 98 kNm due to the load on member 1. In addition, the support at joint 1 exerts a force R_{Z1} and a moment R_{M1}.

The resultant force is given by the algebraic sum of the forces.

$$R_1 = -84 + R_{Z1}$$

The resultant moment is given by the algebraic sum of the moments.

$$M_1 = +98 + R_{M1}$$

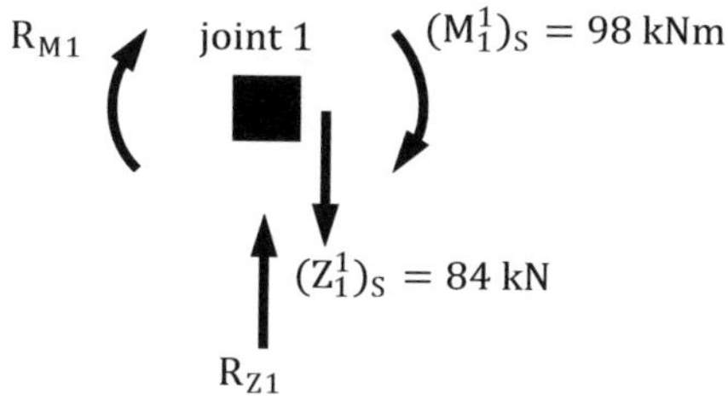

Joint 2: This joint is acted upon by a downward force of 84 kN and an anticlockwise moment of 98 kNm due to the load on member 1 and a downward force of 100 kN and a clockwise moment of 120 kNm due to the load on member 2. In addition, the support at joint 2 exerts a force R_{Z2} .

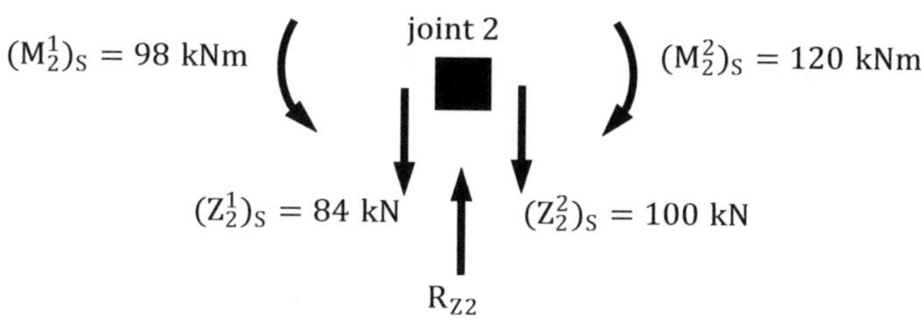

The resultant force, $Z_2 = -84 - 100 + R_{Z2} = -184 + Z_{Z2}$.
The resultant moment, $M_2 = -98 + 120 = +22$ kNm.

Joint 3: This joint is acted upon by a downward force of 35 kN and an anticlockwise moment of 60 kNm due to the load on member 2. In addition, the support at joint 3 exerts an upward force R_{Z3}.

The resultant force, $Z_3 = -35 + R_{Z3}$.
The moment, $M_3 = -60$ kNm.

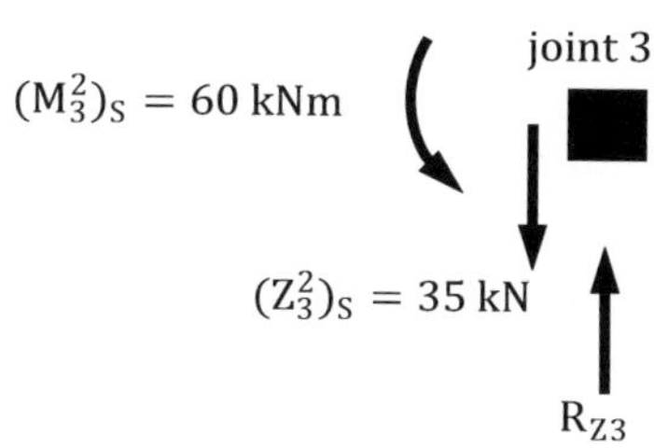

Therefore, the total load vector for the whole structure is

$$F=\begin{bmatrix}F_1\\F_2\\F_3\end{bmatrix}=\begin{bmatrix}Z_1\\M_1\\Z_2\\M_2\\Z_3\\M_3\end{bmatrix}=\begin{bmatrix}-84+R_{Z1}\\+98+R_{M1}\\-184+R_{Z2}\\+22\\-35+R_{Z3}\\-60\end{bmatrix}\tag{4.35}$$

The above result can also be found by direct vector addition of the load vectors due to the loads acting on the members and the reactions of the supports on the structure as follows:

Due to load on member 1 from (4.32), $F_S^1=\begin{bmatrix}(Z_1^1)_S\\(M_1^1)_S\\(Z_2^1)_S\\(M_2^1)_S\\0\\0\end{bmatrix}=\begin{bmatrix}-84\\+98\\-84\\-98\\0\\0\end{bmatrix}$

Due to load on member 2 from (4.34), $F_S^2=\begin{bmatrix}0\\0\\(Z_2^2)_S\\(M_2^2)_S\\(Z_3^2)_S\\(M_3^2)_S\end{bmatrix}=\begin{bmatrix}0\\0\\-100\\+120\\-35\\-60\end{bmatrix}$

Due to support reactions, $F_C=\begin{bmatrix}(Z_1)_C\\(M_1)_C\\(Z_2)_C\\(M_2)_C\\(Z_3)_C\\(M_3)_C\end{bmatrix}=\begin{bmatrix}R_{Z1}\\R_{M1}\\R_{Z2}\\0\\R_{Z3}\\0\end{bmatrix}$

$$F = \begin{bmatrix} Z_1 \\ M_1 \\ Z_2 \\ M_2 \\ Z_3 \\ M_3 \end{bmatrix} = F_S^1 + F_S^2 + F_C = \begin{bmatrix} -84 \\ +98 \\ -84 \\ -98 \\ 0 \\ 0 \end{bmatrix} + \begin{bmatrix} 0 \\ 0 \\ -100 \\ +120 \\ -35 \\ -60 \end{bmatrix} + \begin{bmatrix} R_{Z1} \\ R_{M1} \\ R_{Z2} \\ 0 \\ R_{Z3} \\ 0 \end{bmatrix} = \begin{bmatrix} -84+R_{Z1} \\ +98+R_{M1} \\ -184+R_{Z2} \\ +22 \\ -35+R_{Z3} \\ -60 \end{bmatrix} \tag{4.36}$$

which is the same as (4.35) and this simple method of vector addition will be used in the chapters that will follow.

Substitute K from (4.30) and F from (4.36) in the general relation (4.29) to get:

$$\begin{bmatrix} 600 & -2100 & -600 & -2100 & 0 & 0 \\ -2100 & 9800 & 2100 & 4900 & 0 & 0 \\ -600 & 2100 & 1883 & -1750 & -1283 & -3850 \\ -2100 & 4900 & -1750 & 25200 & 3850 & 7700 \\ 0 & 0 & -1283 & 3850 & 1283 & 3850 \\ 0 & 0 & -3850 & 7700 & 3850 & 15400 \end{bmatrix} \begin{bmatrix} w_1 \\ \theta_1 \\ w_2 \\ \theta_2 \\ w_3 \\ \theta_3 \end{bmatrix} = \begin{bmatrix} -84+R_{Z1} \\ +98+R_{M1} \\ -184+R_{Z2} \\ +22 \\ -35+R_{Z3} \\ -60 \end{bmatrix} \tag{4.26}$$

The next step is to introduce the boundary conditions as follows:

At node 1 where we have a fixed end, both the vertical deflection and rotation are equal to zero, i.e. $w_1 = 0$ hence delete row 1 and column 1, and $\theta_1 = 0$ hence delete row 2 and column 2.

At node 2 where there is a non-yielding support, the vertical deflection is equal to zero, i.e. $w_2 = 0$ hence delete row 3 and column 3.

At node 3 where there is a non-yielding support, the vertical deflection is equal to zero, i.e. $w_3 = 0$ hence delete row 5 and column 5.

The shaded rows and columns are those which are deleted and the resulting 'reduced' matrix is:

$$\begin{bmatrix} 25200 & 7700 \\ 7700 & 15400 \end{bmatrix} \begin{bmatrix} \theta_2 \\ \theta_3 \end{bmatrix} = \begin{bmatrix} +22 \\ -60 \end{bmatrix}$$

The above can be written in the form of two simultaneous equations as:

$$25200\theta_2 + 7700\theta_3 = +22$$

$$7700\theta_2 + 15400\theta_3 = -60$$

The solution of the above equations is:

$$\theta_2 = +0.00244 \text{ rad}$$

$$\theta_3 = -0.00511 \text{ rad}$$

Calculation of reactions at the supports

The reactions at the supports are calculated from (4.26) as follows

From the first row

$600w_1 - 2100\theta_1 - 600w_2 - 2100\theta_2 = -84 + R_{Z1}$

$600 \times 0 - 2100 \times 0 - 600 \times 0 - 2100 \times 0.00244 = -84 + R_{Z1}$

$R_{Z1} = 78.88$ kN

From the second row

$-2100w_1 + 9800\theta_1 + 2100w_2 + 4900\theta_2 = +98 + R_{M1}$

$-2100 \times 0 + 9800 \times 0 + 2100 \times 0 + 4900 \times 0.00244 = +98 + R_{M1}$

$R_{M1} = -86.04$ kNm

From the third row

$-600w_1 + 2100\theta_1 + 1883w_2 - 1750\theta_2 - 1283\theta_3 - 3850\theta_3 = -184 +$ R_{Z2} $-600 \times 0 + 2100 \times 0 + 1883 \times 0 - 1750 \times 0.00244 - 1283 \times 0$ $-3850 \times (-0.00511) = -184 + R_{Z2}$

$R_{Z2} = 199.40$ kN

From the fifth row

$-1283w_2 + 3850\theta_2 + 1283w_3 + 3850\theta_3 = -35 + R_{Z3}$

$-1283 \times 0 + 3850 \times 0.00244 + 1283 \times 0 + 3850 \times (-0.00511)$ $= -35 + R_{Z3}$

$R_{Z3} = 24.72$ kN

Calculation of resultant actions at the ends of the members

The vector of resultant actions on the member at its ends $\bar{F}_r$ is calculated relative to the local $\bar{x}$-axis and is given by: $\bar{F}_r = \bar{F}_d + \bar{F}_f$

where

$\overline{F}_d$ is the vector of actions at the ends of the member due to the resulting displacements and is given by equation (4.26) as $\overline{F}_d = \overline{k}\overline{\delta}$ with $\overline{k}$ from (4.27).

$\overline{F}_f$ is the vector of actions at the ends of the member due to the applied forces acting on the span of the member assuming the member is fixed at its ends (these are commonly called fixed end moments and forces), thus

$$\overline{F}_r = \overline{k}\overline{\delta} + \overline{F}_f$$

Member 1

$$\overline{k}^1 = \begin{bmatrix} 600 & -2100 & -600 & -2100 \\ -2100 & 9800 & 2100 & 4900 \\ -600 & 2100 & 600 & 2100 \\ -2100 & 4900 & 2100 & 9800 \end{bmatrix}$$

$$\overline{\delta} = r\delta$$

where r is transformation matrix and since the member local $\overline{x}$ -axis lies along the global x-axis r = I the unit matrix, thus $\overline{\delta} = \delta$.

$$\overline{\delta}^1 = \delta^1 = \begin{bmatrix} \delta_i^1 \\ \delta_j^1 \end{bmatrix} = \begin{bmatrix} \delta_1 \\ \delta_2 \end{bmatrix} = \begin{bmatrix} w_1 \\ \theta_1 \\ w_2 \\ \theta_2 \end{bmatrix} = \begin{bmatrix} 0 \\ 0 \\ 0 \\ 0.00244 \end{bmatrix}$$

$$\overline{F}_d^1 = \overline{k}^1\overline{\delta}^1 = \begin{bmatrix} 600 & -2100 & -600 & -2100 \\ -2100 & 9800 & 2100 & 4900 \\ -600 & 2100 & 600 & 2100 \\ -2100 & 4900 & 2100 & 9800 \end{bmatrix} \begin{bmatrix} 0 \\ 0 \\ 0 \\ 0.00244 \end{bmatrix} = \begin{bmatrix} -5.12 \\ +11.96 \\ +5.12 \\ +23.91 \end{bmatrix}$$

We had from (4.31)

$$\overline{F}_f^1 = \begin{bmatrix} +84 \\ -98 \\ +84 \\ +98 \end{bmatrix}$$

$$\begin{bmatrix} (\bar{Z}_1^1)_r \\ (\bar{M}_1^1)_r \\ (\bar{Z}_2^1)_r \\ (\bar{M}_2^1)_r \end{bmatrix} = \bar{F}_r^1 = \bar{F}_d^1 + \bar{F}_f^1 = \begin{bmatrix} -5.12 \\ +11.96 \\ +5.12 \\ +23.91 \end{bmatrix} + \begin{bmatrix} +84 \\ -98 \\ +84 \\ +98 \end{bmatrix} = \begin{bmatrix} +78.88 \\ -86.04 \\ +89.12 \\ +121.91 \end{bmatrix}$$

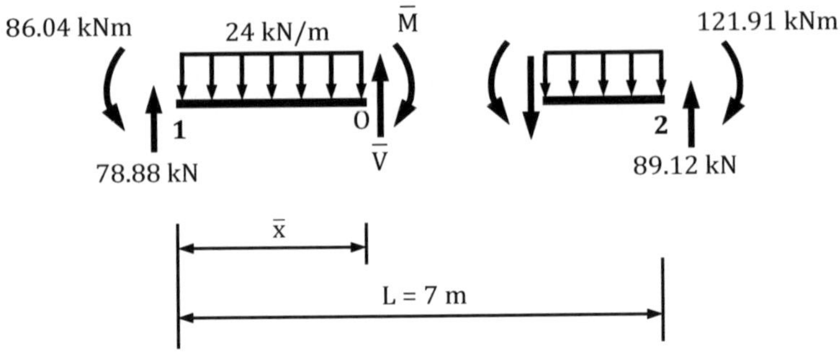

Shear force $\bar{V}$ and bending moment $\bar{M}$ diagrams

Consider a section at a distance $\bar{x}$ from node 1 and apply the equations of equilibrium on the left part of the member.

Summation of the forces in the $\bar{z}$-direction:

$$+78.88 - 24\bar{x} + \bar{V} = 0, \quad \bar{V} = -78.88 + 24\bar{x}$$

Summation of the moments about point O:

$$-86.04 + 78.88\bar{x} - 24\bar{x}\left(\frac{\bar{x}}{2}\right) + \bar{M} = 0, \quad \bar{M} = 86.04 - 78.88\bar{x} + 12\bar{x}^2$$

Calculation of deflection along the beam

The deflection along the beam can be found by solving the standard differential equation for the deflection of beams which is derived in Appendix 2 as given by equation (A2.4) as explained below.

$$EI\frac{d^2\bar{w}}{d\bar{x}^2} = -\bar{M}, \text{ with } EI = 35 \times 10^6 \times 490 \times 10^{-6} = 17150 \text{ kNm}^2$$

$$EI\frac{d^2\bar{w}}{d\bar{x}^2} = -86.04 + 78.88\bar{x} - 12\bar{x}^2$$

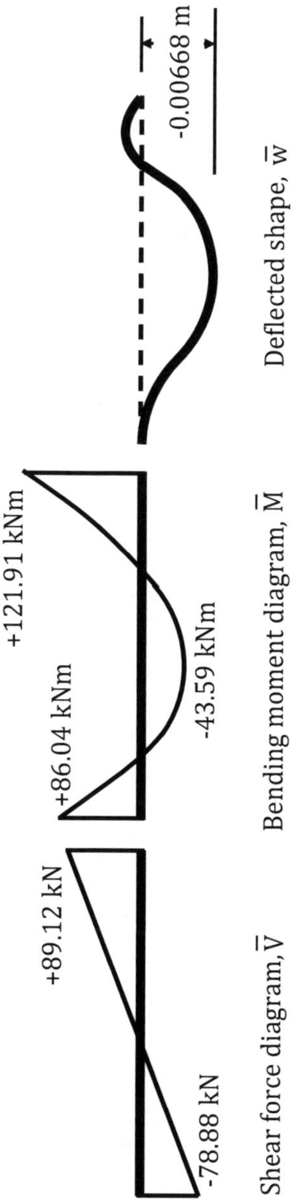
+89.12 kN
-78.88 kN
Shear force diagram, $\overline{V}$
+86.04 kNm
+121.91 kNm
-43.59 kNm
Bending moment diagram, $\overline{M}$
-0.00668 m
Deflected shape, $\overline{w}$

Figure 4.8

Integrate with respect to $\bar{x}$ to get

$$EI\frac{d\bar{w}}{d\bar{x}} = -86.04\bar{x} + 39.44\bar{x}^2 - 4\bar{x}^3 + C_1$$

$$\text{At } \bar{x} = 0, \quad \frac{d\bar{w}}{d\bar{x}} = 0, \quad \text{hence } C_1 = 0$$

Integrate again to get

$$EI\bar{w} = -43.02\bar{x}^2 + 13.15\bar{x}^3 - \bar{x}^4 + C_2$$

$$\text{At } \bar{x} = 0, \quad \bar{w} = 0, \quad \text{hence } C_2 = 0$$

Therefore $$\bar{w} = \frac{-43.02\bar{x}^2 + 13.15\bar{x}^3 - \bar{x}^4}{17150}$$

Member 2

$$\bar{k}^2 = \begin{bmatrix} 1283 & -3850 & -1283 & -3850 \\ -3850 & 15400 & 3850 & 7700 \\ -1283 & 3850 & 1283 & 3850 \\ -3850 & 7700 & 3850 & 15400 \end{bmatrix}$$

$$\delta^2 = \begin{bmatrix} \delta_i^2 \\ \delta_j^2 \end{bmatrix} = \begin{bmatrix} \delta_2 \\ \delta_3 \end{bmatrix} = \begin{bmatrix} w_2 \\ \theta_2 \\ w_3 \\ \theta_3 \end{bmatrix} = \begin{bmatrix} 0 \\ 0.00244 \\ 0 \\ -0.00511 \end{bmatrix}, \ \bar{\delta}^2 = \delta^2 = \begin{bmatrix} 0 \\ 0.00244 \\ 0 \\ -0.00511 \end{bmatrix}$$

$$\bar{F}_d^2 = \bar{k}^2\bar{\delta}^2 = \begin{bmatrix} 1283 & -3850 & -1283 & -3850 \\ -3850 & 15400 & 3850 & 7700 \\ -1283 & 3850 & 1283 & 3850 \\ -3850 & 7700 & 3850 & 15400 \end{bmatrix} \begin{bmatrix} 0 \\ 0.00244 \\ 0 \\ -0.00511 \end{bmatrix} = \begin{bmatrix} +10.28 \\ -1.77 \\ -10.28 \\ -59.91 \end{bmatrix}$$

We had from (4.33)

$$\bar{F}_f^2 = \begin{bmatrix} +100 \\ -120 \\ +35 \\ +60 \end{bmatrix}$$

$$\begin{bmatrix} (\bar{Z}_2^2)_r \\ (\bar{M}_2^2)_r \\ (\bar{Z}_3^2)_r \\ (\bar{M}_3^2)_r \end{bmatrix} = \bar{F}_r^2 = \bar{F}_d^2 + \bar{F}_f^2 = \begin{bmatrix} +10.28 \\ -1.77 \\ -10.28 \\ -59.91 \end{bmatrix} + \begin{bmatrix} +100 \\ -120 \\ +35 \\ +60 \end{bmatrix} = \begin{bmatrix} +110.28 \\ -121.77 \\ +24.72 \\ +0.09 \end{bmatrix}$$

Notice that $(\bar{M}_3^2)_r$ should be zero but the small value of 0.09 is due to rounding off in the computations.

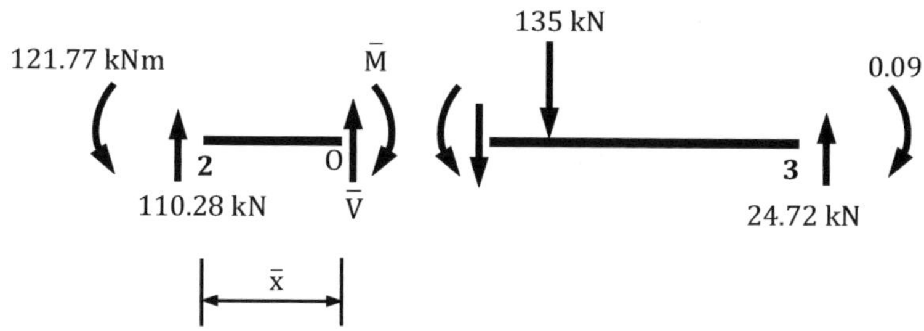

Shear force $\bar{V}$ and bending moment $\bar{M}$ diagrams

Consider a section at a distance $\bar{x}$ from node 2 and apply the equations of equilibrium on the left part of the member

Summation of the forces in the $\bar{z}$-direction is zero

$$\text{For } \bar{x} \le 2\text{ m}: +110.28 + \bar{V} = 0, \quad \bar{V} = -110.28 \text{ kN}$$

$$\text{For } 2\text{ m} \le \bar{x} \le 6\text{ m}: +110.28 - 135 + \bar{V} = 0, \quad \bar{V} = +24.72 \text{ kN}$$

Summation of the moments about point O is zero

$$\text{For } \bar{x} \le 2\text{ m}: -121.77 + 110.28\bar{x} + \bar{M} = 0, \quad \bar{M} = +121.77 - 110.28\bar{x}$$

$$\text{For } 2\text{ m} \le \bar{x} \le 6\text{ m}: -121.77 + 110.28\bar{x} - 135(\bar{x} - 2) + \bar{M} = 0$$

$$\bar{M} = +121.77 - 110.28\bar{x} + 135(\bar{x} - 2)$$

Calculation of deflection along the beam

$$\bar{M} = +121.77 - 110.28\bar{x} + 135[\bar{x} - 2]$$

(The quantity inside the Macaulay's square brackets is ignored if negative, i.e. when $\bar{x} < 2$ m.)

$$EI\frac{d^2\bar{w}}{d\bar{x}^2} = -\bar{M}$$

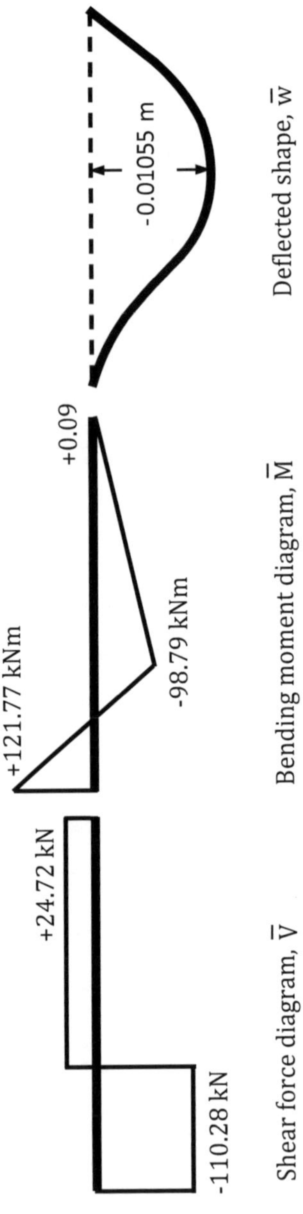
-110.28 kN
+24.72 kN
Shear force diagram, $\overline{V}$
+121.77 kNm
-98.79 kNm
+0.09
Bending moment diagram, $\overline{M}$
-0.01055 m
Deflected shape, $\overline{w}$

Figure 4.9

$$EI\frac{d^2\bar{w}}{d\bar{x}^2} = -121.77 + 110.28\bar{x} - 135[\bar{x} - 2]$$

Integrate twice with respect to $\bar{x}$ to get

$$EI\bar{w} = -60.89\bar{x}^2 + 18.38\bar{x}^3 - 22.5[\bar{x} - 2]^3 + C_1\bar{x} + C_2$$

At $\bar{x} = 0, \bar{w} = 0$, and ignoring the term with the square brackets gives $C_2 = 0$.

At $\bar{x} = 6$ m, $\bar{w} = 0$

$0 = -60.89(6)^2 + 18.38(6)^3 - 22.5[6 - 2]^3 + C_1(6)$,

hence $C_1 = -56.34$ kNm^2.

$$\text{Therefore, } \bar{w} = \frac{-60.89\bar{x}^2 + 18.38\bar{x}^3 - 22.5[\bar{x} - 2]^3 - 56.34\bar{x}}{EI}$$

with EI = $35 \times 10^6 \times 660 \times 10^{-6} = 23100$ kNm^2.

Alternatively, the beams can be divided into 'small' elements and the shear force, bending moment and deflection are determined at the ends of each element and full diagrams are obtained. This will result in more degrees of freedom requiring longer computer time and more storage. But for hand calculations, the method followed in the above example may be more suitable particularly when dealing with relatively small problems and there is no access to specialised software.

4.3 Beams with Elastic Supports

Sometimes the supports of the beam are not rigid but have certain elasticity in the translational or rotational sense. For example, when a beam is supported by a helical spring it will be subjected to a force that is proportional to the amount of deformation of the spring. Another example is when a bolted connection of a steel beam to a steel column is neither completely pinned nor completely fixed but somewhere in between. In such a case it can be assumed that the beam is elastically restrained by a spiral spring whose rotational stiffness is derived from the details of the connection. The treatment of beams with elastic supports is the same as for beams with rigid supports except that the elasticity of the supports is taken into account as shown in the following sections.

4.3.1 Helical Spring

The relationship between the force developed in a helical spring due to an extension w (in the positive z-direction) is $Z_{hs} = +k_{hs}w$ as shown in Fig. 4.10. Where k_{hs} is the stiffness of the spring and the value of the extension w is the displacement at one end of the spring relative to its other end. The force acting on the joint will be in the opposite direction, i.e. $-k_{hs}w$. Notice that when w is negative, the spring will be in compression and the force exerted on the joint will be in the positive z-direction.

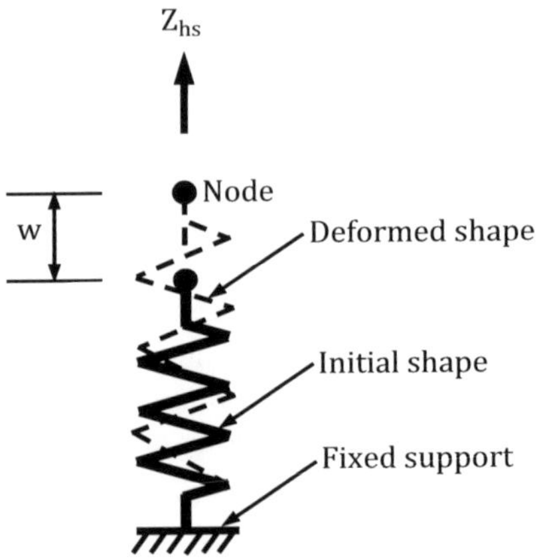

Figure 4.10

The presence of the helical spring will be taken into account when calculating the force acting on the joint as follows. Let the force acting on the joint due to the applied loads on the beam is Z_{beam}, then the total force acting on the joint will be $Z = Z_{beam} - Z_{hs} = Z_{beam} - Z_{hs}w$ and this is placed on the right-hand side of the simultaneous equations since it is part of the load vector. The quantity $-k_{hs}w$ is transferred to the left-hand side of the equation and combined with the term containing the relevant w which means that Z_{hs} will be added to the appropriate coefficient in the standard stiffness matrix.

4.3.2 Spiral Spring

The moment developed in the spiral spring due to a positive rotation, θ is $M_{ss} = +k_{ss}\theta$ as shown in Fig. 4.11 where θ is the rotation at one end of the spring relative to the other end. The moment acting on the joint will be of the same magnitude but in the opposite direction, i.e. $-k_{ss}\theta$.

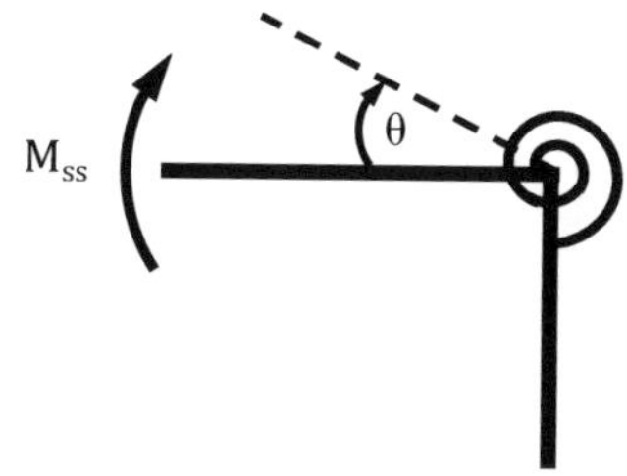

Figure 4.11

The presence of the spiral spring will be taken into account when calculating the moment acting on the joint. Let the moment acting on the joint due to the applied loads on the beam is M_{beam}, then the total moment acting on the joint will be $M = M_{beam} - M_{ss} = M_{beam} - k_{ss}\theta$ and this is on the right-hand side of the simultaneous equations since it is part of the load vector. The quantity $- k_{ss}\theta$ is transferred to the left-hand side of the equation and combined with the term containing the relevant θ which means that k_{ss} will be added to the appropriate coefficient.

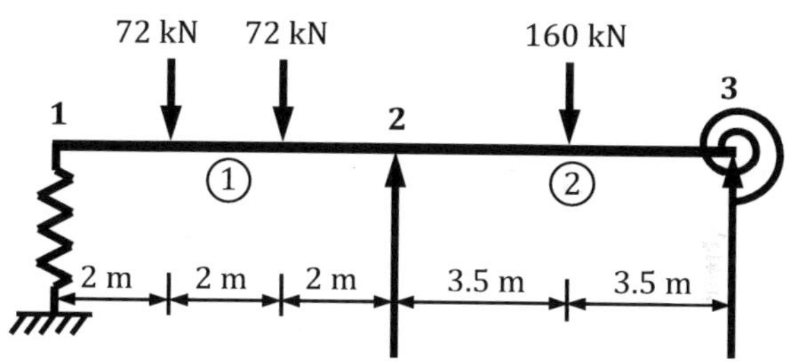

Figure 4.12

Example 2

Draw the shear force and bending moment diagrams for the continuous beam shown in Fig. 4.12. The beam has a roller support at node 2 which is expected to settle by 0.009 m and is hinged to the support at node 3 to provide stability in the x-direction. A helical spring with stiffness k_{hs} = 13000 kN/m provides support in the z-direction to the beam at node 1 and the central spindle of a spiral spring with stiffness k_{ss} = 7000 kNm/radian is fixed to the beam at node 3 to provide rotational resistance. The beam has a uniform cross section with $I = 150 \times 10^{-6}$ m^4 and its modulus of elasticity $E = 210 \times 10^6$ kN/m^2.

Member 1

$L = 6$ m, $I = 150 \times 10^{-6}$ m^4, $E = 210 \times 10^6$ kN/m^2.
From (4.18)

$$k^1 = \begin{bmatrix} 1750 & -5250 & -1750 & -5250 \\ -5250 & 21000 & 5250 & 10500 \\ -1750 & 5250 & 1750 & 5250 \\ -5250 & 10500 & 5250 & 21000 \end{bmatrix}$$

Member 2

$L = 7$ m, $I = 150 \times 10^{-6}$ m^4, $E = 210 \times 10^6$ kN/m^2.
From (4.18)

$$k^2 = \begin{bmatrix} 1102 & -3857 & -1102 & -3857 \\ -3857 & 18000 & 3857 & 9000 \\ -1102 & 3857 & 1102 & 3857 \\ -3857 & 9000 & 3857 & 18000 \end{bmatrix}$$

The overall structure stiffness matrix is

$$K = \begin{bmatrix} 1750 & -5250 & -1750 & -5250 & 0 & 0 \\ -5250 & 21000 & 5250 & 10500 & 0 & 0 \\ -1750 & 5250 & 2852 & 1393 & -1102 & -3857 \\ -5250 & 10500 & 1393 & 39000 & 3857 & 9000 \\ 0 & 0 & -1102 & 3857 & 1102 & 3857 \\ 0 & 0 & -3857 & 9000 & 3857 & 18000 \end{bmatrix} \quad (4.27)$$

Calculation of the load vector

The total load vector for the overall structure will be composed of the applied forces on the beams and the forces and moments exerted by the supports.

Contribution of loads on member 1

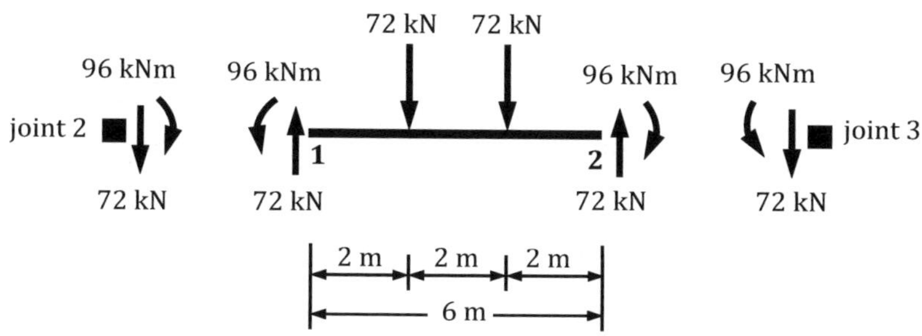

For symmetrical loads applied at the third points of the beam

$$(\bar{Z}_1^1)_f = -W = -(-72) = +72\text{ kN}$$

$$(\bar{Z}_2^1)_f = -W = -(-72) = +72\text{ kN}$$

$$(\bar{M}_1^1)_f = +\frac{2WL}{9} = +\frac{2\times(-72)\times 6}{9} = -96\text{ kNm}$$

$$(\bar{M}_2^1)_f = -\frac{2WL}{9} = -\frac{2\times(-72)\times 6}{9} = +96\text{ kNm}$$

Action vector on member 1

$$\bar{F}_f^1 = \begin{bmatrix} (\bar{Z}_1^1)_f \\ (\bar{M}_1^1)_f \\ (\bar{Z}_2^1)_f \\ (\bar{M}_2^1)_f \end{bmatrix} = \begin{bmatrix} +72 \\ -96 \\ +72 \\ +96 \end{bmatrix} \quad (4.28)$$

Load vector on joints 1 and 2

$$F_S^1 = -\bar{F}_f^1$$

$$F_S^1 = \begin{bmatrix} (Z_1^1)_S \\ (M_1^1)_S \\ (Z_2^1)_S \\ (M_2^1)_S \end{bmatrix} = \begin{bmatrix} -72 \\ +96 \\ -72 \\ -96 \end{bmatrix}$$

And since joint 3 is not affected by the loads on member 1 then

$$F_S^1 = \begin{bmatrix} (Z_1^1)_S \\ (M_1^1)_S \\ (Z_2^1)_S \\ (M_2^1)_S \\ 0 \\ 0 \end{bmatrix} = \begin{bmatrix} -72 \\ +96 \\ -72 \\ -96 \\ 0 \\ 0 \end{bmatrix} \tag{4.29}$$

Contribution of loads on member 2

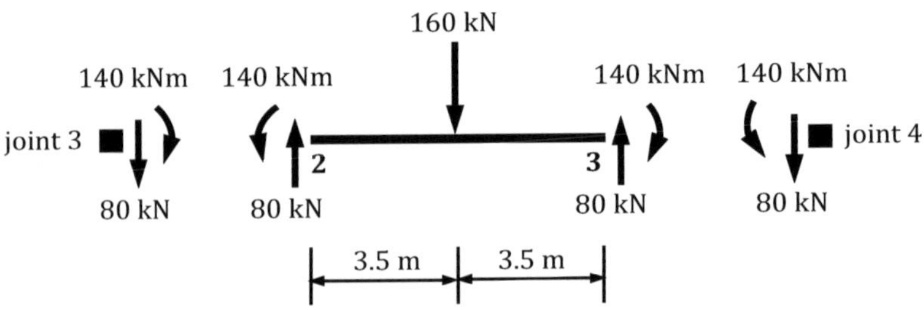

For a beam with a point load at mid-span

$$(\bar{Z}_2^2)_f = -\frac{W}{2} = -\frac{(-160)}{2} = +80 \text{ kN}$$

$$(\bar{Z}_3^2)_f = -\frac{W}{2} = -\frac{(-160)}{2} = +80 \text{ kN}$$

$$(\bar{M}_2^2)_f = +\frac{WL}{8} = +\frac{(-160)\times 7}{8} = -140 \text{ kNm}$$

$$(\bar{M}_3^2)_f = -\frac{WL}{8} = -\frac{(-160)\times 7}{8} = +140 \text{ kNm}$$

Action vector on member 2

$$\bar{F}_f^2 = \begin{bmatrix} (\bar{Z}_2^2)_f \\ (\bar{M}_2^2)_f \\ (\bar{Z}_3^2)_f \\ (\bar{M}_3^2)_f \end{bmatrix} = \begin{bmatrix} +80 \\ -140 \\ +80 \\ +140 \end{bmatrix} \tag{4.30}$$

Load vector on joints 2 and 3

$$F_S^2 = -\bar{F}_f^2$$

$$F_S^2 = \begin{bmatrix} (Z_2^2)_S \\ (M_2^2)_S \\ (Z_3^2)_S \\ (M_3^2)_S \end{bmatrix} = \begin{bmatrix} -80 \\ +140 \\ -80 \\ -140 \end{bmatrix}$$

And since joint 1 is not affected by the loads on member 2 then

$$F_S^2 = \begin{bmatrix} 0 \\ 0 \\ (Z_2^2)_S \\ (M_2^2)_S \\ (Z_3^2)_S \\ (M_3^2)_S \end{bmatrix} = \begin{bmatrix} 0 \\ 0 \\ -80 \\ +140 \\ -80 \\ -140 \end{bmatrix} \tag{4.31}$$

Load vector due to the reactions at the supports is

$$F_C = \begin{bmatrix} (Z_1)_C \\ (M_1)_C \\ (Z_2)_C \\ (M_2)_C \\ (Z_3)_C \\ (M_3)_C \end{bmatrix} = \begin{bmatrix} -k_{hs}w_1 \\ 0 \\ R_{Z2} \\ 0 \\ R_{Z3} \\ -k_{ss}\theta_3 \end{bmatrix} = \begin{bmatrix} -13000w_1 \\ 0 \\ R_{Z2} \\ 0 \\ R_{Z3} \\ -7000\theta_3 \end{bmatrix} \tag{4.32}$$

The total load vector on the structure is given by the sum of values due to the applied loads from (4.29) and (4.31) and those exerted by the reactions from (4.32), thus

$$F=\begin{bmatrix}F_1\\F_2\\F_3\end{bmatrix}=\begin{bmatrix}Z_1\\M_1\\Z_2\\M_2\\Z_3\\M_3\end{bmatrix}=F_S^1+F_S^2+F_C=\begin{bmatrix}-72\\+96\\-72\\-96\\0\\0\end{bmatrix}+\begin{bmatrix}0\\0\\-80\\+140\\-80\\-140\end{bmatrix}+\begin{bmatrix}-13000w_1\\0\\R_{Z2}\\0\\R_{Z3}\\-7000\theta_3\end{bmatrix}$$

$$F=\begin{bmatrix}-72-13000w_1\\+96\\-152+R_{Z2}\\+44\\-80+R_{Z3}\\-140-7000\theta_3\end{bmatrix} \tag{4.33}$$

Substitute K from (4.27) and F from (4.33) in (4.19) to get:

$$\begin{bmatrix}1750&-5250&-1750&-5250&0&0\\-5250&21000&5250&10500&0&0\\-1750&5250&2852&1393&-1102&-3857\\-5250&10500&1393&39000&3857&9000\\0&0&-1102&3857&1102&3857\\0&0&-3857&9000&3857&18000\end{bmatrix}\begin{bmatrix}w_1\\\theta_1\\w_2\\\theta_2\\w_3\\\theta_3\end{bmatrix}=\begin{bmatrix}-72-13000w_1\\+96\\-152+R_{Z2}\\+44\\-80+R_{Z3}\\-140-7000\theta_3\end{bmatrix} \tag{4.34}$$

In rows 1 and 6 transfer the terms $-13000w_1$ and $-7000\theta_3$ respectively from the right-hand side to the left-hand side.

The next step is to introduce the boundary conditions as follows:

At node 2 there is a downward settlement of the support of 0.009 m, hence delete row 3 which correspond to w_2 and substitute the value of $w_2 = -0.009$ m in the rest of the rows.

At node 3 where there is a non-yielding support the translational displacement in the z-direction is zero, i.e. $w_3 = 0$, hence delete row 5 and column 5.

The resulting set of simultaneous equations simplify to

$14750w_1 - 5250\theta_1 - 5250\theta_2 = -87.75$

$-5250w_1 + 21000\theta_1 + 10500\theta_2 = +142.25$

$-5250w_1 + 10500\theta_1 + 39000\theta_2 + 9000\theta_3 = +55.54$

$+ 9000\theta_2 + 25000\theta_3 = -174.71$

The solution of the above equations is:

$w_1 = -0.00363$ m, $\theta_1 = 0.00530$ rad, $\theta_2 = 0.00122$ rad,

$\theta_3 = -0.00743$ rad.

And together with $w_2 = -0.00900$ m and $w_3 = 0$, the full displacement vector will then be:

$$\begin{bmatrix} w_1 \\ \theta_1 \\ w_2 \\ \theta_2 \\ w_3 \\ \theta_3 \end{bmatrix} = \begin{bmatrix} -0.00363 \\ +0.00530 \\ -0.00900 \\ +0.00122 \\ 0 \\ -0.00743 \end{bmatrix}$$

Calculation of reactions at the supports

From the third row of (4.34)

$-1750w_1 + 5250\theta_1 + 2852w_2 + 1393\theta_2 - 1102w_3 - 3857\theta_3 = -152 + R_{Z2}$
$-1750 \times (-0.00363) + 5250 \times 0.00530 + 2852 \times (-0.00900) + 1393 \times 0.00122 - 1102 \times 0 - 3857 \times (-0.00743) = -152 + R_{Z2}$,

$R_{Z2} = +190.83$ kN

From the fifth row of (4.34)

$-1102w_2 + 3857\theta_2 + 1102w_3 + 3857\theta_3 = -80 + R_{Z3}$

$-1102 \times (-0.00900) + 3857 \times 0.00122 + 1102 \times 0 + 3857 \times (-0.00743) = -80 + R_{Z3}$, $R_{Z3} = +65.96$ kN

Force developed in the helical spring,

$Z_{hs} = k_{hs}w_1 = 13000 \times (-0.00363) = -47.19$ kN, i.e. compression

Moment developed in the spiral spring,

$M_{ss} = k_{ss}\theta_3 = 7000 \times (-0.00743) = -52.01$ kNm, i.e. anticlockwise

Calculation of actions on the members

The vector of resultant actions at the ends of the member is calculated in a similar way as in the previous example and is given by:

$$\overline{F}_r = \overline{F}_d + \overline{F}_f$$

Member 1

$$\overline{k}^1 = \begin{bmatrix} 1750 & -5250 & -1750 & -5250 \\ -5250 & 21000 & 5250 & 10500 \\ -1750 & 5250 & 1750 & 5250 \\ -5250 & 10500 & 5250 & 21000 \end{bmatrix}$$

$$\overline{\delta}^1 = \delta^1 = \begin{bmatrix} \delta_i^1 \\ \delta_j^1 \end{bmatrix} = \begin{bmatrix} \delta_1 \\ \delta_2 \end{bmatrix} = \begin{bmatrix} w_1 \\ \theta_1 \\ w_2 \\ \theta_2 \end{bmatrix} = \begin{bmatrix} -0.00363 \\ +0.00530 \\ -0.00900 \\ +0.00122 \end{bmatrix}$$

$$\overline{F}_d^1 = \overline{k}^1 \overline{\delta}^1 = \begin{bmatrix} 1750 & -5250 & -1750 & -5250 \\ -5250 & 21000 & 5250 & 10500 \\ -1750 & 5250 & 1750 & 5250 \\ -5250 & 10500 & 5250 & 21000 \end{bmatrix} \begin{bmatrix} -0.00363 \\ +0.00530 \\ -0.00900 \\ +0.00122 \end{bmatrix} = \begin{bmatrix} -24.83 \\ +95.92 \\ +24.83 \\ +53.08 \end{bmatrix}$$

From (4.28), $\overline{F}_f^1 = \begin{bmatrix} +72 \\ -96 \\ +72 \\ +96 \end{bmatrix}$

$$\begin{bmatrix} (\overline{Z}_1^1)_r \\ (\overline{M}_1^1)_r \\ (\overline{Z}_2^1)_r \\ (\overline{M}_2^1)_r \end{bmatrix} = \overline{F}_r^1 = \overline{F}_d^1 + \overline{F}_f^1 = \begin{bmatrix} -24.83 \\ +95.92 \\ +24.83 \\ +53.08 \end{bmatrix} + \begin{bmatrix} +72 \\ -96 \\ +72 \\ +96 \end{bmatrix} = \begin{bmatrix} +47.17 \\ -0.08 \\ +96.83 \\ +149.08 \end{bmatrix}$$

A useful check may be made by noting that the force at node 1 of the beam is +47.17 kN while that on the helical spring was found to be −47.19 kN. The bending moment on the beam at the simple support at node 1 which should be zero has the small value of 0.08 kNm due to rounding off.

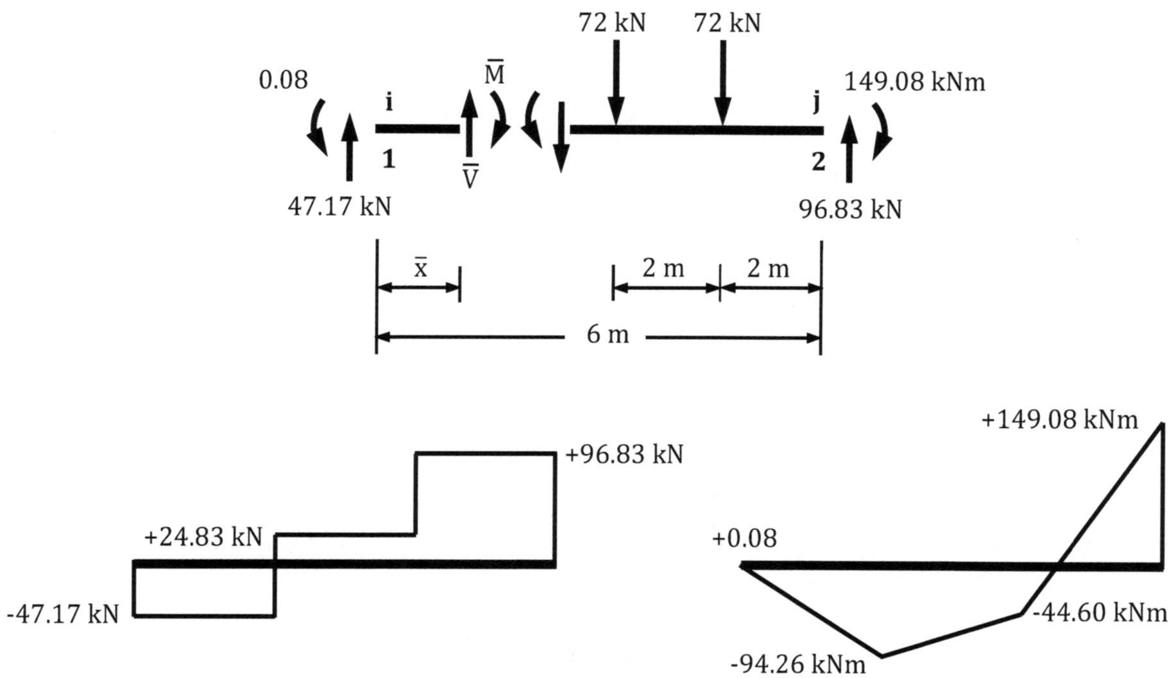

Shear force diagram, $\overline{V}$

Bending moment diagram, $\overline{M}$

Figure 4.13

Member 2

$$\bar{k}^2 = \begin{bmatrix} 1102 & -3857 & -1102 & -3857 \\ -3857 & 18000 & 3857 & 9000 \\ -1102 & 3857 & 1102 & 3857 \\ -3857 & 9000 & 3857 & 18000 \end{bmatrix}$$

$$\bar{\delta}^2 = \delta^2 = \begin{bmatrix} \delta_i^2 \\ \delta_j^2 \end{bmatrix} = \begin{bmatrix} \delta_2 \\ \delta_3 \end{bmatrix} = \begin{bmatrix} w_2 \\ \theta_2 \\ w_3 \\ \theta_3 \end{bmatrix} = \begin{bmatrix} -0.00900 \\ +0.00122 \\ 0 \\ -0.00743 \end{bmatrix}$$

$$\bar{F}_d^2 = \bar{k}^2\bar{\delta}^2 = \begin{bmatrix} 1102 & -3857 & -1102 & -3857 \\ -3857 & 18000 & 3857 & 9000 \\ -1102 & 3857 & 1102 & 3857 \\ -3857 & 9000 & 3857 & 18000 \end{bmatrix} \begin{bmatrix} -0.00900 \\ +0.00122 \\ 0 \\ -0.00743 \end{bmatrix} = \begin{bmatrix} +14.03 \\ -10.20 \\ -14.03 \\ -88.05 \end{bmatrix}$$

$$\text{From (4.30), } \bar{F}_f^2 = \begin{bmatrix} +80 \\ -140 \\ +80 \\ +140 \end{bmatrix}$$

$$\begin{bmatrix} (\bar{Z}_2^2)_r \\ (\bar{M}_2^2)_r \\ (\bar{Z}_3^2)_r \\ (\bar{M}_3^2)_r \end{bmatrix} = \bar{F}_r^2 = \bar{F}_d^2 + \bar{F}_f^2 = \begin{bmatrix} +14.03 \\ -10.20 \\ -14.03 \\ -88.05 \end{bmatrix} + \begin{bmatrix} +80 \\ -140 \\ +80 \\ +140 \end{bmatrix} = \begin{bmatrix} +94.03 \\ -150.20 \\ +65.97 \\ +51.95 \end{bmatrix}$$

Note that the moment at node 3 of the beam is +51.95 kNm while that on the spiral spring was found to be –52.01 kNm.

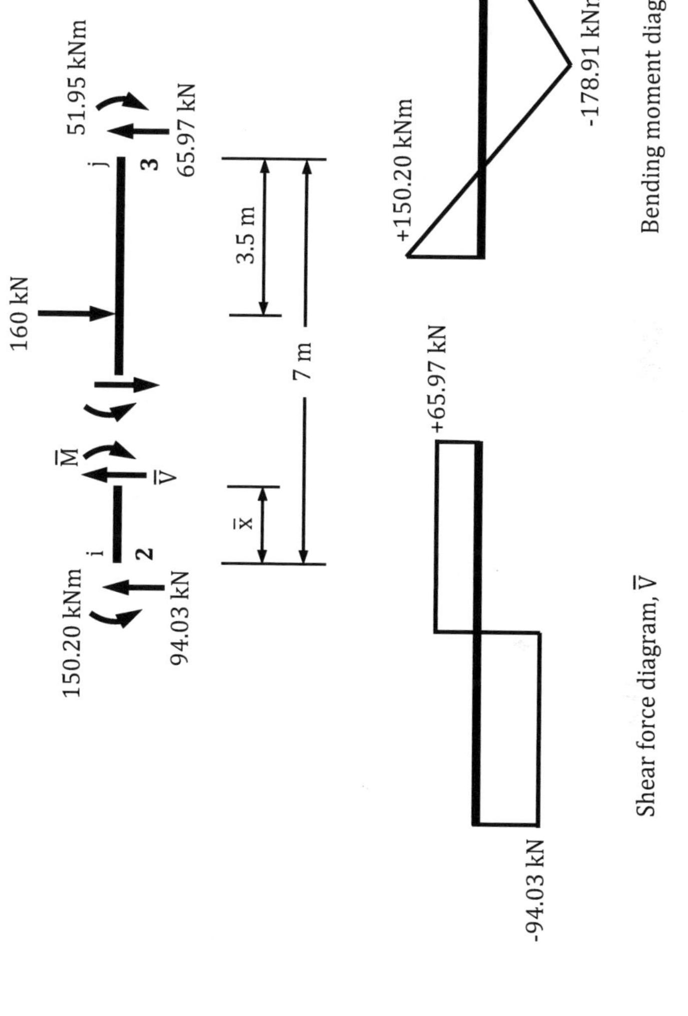
150.20 kNm
94.03 kN
i
2
M̄
V̄
160 kN
j
3
51.95 kNm
65.97 kN
x̄
3.5 m
7 m
+65.97 kN
-94.03 kN
Shear force diagram, V̄
+150.20 kNm
-178.91 kNm
+51.95 kNm
Bending moment diagram, M̄

Figure 4.14

Problems

P4.1. The continuous beam shown in Fig. P4.1 is simply supported on rollers at nodes 1 and 2 and fixed at node 3. Calculate and draw the shear force and bending moment diagrams and the deflected shape of the beam.

The beam is made of concrete with modulus of elasticity $E = 25 \times 10^6$ kN/m^2 and has the following properties:

Member 1: $L = 6$ m, $I = 300 \times 10^{-6}$ m^4

Member 2: $L = 4$ m, $I = 200 \times 10^{-6}$ m^4

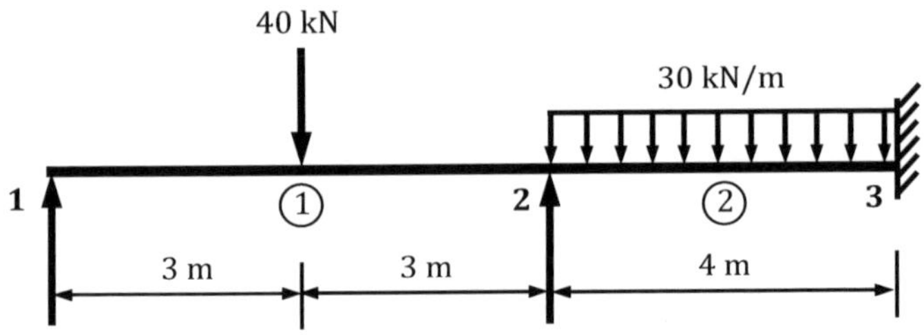

Figure P4.1

Answer:

$w_1 = 0,\ \theta_1 = 0.00629$ rad, $w_2 = 0,\ \theta_2 = -0.00057$ rad, $w_3 = 0,\ \theta_3 = 0$

$$\text{Member 1: } \begin{bmatrix} (\bar{Z}_1^1)_r \\ (\bar{M}_1^1)_r \\ (\bar{Z}_2^1)_r \\ (\bar{M}_2^1)_r \end{bmatrix} = \begin{bmatrix} +12.86 \\ 0 \\ +27.14 \\ +42.86 \end{bmatrix}, \text{ Member 2: } \begin{bmatrix} (\bar{Z}_2^2)_r \\ (\bar{M}_2^2)_r \\ (\bar{Z}_3^2)_r \\ (\bar{M}_3^2)_r \end{bmatrix} = \begin{bmatrix} +61.07 \\ -42.86 \\ +58.93 \\ +38.57 \end{bmatrix}$$

P4.2. Draw the shear force and bending moment diagrams for the continuous beam shown in Fig. P4.2. The beam is fixed at node 1 and has roller supports at nodes 2 and 3. It is expected that when the loads are applied the support at node 3 will settle by 0.006 m. The beam has a uniform cross section with a value of $I = 80 \times 10^{-6}$ m^4 and its modulus of elasticity, $E = 210 \times 10^6$ kN/m^2.

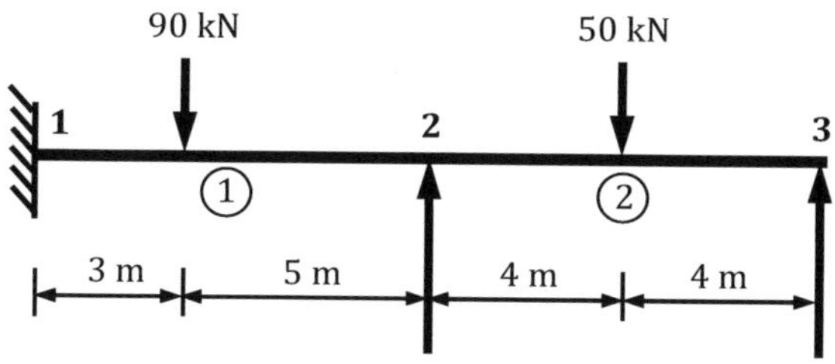

Figure P4.2

Answer:

$w_1 = 0$, $\theta_1 = 0$, $w_2 = 0$, $\theta_2 = 0.00112$ rad, $w_3 = -0.00600$ m, $\theta_3 = -0.00539$ rad.

$$\text{Member 1:} \begin{bmatrix} (\bar{Z}_1^1)_r \\ (\bar{M}_1^1)_r \\ (\bar{Z}_2^1)_r \\ (\bar{M}_2^1)_r \end{bmatrix} = \begin{bmatrix} +59.76 \\ -100.77 \\ +30.24 \\ +72.68 \end{bmatrix}, \text{ Member 2:} \begin{bmatrix} (\bar{Z}_2^2)_r \\ (\bar{M}_2^2)_r \\ (\bar{Z}_3^2)_r \\ (\bar{M}_3^2)_r \end{bmatrix} = \begin{bmatrix} +34.09 \\ -72.68 \\ +15.92 \\ 0 \end{bmatrix}$$

P4.3. Calculate and draw the shear force and bending moment diagrams for the continuous beam shown in Fig. P4.3. The beam is hinged at support 1 and is resting on rollers at nodes 2 and 3. At node 4 the beam is supported by a helical spring with a stiffness $k_{hs} = 7000$ kN/m. The beam is made of concrete with modulus of elasticity $E = 30 \times 10^6$ kN/m^2 and has the following properties:
Member 1: $L = 6$ m, $I = 400 \times 10^{-6}$ m^4
Member 2: $L = 10$ m, $I = 900 \times 10^{-6}$ m^4
Member 3: $L = 8$ m, $I = 700 \times 10^{-6}$ m^4

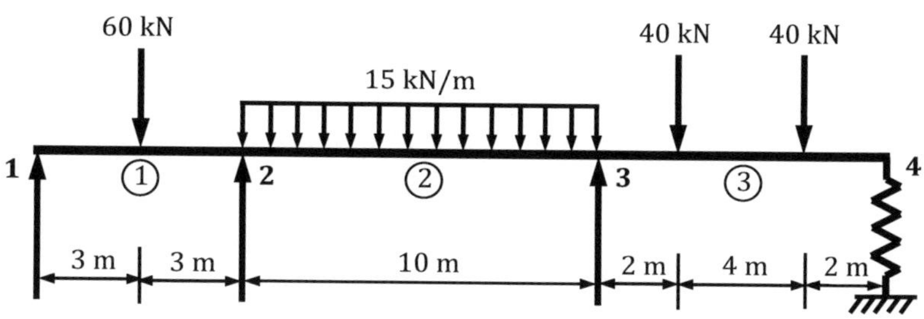

Figure P4.3

Answer:

$w_1 = 0$, $\theta_1 = 0.00344$ rad, $w_2 = 0$, $\theta_2 = 0.00437$ rad, $w_3 = 0$, $\theta_3 = -0.00295$ rad, $\theta_4 = -0.00363$ m, $\theta_4 = -0.00356$ rad.

Force developed in the spring = −25.41 kN (compression)

$$\text{Member 1: } \begin{bmatrix} (\bar{Z}_1^1)_r \\ (\bar{M}_1^1)_r \\ (\bar{Z}_2^1)_r \\ (\bar{M}_2^1)_r \end{bmatrix} = \begin{bmatrix} +14.38 \\ 0 \\ +45.62 \\ +93.72 \end{bmatrix}, \text{ Member 2: } \begin{bmatrix} (\bar{Z}_2^2)_r \\ (\bar{M}_2^2)_r \\ (\bar{Z}_3^2)_r \\ (\bar{M}_3^2)_r \end{bmatrix} = \begin{bmatrix} +72.69 \\ -93.72 \\ +77.31 \\ +116.78 \end{bmatrix},$$

$$\text{Member 3: } \begin{bmatrix} (\bar{Z}_3^3)_r \\ (\bar{M}_3^3)_r \\ (\bar{Z}_4^3)_r \\ (\bar{M}_4^3)_r \end{bmatrix} = \begin{bmatrix} +54.60 \\ -116.78 \\ +25.40 \\ 0 \end{bmatrix}.$$

P4.4. Calculate and draw the shear force and bending moment diagrams for the continuous beam shown in Fig. P4.4. The beam is hinged to the support at node 1 and supported by rollers at nodes 2 and 3. The central spindle of a spiral spring with k_{ss} = 11000 kNm/rad is fixed to the beam at node 1 to provide rotational resistance at that node. The material of the beam is steel of modulus of elasticity E = 210 × 10^6 kN/m^2 and has a uniform cross section with a value of I = 190 × 10^{-6} m^4.

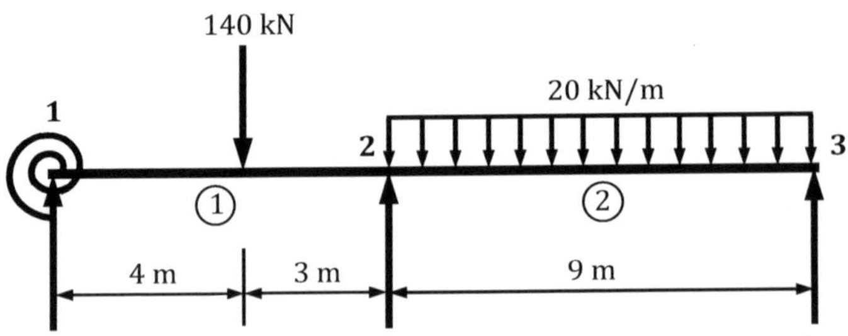

Figure P4.4

Answer:

$w_1 = 0$, $\theta_1 = 0.00272$ rad, $w_2 = 0$, $\theta_2 = 0.00095$ rad, $w_3 = 0$, $\theta_3 = -0.00809$ rad.

Moment developed in the spring = 29.92 kNm (clockwise)

$$\text{Member 1:} \begin{bmatrix} (\bar{Z}_1^1)_r \\ (\bar{M}_1^1)_r \\ (\bar{Z}_2^1)_r \\ (\bar{M}_2^1)_r \end{bmatrix} = \begin{bmatrix} +37.16 \\ -29.95 \\ +102.84 \\ +189.86 \end{bmatrix},$$

$$\text{Member 2:} \begin{bmatrix} (\bar{Z}_2^2)_r \\ (\bar{M}_2^2)_r \\ (\bar{Z}_3^2)_r \\ (\bar{M}_3^2)_r \end{bmatrix} = \begin{bmatrix} +111.10 \\ -189.86 \\ +68.91 \\ 0 \end{bmatrix}.$$

Chapter 5

Rigidly Connected Plane Frames

The members of these frames are connected in such a way that their joints have sufficient stiffness to resist moments. This implies that the angle between two rigidly connected members is not changed, i.e. members meeting at a joint do not rotate relative to each other. However, the joint as a whole will rotate when loads are applied to the structure. There are mainly two methods for achieving rigidity of the joints in steel frames: either by welding the members together therefore establishing continuity across the joint or more commonly by designing a bolted 'moment resisting' connection. The assumption made in the analysis of frames as to whether they are regarded pin-connected (as explained in Chapter 3) or rigidly connected depends upon the way they will be constructed. Reinforced concrete frames are designed as rigid frames when continuity is achieved by proper detailing of steel reinforcement at the joints between beams and columns.

Rigidly connected frames are often used for single storey industrial or leisure buildings of medium spans when a column free space is required as shown in Fig. 5.1. Another application of this type of frames is in design of multistorey buildings shown in Fig. 5.2 when resistance to wind loading is assumed to be dependent partly or wholly on the rigidity of the joints rather than by using cross bracing of some type.

Analysis of Structures by Matrix Methods
Fathi Al-Shawi

ISBN 978-981-4968-19-5 (Hardcover), 978-1-003-29130-5 (eBook)
www.jennystanford.com

Figure 5.1 Single storey portal frame.

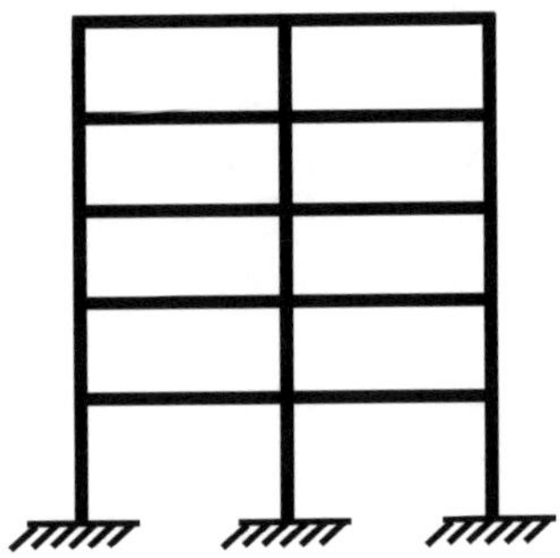

Figure 5.2 Multistorey frame.

5.1 Derivation of Stiffness Matrix

The members of rigidly connected frames are subjected to the combined effect of shear forces and bending moments as well as axial forces. The treatment of members subjected to shear forces and bending moments is explained in Chapter 4 and the additional effect of axial forces is covered in Chapter 2.

In Chapter 4 (Section 4.1) the stiffness matrix for the bending about the $\bar{y}$-axis of a beam lying in the $\overline{xz}$ plane was derived and is given by (4.25) as:

$$\begin{bmatrix} \bar{Z}_i \\ \bar{M}_i \\ \bar{Z}_j \\ \bar{M}_j \end{bmatrix} = \begin{bmatrix} \frac{12EI}{L^3} & -\frac{6EI}{L^2} & -\frac{12EI}{L^3} & -\frac{6EI}{L^2} \\ -\frac{6EI}{L^2} & \frac{4EI}{L} & \frac{6EI}{L^2} & \frac{2EI}{L} \\ -\frac{12EI}{L^3} & \frac{6EI}{L^2} & \frac{12EI}{L^3} & \frac{6EI}{L^2} \\ -\frac{6EI}{L^2} & \frac{2EI}{L} & \frac{6EI}{L^2} & \frac{4EI}{L} \end{bmatrix} \begin{bmatrix} \bar{w}_i \\ \bar{\theta}_i \\ \bar{w}_j \\ \bar{\theta}_j \end{bmatrix} \tag{5.1}$$

In Chapter 2 (Section 2.1.1) the stiffness matrix for a bar element lying along the $\bar{x}$-axis was derived and is given by (2.4) as:

$$\begin{bmatrix} \bar{X}_i \\ \bar{X}_j \end{bmatrix} = \begin{bmatrix} \frac{EA}{L} & -\frac{EA}{L} \\ -\frac{EA}{L} & \frac{EA}{L} \end{bmatrix} \begin{bmatrix} \bar{u}_i \\ \bar{u}_j \end{bmatrix} \tag{5.2}$$

Combining (5.1) and (5.2) we get

$$\begin{bmatrix} \bar{X}_i \\ \bar{Z}_i \\ \bar{M}_i \\ \bar{X}_j \\ \bar{Z}_J \\ \bar{M}_J \end{bmatrix} = \begin{bmatrix} \frac{EA}{L} & 0 & 0 & -\frac{EA}{L} & 0 & 0 \\ 0 & \frac{12EI}{L^3} & -\frac{6EI}{L^2} & 0 & -\frac{12EI}{L^3} & -\frac{6EI}{L^2} \\ 0 & -\frac{6EI}{L^2} & \frac{4EI}{L} & 0 & \frac{6EI}{L^2} & \frac{2EI}{L} \\ -\frac{EA}{L} & 0 & 0 & \frac{EA}{L} & 0 & 0 \\ 0 & -\frac{12EI}{L^3} & \frac{6EI}{L^2} & 0 & \frac{12EI}{L^3} & \frac{6EI}{L^2} \\ 0 & -\frac{6EI}{L^2} & \frac{2EI}{L} & 0 & \frac{6EI}{L^2} & \frac{4EI}{L} \end{bmatrix} = \begin{bmatrix} \bar{u}_i \\ \bar{w}_i \\ \bar{\theta}_i \\ \bar{u}_j \\ \bar{w}_j \\ \bar{\theta}_j \end{bmatrix} \tag{5.3}$$

or

$$\bar{F} = \bar{k}\bar{\delta} \tag{5.4}$$

where the stiffness matrix relative to local coordinates is:

$$\bar{k} = \begin{bmatrix} \frac{EA}{L} & 0 & 0 & -\frac{EA}{L} & 0 & 0 \\ 0 & \frac{12EI}{L^3} & -\frac{6EI}{L^2} & 0 & -\frac{12EI}{L^3} & -\frac{6EI}{L^2} \\ 0 & -\frac{6EI}{L^2} & \frac{4EI}{L} & 0 & \frac{6EI}{L^2} & \frac{2EI}{L} \\ -\frac{EA}{L} & 0 & 0 & \frac{EA}{L} & 0 & 0 \\ 0 & -\frac{12EI}{L^3} & \frac{6EI}{L^2} & 0 & \frac{12EI}{L^3} & \frac{6EI}{L^2} \\ 0 & -\frac{6EI}{L^2} & \frac{2EI}{L} & 0 & \frac{6EI}{L^2} & \frac{4EI}{L} \end{bmatrix} \tag{5.5}$$

the displacement vector $\bar{\delta} = \begin{bmatrix} \bar{\delta}_i \\ \bar{\delta}_j \end{bmatrix} = \begin{bmatrix} \bar{u}_i \\ \bar{w}_i \\ \bar{\theta}_i \\ \bar{u}_j \\ \bar{w}_j \\ \bar{\theta}_j \end{bmatrix}$ and the action vector

$$\bar{F} = \begin{bmatrix} \bar{F}_i \\ \bar{F}_j \end{bmatrix} \begin{bmatrix} \bar{X}_i \\ \bar{Z}_i \\ \bar{M}_i \\ \bar{X}_j \\ \bar{Z}_j \\ \bar{M}_j \end{bmatrix}.$$

5.2 Transformation from Local to Global Coordinates

The stiffness matrix in (5.5) is written in terms of local coordinates and when the member local $\bar{x}$-axis does not coincide with the global x-axis, this matrix needs to be transformed from local to global coordinates as explained below.

5.2.1 Transformation of Displacements

In Chapter 3, the displacements at node i, $\bar{u}_i$ and $\bar{w}_i$ along the local $\bar{x}$- and $\bar{z}$-axes respectively were transformed to u_i and w_i along the global x- and z-axes respectively resulted in the following relationships

$$\bar{u}_i = u_i \cos\varphi_{\bar{y}} - w_i \sin\varphi_{\bar{y}}$$

$$\bar{w}_i = u_i \sin\varphi_{\bar{y}} + w_i \cos\varphi_{\bar{y}}$$

The member has taken up its final position by a rotation about the $\bar{y}$-axis only and this means that the $\bar{y}$- and y-axes are still coincident and the rotational displacement relative to the local coordinates is not changed as shown in Fig. 5.3, thus

$$\bar{\theta}_i = \theta_i$$

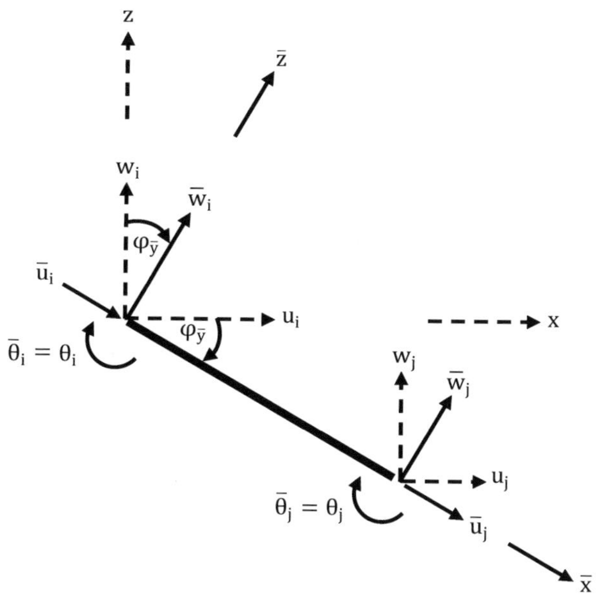

Figure 5.3

In matrix form

$$\begin{bmatrix} \bar{u}_i \\ \bar{w}_i \\ \bar{\theta}_i \end{bmatrix} = \begin{bmatrix} \cos\varphi_{\bar{y}} & -\sin\varphi_{\bar{y}} & 0 \\ \sin\varphi_{\bar{y}} & \cos\varphi_{\bar{y}} & 0 \\ 0 & 0 & 1 \end{bmatrix} \begin{bmatrix} u_i \\ w_i \\ \theta_i \end{bmatrix} = \rho_{\bar{y}} \begin{bmatrix} u_i \\ w_i \\ \theta_i \end{bmatrix}$$

Similarly, at node j

$$\begin{bmatrix} \bar{u}_j \\ \bar{w}_j \\ \bar{\theta}_j \end{bmatrix} = \rho_{\bar{y}} \begin{bmatrix} u_j \\ w_j \\ \theta_j \end{bmatrix}$$

For nodes I and j, $\begin{bmatrix} \bar{u}_i \\ \bar{w}_i \\ \bar{\theta}_i \\ \bar{u}_j \\ \bar{w}_j \\ \bar{\theta}_j \end{bmatrix} = \begin{bmatrix} \rho_{\bar{y}} & O \\ O & \rho_{\bar{y}} \end{bmatrix} \begin{bmatrix} u_i \\ w_i \\ \theta_i \\ u_j \\ w_j \\ \theta_j \end{bmatrix}$ with $O = \begin{bmatrix} 0 & 0 & 0 \\ 0 & 0 & 0 \\ 0 & 0 & 0 \end{bmatrix}$

The full transformation is

$$\begin{bmatrix} \bar{u}_i \\ \bar{w}_i \\ \bar{\theta}_i \\ \bar{u}_j \\ \bar{w}_j \\ \bar{\theta}_j \end{bmatrix} = \begin{bmatrix} \cos\varphi_{\bar{y}} & -\sin\varphi_{\bar{y}} & 0 & 0 & 0 & 0 \\ \sin\varphi_{\bar{y}} & \cos\varphi_{\bar{y}} & 0 & 0 & 0 & 0 \\ 0 & 0 & 1 & 0 & 0 & 0 \\ 0 & 0 & 0 & \cos\varphi_{\bar{y}} & -\sin\varphi_{\bar{y}} & 0 \\ 0 & 0 & 0 & \sin\varphi_{\bar{y}} & \cos\varphi_{\bar{y}} & 0 \\ 0 & 0 & 0 & 0 & 0 & 1 \end{bmatrix} \begin{bmatrix} u_i \\ w_i \\ \theta_i \\ u_j \\ w_j \\ \theta_j \end{bmatrix}$$

or $\bar{\delta} = r\delta$ where r is the transformation matrix which is given by

$$r = \begin{bmatrix} \cos\varphi_{\bar{y}} & -\sin\varphi_{\bar{y}} & 0 & 0 & 0 & 0 \\ \sin\varphi_{\bar{y}} & \cos\varphi_{\bar{y}} & 0 & 0 & 0 & 0 \\ 0 & 0 & 1 & 0 & 0 & 0 \\ 0 & 0 & 0 & \cos\varphi_{\bar{y}} & -\sin\varphi_{\bar{y}} & 0 \\ 0 & 0 & 0 & \sin\varphi_{\bar{y}} & \cos\varphi_{\bar{y}} & 0 \\ 0 & 0 & 0 & 0 & 0 & 1 \end{bmatrix}$$

5.2.2 Transformation of Actions

From Chapter 3, we had at node i

$$\bar{X}_i = X_i\cos\varphi_{\bar{y}} - Z_i\sin\varphi_{\bar{y}}$$

$$\bar{Z}_i = X_i\sin\varphi_{\bar{y}} + Z_i\cos\varphi_{\bar{y}}$$

The location of the local $\bar{y}$-axis is not changed since it is still coincident with the global y-axis, it follows that the moment is not changed as shown in Fig 5.4. Thus

$$\bar{M}_i = M_i$$

In matrix form

$$\begin{bmatrix} \bar{X}_i \\ \bar{Z}_i \\ \bar{M}_i \end{bmatrix} = \begin{bmatrix} \cos\varphi_{\bar{y}} & -\sin\varphi_{\bar{y}} & 0 \\ -\sin\varphi_{\bar{y}} & \cos\varphi_{\bar{y}} & 0 \\ 0 & 0 & 1 \end{bmatrix} \begin{bmatrix} X_i \\ Z_i \\ M_i \end{bmatrix} = \rho_{\bar{y}} \begin{bmatrix} X_i \\ Z_i \\ M_i \end{bmatrix}$$

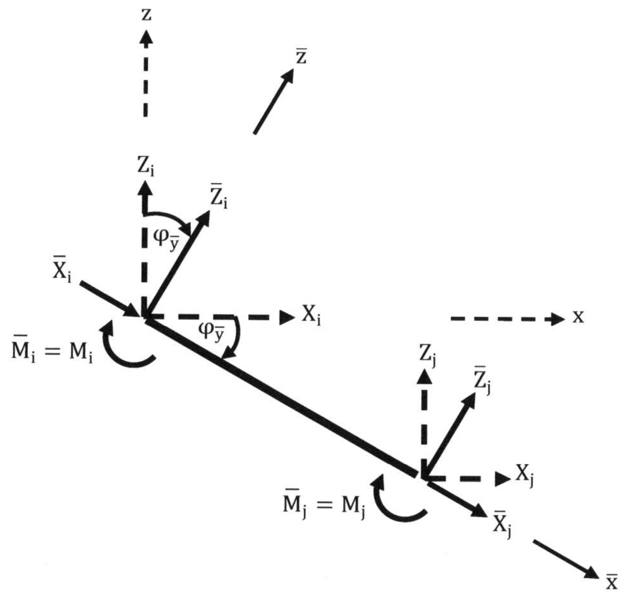

Figure 5.4

Similarly, for node j

$$\begin{bmatrix} \bar{X}_j \\ \bar{Z}_j \\ \bar{M}_j \end{bmatrix} = \rho_{\bar{y}} \begin{bmatrix} X_j \\ Z_j \\ M_j \end{bmatrix}$$

$$\begin{bmatrix} \bar{X}_i \\ \bar{Z}_i \\ \bar{M}_i \\ \bar{X}_j \\ \bar{Z}_j \\ \bar{M}_j \end{bmatrix} = \begin{bmatrix} \rho_{\bar{y}} & O \\ O & \rho_{\bar{y}} \end{bmatrix} \begin{bmatrix} X_i \\ Z_i \\ M_i \\ X_j \\ Z_j \\ M_j \end{bmatrix}$$

The full transformation is

$$\begin{bmatrix} \bar{X}_i \\ \bar{Z}_i \\ \bar{M}_i \\ \bar{X}_j \\ \bar{Z}_j \\ \bar{M}_j \end{bmatrix} = \begin{bmatrix} \cos\varphi_{\bar{y}} & -\sin\varphi_{\bar{y}} & 0 & 0 & 0 & 0 \\ \sin\varphi_{\bar{y}} & \cos\varphi_{\bar{y}} & 0 & 0 & 0 & 0 \\ 0 & 0 & 1 & 0 & 0 & 0 \\ 0 & 0 & 0 & \cos\varphi_{\bar{y}} & -\sin\varphi_{\bar{y}} & 0 \\ 0 & 0 & 0 & \sin\varphi_{\bar{y}} & \cos\varphi_{\bar{y}} & 0 \\ 0 & 0 & 0 & 0 & 0 & 1 \end{bmatrix} \begin{bmatrix} X_i \\ Z_i \\ M_i \\ X_j \\ Z_j \\ M_j \end{bmatrix}$$

or $\bar{F} = rF$.

Notice that matrix r for the transformation of actions from local coordinates to global coordinates is the same as that for the transformation of displacements because both of them are vectors having the same respective directions relative to the relevant coordinate axes.

The transformation matrix r can be written in a more convenient form by expressing $\sin\varphi_{\bar{y}}$ and $\cos\varphi_{\bar{y}}$ in terms of the coordinates at the ends of the member as shown in Fig. 3.7 of Chapter 3 as

$$\sin\varphi_{\bar{y}} = -\frac{z_j - z_i}{L} = -\frac{z_{ij}}{L}, \quad \cos\varphi_{\bar{y}} = \frac{x_j - x_i}{L} = \frac{x_{ij}}{L}$$

$$L = \sqrt{(x_j - x_i)^2 + (z_j - z_i)^2} = \sqrt{x_{ij}^2 + z_{ij}^2}$$

$$r = \begin{bmatrix} x_{ij}/L & z_{ij}/L & 0 & 0 & 0 & 0 \\ -z_{ij}/L & x_{ij}/L & 0 & 0 & 0 & 0 \\ 0 & 0 & 1 & 0 & 0 & 0 \\ 0 & 0 & 0 & x_{ij}/L & z_{ij}/L & 0 \\ 0 & 0 & 0 & -z_{ij}/L & x_{ij}/L & 0 \\ 0 & 0 & 0 & 0 & 0 & 1 \end{bmatrix} \tag{5.6}$$

Notice that the transformation matrix from local coordinates to global coordinates, r, is similar to that for axial straining with the additional transformation of rotations and moments about the $\bar{y}$-axis.

The equation relative to local coordinates is $\overline{F} = \overline{k}\overline{\delta}$ and with the substitution of $\overline{F} = rF$ and $\overline{\delta} = r\delta$ we get $rF = \overline{k}r\delta$, premultiply both sides by r^{-1} we get $r^{-1}rF = r^{-1}\overline{k}r\delta$, and since $r^{-1}r = I$ (the unit matrix), $F = r^{-1}\overline{k}r\delta$.

Also note that one of the properties of the transformation matrix is that its inverse is equal to its transpose, i.e. $r^{-1} = r^T$, thus $F = r^T\overline{k}r\delta$, and this can be written as $F = k\delta$, where $k = r^T\overline{k}r$ and $\overline{k}$ and r as given in (5.5) and (5.6) respectively.

The stiffness matrix k and the column vectors of displacements δ and actions F are all relative to global coordinates.

The stiffness matrix relative to global coordinates is given by

$$k = r^T\overline{k}r = \begin{bmatrix} x_{ij}/L & -z_{ij}/L & 0 & 0 & 0 & 0 \\ z_{ij}/L & x_{ij}/L & 0 & 0 & 0 & 0 \\ 0 & 0 & 1 & 0 & 0 & 0 \\ 0 & 0 & 0 & x_{ij}/L & -z_{ij}/L & 0 \\ 0 & 0 & 0 & z_{ij}/L & x_{ij}/L & 0 \\ 0 & 0 & 0 & 0 & 0 & 1 \end{bmatrix}$$

$$\begin{bmatrix} \frac{EA}{L} & 0 & 0 & -\frac{EA}{L} & 0 & 0 \\ 0 & \frac{12EI}{L^3} & -\frac{6EI}{L^2} & 0 & -\frac{12EI}{L^3} & -\frac{6EI}{L^2} \\ 0 & -\frac{6EI}{L^2} & \frac{4EI}{L} & 0 & \frac{6EI}{L^2} & \frac{2EI}{L} \\ -\frac{EA}{L} & 0 & 0 & \frac{EA}{L} & 0 & 0 \\ 0 & -\frac{12EI}{L^3} & \frac{6EI}{L^2} & 0 & \frac{12EI}{L^3} & \frac{6EI}{L^2} \\ 0 & -\frac{6EI}{L^2} & \frac{2EI}{L} & 0 & \frac{6EI}{L^2} & \frac{4EI}{L} \end{bmatrix}$$

$$
\begin{bmatrix}
x_{ij}/L & z_{ij}/L & 0 & 0 & 0 & 0 \\
-z_{ij}/L & x_{ij}/L & 0 & 0 & 0 & 0 \\
0 & 0 & 1 & 0 & 0 & 0 \\
0 & 0 & 0 & x_{ij}/L & z_{ij}/L & 0 \\
0 & 0 & 0 & -z_{ij}/L & x_{ij}/L & 0 \\
0 & 0 & 0 & 0 & 0 & 1
\end{bmatrix}
$$

$$
k = \left[\begin{matrix}
\frac{EAx_{ij}^2}{L^3}+\frac{12EIz_{ij}^2}{L^5} & \frac{EAx_{ij}z_{ij}}{L^3}-\frac{12EIx_{ij}z_{ij}}{L^5} & \frac{6EIz_{ij}}{L^3} \\
\frac{EAx_{ij}z_{ij}}{L^3}-\frac{12EIx_{ij}z_{ij}}{L^5} & \frac{EAz_{ij}^2}{L^3}+\frac{12EIx_{ij}^2}{L^5} & -\frac{6EIx_{ij}}{L^3} \\
\frac{6EIz_{ij}}{L^3} & -\frac{6EIx_{ij}}{L^3} & \frac{4EI}{L} \\
-\frac{EAx_{ij}^2}{L^3}-\frac{12EIz_{ij}^2}{L^5} & -\frac{EAx_{ij}z_{ij}}{L^3}+\frac{12EIx_{ij}z_{ij}}{L^5} & -\frac{6EIz_{ij}}{L^3} \\
-\frac{EAx_{ij}z_{ij}}{L^3}+\frac{12EIx_{ij}z_{ij}}{L^5} & -\frac{EAz_{ij}^2}{L^3}-\frac{12EIx_{ij}^2}{L^5} & \frac{6EIx_{ij}}{L^3} \\
\frac{6EIz_{ij}}{L^3} & -\frac{6EIx_{ij}}{L^3} & \frac{2EI}{L}
\end{matrix}\right.
\tag{5.7}
$$

$$
\left.\begin{matrix}
-\frac{EAx_{ij}^2}{L^3}-\frac{12EIz_{ij}^2}{L^5} & -\frac{EAx_{ij}z_{ij}}{L^3}+\frac{12EIx_{ij}z_{ij}}{L^5} & \frac{6EIz_{ij}}{L^3} \\
-\frac{EAx_{ij}z_{ij}}{L^3}+\frac{12EIx_{ij}z_{ij}}{L^5} & -\frac{EAz_{ij}^2}{L^3}-\frac{12EIx_{ij}^2}{L^5} & -\frac{6EIx_{ij}}{L^3} \\
-\frac{6EIz_{ij}}{L^3} & \frac{6EIx_{ij}}{L^3} & \frac{2EI}{L} \\
\frac{EAx_{ij}^2}{L^3}+\frac{12EIz_{ij}^2}{L^5} & \frac{EAx_{ij}z_{ij}}{L^3}-\frac{12EIx_{ij}z_{ij}}{L^5} & -\frac{6EIz_{ij}}{L^3} \\
\frac{EAx_{ij}z_{ij}}{L^3}-\frac{12EIx_{ij}z_{ij}}{L^5} & \frac{EAz_{ij}^2}{L^3}+\frac{12EIx_{ij}^2}{L^5} & \frac{6EIx_{ij}}{L^3} \\
-\frac{6EIz_{ij}}{L^3} & \frac{6EIx_{ij}}{L^3} & \frac{4EI}{L}
\end{matrix}\right]
$$

$$k_{ii} = \begin{bmatrix} \frac{EAx_{ij}^2}{L^3} + \frac{12EIz_{ij}^2}{L^5} & \frac{EAx_{ij}z_{ij}}{L^3} - \frac{12EIx_{ij}z_{ij}}{L^5} & \frac{6EIz_{ij}}{L^3} \\ \frac{EAx_{ij}z_{ij}}{L^3} - \frac{12EIx_{ij}z_{ij}}{L^5} & \frac{EAz_{ij}^2}{L^3} + \frac{12EIx_{ij}^2}{L^5} & -\frac{6EIx_{ij}}{L^3} \\ \frac{6EIz_{ij}}{L^3} & -\frac{6EIx_{ij}}{L^3} & \frac{4EI}{L} \end{bmatrix}$$

$$k_{ij} = \begin{bmatrix} -\frac{EAx_{ij}^2}{L^3} - \frac{12EIz_{ij}^2}{L^5} & -\frac{EAx_{ij}z_{ij}}{L^3} + \frac{12EIx_{ij}z_{ij}}{L^5} & \frac{6EIz_{ij}}{L^3} \\ -\frac{EAx_{ij}z_{ij}}{L^3} + \frac{12EIx_{ij}z_{ij}}{L^5} & -\frac{EAz_{ij}^2}{L^3} - \frac{12EIx_{ij}^2}{L^5} & -\frac{6EIx_{ij}}{L^3} \\ -\frac{6EIz_{ij}}{L^3} & \frac{6EIx_{ij}}{L^3} & \frac{2EI}{L} \end{bmatrix}$$

$$k_{ji} = \begin{bmatrix} -\frac{EAx_{ij}^2}{L^3} - \frac{12EIz_{ij}^2}{L^5} & -\frac{EAx_{ij}z_{ij}}{L^3} + \frac{12EIx_{ij}z_{ij}}{L^5} & -\frac{6EIz_{ij}}{L^3} \\ -\frac{EAx_{ij}z_{ij}}{L^3} + \frac{12EIx_{ij}z_{ij}}{L^5} & -\frac{EAz_{ij}^2}{L^3} - \frac{12EIx_{ij}^2}{L^5} & \frac{6EIx_{ij}}{L^3} \\ \frac{6EIz_{ij}}{L^3} & -\frac{6EIx_{ij}}{L^3} & \frac{2EI}{L} \end{bmatrix}$$

$$k_{jj} = \begin{bmatrix} \frac{EAx_{ij}^2}{L^3} + \frac{12EIz_{ij}^2}{L^5} & \frac{EAx_{ij}z_{ij}}{L^3} - \frac{12EIx_{ij}z_{ij}}{L^5} & -\frac{6EIz_{ij}}{L^3} \\ \frac{EAx_{ij}z_{ij}}{L^3} - \frac{12EIx_{ij}z_{ij}}{L^5} & \frac{EAz_{ij}^2}{L^3} + \frac{12EIx_{ij}^2}{L^5} & \frac{6EIx_{ij}}{L^3} \\ -\frac{6EIz_{ij}}{L^3} & \frac{6EIx_{ij}}{L^3} & \frac{4EI}{L} \end{bmatrix}$$

Example 1

The rigidly jointed plane frame shown in Fig. 5.5 is fixed at base A and pin-connected to base D. The properties of the members of the frame are as follows: member AB, $I_1 = 0.003\ m^4$, $A_1 = 0.14\ m^2$, member BC, $I_2 = 0.005\ m^4$, $A_2 = 0.18\ m^2$, and member CD, $I_3 = 0.008\ m^4$, $A_3 = 0.23\ m^2$. The modulus of elasticity of all members, $E = 25 \times 10^6\ kN/m^2$. Analyse the frame for the loading shown and draw the axial force, shear force and bending moment diagrams.

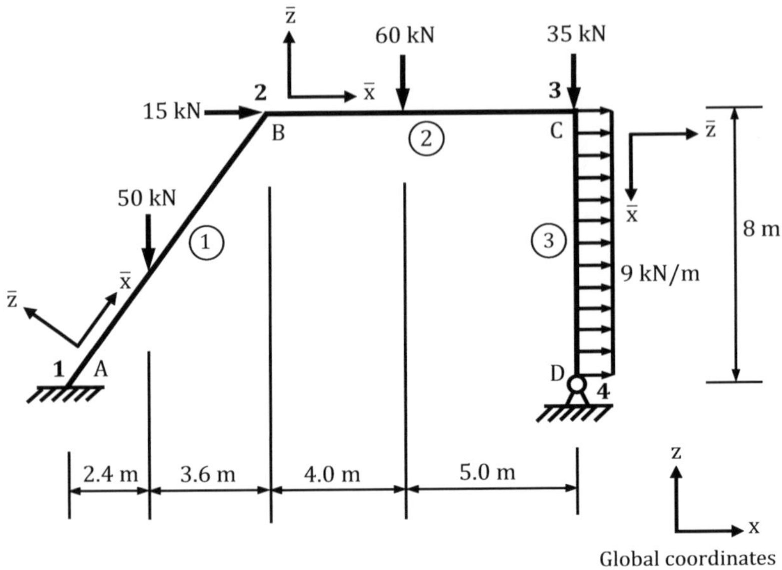

Figure 5.5

Calculation of stiffness matrices of the members

Member 1

Member address in k:	i	j
Structure address in K:	1	2

$E = 25 \times 10^6\ kN/m^2$, $A = 0.14\ m^2$, $I = 0.003\ m^4$.

$x_i = 0$, $\quad x_j = 6\ m$, $\quad x_{ij} = x_j - x_i = 6 - 0 = 6\ m$

$z_i = 0$, $\quad z_j = 8\ m$, $\quad z_{ij} = z_j - z_i = 8 - 0 = 8\ m$

$$L = \sqrt{x_{ij}^2 + z_{ij}^2} = \sqrt{6^2 + 8^2} = 10\ m$$

From (5.7)

$$\overbrace{\overbrace{\begin{matrix} u_i & w_i & \theta_i \\ u_1 & w_1 & \theta_1 \end{matrix}}^{\delta_i=\delta_1} \quad \overbrace{\begin{matrix} u_j & w_j & \theta_j \\ u_2 & w_2 & \theta_2 \end{matrix}}^{\delta_j=\delta_2}}^{\delta^1}$$

$$k^1 = \begin{bmatrix} 126576 & 167568 & 3600 & -126576 & -167568 & 3600 \\ 167568 & 224324 & -2700 & -167568 & -224324 & -2700 \\ 3600 & -2700 & 30000 & -3600 & 2700 & 15000 \\ -126576 & -167568 & -3600 & 126576 & 167568 & -3600 \\ -167568 & -224324 & 2700 & 167568 & 224324 & 2700 \\ 3600 & -2700 & 15000 & -3600 & 2700 & 30000 \end{bmatrix} \begin{matrix} u_1 \\ w_1 \\ \theta_1 \\ u_2 \\ w_2 \\ \theta_2 \end{matrix}$$

Member 2

Member address in k: i j
Structure address in K: 2 3

$E = 25 \times 10^6\ \text{kN/m}^2$, $A = 0.18\ \text{m}^2$, $I = 0.005\ \text{m}^4$.

$x_i = 6$ m, $x_j = 15$ m, $x_{ij} = x_j - x_i = 15 - 6 = 9$ m

$z_i = 8$ m, $z_j = 8$ m, $z_{ij} = z_j - z_i = 8 - 8 = 0$

$$L = \sqrt{x_{ij}^2 + z_{ij}^2} = \sqrt{9^2 + 0^2} = 9\ \text{m}$$

$$\overbrace{\overbrace{\begin{matrix} u_i & w_i & \theta_i \\ u_2 & w_2 & \theta_2 \end{matrix}}^{\delta_i=\delta_2} \quad \overbrace{\begin{matrix} u_j & w_j & \theta_j \\ u_3 & w_3 & \theta_3 \end{matrix}}^{\delta_j=\delta_3}}^{\delta^2}$$

$$k^2 = \begin{bmatrix} 500000 & 0 & 0 & -500000 & 0 & 0 \\ 0 & 2058 & -9259 & 0 & -2058 & -9259 \\ 0 & -9259 & 55556 & 0 & 9259 & 27778 \\ -500000 & 0 & 0 & 500000 & 0 & 0 \\ 0 & -2058 & 9259 & 0 & 2058 & 9259 \\ 0 & -9259 & 27778 & 0 & 9259 & 55556 \end{bmatrix} \begin{matrix} u_2 \\ w_2 \\ \theta_2 \\ u_3 \\ w_3 \\ \theta_3 \end{matrix}$$

Notice that the member local $\bar{x}$-axis coincides with the global x-axis with the transformation matrix r = I (the unit matrix) then $k^2 = \bar{k}^2$ which means that transformation is not necessary in such a case.

Member 3

Member address in k: i j

Structure address in K: 3 4

$E = 25 \times 10^6$ kN/m^2, A = 0.23 m^2, I = 0.008 m^4.

$x_i = 15$ m, $x_j = 15$ m, $x_{ij} = x_j - x_i = 15 - 15 = 0$ m

$z_i = 8$ m, $z_j = 0$, $z_{ij} = z_j - z_i = 0 - 8 = -8$ m

$$L = \sqrt{x_{ij}^2 + z_{ij}^2} = \sqrt{0^2 + (-8)^2} = 8 \text{ m}$$

$$\overbrace{\overbrace{\begin{matrix} u_i & w_i & \theta_i \\ u_3 & w_3 & \theta_3 \end{matrix}}^{\delta_i = \delta_3} \quad \overbrace{\begin{matrix} u_j & w_j & \theta_j \\ u_4 & w_4 & \theta_4 \end{matrix}}^{\delta_j = \delta_4}}^{\delta^3}$$

$$k^3 = \begin{bmatrix} 4668 & 0 & -18750 & -4688 & 0 & -18750 \\ 0 & 718750 & 0 & 0 & -718750 & 0 \\ -18750 & 0 & 100000 & 18750 & 0 & 50000 \\ -4668 & 0 & 18750 & 4688 & 0 & 18750 \\ 0 & -718750 & 0 & 0 & 718750 & 0 \\ -18750 & 0 & 50000 & 18750 & 0 & 100000 \end{bmatrix} \begin{matrix} u_3 \\ w_3 \\ \theta_3 \\ u_4 \\ w_4 \\ \theta_4 \end{matrix}$$

Assembly of the overall structure stiffness matrix

Since the frame has four nodes, the overall structure stiffness matrix is made of 4×4 sub-matrices each of which is a 3×3 matrix.

$$K = \begin{bmatrix} K_{11} & K_{12} & K_{13} & K_{14} \\ K_{21} & K_{22} & K_{23} & K_{24} \\ K_{31} & K_{32} & K_{33} & K_{34} \\ K_{41} & K_{42} & K_{43} & K_{44} \end{bmatrix}$$

By inspection

$$K = \begin{bmatrix} k_{ii}^1 & k_{ij}^1 & 0 & 0 \\ k_{ji}^1 & k_{jj}^1 + k_{ii}^2 & k_{ij}^2 & 0 \\ 0 & k_{ji}^2 & k_{jj}^2 + k_{ii}^3 & k_{ij}^3 \\ 0 & 0 & k_{ji}^3 & k_{jj}^3 \end{bmatrix}$$

K = (5.8)

	u_1	w_1	θ_1	u_2	w_2	θ_2	u_3	w_3	θ_3	u_4	w_4	θ_4
u_1	126576	167568	3600	−126576	−167568	3600	0	0	0	0	0	0
w_1	167568	224324	−2700	−167568	−224324	−2700	0	0	0	0	0	0
θ_1	3600	−2700	30000	−3600	2700	15000	0	0	0	0	0	0
u_2	−126576	−167568	−3600	626576	167568	−3600	−500000	0	0	0	0	0
w_2	−167568	−224324	2700	167568	226382	−6559	0	−2058	−9259	0	0	0
θ_2	3600	−2700	15000	−3600	−6559	85556	0	9259	27778	0	0	0
u_3	0	0	0	−500000	0	0	504688	0	−18750	−4688	0	−18750
w_3	0	0	0	0	−2058	9259	0	720808	9259	0	−718750	0
θ_3	0	0	0	0	−9259	27778	−18750	9259	155556	18750	0	50000
u_4	0	0	0	0	0	0	−4688	0	18750	4688	0	18750
w_4	0	0	0	0	0	0	0	−718750	0	0	718750	0
θ_4	0	0	0	0	0	0	−18750	0	50000	18750	0	100000

Load vector

This is calculated relative to global coordinates from loads acting on the members and the directly applied loads on the joints. In addition, the joints are subjected to loads resulting from the reactions of the supports on the structure.

All forces and moments on the members and joints will be shown in the positive directions and if the calculations give negative answers to any of these then their actual directions will be opposite to those shown.

The structure load vector due to the external loads acting directly on the members

Member 1

$$x_{ij} = 6\text{ m}, z_{ij} = 8\text{ m}, L = 10\text{ m}, a = 2.4 \times \frac{10}{6} = 4\text{ m}, b = 3.6 \times \frac{10}{6} = 6\text{ m},$$

$$\sin\alpha = \frac{8}{10} = 0.8, \cos\alpha = \frac{6}{10} = 0.6, \quad \alpha + \beta = 90^0.$$

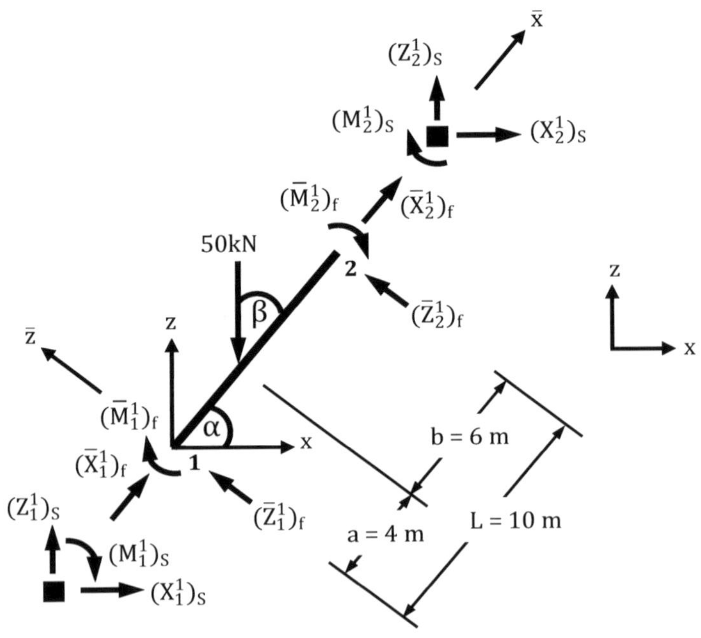

Figure 5.6

The load of 50 kN is resolved into two components, X and W, along and at right angles to the axis of the member respectively and are given by

$$X = -50\cos\beta = -50 \times 0.8 = -40 \text{ kN}$$
$$W = -50\sin\beta = -50 \times 0.6 = -30 \text{ kN}$$

Actions on member 1

$$(\bar{X}_1^1)_f = -\frac{Xb}{L} = -\frac{(-40)\times 6}{10} = +24.00 \text{ kN}$$

$$(\bar{X}_2^1)_f = -\frac{Xa}{L} = -\frac{(-40)\times 4}{10} = +16.00 \text{ kN}$$

$$(\bar{Z}_1^1)_f = -\frac{Wb\left(L^2+ab-a^2\right)}{L^3} = -\frac{(-30)\times 6\times\left(10^2+4\times 6-4^2\right)}{10^3} = +19.44 \text{ kN}$$

$$(\bar{Z}_2^1)_f = -\frac{Wa\left(L^2+ab-b^2\right)}{L^3} = -\frac{(-30)\times 4\times\left(10^2+4\times 6-6^2\right)}{10^3} = +10.56 \text{ kN}$$

$$(\bar{M}_1^1)_f = +\frac{Wab^2}{L^2} = +\frac{(-30)\times 4\times 6^2}{10^2} = -43.20 \text{ kNm}$$

$$(\bar{M}_2^1)_f = -\frac{Wa^2b}{L^2} = -\frac{(-30)\times 4^2\times 6}{10^2} = +28.80 \text{ kNm}$$

$$\bar{F}_f^1 = \begin{bmatrix} (\bar{X}_1^1)_f \\ (\bar{Z}_1^1)_f \\ (\bar{M}_1^1)_f \\ (\bar{X}_2^1)_f \\ (\bar{Z}_2^1)_f \\ (\bar{M}_2^1)_f \end{bmatrix} = \begin{bmatrix} +24.00 \\ +19.44 \\ -43.20 \\ +16.00 \\ +10.56 \\ +28.80 \end{bmatrix} \tag{5.9}$$

(The above actions are relative to the local coordinates of the member and will be use later in the calculation of the resultant end actions on member 1.)

Loads on joints 1 and 2

For the calculation of loads on the joints relative to global axes it is more convenient to resolve the actions in (5.9) into components along the global axes and from equilibrium the loads on the joints will be equal to these but acting in the opposite direction and hence with a reversed sign, thus

For joint 1:

$$(X_1^1)_S = -\left(\left(\overline{X}_1^1\right)_f \cos\alpha - \left(\overline{Z}_1^1\right)_f \sin\alpha\right) = -(24.00\times 0.6 - 19.44\times 0.8) = +1.15\ \text{kN}$$

$$(Z_1^1)_S = -\left(\left(\overline{X}_1^1\right)_f \sin\alpha + \left(\overline{Z}_1^1\right)_f \cos\alpha\right) = -(24.00\times 0.8 + 19.44\times 0.6) = -30.86\ \text{kN}$$

$$(M_1^1)_S = -\left(\overline{M}_1^1\right)_f = -(-43.20) = +43.20\ \text{kNm}$$

Similarly, for joint 2

$$(X_2^1)_S = -\left(\left(\overline{X}_2^1\right)_f \cos\alpha - \left(\overline{Z}_2^1\right)_f \sin\alpha\right) = -(16.00\times 0.6 - 10.56\times 0.8) = -1.15\ \text{kN}$$

$$(Z_2^1)_S = -\left(\left(\overline{X}_2^1\right)_f \sin\alpha + \left(\overline{Z}_2^1\right)_f \cos\alpha\right) = -(16.00\times 0.8 + 10.56\times 0.6) = -19.14\ \text{kN}$$

$$(M_2^1)_S = -\left(\overline{M}_2^1\right)_f = -28.80\ \text{kNm}$$

The above equations are written in matrix form as:

$$F_S^1 = \begin{bmatrix} (X_1^1)_S \\ (Z_1^1)_S \\ (M_1^1)_S \\ (X_2^1)_S \\ (Z_2^1)_S \\ (M_2^1)_S \end{bmatrix} = \begin{bmatrix} +1.15 \\ -30.86 \\ +43.20 \\ -1.15 \\ -19.14 \\ -28.80 \end{bmatrix} \qquad (5.10)$$

Alternatively, and to make the computations more systematic, the load vector on the joints of the structure, F_S, which is relative to global coordinates can be calculated from the actions at the ends of the member which are relative to the local coordinates of the member by using the transformation matrix as follows:

Consider the equilibrium of a section cut at the junction of the member and the joint $F_S + F_f = 0$ or $F_S = -F_f$ where F_f is the action vector on end of the member relative to global coordinates. The action vector relative to local coordinates is $\overline{F}_f$, therefore, $\overline{F}_f = rF_f$ or $F_f = r^{-1}\overline{F}_f = r^T\overline{F}_f$ (since $r^{-1} = r^T$) and hence $F_S = -r^T\overline{F}_f$.

We have from (5.6) and with $x_{ij} = 6$ m, $z_{ij} = 8$ m, L = 10 m

$$r^1 = \begin{bmatrix} 0.6 & 0.8 & 0 & 0 & 0 & 0 \\ -0.8 & 0.6 & 0 & 0 & 0 & 0 \\ 0 & 0 & 1.0 & 0 & 0 & 0 \\ 0 & 0 & 0 & 0.6 & 0.8 & 0 \\ 0 & 0 & 0 & -0.8 & 0.6 & 0 \\ 0 & 0 & 0 & 0 & 0 & 1.0 \end{bmatrix}$$

$$F_S^1 = -\left(r^1\right)^T \bar{F}_f^1 = -\begin{bmatrix} 0.6 & -0.8 & 0 & 0 & 0 & 0 \\ 0.8 & 0.6 & 0 & 0 & 0 & 0 \\ 0 & 0 & 1.0 & 0 & 0 & 0 \\ 0 & 0 & 0 & 0.6 & -0.8 & 0 \\ 0 & 0 & 0 & 0.8 & 0.6 & 0 \\ 0 & 0 & 0 & 0 & 0 & 1.0 \end{bmatrix} \begin{bmatrix} +24.00 \\ +19.44 \\ -43.20 \\ +16.00 \\ +10.56 \\ +28.80 \end{bmatrix} = \begin{bmatrix} +1.15 \\ -30.86 \\ +43.20 \\ -1.15 \\ -19.14 \\ -28.80 \end{bmatrix}$$

which is the same as the load vector obtained in (5.10).

Member 2

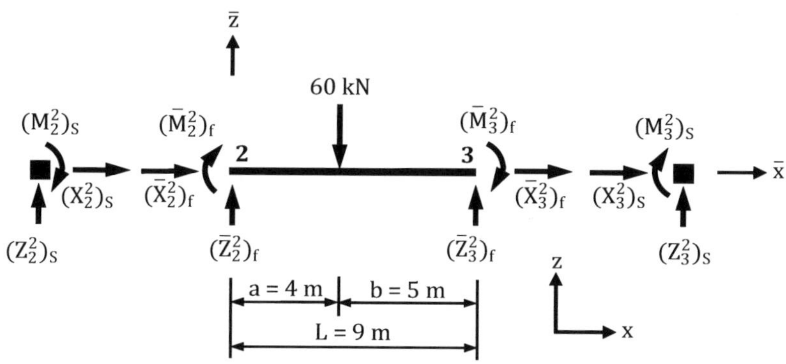

Figure 5.7

Actions on member 2 (W = –60 kN)

$$(\bar{X}_2^2)_f = 0$$

$$(\bar{X}_3^2)_f = 0$$

$$(\bar{Z}_2^2)_f = -\frac{Wb}{L^3}\left(L^2 + ab - a^2\right) = -\frac{(-60)\times 5}{9^3}\left(9^2 + 4\times 5 - 4^2\right) = +34.98 \text{ kN}$$

$$(\bar{Z}_3^2)_f = -\frac{Wa}{L^3}\left(L^2 + ab - b^2\right) = -\frac{(-60)\times 4}{9^3}\left(9^2 + 4\times 5 - 5^2\right) = +25.02\text{ kN}$$

$$(\bar{M}_2^2)_f = +\frac{Wab^2}{L^2} = +\frac{(-60)\times 4\times 5^2}{9^2} = -74.07\text{ kNm}$$

$$(\bar{M}_3^2)_f = -\frac{Wa^2b}{L^2} = -\frac{(-60)\times 4^2\times 5}{9^2} = +59.26\text{ kNm}$$

$$\bar{F}_f^2 = \begin{bmatrix} (\bar{X}_2^2)_f \\ (\bar{Z}_2^2)_f \\ (\bar{M}_2^2)_f \\ (\bar{X}_3^2)_f \\ (\bar{Z}_3^2)_f \\ (\bar{M}_3^2)_f \end{bmatrix} = \begin{bmatrix} 0 \\ +34.98 \\ -74.07 \\ 0 \\ +25.02 \\ +59.26 \end{bmatrix} \tag{5.11}$$

(The above actions are relative to the local coordinates of the member and will be used later in the calculation of the resultant end actions on member 2.)

Loads on the joints 2 and 3

$$(X_2^2)_S = -\left(\bar{X}_2^2\right)_f = 0$$

$$(Z_2^2)_S = -\left(\bar{Z}_2^2\right)_f = -34.98\text{ kN}$$

$$(M_2^2)_S = -\left(\bar{M}_2^2\right)_f = -(-74.07) = +74.07\text{ kNm}$$

$$(X_3^2)_S = -\left(\bar{X}_3^2\right)_f = 0$$

$$(Z_3^2)_S = -\left(\bar{Z}_3^2\right)_f = -25.02\text{ kN}$$

$$(M_3^2)_S = -\left(\bar{M}_3^2\right)_f = -59.26\text{ kNm}$$

$$F_S^2 = \begin{bmatrix} (X_2^2)_S \\ (Z_2^2)_S \\ (M_2^2)_S \\ (X_3^2)_S \\ (Z_3^2)_S \\ (M_3^2)_S \end{bmatrix} = \begin{bmatrix} 0 \\ -34.98 \\ +74.07 \\ 0 \\ -25.02 \\ -59.26 \end{bmatrix} \tag{5.12}$$

The above vector could have been obtained by matrix operation as follow:

$x_{ij} = 9$ m, $z_{ij} = 0$, L = 9 m and from (5.6)

$$r^2 = \begin{bmatrix} 1 & 0 & 0 & 0 & 0 & 0 \\ 0 & 1 & 0 & 0 & 0 & 0 \\ 0 & 0 & 1 & 0 & 0 & 0 \\ 0 & 0 & 0 & 1 & 0 & 0 \\ 0 & 0 & 0 & 0 & 1 & 0 \\ 0 & 0 & 0 & 0 & 0 & 1 \end{bmatrix} \text{(the unit matrix)}$$

$$F_S^2 = -(r^2)^T \bar{F}_f^2$$

$$F_S^2 = -\begin{bmatrix} 1 & 0 & 0 & 0 & 0 & 0 \\ 0 & 1 & 0 & 0 & 0 & 0 \\ 0 & 0 & 1 & 0 & 0 & 0 \\ 0 & 0 & 0 & 1 & 0 & 0 \\ 0 & 0 & 0 & 0 & 1 & 0 \\ 0 & 0 & 0 & 0 & 0 & 1 \end{bmatrix} \begin{bmatrix} 0 \\ +34.98 \\ -74.07 \\ 0 \\ +25.02 \\ +59.26 \end{bmatrix} = \begin{bmatrix} 0 \\ -34.98 \\ +74.07 \\ 0 \\ -25.02 \\ -59.26 \end{bmatrix}$$

Member 3

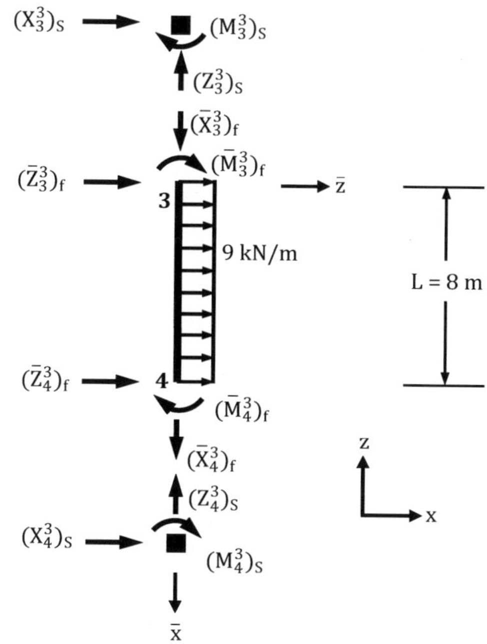

Figure 5.8

Actions on member 3 (n = +9 kN/m)

$$(\bar{X}_3^3)_f = 0$$

$$(\bar{X}_4^3)_f = 0$$

$$(\bar{Z}_3^3)_f = -\frac{nL}{2} = -\frac{9\times 8}{2} = -36\text{ kN}$$

$$(\bar{Z}_4^3)_f = -\frac{nL}{2} = -\frac{9\times 8}{2} = -36\text{ kN}$$

$$(\bar{M}_3^3)_f = +\frac{nL^2}{12} = +\frac{9\times 8^2}{12} = +48\text{ kNm}$$

$$(\bar{M}_4^3)_f = -\frac{nL^2}{12} = -\frac{9\times 8^2}{12} = -48\text{ kNm}$$

$$\bar{F}_f^3 = \begin{bmatrix} (\bar{X}_3^3)_f \\ (\bar{Z}_3^3)_f \\ (\bar{M}_3^3)_f \\ (\bar{X}_4^3)_f \\ (\bar{Z}_4^3)_f \\ (\bar{M}_4^3)_f \end{bmatrix} = \begin{bmatrix} 0 \\ -36 \\ +48 \\ 0 \\ -36 \\ -48 \end{bmatrix} \tag{5.13}$$

(The above actions are relative to the local coordinates of the member and will be use later in the calculation of the resultant end actions on member 3.)

Loads on joints 3 and 4

$$(X_3^3)_S = -\left(\bar{Z}_3^3\right)_f = -(-36) = +36\text{ kN}$$

$$(Z_3^3)_S = -\left(\bar{X}_3^3\right)_f = 0$$

$$(M_3^3)_S = -\left(\bar{M}_3^3\right)_f = -48\text{ kNm}$$

$$(X_4^3)_S = -\left(\bar{Z}_4^3\right)_f = -(-36) = +36\text{ kN}$$

$$(Z_4^3)_S = -\left(\bar{X}_4^3\right)_f = 0$$

$$(M_4^3)_S = -\left(\bar{M}_4^3\right)_f = -(-48) = +48\text{ kNm}$$

$$F_S^3 = \begin{bmatrix} (X_3^3)_S \\ (Z_3^3)_S \\ (M_3^3)_S \\ (X_4^3)_S \\ (Z_4^3)_S \\ (M_4^3)_S \end{bmatrix} = \begin{bmatrix} +36 \\ 0 \\ -48 \\ +36 \\ 0 \\ +48 \end{bmatrix} \tag{5.14}$$

The above vector could have been obtained by matrix operation as follow:

$x_{ij} = 0$, $z_{ij} = -8$ m, $L = 8$ m and from (5.6)

$$r^3 = \begin{bmatrix} 0 & -1 & 0 & 0 & 0 & 0 \\ 1 & 0 & 0 & 0 & 0 & 0 \\ 0 & 0 & 1 & 0 & 0 & 0 \\ 0 & 0 & 0 & 0 & -1 & 0 \\ 0 & 0 & 0 & 1 & 0 & 0 \\ 0 & 0 & 0 & 0 & 0 & 1 \end{bmatrix}$$

$$F_S^3 = -(r^3)^T \overline{F}_f^3$$

$$F_S^3 = -\begin{bmatrix} 0 & 1 & 0 & 0 & 0 & 0 \\ -1 & 0 & 0 & 0 & 0 & 0 \\ 0 & 0 & 1 & 0 & 0 & 0 \\ 0 & 0 & 0 & 0 & 1 & 0 \\ 0 & 0 & 0 & -1 & 0 & 0 \\ 0 & 0 & 0 & 0 & 0 & 1 \end{bmatrix} \begin{bmatrix} 0 \\ -36 \\ +48 \\ 0 \\ -36 \\ -48 \end{bmatrix} = \begin{bmatrix} +36 \\ 0 \\ -48 \\ +36 \\ 0 \\ +48 \end{bmatrix}$$

The total load vector on the joints due to the external loads acting directly on the members is:

$$F_S = \begin{bmatrix} (F_1)_S \\ (F_2)_S \\ (F_3)_S \\ (F_4)_S \end{bmatrix} = F_S^1 + F_S^2 + F_S^3$$

From (5.10), (5.12), and (5.14)

$$F_S = \begin{bmatrix} (F_1)_S \\ (F_2)_S \\ (F_3)_S \\ (F_4)_S \end{bmatrix} = \begin{bmatrix} (X_1)_S \\ (Z_1)_S \\ (M_1)_S \\ (X_2)_S \\ (Z_2)_S \\ (M_2)_S \\ (X_3)_S \\ (Z_3)_S \\ (M_3)_S \\ (X_4)_S \\ (Z_4)_S \\ (M_4)_S \end{bmatrix} = \begin{bmatrix} +1.15 \\ -30.86 \\ +43.20 \\ -1.15 \\ -19.14 \\ -28.80 \\ 0 \\ 0 \\ 0 \\ 0 \\ 0 \\ 0 \end{bmatrix} + \begin{bmatrix} 0 \\ 0 \\ 0 \\ 0 \\ -34.98 \\ +74.07 \\ 0 \\ -25.02 \\ -59.26 \\ 0 \\ 0 \\ 0 \end{bmatrix} + \begin{bmatrix} 0 \\ 0 \\ 0 \\ 0 \\ 0 \\ 0 \\ +36 \\ 0 \\ -48 \\ +36 \\ 0 \\ +48 \end{bmatrix} = \begin{bmatrix} +1.15 \\ -30.86 \\ +43.20 \\ -1.15 \\ -54.12 \\ +45.27 \\ +36.00 \\ -25.02 \\ -107.26 \\ +36.00 \\ 0 \\ +48.00 \end{bmatrix} \tag{5.15a}$$

The structure load vector due to the external loads applied directly at the nodes

A load of +15 kN applied at node 2 in the x-direction and a load of –35 at node 3 in the z-direction, thus

$$F_N = \begin{bmatrix} (F_1)_N \\ (F_2)_N \\ (F_3)_N \\ (F_4)_N \end{bmatrix} = \begin{bmatrix} (X_1)_N \\ (Z_1)_N \\ (M_1)_N \\ (X_2)_N \\ (Z_2)_N \\ (M_2)_N \\ (X_3)_N \\ (Z_3)_N \\ (M_3)_N \\ (X_4)_N \\ (Z_4)_N \\ (M_4)_N \end{bmatrix} = \begin{bmatrix} 0 \\ 0 \\ 0 \\ +15 \\ 0 \\ 0 \\ 0 \\ -35 \\ 0 \\ 0 \\ 0 \\ 0 \end{bmatrix} \tag{5.15b}$$

The structure load vector due to the reactions at the supports

$$F_C = \begin{bmatrix} (F_1)_C \\ (F_2)_C \\ (F_3)_C \\ (F_4)_C \end{bmatrix} = \begin{bmatrix} (X_1)_C \\ (Z_1)_C \\ (M_1)_C \\ (X_2)_C \\ (Z_2)_C \\ (M_2)_C \\ (X_3)_C \\ (Z_3)_C \\ (M_3)_C \\ (X_4)_C \\ (Z_4)_C \\ (M_4)_C \end{bmatrix} = \begin{bmatrix} R_{X1} \\ R_{Z1} \\ R_{M1} \\ 0 \\ 0 \\ 0 \\ 0 \\ 0 \\ 0 \\ R_{X4} \\ R_{Z4} \\ 0 \end{bmatrix} \tag{5.16}$$

Total load vector on the joints of the structure is obtained from the algebraic addition of (5.15a), (5.15b), and (5.16) as:

$$F = F_S + F_N + F_C = \begin{bmatrix} F_1 \\ F_2 \\ F_3 \\ F_4 \end{bmatrix} = \begin{bmatrix} X_1 \\ Z_1 \\ M_1 \\ X_2 \\ Z_2 \\ M_2 \\ X_3 \\ Z_3 \\ M_3 \\ X_4 \\ Z_4 \\ M_4 \end{bmatrix} = \begin{bmatrix} +1.15 \\ -30.86 \\ +43.20 \\ -1.15 \\ -54.12 \\ +45.27 \\ +36.00 \\ -25.02 \\ -107.26 \\ +36.00 \\ 0 \\ +48.00 \end{bmatrix} + \begin{bmatrix} 0 \\ 0 \\ 0 \\ +15 \\ 0 \\ 0 \\ 0 \\ -35 \\ 0 \\ 0 \\ 0 \\ 0 \end{bmatrix} + \begin{bmatrix} R_{X1} \\ R_{Z1} \\ R_{M1} \\ 0 \\ 0 \\ 0 \\ 0 \\ 0 \\ 0 \\ R_{X4} \\ R_{Z4} \\ 0 \end{bmatrix} = \begin{bmatrix} +1.15+R_{X1} \\ -30.86+R_{Z1} \\ +43.20+R_{M1} \\ +13.85 \\ -54.12 \\ +45.27 \\ +36.00 \\ -60.02 \\ -107.26 \\ +36.00+R_{X4} \\ R_{Z4} \\ +48.00 \end{bmatrix}$$

(5.17)

From (5.8) and (5.17) we get (in a tabular form)

$$
\begin{array}{cccccccccccc}
u_1 & w_1 & \theta_1 & u_2 & w_2 & \theta_2 & u_3 & w_3 & \theta_3 & u_4 & w_4 & \theta_4 \\
\end{array}
$$

$$
\left[\begin{array}{ccc|ccc|ccc|ccc}
126576 & 167568 & 3600 & -126576 & -167568 & 3600 & 0 & 0 & 0 & 0 & 0 & 0 \\
167568 & 224324 & -2700 & -167568 & -224324 & -2700 & 0 & 0 & 0 & 0 & 0 & 0 \\
3600 & -2700 & 30000 & -3600 & 2700 & 15000 & 0 & 0 & 0 & 0 & 0 & 0 \\
\hline
-126576 & -167568 & -3600 & 626576 & 167568 & -3600 & -500000 & 0 & 0 & 0 & 0 & 0 \\
-167568 & -224324 & 2700 & 167568 & 226382 & -6559 & 0 & -2058 & -9259 & 0 & 0 & 0 \\
3600 & -2700 & 15000 & -3600 & -6559 & 85556 & 0 & 9259 & 27778 & 0 & 0 & 0 \\
\hline
0 & 0 & 0 & -500000 & 0 & 0 & 504688 & 0 & -18750 & -4688 & 0 & -18750 \\
0 & 0 & 0 & 0 & -2058 & 9259 & 0 & 720808 & 9259 & 0 & -718750 & 0 \\
0 & 0 & 0 & 0 & -9259 & 27778 & -18750 & 9259 & 155556 & 18750 & 0 & 50000 \\
\hline
0 & 0 & 0 & 0 & 0 & 0 & -4688 & 0 & 18750 & 4688 & 0 & 18750 \\
0 & 0 & 0 & 0 & 0 & 0 & 0 & -718750 & 0 & 0 & 718750 & 0 \\
0 & 0 & 0 & 0 & 0 & 0 & -18750 & 0 & 50000 & 18750 & 0 & 100000
\end{array}\right]
\left[\begin{array}{c}
u_1 \\ w_1 \\ \theta_1 \\ \hline u_2 \\ w_2 \\ \theta_2 \\ \hline u_3 \\ w_3 \\ \theta_3 \\ \hline u_4 \\ w_4 \\ \theta_4
\end{array}\right]
=
\left[\begin{array}{c}
+1.15+R_{X1} \\ -30.86+R_{Z1} \\ +43.20+R_{M1} \\ \hline +13.85 \\ -54.12 \\ +45.27 \\ \hline +36.00 \\ -60.02 \\ -107.26 \\ \hline +36.00+R_{X4} \\ R_{Z4} \\ +48.00
\end{array}\right]
\tag{5.18}
$$

Apply the boundary conditions

For the fixed support at A: $u_1 = 0$, $w_1 = 0$, and $\theta_1 = 0$. Hence delete rows 1, 2, and 3, and columns 1, 2, and 3.

For the pinned support at D: $u_4 = 0$ and $w_4 = 0$. Hence delete rows 10 and 11 and columns 10 and 11.

The resulting set of equations are:

$$\begin{bmatrix} 626576 & 167568 & -3600 & -500000 & 0 & 0 & 0 \\ 167568 & 226382 & -6559 & 0 & -2058 & -9259 & 0 \\ -3600 & -6559 & 85556 & 0 & 9259 & 27778 & 0 \\ -500000 & 0 & 0 & 504688 & 0 & -18750 & -18750 \\ 0 & -2058 & 9259 & 0 & 720808 & 9259 & 0 \\ 0 & -9259 & 27778 & -18750 & 9259 & 155556 & 50000 \\ 0 & 0 & 0 & -18750 & 0 & 50000 & 100000 \end{bmatrix}$$

$$\begin{bmatrix} u_2 \\ w_2 \\ \theta_2 \\ u_3 \\ w_3 \\ \theta_3 \\ \theta_4 \end{bmatrix} = \begin{bmatrix} +13.85 \\ -54.12 \\ +45.27 \\ +36.00 \\ -60.02 \\ -107.26 \\ +48.00 \end{bmatrix}$$

The solution of the above set is:
$u_2 = +0.026181$ m, $w_2 = -0.019634$ m, $\theta_2 = +0.000328$ rad, $u_3 = +0.026198$ m, $w_3 = -0.000136$ m, $\theta_3 = -0.000577$ rad, $\theta_4 = +0.005681$ rad.

$$\delta = \begin{bmatrix} \delta_1 \\ \delta_2 \\ \delta_3 \\ \delta_4 \end{bmatrix}, \text{ where } \delta_1 = \begin{bmatrix} u_1 \\ w_1 \\ \theta_1 \end{bmatrix} = \begin{bmatrix} 0 \\ 0 \\ 0 \end{bmatrix}, \ \delta_2 = \begin{bmatrix} u_2 \\ w_2 \\ \theta_2 \end{bmatrix} = \begin{bmatrix} +0.026181 \\ -0.019634 \\ +0.000328 \end{bmatrix}$$

$$\delta_3 = \begin{bmatrix} u_3 \\ w_3 \\ \theta_3 \end{bmatrix} = \begin{bmatrix} +0.026199 \\ -0.000136 \\ -0.000577 \end{bmatrix}, \ \delta_4 = \begin{bmatrix} u_4 \\ w_4 \\ \theta_4 \end{bmatrix} = \begin{bmatrix} 0 \\ 0 \\ +0.005681 \end{bmatrix}$$

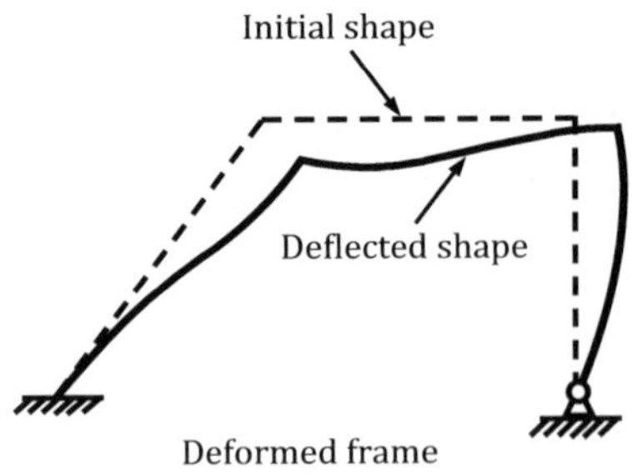

Calculation of reactions at the supports

From the first row of (5.18)

$126576u_1 + 167568w_1 + 3600\theta_1 - 126576u_2 - 167568w_2 + 3600\theta_2 = +1.15 + R_{X1}$

$126576 \times 0 + 167568 \times 0 + 3600 \times 0 - 126576 \times 0.026181 - 167568 \times (-0.019634) + 3600 \times 0.000328 = +1.15 \times R_{X1}$, $R_{X1} = -23.83$ kN

From the second row

$167568u_1 + 224324w_1 - 2700\theta_1 - 167568u_2 - 224324w_2 - 2700\theta_2 = -30.86 + R_{Z1} 167568 \times 0 + 224324 \times 0 - 2700 \times 0 - 167568 \times 0.026181 - 224324 \times (-0.019634) - 2700 \times 0.000328 = -30.86 + R_{Z1}$, $R_{Z1} = +47.25$ kN

From the third row

$3600u_1 - 2700w_1 + 300000\theta_1 - 3600u_2 + 2700w_2 + 150000\theta_2 = +43.20 + R_{M1}$

$3600 \times 0 - 2700 \times 0 + 30000 \times 0 - 3600 \times 0.026181 + 2700 \times (-0.019634) + 15000 \times 0.000328 = +43.20 + R_{M1}$, $R_{M1} = -185.54$ kNm

Similarly, $R_{X4} = -63.12$ kN and $R_{Z4} = +97.75$ kN are obtained from rows ten and eleven, respectively.

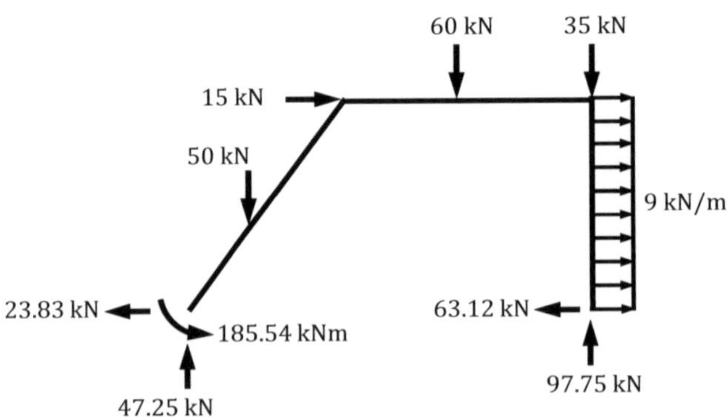

Figure 5.9 External reactions on the frame.

Calculation of actions on the members

Sign convention

When calculating the internal actions (axial force, shear force, and bending moment) along the member it is usual to start at the left end and working towards the right end of the member. A section is cut at a distance $\bar{x}$ from the left end (node i) and the sign of the internal actions are based on their directions at the right (not the left) end of the cut portion of the member. The positive axial force at the right end means that the member is in tension and the positive shear force is tending to move the section in the positive $\bar{z}$ direction. The positive bending moment at the right end of the member causes tension in the top face and compression in the bottom face of the member as shown in Fig. 5.10.

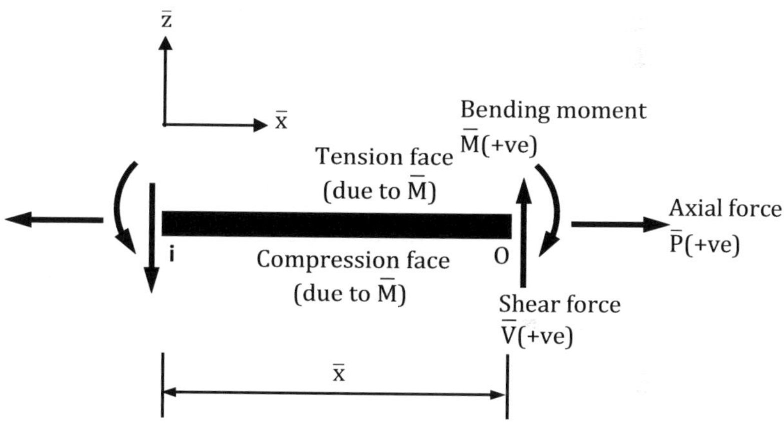

Figure 5.10 Internal actions developed in the member at section O.

The actions on the member are calculated relative to local coordinates and the resultant action, $\bar{F}_r = \bar{F}_d + \bar{F}_f$.

Where the action due to displacements, $\bar{F}_d = \bar{k}\bar{\delta}$, and $\bar{F}_f$ is the action due to the applied loads.

Member 1

The stiffness matrix relative to local coordinates $\bar{k}$ is given in (5.5), thus

$$\bar{k}^1 = \begin{bmatrix} 350000 & 0 & 0 & -350000 & 0 & 0 \\ 0 & 900 & -4500 & 0 & -900 & -4500 \\ 0 & -4500 & 30000 & 0 & 4500 & 15000 \\ -350000 & 0 & 0 & 350000 & 0 & 0 \\ 0 & -900 & 4500 & 0 & 900 & 4500 \\ 0 & -4500 & 15000 & 0 & 4500 & 30000 \end{bmatrix}$$

$x_{ij} = 6$ m, $z_{ij} = 8$ m, L = 10 m
The transformation matrix r is given by (5.6), thus

$$r^1 = \begin{bmatrix} 0.6 & 0.8 & 0 & 0 & 0 & 0 \\ -0.8 & 0.6 & 0 & 0 & 0 & 0 \\ 0 & 0 & 1.0 & 0 & 0 & 0 \\ 0 & 0 & 0 & 0.6 & 0.8 & 0 \\ 0 & 0 & 0 & -0.8 & 0.6 & 0 \\ 0 & 0 & 0 & 0 & 0 & 1.0 \end{bmatrix}$$

$$\delta^1 = \begin{bmatrix} \delta_i^1 \\ \delta_j^1 \end{bmatrix} = \begin{bmatrix} \delta_1 \\ \delta_2 \end{bmatrix} = \begin{bmatrix} u_1 \\ w_1 \\ \theta_1 \\ u_2 \\ w_2 \\ \theta_2 \end{bmatrix} = \begin{bmatrix} 0 \\ 0 \\ 0 \\ +0.026181 \\ -0.019634 \\ +0.000328 \end{bmatrix}$$

$$\bar{\delta}^1 = r^1\delta^1 = \begin{bmatrix} 0.6 & 0.8 & 0 & 0 & 0 & 0 \\ -0.8 & 0.6 & 0 & 0 & 0 & 0 \\ 0 & 0 & 1.0 & 0 & 0 & 0 \\ 0 & 0 & 0 & 0.6 & 0.8 & 0 \\ 0 & 0 & 0 & -0.8 & 0.6 & 0 \\ 0 & 0 & 0 & 0 & 0 & 1.0 \end{bmatrix} \begin{bmatrix} 0 \\ 0 \\ 0 \\ +0.026181 \\ -0.019634 \\ +0.000328 \end{bmatrix} = \begin{bmatrix} 0 \\ 0 \\ 0 \\ +0.000001 \\ -0.032725 \\ +0.000328 \end{bmatrix}$$

$$\bar{F}_d^1 = \bar{k}^1 \bar{\delta}^1 = \begin{bmatrix} 350000 & 0 & 0 & -350000 & 0 & 0 \\ 0 & 900 & -4500 & 0 & -900 & -4500 \\ 0 & -4500 & 30000 & 0 & 4500 & 15000 \\ -350000 & 0 & 0 & 350000 & 0 & 0 \\ 0 & -900 & 4500 & 0 & 900 & 4500 \\ 0 & -4500 & 15000 & 0 & 4500 & 30000 \end{bmatrix} \begin{bmatrix} 0 \\ 0 \\ 0 \\ +0.000001 \\ -0.032725 \\ +0.000328 \end{bmatrix}$$

$$\bar{F}_d^1 = \begin{bmatrix} -0.35 \\ +27.98 \\ -142.34 \\ +0.35 \\ -27.98 \\ -137.42 \end{bmatrix} \text{ and from (5.9), } \bar{F}_f^1 = \begin{bmatrix} +24.00 \\ +19.44 \\ -43.20 \\ +16.00 \\ +10.56 \\ +28.80 \end{bmatrix}$$

$$\bar{F}_r^1 = \bar{F}_d^1 + \bar{F}_f^1$$

$$\bar{F}_r^1 = \begin{bmatrix} -0.35 \\ +27.98 \\ -142.34 \\ +0.35 \\ -27.98 \\ -137.42 \end{bmatrix} + \begin{bmatrix} +24.00 \\ +19.44 \\ -43.20 \\ +16.00 \\ +10.56 \\ +28.80 \end{bmatrix} = \begin{bmatrix} +23.65 \\ +47.42 \\ -185.54 \\ +16.35 \\ -17.42 \\ -108.62 \end{bmatrix}, \text{ i.e. } \begin{bmatrix} (\bar{X}_1^1)_r \\ (\bar{Z}_1^1)_r \\ (\bar{M}_1^1)_r \\ (\bar{X}_2^1)_r \\ (\bar{Z}_2^1)_r \\ (\bar{M}_2^1)_r \end{bmatrix} = \begin{bmatrix} +23.65 \\ +47.42 \\ -185.54 \\ +16.35 \\ -17.42 \\ -108.62 \end{bmatrix}$$

Internal actions (axial force, shear force, and bending moment)

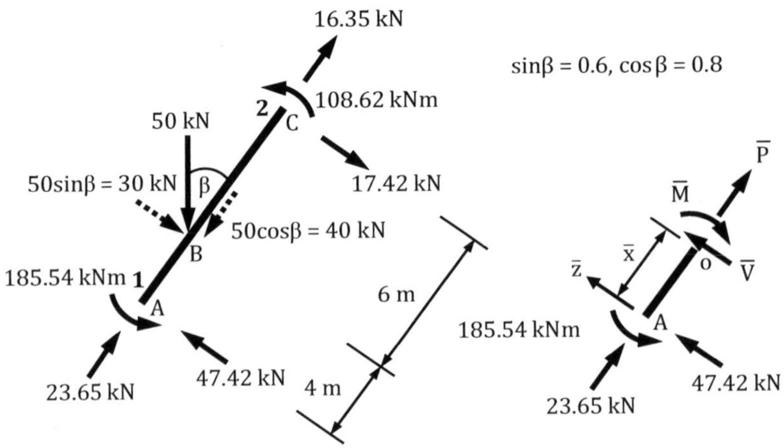

Figure 5.11

For part AB:

$$\sum \overline{X} = 0, \quad +23.65 + \overline{P} = 0, \quad \overline{P} = -23.65\,\text{kN}, \quad \text{i.e. compression}$$

$$\sum \overline{Z} = 0, \quad +47.42 + \overline{V} = 0, \quad \overline{V} = -47.42\,\text{kN}$$

$$\sum \overline{M}_o = 0, \quad -185.54 + 47.42\overline{x} + \overline{M} = 0, \quad \overline{M} = +185.54 - 47.42\overline{x}$$

For part BC:

$$\sum \overline{X} = 0, \quad +23.65 - 40 + \overline{P} = 0, \quad \overline{P} = +16.35\,\text{kN}, \ \text{i.e. tension}$$

$$\sum \overline{Z} = 0, \quad +47.42 - 30 + \overline{V} = 0, \quad \overline{V} = -17.42\,\text{kN}$$

$$\sum \overline{M}_o = 0, \quad -185.54 + 47.42\overline{x} - 30(\overline{x} - 4) + \overline{M} = 0,$$

$$\overline{M} = +185.54 - 47.42\overline{x} + 30(\overline{x} - 4)$$

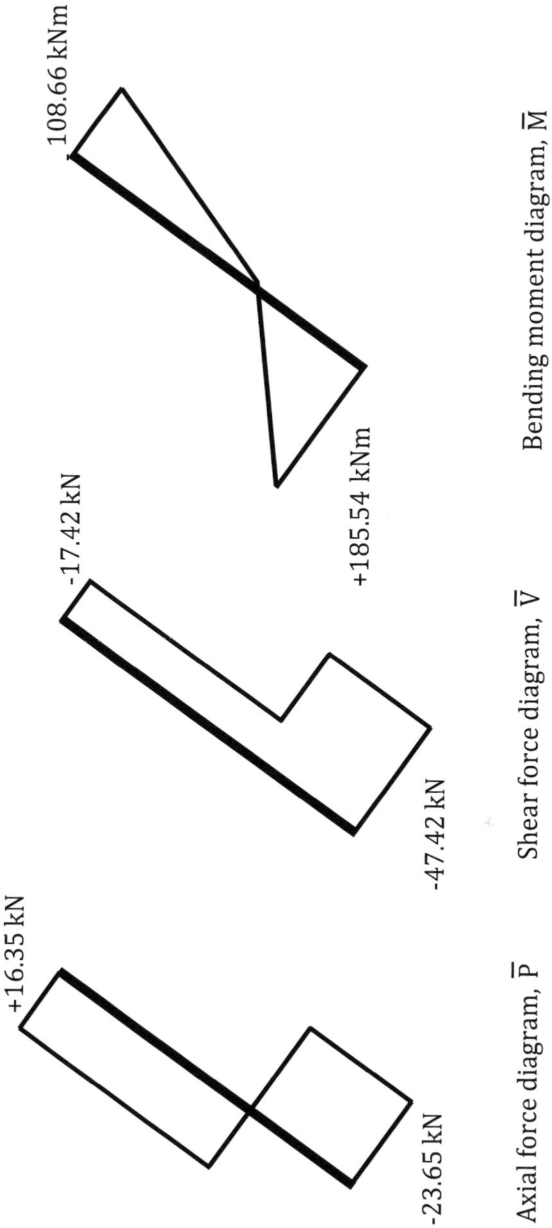
+16.35 kN
-23.65 kN
Axial force diagram, $\overline{P}$
-17.42 kN
-47.42 kN
Shear force diagram, $\overline{V}$
108.66 kNm
+185.54 kNm
Bending moment diagram, $\overline{M}$

Figure 5.12

Member 2

$$\delta^2 = \begin{bmatrix} \delta_i^2 \\ \delta_j^2 \end{bmatrix} = \begin{bmatrix} \delta_2 \\ \delta_3 \end{bmatrix} = \begin{bmatrix} u_2 \\ w_2 \\ \theta_2 \\ u_3 \\ w_3 \\ \theta_3 \end{bmatrix} = \begin{bmatrix} +0.026181 \\ -0.019634 \\ +0.000328 \\ +0.026199 \\ -0.000136 \\ -0.000577 \end{bmatrix},$$

$$\bar{\delta}^2 = r^2\delta^2 = I\delta^2 = \begin{bmatrix} +0.026181 \\ -0.019634 \\ +0.000328 \\ +0.026199 \\ -0.000136 \\ -0.000577 \end{bmatrix}$$

From (5.5)

$$\bar{k}^2 = \begin{bmatrix} 500000 & 0 & 0 & -500000 & 0 & 0 \\ 0 & 2058 & -9259 & 0 & -2058 & -9259 \\ 0 & -9259 & 55556 & 0 & 9259 & 27778 \\ -500000 & 0 & 0 & 500000 & 0 & 0 \\ 0 & -2058 & 9259 & 0 & 2058 & 9259 \\ 0 & -9259 & 27778 & 0 & 9259 & 55556 \end{bmatrix}$$

$$\bar{F}_d^2 = \bar{k}^2\bar{\delta}^2 = \begin{bmatrix} 500000 & 0 & 0 & -500000 & 0 & 0 \\ 0 & 2058 & -9259 & 0 & -2058 & -9259 \\ 0 & -9259 & 55556 & 0 & 9259 & 27778 \\ -500000 & 0 & 0 & 500000 & 0 & 0 \\ 0 & -2058 & 9259 & 0 & 2058 & 9259 \\ 0 & -9259 & 27778 & 0 & 9259 & 55556 \end{bmatrix}$$

$$\begin{bmatrix} +0.026181 \\ -0.019634 \\ +0.000328 \\ +0.026199 \\ -0.000136 \\ -0.000577 \end{bmatrix}$$

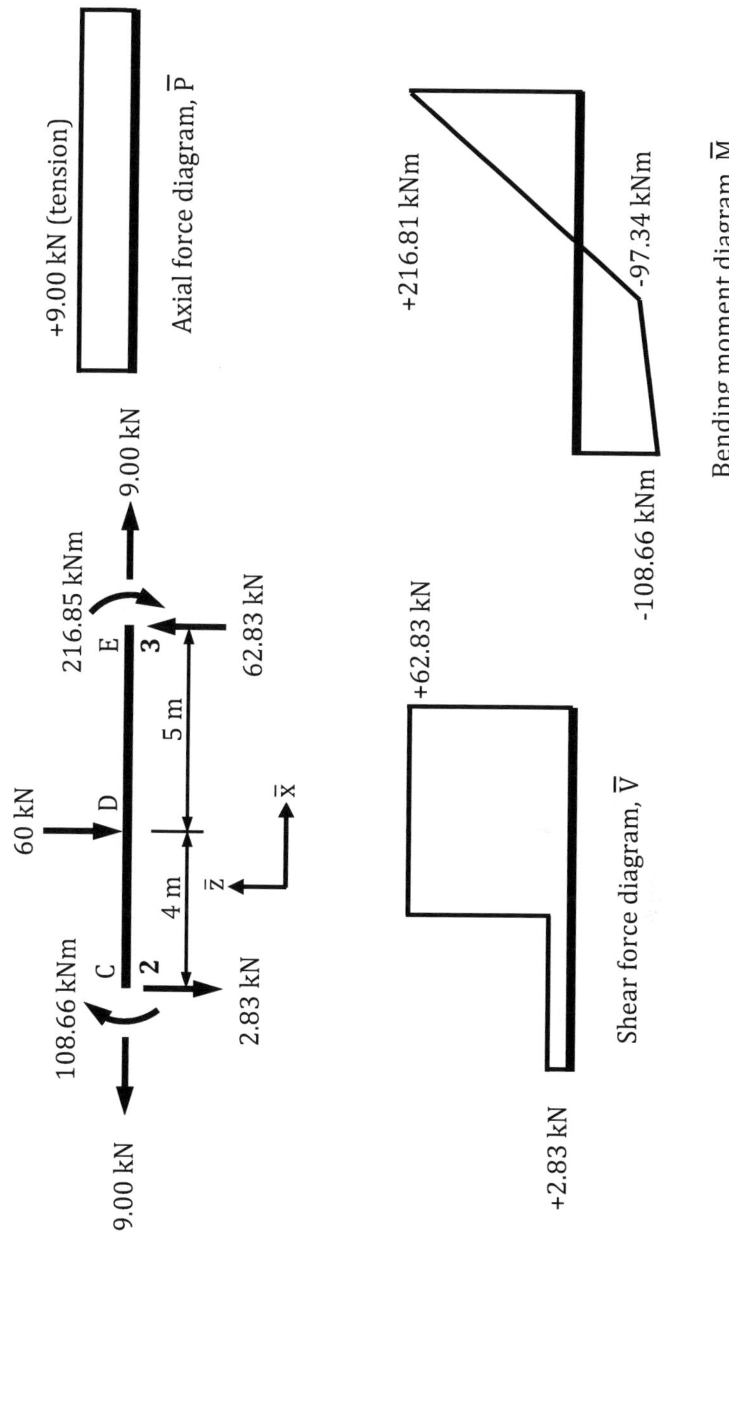

+9.00 kN (tension)
Axial force diagram, $\bar{P}$
60 kN
108.66 kNm
216.85 kNm
C
D
E
2
3
4 m
5 m
9.00 kN
9.00 kN
2.83 kN
62.83 kN
$\bar{z}$
$\bar{x}$
+216.81 kNm
-97.34 kNm
-108.66 kNm
Bending moment diagram, $\bar{M}$
+62.83 kN
+2.83 kN
Shear force diagram, $\bar{V}$

Figure 5.13

$$\bar{F}_d^2 = \begin{bmatrix} -9.00 \\ -37.81 \\ +182.73 \\ +9.00 \\ +37.81 \\ +157.59 \end{bmatrix} \text{ and from (5.11) } \bar{F}_f^2 = \begin{bmatrix} 0 \\ +34.98 \\ -74.07 \\ 0 \\ +25.02 \\ +59.26 \end{bmatrix}$$

$$\bar{F}_r^2 = \bar{F}_d^2 + \bar{F}_f^2 = \begin{bmatrix} -9.00 \\ -37.81 \\ +182.73 \\ +9.00 \\ +37.81 \\ +157.59 \end{bmatrix} + \begin{bmatrix} 0 \\ +34.98 \\ -74.07 \\ 0 \\ +25.02 \\ +59.26 \end{bmatrix} = \begin{bmatrix} -9.00 \\ -2.83 \\ +108.66 \\ +9.00 \\ +62.83 \\ +216.85 \end{bmatrix}$$

Member 3

$$\delta^3 = \begin{bmatrix} \delta_i^3 \\ \delta_j^3 \end{bmatrix} = \begin{bmatrix} \delta_3 \\ \delta_4 \end{bmatrix} = \begin{bmatrix} u_3 \\ w_3 \\ \theta_3 \\ u_4 \\ w_4 \\ \theta_4 \end{bmatrix} = \begin{bmatrix} +0.026199 \\ -0.000136 \\ -0.000577 \\ 0 \\ 0 \\ +0.005681 \end{bmatrix}$$

From (5.6) with $x_{ij} = 0$, $z_{ij} = -8$ m, and $L = 8$ m

$$r^3 = \begin{bmatrix} 0 & -1 & 0 & 0 & 0 & 0 \\ 1 & 0 & 0 & 0 & 0 & 0 \\ 0 & 0 & 1 & 0 & 0 & 0 \\ 0 & 0 & 0 & 0 & -1 & 0 \\ 0 & 0 & 0 & 1 & 0 & 0 \\ 0 & 0 & 0 & 0 & 0 & 1 \end{bmatrix}$$

$$\bar{\delta}^3 = r^3\delta^3 = \begin{bmatrix} 0 & -1 & 0 & 0 & 0 & 0 \\ 1 & 0 & 0 & 0 & 0 & 0 \\ 0 & 0 & 1 & 0 & 0 & 0 \\ 0 & 0 & 0 & 0 & -1 & 0 \\ 0 & 0 & 0 & 1 & 0 & 0 \\ 0 & 0 & 0 & 0 & 0 & 1 \end{bmatrix} \begin{bmatrix} +0.026199 \\ -0.000136 \\ -0.000577 \\ 0 \\ 0 \\ +0.005681 \end{bmatrix} = \begin{bmatrix} +0.000136 \\ +0.026199 \\ -0.000577 \\ 0 \\ 0 \\ +0.005681 \end{bmatrix}$$

From (5.5)

$$\bar{k}^3 = \begin{bmatrix} 718750 & 0 & 0 & -718750 & 0 & 0 \\ 0 & 4688 & -18750 & 0 & -4688 & -18750 \\ 0 & -18750 & 100000 & 0 & 18750 & 50000 \\ -718750 & 0 & 0 & 718750 & 0 & 0 \\ 0 & -4688 & 18750 & 0 & 4688 & 18750 \\ 0 & -18750 & 50000 & 0 & 18750 & 100000 \end{bmatrix}$$

$$\bar{F}_d^3 = \bar{k}^3\bar{\delta}^3 = \begin{bmatrix} 718750 & 0 & 0 & -718750 & 0 & 0 \\ 0 & 4688 & -18750 & 0 & -4688 & -18750 \\ 0 & -18750 & 100000 & 0 & 18750 & 50000 \\ -718750 & 0 & 0 & 718750 & 0 & 0 \\ 0 & -4688 & 18750 & 0 & 4688 & 18750 \\ 0 & -18750 & 50000 & 0 & 18750 & 100000 \end{bmatrix} \begin{bmatrix} +0.000136 \\ +0.026199 \\ -0.000577 \\ 0 \\ 0 \\ +0.005681 \end{bmatrix}$$

$$\bar{F}_d^3 = \begin{bmatrix} +97.75 \\ +27.12 \\ -264.88 \\ -97.75 \\ -27.12 \\ +48.02 \end{bmatrix} \text{ and from (5.13) } \bar{F}_f^3 = \begin{bmatrix} 0 \\ -36 \\ +48 \\ 0 \\ -36 \\ -48 \end{bmatrix}$$

$$\overline{F}_r^3 = \overline{F}_d^3 + \overline{F}_f^3 = \begin{bmatrix} +97.75 \\ +27.12 \\ -264.88 \\ -97.75 \\ -27.12 \\ +48.02 \end{bmatrix} + \begin{bmatrix} 0 \\ -36 \\ +48 \\ 0 \\ -36 \\ -48 \end{bmatrix} = \begin{bmatrix} +97.75 \\ -8.88 \\ -216.88 \\ -97.75 \\ -63.12 \\ +0.02 \end{bmatrix}$$

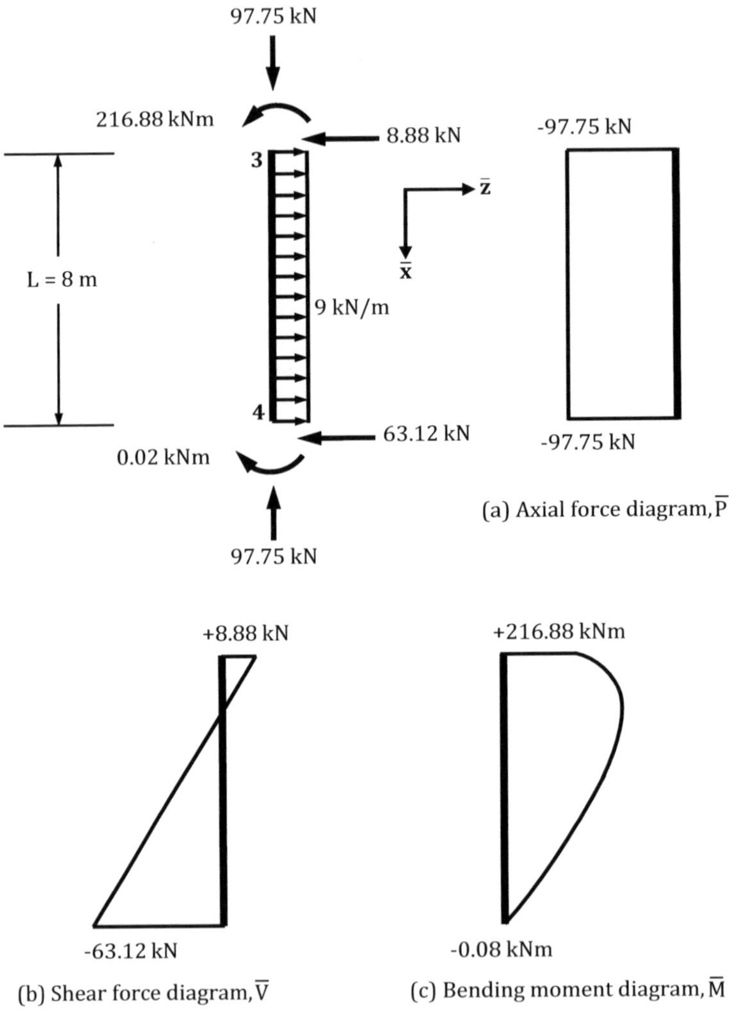

Figure 5.14

5.3 Members with a Pin at One End

Sometimes a hinge is inserted at the end of a member in a rigidly connected frame to achieve certain structural behaviour. Such a member has a special stiffness matrix whose derivation is explained below.

(a) Beam member with a pin at node j

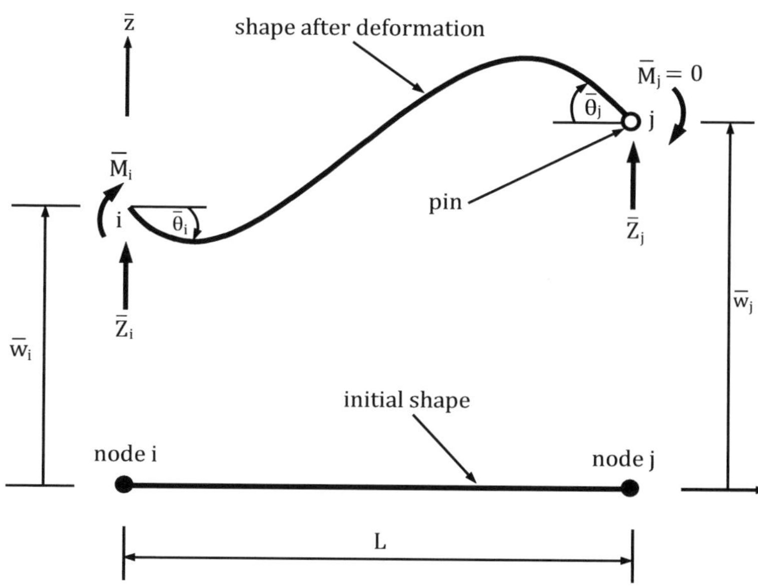

Figure 5.15 Beam element.

For a beam member we have the following relationships:

$$\bar{Z}_i = \frac{12EI}{L^3}\bar{w}_i - \frac{6EI}{L^2}\bar{\theta}_i - \frac{12EI}{L^3}\bar{w}_j - \frac{6EI}{L^2}\bar{\theta}_j \tag{5.19}$$

$$\bar{M}_i = -\frac{6EI}{L^2}\bar{w}_i + \frac{4EI}{L}\bar{\theta}_i + \frac{6EI}{L^2}\bar{w}_j + \frac{2EI}{L}\bar{\theta}_j \tag{5.20}$$

$$\bar{Z}_j = -\frac{12EI}{L^3}\bar{w}_i + \frac{6EI}{L^2}\bar{\theta}_i + \frac{12EI}{L^3}\bar{w}_j + \frac{6EI}{L^2}\bar{\theta}_j \tag{5.21}$$

$$\bar{M}_j = -\frac{6EI}{L^2}\bar{w}_i + \frac{2EI}{L}\bar{\theta}_i + \frac{6EI}{L^2}\bar{w}_j + \frac{4EI}{L}\bar{\theta}_j \tag{5.22}$$

At node j where there is a pin the moment is zero, i.e. $\bar{M}_j = 0$. Therefore (5.22) becomes:

$$0 = -\frac{6EI}{L^2}\bar{w}_i + \frac{2EI}{L}\bar{\theta}_i + \frac{6EI}{L^2}\bar{w}_j + \frac{4EI}{L}\bar{\theta}_j$$

$$\bar{\theta}_j = +\frac{3}{2L}\bar{w}_i - \frac{1}{2}\bar{\theta}_i - \frac{3}{2L}\bar{w}_j$$

Substitute the above value of $\bar{\theta}_j$ into (5.19) to (5.21) and simplify to get:

$$\bar{Z}_i = +\frac{3EI}{L^3}\bar{w}_i - \frac{3EI}{L^2}\bar{\theta}_i - \frac{3EI}{L^3}\bar{w}_j$$

$$\bar{M}_i = -\frac{3EI}{L^2}\bar{w}_i + \frac{3EI}{L}\bar{\theta}_i + \frac{3EI}{L^2}\bar{w}_j$$

$$\bar{Z}_j = -\frac{3EI}{L^3}\bar{w}_i + \frac{3EI}{L^2}\bar{\theta}_i + \frac{3EI}{L^3}\bar{w}_j$$

Writing the above equations in matrix form leads to:

$$\begin{bmatrix} \bar{Z}_i \\ \bar{M}_i \\ \bar{Z}_j \end{bmatrix} = \begin{bmatrix} \frac{3EI}{L^3} & -\frac{3EI}{L^2} & -\frac{3EI}{L^3} \\ -\frac{3EI}{L^2} & \frac{3EI}{L} & \frac{3EI}{L^2} \\ -\frac{3EI}{L^3} & \frac{3EI}{L^2} & \frac{3EI}{L^3} \end{bmatrix} \begin{bmatrix} \bar{w}_i \\ \bar{\theta}_i \\ \bar{w}_j \end{bmatrix}$$

$$\begin{bmatrix} \bar{Z}_i \\ \bar{M}_i \\ \bar{Z}_j \\ 0 \end{bmatrix} = \begin{bmatrix} \frac{3EI}{L^3} & -\frac{3EI}{L^2} & -\frac{3EI}{L^3} & 0 \\ -\frac{3EI}{L^2} & \frac{3EI}{L} & \frac{3EI}{L^2} & 0 \\ -\frac{3EI}{L^3} & \frac{3EI}{L^2} & \frac{3EI}{L^3} & 0 \\ 0 & 0 & 0 & 0 \end{bmatrix} \begin{bmatrix} \bar{w}_i \\ \bar{\theta}_i \\ \bar{w}_j \\ \bar{\theta}_j \end{bmatrix} \quad (5.23)$$

The inclusion of $\bar{\theta}_j$ in the above matrix is to maintain consistency of using a 4×4 matrix and also to indicate that it is not equal to zero since the pin at node j will rotate.

The full stiffness matrix with the added axial force effect is:

$$\begin{bmatrix} \bar{X}_i \\ \bar{Z}_i \\ \bar{M}_i \\ \bar{X}_j \\ \bar{Z}_j \\ 0 \end{bmatrix} = \begin{bmatrix} \frac{EA}{L} & 0 & 0 & -\frac{EA}{L} & 0 & 0 \\ 0 & \frac{3EI}{L^3} & -\frac{3EI}{L^2} & 0 & -\frac{3EI}{L^3} & 0 \\ 0 & -\frac{3EI}{L^2} & \frac{3EI}{L} & 0 & \frac{3EI}{L^2} & 0 \\ -\frac{EA}{L} & 0 & 0 & \frac{EA}{L} & 0 & 0 \\ 0 & -\frac{3EI}{L^3} & \frac{3EI}{L^2} & 0 & \frac{3EI}{L^3} & 0 \\ 0 & 0 & 0 & 0 & 0 & 0 \end{bmatrix} \begin{bmatrix} \bar{u}_i \\ \bar{w}_i \\ \bar{\theta}_i \\ \bar{u}_j \\ \bar{w}_j \\ \bar{\theta}_j \end{bmatrix}$$

$$\text{i.e., } \bar{k} = \begin{bmatrix} \frac{EA}{L} & 0 & 0 & -\frac{EA}{L} & 0 & 0 \\ 0 & \frac{3EI}{L^3} & -\frac{3EI}{L^2} & 0 & -\frac{3EI}{L^3} & 0 \\ 0 & -\frac{3EI}{L^2} & \frac{3EI}{L} & 0 & \frac{3EI}{L^2} & 0 \\ -\frac{EA}{L} & 0 & 0 & \frac{EA}{L} & 0 & 0 \\ 0 & -\frac{3EI}{L^3} & \frac{3EI}{L^2} & 0 & \frac{3EI}{L^3} & 0 \\ 0 & 0 & 0 & 0 & 0 & 0 \end{bmatrix} \qquad (5.24)$$

To transform the above matrix into global coordinates

$k = r^T \bar{k} r$ and r is given in (5.6), thus

$$k = \begin{bmatrix} \frac{EAx_{ij}^2}{L^3}+\frac{3EIz_{ij}^2}{L^5} & \frac{EAx_{ij}z_{ij}}{L^3}-\frac{3EIx_{ij}z_{ij}}{L^5} & \frac{3EIz_{ij}}{L^3} & -\frac{EAx_{ij}^2}{L^3}-\frac{3EIz_{ij}^2}{L^5} & -\frac{EAx_{ij}z_{ij}}{L^3}+\frac{3EIx_{ij}z_{ij}}{L^5} & 0 \\ \frac{EAx_{ij}z_{ij}}{L^3}-\frac{3EIx_{ij}z_{ij}}{L^5} & \frac{EAz_{ij}^2}{L^3}+\frac{3EIx_{ij}^2}{L^5} & -\frac{3EIx_{ij}}{L^3} & -\frac{EAx_{ij}z_{ij}}{L^3}+\frac{3EIx_{ij}z_{ij}}{L^5} & -\frac{EAz_{ij}^2}{L^3}-\frac{3EIx_{ij}^2}{L^5} & 0 \\ \frac{3EIz_{ij}}{L^3} & -\frac{3EIx_{ij}}{L^3} & \frac{3EI}{L} & -\frac{3EIz_{ij}}{L^3} & \frac{3EIx_{ij}}{L^3} & 0 \\ -\frac{EAx_{ij}^2}{L^3}-\frac{3EIz_{ij}^2}{L^5} & -\frac{EAx_{ij}z_{ij}}{L^3}+\frac{3EIx_{ij}z_{ij}}{L^5} & -\frac{3EIz_{ij}}{L^3} & \frac{EAx_{ij}^2}{L^3}+\frac{3EIz_{ij}^2}{L^5} & \frac{EAx_{ij}z_{ij}}{L^3}-\frac{3EIx_{ij}z_{ij}}{L^5} & 0 \\ -\frac{EAx_{ij}z_{ij}}{L^3}+\frac{3EIx_{ij}z_{ij}}{L^5} & -\frac{EAz_{ij}^2}{L^3}-\frac{3EIx_{ij}^2}{L^5} & \frac{3EIx_{ij}}{L^3} & \frac{EAx_{ij}z_{ij}}{L^3}-\frac{3EIx_{ij}z_{ij}}{L^5} & \frac{EAz_{ij}^2}{L^3}+\frac{3EIx_{ij}^2}{L^5} & 0 \\ 0 & 0 & 0 & 0 & 0 & 0 \end{bmatrix} \tag{5.25}$$

$$\delta = \begin{bmatrix} \delta_i \\ \delta_j \end{bmatrix} = \begin{bmatrix} u_i \\ w_i \\ \theta_i \\ u_j \\ w_j \\ \theta_j \end{bmatrix} \text{ and } F = \begin{bmatrix} F_i \\ F_j \end{bmatrix} = \begin{bmatrix} X_i \\ Z_i \\ M_i \\ X_j \\ Z_j \\ 0 \end{bmatrix}$$

(b) Beam member with a pin at node i

At node i where there is a pin the moment is zero, i.e. $\bar{M}_i = 0$ and following the same procedure as in (a) above leads to the following relationship

$$\begin{bmatrix} \bar{X}_i \\ \bar{Z}_i \\ 0 \\ \bar{X}_j \\ \bar{Z}_j \\ \bar{M}_j \end{bmatrix} = \begin{bmatrix} \frac{EA}{L} & 0 & 0 & -\frac{EA}{L} & 0 & 0 \\ 0 & \frac{3EI}{L^3} & 0 & 0 & -\frac{3EI}{L^3} & -\frac{3EI}{L^2} \\ 0 & 0 & 0 & 0 & 0 & 0 \\ -\frac{EA}{L} & 0 & 0 & \frac{EA}{L} & 0 & 0 \\ 0 & -\frac{3EI}{L^3} & 0 & 0 & \frac{3EI}{L^3} & \frac{3EI}{L^2} \\ 0 & -\frac{3EI}{L^2} & 0 & 0 & \frac{3EI}{L^2} & \frac{3EI}{L} \end{bmatrix} \begin{bmatrix} \bar{u}_i \\ \bar{w}_i \\ \bar{\theta}_i \\ \bar{u}_j \\ \bar{w}_j \\ \bar{\theta}_j \end{bmatrix} \quad (5.26)$$

The above matrix can be transformed, if required, from local to global coordinates in the usual way and the resulting matrix is:

$$k = \left[\begin{matrix} \frac{EAx_{ij}^2}{L^3} + \frac{3EIz_{ij}^2}{L^5} & \frac{EAx_{ij}z_{ij}}{L^3} - \frac{3EIx_{ij}z_{ij}}{L^5} & 0 \\ \frac{EAx_{ij}z_{ij}}{L^3} - \frac{3EIx_{ij}z_{ij}}{L^5} & \frac{EAz_{ij}^2}{L^3} + \frac{3EIx_{ij}^2}{L^5} & 0 \\ 0 & 0 & 0 \\ -\frac{EAx_{ij}^2}{L^3} - \frac{3EIz_{ij}^2}{L^5} & -\frac{EAx_{ij}z_{ij}}{L^3} + \frac{3EIx_{ij}z_{ij}}{L^5} & 0 \\ -\frac{EAx_{ij}z_{ij}}{L^3} + \frac{3EIx_{ij}z_{ij}}{L^5} & -\frac{EAz_{ij}^2}{L^3} - \frac{3EIx_{ij}^2}{L^5} & 0 \\ \frac{3EIz_{ij}}{L^3} & -\frac{3EIx_{ij}}{L^3} & 0 \end{matrix}\right. \quad (5.27)$$

$$\left.\begin{matrix} -\frac{EAx_{ij}^2}{L^3} - \frac{3EIz_{ij}^2}{L^5} & -\frac{EAx_{ij}z_{ij}}{L^3} + \frac{3EIx_{ij}z_{ij}}{L^5} & \frac{3EIz_{ij}}{L^3} \\ -\frac{EAx_{ij}z_{ij}}{L^3} + \frac{3EIx_{ij}z_{ij}}{L^5} & -\frac{EAz_{ij}^2}{L^3} - \frac{3EIx_{ij}^2}{L^5} & -\frac{3EIx_{ij}}{L^3} \\ 0 & 0 & 0 \\ \frac{EAx_{ij}^2}{L^3} + \frac{3EIz_{ij}^2}{L^5} & \frac{EAx_{ij}z_{ij}}{L^3} - \frac{3EIx_{ij}z_{ij}}{L^5} & -\frac{3EIz_{ij}}{L^3} \\ \frac{EAx_{ij}z_{ij}}{L^3} - \frac{3EIx_{ij}z_{ij}}{L^5} & \frac{EAz_{ij}^2}{L^3} + \frac{3EIx_{ij}^2}{L^5} & \frac{3EIx_{ij}}{L^3} \\ -\frac{3EIz_{ij}}{L^3} & \frac{3EIx_{ij}}{L^3} & \frac{3EI}{L} \end{matrix}\right]$$

$$\delta = \begin{bmatrix} \delta_i \\ \delta_j \end{bmatrix} = \begin{bmatrix} u_i \\ w_i \\ \theta_i \\ u_j \\ w_j \\ \theta_j \end{bmatrix} \text{ and } F = \begin{bmatrix} F_i \\ F_j \end{bmatrix} = \begin{bmatrix} X_i \\ Z_i \\ 0 \\ X_j \\ Z_j \\ M_j \end{bmatrix}$$

For a member with pins at both ends then the stiffness matrix is the same as that for an axially loaded member as explained in Chapter 3, i.e. a 4×4 matrix but with the addition of zero rows and zero columns to get a 6×6 matrix thus making it consistent with the stiffness matrices of the rest of the members of the structure. The nodal forces acting on the joints due to the loads applied to the member are calculated in the normal manner except that the fixed end moments are zero.

Example 2

The frame shown in Fig. 5.16 is pinned at support A and fixed to the supports at points B and D. Members AC and CD are rigidly connected together at joint C while member BC is pin-connected to joint C. Analyse the frame for the loading shown for the following data:

$E = 210 \times 10^6$ kN/m^2, $I_1 = 66 \times 10^{-6}$ m^4, $A_1 = 0.003$ m^2, $I_2 = 75 \times 10^{-6}$ m^4, $A_2 = 0.004$ m^2, $I_3 = 16 \times 10^{-6}$ m^4, $A_3 = 0.001$ m^2.

For member 1, which has a pin at node 1, the standard matrix can be used because the pin occurs at the support. The boundary conditions take account of the pin, i.e. $u_1 = 0$ and $w_1 = 0$ but the rotation θ_1 is unknown and is determined in the usual way from the set of the resulting simultaneous equations (see example 1 above). Alternatively, it can be treated as a member with a pin at node i as is shown in the following analysis.

For member 2, which has a pin at node 3 (node j), it is necessary to use the modified matrix because the pin occurs at an internal node.

Member 3 is treated as a standard member since both its ends are continuous.

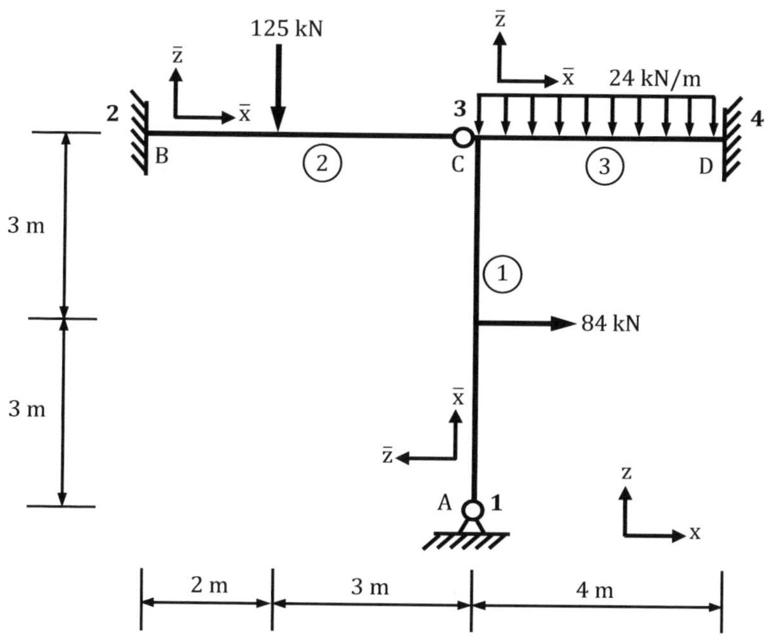

Figure 5.16

Member 1: (i,j) = (1,3) with a pin at node 1

$E = 210 \times 10^6 \text{ kN/m}^2$, $I_1 = 66 \times 10^{-6} \text{ m}^4$, and $A_1 = 0.003 \text{ m}^2$.

$x_i = 0,\quad x_j = 0,\quad x_{ij} = x_j - x_i = 0 - 0 = 0$

$z_i = 0,\quad z_j = 6 \text{ m}, z_{ij} = z_j - z_i = 6 - 0 = 6 \text{ m}$

$$L = \sqrt{x_{ij}^2 + z_{ij}^2} = \sqrt{0^2 + 6^2} = 6 \text{ m}$$

For a member with a pin at node i the stiffness matrix is given by (5.27), thus

$$k^1 = \begin{bmatrix} 193 & 0 & 0 & -193 & 0 & 1155 \\ 0 & 105000 & 0 & 0 & -105000 & 0 \\ 0 & 0 & 0 & 0 & 0 & 0 \\ -193 & 0 & 0 & 193 & 0 & -1155 \\ 0 & -105000 & 0 & 0 & 105000 & 0 \\ 1155 & 0 & 0 & -1155 & 0 & 6930 \end{bmatrix}$$

Member 2: (i,j) = (2,3) with a pin at node 3

$E = 210 \times 10^6\ \text{kN/m}^2$, $I_2 = 75 \times 10^{-6}\ \text{m}^4$, and $A_2 = 0.004\ \text{m}^2$.

The local $\bar{x}$-axis of this member lies along the global x-axis, therefore $k = \bar{k}$ as given by (5.24) for a member with a pin at node j, thus

$$k^2 = \begin{bmatrix} 168000 & 0 & 0 & -168000 & 0 & 0 \\ 0 & 378 & -1890 & 0 & -378 & 0 \\ 0 & -1890 & 9450 & 0 & 1890 & 0 \\ -168000 & 0 & 0 & 168000 & 0 & 0 \\ 0 & -378 & 1890 & 0 & 378 & 0 \\ 0 & 0 & 0 & 0 & 0 & 0 \end{bmatrix}$$

Member 3: (i,j) = (3,4)

$E = 210 \times 10^6\ \text{kN/m}^2$, $I_3 = 16 \times 10^{-6}\ \text{m}^4$, and $A_3 = 0.001\ \text{m}^2$.

The standard stiffness matrix is given by (5.7), thus

$$k^3 = \begin{bmatrix} 52500 & 0 & 0 & -52500 & 0 & 0 \\ 0 & 630 & -1260 & 0 & -630 & -1260 \\ 0 & -1260 & 3360 & 0 & 1260 & 1680 \\ -52500 & 0 & 0 & 52500 & 0 & 0 \\ 0 & -630 & 1260 & 0 & 630 & 1260 \\ 0 & -1260 & 1680 & 0 & 1260 & 3360 \end{bmatrix}$$

By inspection the overall structure matrix is given by

K =

δ_1	δ_2	δ_3	δ_4
$K_{11} = k_{ii}^1$	0	$K_{13} = k_{ij}^1$	0
0	$K_{22} = k_{ii}^2$	$K_{23} = k_{ij}^2$	0
$K_{31} = k_{ji}^1$	$K_{32} = k_{ji}^2$	$K_{33} = k_{jj}^1 + k_{jj}^2 + k_{ii}^3$	$K_{34} = k_{ij}^3$
0	0	$K_{43} = k_{ji}^3$	$K_{44} = k_{jj}^3$

K =

u$_1$	w$_1$	θ$_1$	u$_2$	w$_2$	θ$_2$	u$_3$	w$_3$	θ$_3$	u$_4$	w$_4$	θ$_4$
193	0	0	0	0	0	−193	0	1155	0	0	0
0	105000	0	0	0	0	0	−105000	0	0	0	0
0	0	0	0	0	0	0	0	0	0	0	0
0	0	0	168000	0	0	−168000	0	0	0	0	0
0	0	0	0	378	−1890	0	−378	0	0	0	0
0	0	0	0	−1890	9450	0	1890	0	0	0	0
−193	0	0	−168000	0	0	220693	0	−1155	-52500	0	0
0	−105000	0	0	−378	1890	0	106008	−1260	0	−630	−1260
1155	0	0	0	0	0	−1155	−1260	10290	0	1260	1680
0	0	0	0	0	0	-52500	0	0	52500	0	0
0	0	0	0	0	0	0	−630	1260	0	630	1260
0	0	0	0	0	0	0	−1260	1680	0	1260	3360

(5.28)

Applying the boundary conditions of $u_1 = 0$, $w_1 = 0$, $u_2 = 0$, $w_2 = 0$, $\theta_2 = 0$, $u_4 = 0$, $w_4 = 0$, $\theta_4 = 0$ and noting that all the coefficients in row 3 and column 3 are equal to zero, therefore, delete the corresponding rows and columns from the above matrix to get

$$K = \begin{bmatrix} 220693 & 0 & -1155 \\ 0 & 106008 & -1260 \\ -1155 & -1260 & 10290 \end{bmatrix} \tag{5.28a}$$

Load vector

Member 1

Contribution of loads acting on member 1 to the loads on joints 1 and 3:

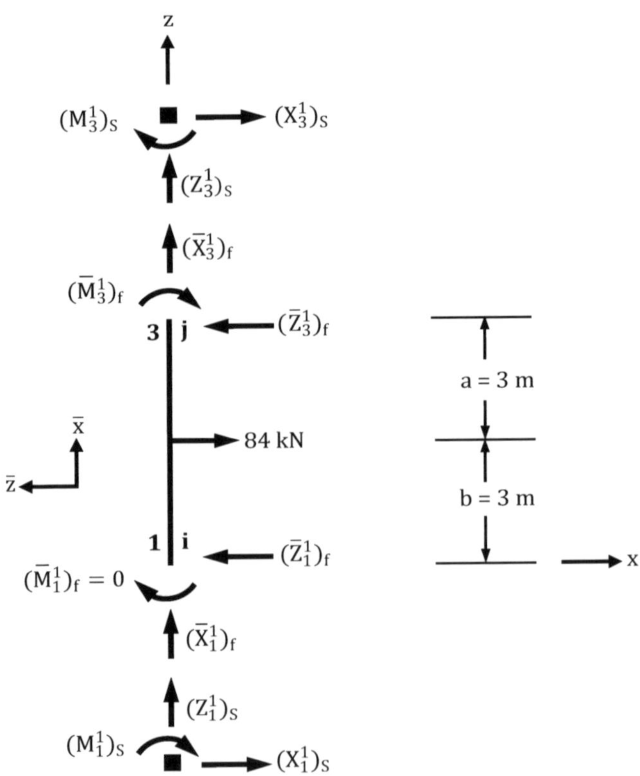

Figure 5.17

Actions on member 1 (W = –84 kN, L = 6 m, a = 3 m, and b = 3 m)

$$(\overline{X}_1^1)_f = 0$$

$$(\overline{X}_3^1)_f = 0$$

$$(\overline{Z}_1^1)_f = -\frac{Wa}{2L^3}\left(3La - a^2\right) = -\frac{(-84)\times 3}{2\times 6^3}\left(3\times 6\times 3 - 3^2\right) = +26.25\text{ kN}$$

$$(\overline{Z}_3^1)_f = -\frac{Wb}{2L^3}\left(3L^2 - b^2\right) = -\frac{(-84)\times 3}{2\times 6^3}\left(3\times 6^2 - 3^2\right) = +57.75\text{ kN}$$

$(\overline{M}_1^1)_f = 0$, because there is a pin at end i of this member.

$$(\overline{M}_3^1)_f = -\frac{Wab}{2L^2}(L+b) = -\frac{(-84)\times 3\times 3}{2\times 6^2}(6+3) = +94.50\text{ kNm}$$

$$\overline{F}_f^1 = \begin{bmatrix} (\overline{X}_1^1)_f \\ (\overline{Z}_1^1)_f \\ (\overline{M}_1^1)_f \\ (\overline{X}_3^1)_f \\ (\overline{Z}_3^1)_f \\ (\overline{M}_3^1)_f \end{bmatrix} = \begin{bmatrix} 0 \\ +26.25 \\ 0 \\ 0 \\ +57.75 \\ +94.50 \end{bmatrix} \tag{5.29}$$

Loads on joints 1 and 3

$$(X_1^1)_S = \left(\overline{Z}_1^1\right)_f = +26.25\text{ kN}$$

$$(Z_1^1)_S = -\left(\overline{X}_1^1\right)_f = 0$$

$$(M_1^1)_S = -\left(\overline{M}_1^1\right)_f = 0$$

$$(X_3^1)_S = \left(\overline{Z}_3^1\right)_f = +57.75\text{ kN}$$

$$(Z_3^1)_S = -\left(\overline{X}_3^1\right)_f = 0$$

$$(M_3^1)_S = -\left(\overline{M}_3^1\right)_f = -94.50\text{ kNm}$$

$$\bar{F}_S^1 = \begin{bmatrix} (\bar{X}_1^1)_S \\ (\bar{Z}_1^1)_S \\ (\bar{M}_1^1)_S \\ (\bar{X}_3^1)_S \\ (\bar{Z}_3^1)_S \\ (\bar{M}_3^1)_S \end{bmatrix} = \begin{bmatrix} +26.25 \\ 0 \\ 0 \\ +57.75 \\ 0 \\ -94.50 \end{bmatrix} \tag{5.30}$$

Member 2

Contribution of loads acting on member 2 to the loads on joints 2 and 3:

Actions on member 2 (W = –125 kN, L = 5 m, a = 2 m, and b = 3 m)

$$(\bar{X}_2^2)_f = 0$$

$$(\bar{X}_3^2)_f = 0$$

$$(\bar{Z}_2^2)_f = -\frac{Wb}{2L^3}\left(3L^2 - b^2\right) = -\frac{(-125)\times 3}{2\times 5^3}\left(3\times 5^2 - 3^2\right) = +99 \text{ kN}$$

$$(\bar{Z}_3^2)_f = -\frac{Wa}{2L^3}\left(3La - a^2\right) = -\frac{(-125)\times 2}{2\times 5^3}\left(3\times 5\times 2 - 2^2\right) = +26 \text{ kN}$$

$$(\bar{M}_2^2)_f = +\frac{Wab}{2L^2}(L+b) = +\frac{(-125)\times 2\times 3}{2\times 5^2}(5+3) = -120 \text{ kNm}$$

$(\bar{M}_3^2)_f = 0$, because there is a pin at end j of this member.

$$\bar{F}_f^2 = \begin{bmatrix} (\bar{X}_2^2)_f \\ (\bar{Z}_2^2)_f \\ (\bar{M}_2^2)_f \\ (\bar{X}_3^2)_f \\ (\bar{Z}_3^2)_f \\ (\bar{M}_3^2)_f \end{bmatrix} = \begin{bmatrix} 0 \\ +99.00 \\ -120.00 \\ 0 \\ +26.00 \\ 0 \end{bmatrix} \tag{5.31}$$

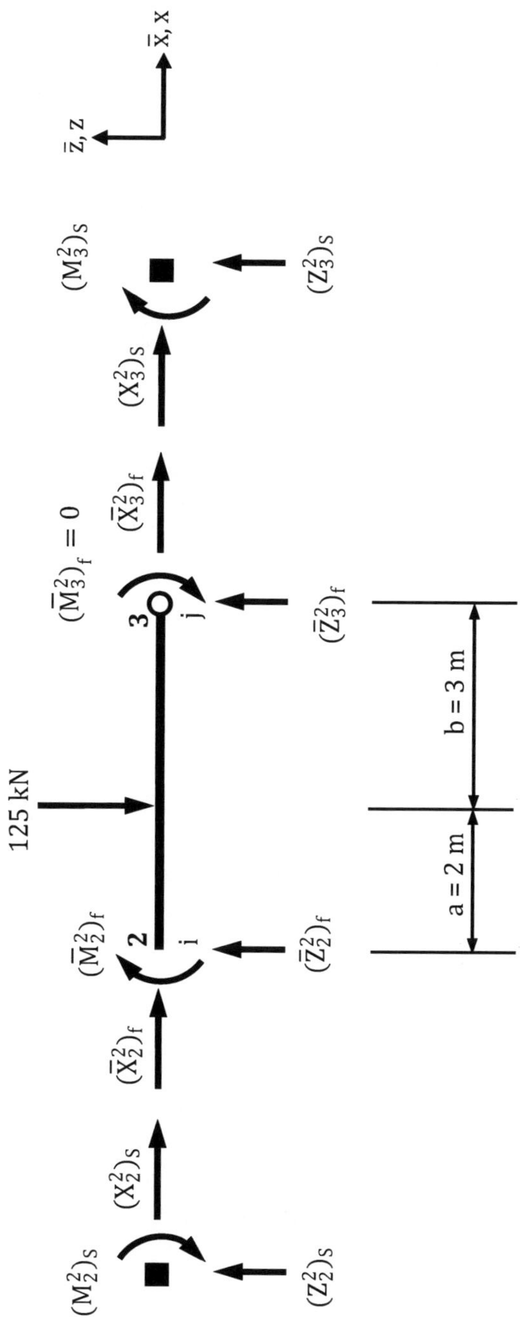

Figure 5.18

Loads on joints 2 and 3

$$F_S^2 = \begin{bmatrix} (X_2^2)_S \\ (Z_2^2)_S \\ (M_2^2)_S \\ (X_3^2)_S \\ (Z_3^2)_S \\ (M_3^2)_S \end{bmatrix} = -\bar{F}_f^2 = \begin{bmatrix} 0 \\ -99.00 \\ +120.00 \\ 0 \\ -26.00 \\ 0 \end{bmatrix} \tag{5.32}$$

Member 3

Contribution of loads acting on member 3 to the loads on joints 3 and 4:

Actions on member 3 (n = –24 kN/m)

$$(\bar{X}_3^3)_f = 0$$

$$(\bar{X}_4^3)_f = 0$$

$$(\bar{Z}_3^3)_f = -\frac{nL}{2} = -\frac{(-24)\times 4}{2} = +48.00\text{ kN}$$

$$(\bar{Z}_4^3)_f = -\frac{nL}{2} = -\frac{(-24)\times 4}{2} = +48.00\text{ kN}$$

$$(\bar{M}_3^3)_f = +\frac{nL^2}{12} = +\frac{(-24)\times 4^2}{12} = -32.00\text{ kNm}$$

$$(\bar{M}_4^3)_f = -\frac{nL^2}{12} = -\frac{(-24)\times 4^2}{12} = +32.00\text{ kNm}$$

$$\bar{F}_f^3 = \begin{bmatrix} (\bar{X}_3^3)_f \\ (\bar{Z}_3^3)_f \\ (\bar{M}_3^3)_f \\ (\bar{X}_4^3)_f \\ (\bar{Z}_4^3)_f \\ (\bar{M}_4^3)_f \end{bmatrix} = \begin{bmatrix} 0 \\ +48.00 \\ -32.00 \\ 0 \\ +48.00 \\ +32.00 \end{bmatrix} \tag{5.33}$$

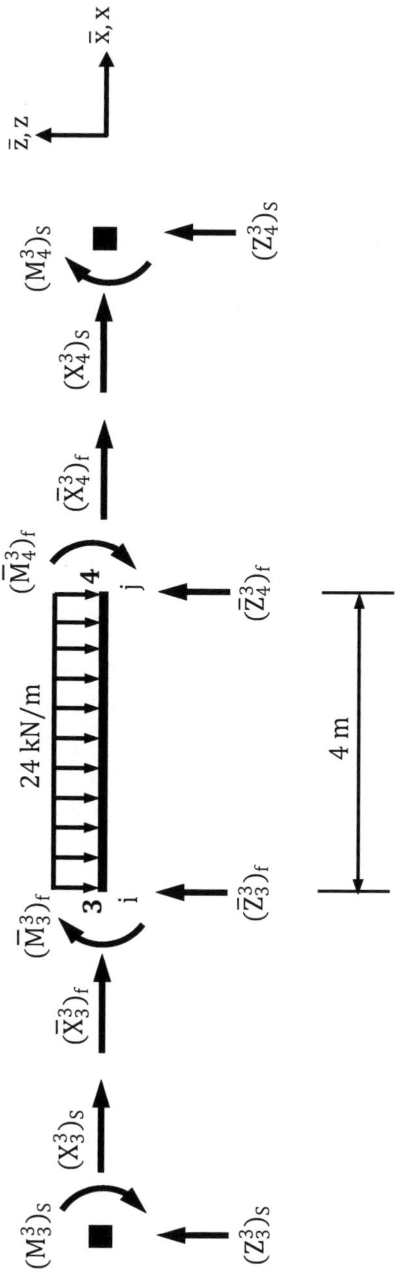
$(M_3^3)_S$
$(X_3^3)_S$
$(Z_3^3)_S$
$(\bar{X}_3^3)_f$
$(\bar{M}_3^3)_f$
3
i
$(\bar{Z}_3^3)_f$
24 kN/m
4
j
$(\bar{M}_4^3)_f$
$(\bar{Z}_4^3)_f$
4 m
$(\bar{X}_4^3)_f$
$(X_4^3)_S$
$(M_4^3)_S$
$(Z_4^3)_S$
$\bar{z}, z$
$\bar{x}, x$

Figure 5.19

Loads on joints 3 and 4

$$F_S^3 = \begin{bmatrix} (X_3^3)_S \\ (Z_3^3)_S \\ (M_3^3)_S \\ (X_4^3)_S \\ (Z_4^3)_S \\ (M_4^3)_S \end{bmatrix} = -\bar{F}_f^3 = \begin{bmatrix} 0 \\ -48.00 \\ +32.00 \\ 0 \\ -48.00 \\ -32.00 \end{bmatrix} \quad (5.34)$$

From (5.30), (5.32), and (5.34)

$$\bar{F}_S^1 = \begin{bmatrix} (\bar{X}_1^1)_S \\ (\bar{Z}_1^1)_S \\ (\bar{M}_1^1)_S \\ (\bar{X}_3^1)_S \\ (\bar{Z}_3^1)_S \\ (\bar{M}_3^1)_S \end{bmatrix} = \begin{bmatrix} +26.25 \\ 0 \\ 0 \\ +57.75 \\ 0 \\ -94.50 \end{bmatrix}, \qquad F_S^2 = \begin{bmatrix} (X_2^2)_S \\ (Z_2^2)_S \\ (M_2^2)_S \\ (X_3^2)_S \\ (Z_3^2)_S \\ (M_3^2)_S \end{bmatrix} = \begin{bmatrix} 0 \\ -99.00 \\ +120.00 \\ 0 \\ -26.00 \\ 0 \end{bmatrix},$$

$$F_S^3 = \begin{bmatrix} (X_3^3)_S \\ (Z_3^3)_S \\ (M_3^3)_S \\ (X_4^3)_S \\ (Z_4^3)_S \\ (M_4^3)_S \end{bmatrix} = \begin{bmatrix} 0 \\ -48.00 \\ +32.00 \\ 0 \\ -48.00 \\ -32.00 \end{bmatrix}$$

$$F_S = \begin{bmatrix} (X_1)_S \\ (Z_1)_S \\ (M_1)_S \\ (X_2)_S \\ (Z_2)_S \\ (M_2)_S \\ (X_3)_S \\ (Z_3)_S \\ (M_3)_S \\ (X_4)_S \\ (Z_4)_S \\ (M_4)_S \end{bmatrix} = F_S^1 + F_S^2 + F_S^3 = \begin{bmatrix} +26.25 \\ 0 \\ 0 \\ 0 \\ 0 \\ 0 \\ +57.75 \\ 0 \\ -94.50 \\ 0 \\ 0 \\ 0 \end{bmatrix} + \begin{bmatrix} 0 \\ 0 \\ 0 \\ 0 \\ -99.00 \\ +120.00 \\ 0 \\ -26.00 \\ 0 \\ 0 \\ 0 \\ 0 \end{bmatrix} + \begin{bmatrix} 0 \\ 0 \\ 0 \\ 0 \\ 0 \\ 0 \\ 0 \\ -48.00 \\ +32.00 \\ 0 \\ -48.00 \\ -32.00 \end{bmatrix}$$

$$F_S = \begin{bmatrix} +26.25 \\ 0 \\ 0 \\ 0 \\ -99.00 \\ +120.00 \\ +57.75 \\ -74.00 \\ -62.50 \\ 0 \\ -48.00 \\ -32.00 \end{bmatrix} \tag{5.35}$$

Since there are no externally applied direct actions at the nodes then:

$$F_N = 0 \tag{5.36}$$

The load vector due to the reactions at the supports

$$F_C = \begin{bmatrix} (F_1)_C \\ (F_2)_C \\ (F_3)_C \\ (F_4)_C \end{bmatrix} = \begin{bmatrix} (X_1)_C \\ (Z_1)_C \\ (M_1)_C \\ (X_2)_C \\ (Z_2)_C \\ (M_2)_C \\ (X_3)_C \\ (Z_3)_C \\ (M_3)_C \\ (X_4)_C \\ (Z_4)_C \\ (M_4)_C \end{bmatrix} = \begin{bmatrix} R_{X1} \\ R_{Z1} \\ 0 \\ R_{X2} \\ R_{Z2} \\ R_{M2} \\ 0 \\ 0 \\ 0 \\ R_{X4} \\ R_{Z4} \\ R_{M4} \end{bmatrix} \tag{5.37}$$

Note that the reaction moment at the pinned support A, $(M_1)_c = 0$.

Total load vector, F, is given by:

$$F = F_S + F_N + F_C$$

From (5.35) to (5.37)

$$F = \begin{bmatrix} +26.25 \\ 0 \\ 0 \\ 0 \\ -99.00 \\ +120.00 \\ +57.75 \\ -74.00 \\ -62.50 \\ 0 \\ -48.00 \\ -32.00 \end{bmatrix} + \begin{bmatrix} 0 \\ 0 \\ 0 \\ 0 \\ 0 \\ 0 \\ 0 \\ 0 \\ 0 \\ 0 \\ 0 \\ 0 \end{bmatrix} + \begin{bmatrix} R_{X1} \\ R_{Z1} \\ 0 \\ R_{X2} \\ R_{Z2} \\ R_{M2} \\ 0 \\ 0 \\ 0 \\ R_{X4} \\ R_{Z4} \\ R_{M4} \end{bmatrix} = \begin{bmatrix} R_{X1} + 26.25 \\ R_{Z1} \\ 0 \\ R_{X2} \\ R_{Z2} - 99.00 \\ R_{M2} + 120.00 \\ +57.75 \\ -74.00 \\ -62.50 \\ R_{X4} \\ R_{Z4} - 48.00 \\ R_{M4} - 32.00 \end{bmatrix} \tag{5.38}$$

From (5.28) and (5.38)

u_1	w_1	θ_1	u_2	w_2	θ_2	u_3	w_3	θ_3	u_4	w_4	θ_4				
193	0	0	0	0	0	−193	0	1155	0	0	0	u_1		$R_{X1}+26.25$	
0	105000	0	0	0	0	0	−105000	0	0	0	0	w_1		R_{Z1}	
0	0	0	0	0	0	0	0	0	0	0	0	θ_1		0	
0	0	0	168000	0	0	−168000	0	0	0	0	0	u_2		R_{X2}	
0	0	0	0	378	−1890	0	-378	0	0	0	0	w_2		$R_{Z2}-99.00$	
0	0	0	0	-1890	9450	0	1890	0	0	0	0	θ_2	=	$R_{M2}+120.00$	(5.39)
-193	0	0	-168000	0	0	220693	0	-1155	-52500	0	0	u_3		+57.75	
0	-105000	0	0	-378	1890	0	106008	-1260	0	-630	-1260	w_3		-74.00	
1155	0	0	0	0	0	-1155	-1260	10290	0	1260	1680	θ_3		-62.50	
0	0	0	0	0	0	-52500	0	0	52500	0	0	u_4		R_{X4}	
0	0	0	0	0	0	0	-630	1260	0	630	1260	w_4		$R_{Z4}-48.00$	
0	0	0	0	0	0	0	-1260	1680	0	1260	3360	θ_4		$R_{M4}-32.00$	

The boundary conditions are: $u_1 = 0$, $w_1 = 0$, $u_2 = 0$, $w_2 = 0$, $\theta_2 = 0$, $u_4 = 0$, $w_4 = 0$, and $\theta_4 = 0$, so delete rows and columns 1, 2, 4, 5, 6, 10, 11, and 12. Also delete the third row and third column because the rotation at node 1 is not included since the stiffness matrix for member 1 is derived as a member with a pin at one end.

The resulting reduced set is:

$$\begin{bmatrix} 220693 & 0 & -1155 \\ 0 & 106008 & -1260 \\ -1155 & -1260 & 10290 \end{bmatrix} \begin{bmatrix} u_3 \\ w_3 \\ \theta_3 \end{bmatrix} = \begin{bmatrix} +57.75 \\ -74.00 \\ -62.50 \end{bmatrix}$$

The solution of the above set is: $u_3 = +0.000230$ m, $w_3 = -0.000771$ m, and $\theta_3 = -0.006143$ rad (which is the rotation of the rigid joint at node 3).

Reactions at the supports of the frame are calculated from (5.39) and using the above values of u_3, w_3, and θ_3.

Actions on member 1

$E = 210 \times 10^6$ kN/m², $I_1 = 66 \times 10^{-6}$ m⁴, $A_1 = 0.003$ m², $L = 6$ m, $x_{ij} = 0$, and $z_{ij} = 6$ m.

From (5.26) for a member with a pin at node i

$$\bar{k} = \begin{bmatrix} \frac{EA}{L} & 0 & 0 & -\frac{EA}{L} & 0 & 0 \\ 0 & \frac{3EI}{L^3} & 0 & 0 & -\frac{3EI}{L^3} & -\frac{3EI}{L^2} \\ 0 & 0 & 0 & 0 & 0 & 0 \\ -\frac{EA}{L} & 0 & 0 & \frac{EA}{L} & 0 & 0 \\ 0 & -\frac{3EI}{L^3} & 0 & 0 & \frac{3EI}{L^3} & \frac{3EI}{L^2} \\ 0 & -\frac{3EI}{L^2} & 0 & 0 & \frac{3EI}{L^2} & \frac{3EI}{L} \end{bmatrix}$$

For the calculation of actions on members with a pin at one end it is more convenient to delete the row and column corresponding to the rotation of the pin since the moment at the pinned end is zero. Hence delete the third row and third column which correspond to the pin at node i of this member.

$$\bar{k}^1 = \begin{bmatrix} 105000 & 0 & -105000 & 0 & 0 \\ 0 & 193 & 0 & -193 & -1155 \\ -105000 & 0 & 105000 & 0 & 0 \\ 0 & -193 & 0 & 193 & 1155 \\ 0 & -1155 & 0 & 1155 & 6930 \end{bmatrix}$$

From (5.6)

$$r = \begin{bmatrix} \frac{x_{ij}}{L} & \frac{z_{ij}}{L} & 0 & 0 & 0 & 0 \\ -\frac{z_{ij}}{L} & \frac{x_{ij}}{L} & 0 & 0 & 0 & 0 \\ 0 & 0 & 1 & 0 & 0 & 0 \\ 0 & 0 & 0 & \frac{x_{ij}}{L} & \frac{z_{ij}}{L} & 0 \\ 0 & 0 & 0 & -\frac{z_{ij}}{L} & \frac{x_{ij}}{L} & 0 \\ 0 & 0 & 0 & 0 & 0 & 1 \end{bmatrix}$$

$$r^1 = \begin{bmatrix} 0 & 1 & 0 & 0 & 0 \\ -1 & 0 & 0 & 0 & 0 \\ 0 & 0 & 0 & 1 & 0 \\ 0 & 0 & -1 & 0 & 0 \\ 0 & 0 & 0 & 0 & 1 \end{bmatrix}$$

Notice that the third row and third column have been deleted since they correspond to θ_1 which is not included for a member with a pin at one end.

$$\delta^1 = \begin{bmatrix} \delta_i^1 \\ \delta_j^1 \end{bmatrix} = \begin{bmatrix} u_1 \\ w_1 \\ u_3 \\ w_3 \\ \theta_3 \end{bmatrix} = \begin{bmatrix} 0 \\ 0 \\ +0.000230 \\ -0.000771 \\ -0.006143 \end{bmatrix}$$

$$\bar{\delta}^1 = r^1\delta^1 = \begin{bmatrix} 0 & 1 & 0 & 0 & 0 \\ -1 & 0 & 0 & 0 & 0 \\ 0 & 0 & 0 & 1 & 0 \\ 0 & 0 & -1 & 0 & 0 \\ 0 & 0 & 0 & 0 & 1 \end{bmatrix}\begin{bmatrix} 0 \\ 0 \\ +0.000230 \\ -0.000771 \\ -0.006143 \end{bmatrix} = \begin{bmatrix} 0 \\ 0 \\ -0.000771 \\ -0.000230 \\ -0.006143 \end{bmatrix}$$

$$\bar{F}_d^1 = \bar{k}^1\bar{\delta}^1 = \begin{bmatrix} 105000 & 0 & -105000 & 0 & 0 \\ 0 & 193 & 0 & -193 & -1155 \\ -105000 & 0 & 105000 & 0 & 0 \\ 0 & -193 & 0 & 193 & 1155 \\ 0 & -1155 & 0 & 1155 & 6930 \end{bmatrix}\begin{bmatrix} 0 \\ 0 \\ -0.000771 \\ -0.000230 \\ -0.006143 \end{bmatrix}$$

$$\bar{F}_d^1 = \begin{bmatrix} +80.96 \\ +7.14 \\ -80.96 \\ -7.14 \\ -42.84 \end{bmatrix} \text{ and from (5.29), } \bar{F}_f^1 = \begin{bmatrix} 0 \\ +26.25 \\ 0 \\ 0 \\ +57.75 \\ +94.50 \end{bmatrix}$$

$$\bar{F}_r^1 = \bar{F}_d^1 + \bar{F}_f^1 = \begin{bmatrix} +80.96 \\ +7.14 \\ -80.96 \\ -7.14 \\ -42.84 \end{bmatrix} + \begin{bmatrix} 0 \\ +26.25 \\ 0 \\ +57.75 \\ +94.50 \end{bmatrix} = \begin{bmatrix} +80.96 \\ +33.39 \\ -80.96 \\ +50.61 \\ +51.66 \end{bmatrix}$$

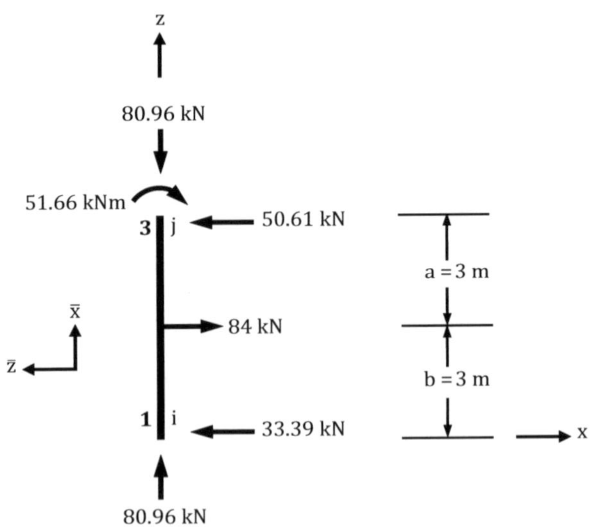

The actions on members 2 and 3 are calculated in a similar manner. Once the actions at the ends of the members are calculated, the axial force, shear force, and bending moment diagrams can be determined and these are shown in Fig. 5.20.

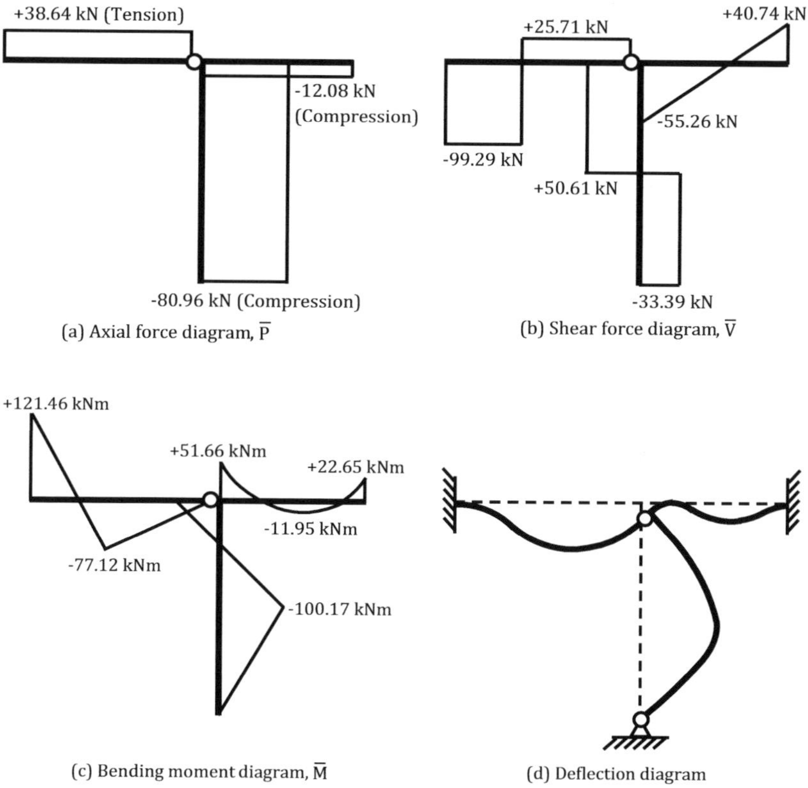

(a) Axial force diagram, $\overline{P}$

(b) Shear force diagram, $\overline{V}$

(c) Bending moment diagram, $\overline{M}$

(d) Deflection diagram

Figure 5.20

Problems

P5.1. The rigidly jointed plane frame shown in Fig. P5.1 is fixed at its bases A and D. The properties of the members of the frame are as follows: member AB, $I_1 = 0.0028$ m^4, $A_1 = 0.19$ m^2, member BC, $I_2 = 0.0021$ m^4, $A_2 = 0.16$ m^2, and member CD, $I_3 = 0.0024$ m^4, $A_3 = 0.17$ m^2. The modulus of elasticity of all members, E = 30×10^6 kN/m^2. Analyse the frame for the loading shown and draw the axial force, shear force, and bending moment diagrams.

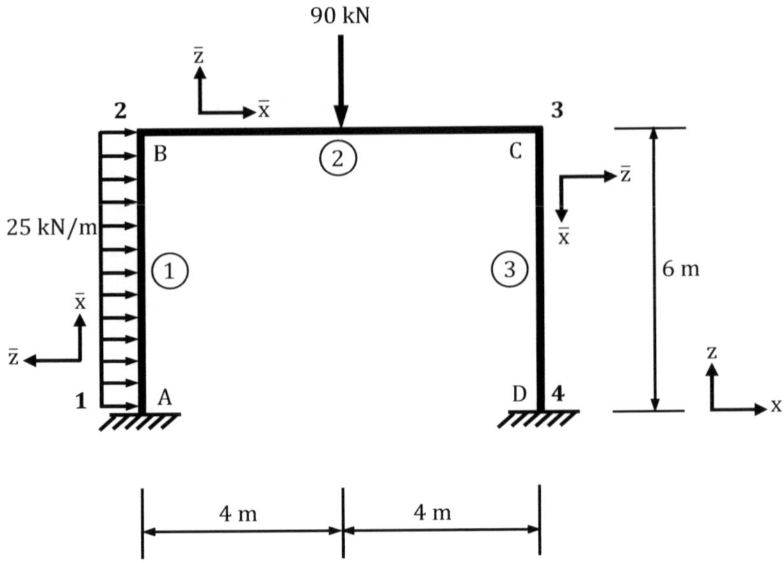

Figure P5.1

Answer:

$u_1 = 0$, $w_1 = 0$, $\theta_1 = 0$, $u_2 = +0.01260$ m, $w_2 = -0.00003$ m, $\theta_2 = +0.00213$ rad,

$u_3 = +0.01252$ m, $w_3 = -0.00007$ m, $\theta_3 = +0.00034$ rad, $u_4 = 0$, $w_4 = 0$, $\theta_4 = 0$,

$R_{X1} = -103.99$ kN, $R_{Z1} = +30.48$ kN, $R_{M1} = -191.76$ kNm

$R_{X4} = -46.01$ kN, $R_{Z4} = +59.52$ kN, $R_{M4} = -142.11$ kNm

$$\text{Member 1: } \bar{F}_r^1 = \begin{bmatrix} (\bar{X}_1^1)_r \\ (\bar{Z}_1^1)_r \\ (\bar{M}_1^1)_r \\ (\bar{X}_2^1)_r \\ (\bar{Z}_2^1)_r \\ (\bar{M}_2^1)_r \end{bmatrix} = \begin{bmatrix} +30.48 \\ +103.99 \\ -191.76 \\ -30.48 \\ +46.01 \\ +17.84 \end{bmatrix},$$

$$\text{Member 2:}\quad \overline{F}_r^2 = \begin{bmatrix} (\overline{X}_2^2)_r \\ (\overline{Z}_2^2)_r \\ (\overline{M}_2^2)_r \\ (\overline{X}_3^2)_r \\ (\overline{Z}_3^2)_r \\ (\overline{M}_3^2)_r \end{bmatrix} = \begin{bmatrix} +46.01 \\ +30.48 \\ -17.84 \\ -46.01 \\ +59.52 \\ +133.97 \end{bmatrix},$$

$$\text{Member 3:}\quad \overline{F}_r^3 = \begin{bmatrix} (\overline{X}_3^3)_r \\ (\overline{Z}_3^3)_r \\ (\overline{M}_3^3)_r \\ (\overline{X}_4^3)_r \\ (\overline{Z}_4^3)_r \\ (\overline{M}_4^3)_r \end{bmatrix} = \begin{bmatrix} +59.52 \\ +46.01 \\ -133.97 \\ -59.52 \\ -46.01 \\ -142.11 \end{bmatrix}$$

P5.2. The rigidly jointed plane frame shown in Fig. P5.2 is fixed at base A and pinned at base D. The properties of the members of the frame are as follows: member AB, $I_1 = 0.0002\ m^4$, $A_1 = 0.009\ m^2$, member BC, $I_2 = 0.0001\ m^4$, $A_2 = 0.005\ m^2$ and

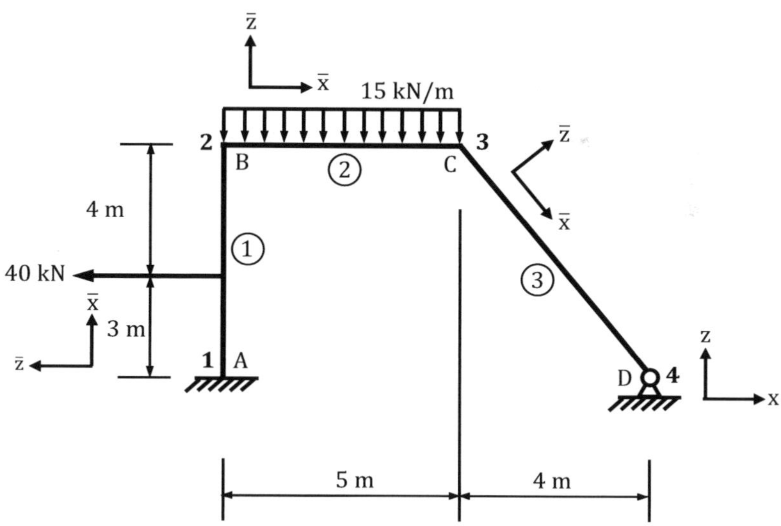

Figure P5.2

member CD, $I_3 = 0.00009\ m^4$, $A_3 = 0.004\ m^2$. The modulus of elasticity of all members, $E = 210 \times 10^6\ kN/m^2$. Analyse the frame for the loading shown and draw the axial force, shear force, and bending moment diagrams. In this problem use the standard stiffness matrix for member 3 and not the stiffness matrix for a member with a pin at node j. The treatment of the hinged base at node 4 is similar to that followed in example 1.

Answer:

$u_1 = 0$, $w_1 = 0$, $\theta_1 = 0$, $u_2 = -0.01530$ m, $w_2 = -0.00020$ m, $\theta_2 = +0.00072$ rad,

$u_3 = -0.01535$ m, $w_3 = -0.00904$ m, $\theta_3 = -0.00035$ rad, $u_4 = 0$, $w_4 = 0$, $\theta_4 = -0.00314$ rad.

$R_{X1} = +50.45$ kN, $R_{Z1} = +53.44$ kN, $R_{M1} = +126.53$ kNm

$R_{X4} = -10.45$ kN, $R_{Z4} = +21.56$ kN, $R_{M4} = 0$

$$\text{Member 1:}\quad \bar{F}_r^1 = \begin{bmatrix} (\bar{X}_1^1)_r \\ (\bar{Z}_1^1)_r \\ (\bar{M}_1^1)_r \\ (\bar{X}_2^1)_r \\ (\bar{Z}_2^1)_r \\ (\bar{M}_2^1)_r \end{bmatrix} = \begin{bmatrix} +53.44 \\ -50.45 \\ +126.53 \\ -53.44 \\ +10.45 \\ +66.61 \end{bmatrix},$$

$$\text{Member 2:}\quad \bar{F}_r^2 = \begin{bmatrix} (\bar{X}_2^2)_r \\ (\bar{Z}_2^2)_r \\ (\bar{M}_2^2)_r \\ (\bar{X}_3^2)_r \\ (\bar{Z}_3^2)_r \\ (\bar{M}_3^2)_r \end{bmatrix} = \begin{bmatrix} +10.45 \\ +53.44 \\ -66.61 \\ -10.45 \\ +21.56 \\ -13.09 \end{bmatrix},$$

$$\text{Member 3:}\quad \bar{F}_r^3 = \begin{bmatrix} (\bar{X}_3^3)_r \\ (\bar{Z}_3^3)_r \\ (\bar{M}_3^3)_r \\ (\bar{X}_4^3)_r \\ (\bar{Z}_4^3)_r \\ (\bar{M}_4^3)_r \end{bmatrix} = \begin{bmatrix} +23.90 \\ -1.62 \\ +13.09 \\ -23.90 \\ +1.62 \\ 0 \end{bmatrix}$$

P5.3. The frame shown in Fig. P5.3 is fixed to the supports at A and D. Members BC and CD are rigidly connected together at joint C while members AB and BC are pinned at joint B. Analyse the frame for the loading show and draw the axial force, shear force, and bending moment diagrams. The modulus of elasticity of frame is $E = 210 \times 10^6$ kN/m^2 and all the members have the same cross-sectional area A = 0.008 m^2 and second moment of area I = 0.0003 m^4.

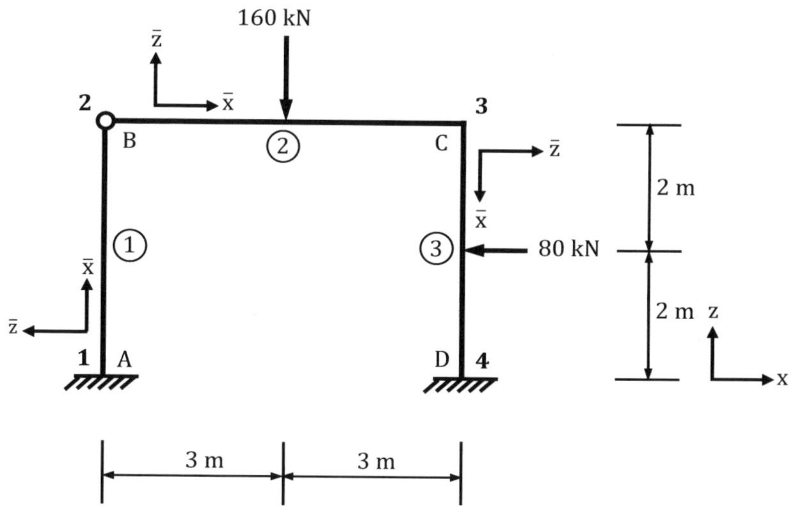

Figure P5.3

Answer:

$u_1 = 0$, $w_1 = 0$, $\theta_1 = 0$, $u_2 = -0.00840$ m, $w_2 = -0.00016$ m, $u_3 = -0.00849$ m,

$w_3 = -0.00022$ m, $\theta_3 = -0.00360$ rad, $u_4 = 0$, $w_4 = 0$, $\theta_4 = 0$.

$R_{X1} = +24.80$ kN, $R_{Z1} = +68.95$ kN, $R_{M1} = +99.21$ kNm,

$R_{X4} = +55.20$ kN, $R_{Z4} = +91.05$ kN, $R_{M4} = +127.10$ kNm.

$$\text{Member 1:}\quad \bar{F}_r^1 = \begin{bmatrix} (\bar{X}_1^1)_r \\ (\bar{Z}_1^1)_r \\ (\bar{M}_1^1)_r \\ (\bar{X}_2^1)_r \\ (\bar{Z}_2^1)_r \\ (\bar{M}_2^1)_r \end{bmatrix} = \begin{bmatrix} +68.95 \\ -24.80 \\ +99.21 \\ -68.95 \\ +24.80 \\ 0 \end{bmatrix},$$

$$\text{Member 2:}\quad \bar{F}_r^2 = \begin{bmatrix} (\bar{X}_2^2)_r \\ (\bar{Z}_2^2)_r \\ (\bar{M}_2^2)_r \\ (\bar{X}_3^2)_r \\ (\bar{Z}_3^2)_r \\ (\bar{M}_3^2)_r \end{bmatrix} = \begin{bmatrix} +24.80 \\ +68.95 \\ 0 \\ -24.80 \\ +91.05 \\ +66.31 \end{bmatrix},$$

$$\text{Member 3:}\quad \bar{F}_r^3 = \begin{bmatrix} (\bar{X}_3^3)_r \\ (\bar{Z}_3^3)_r \\ (\bar{M}_3^3)_r \\ (\bar{X}_4^3)_r \\ (\bar{Z}_4^3)_r \\ (\bar{M}_4^3)_r \end{bmatrix} = \begin{bmatrix} +91.05 \\ +24.80 \\ -66.31 \\ -91.05 \\ +55.20 \\ +127.10 \end{bmatrix}$$

P5.4. The frame shown in Fig. P5.4 is pinned to the support at A is fixed to the supports at C and D. Members AB and BC are rigidly connected together at joint B while member BD is pinned to joint B. Analyse the frame for the loading shown

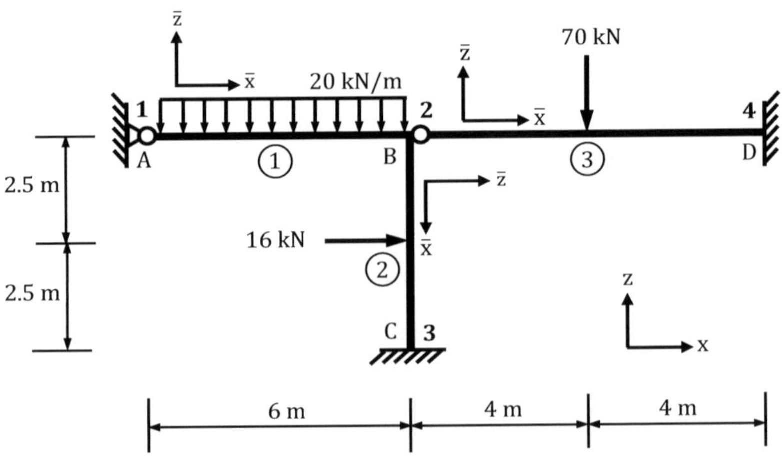

Figure P5.4

and draw the axial force, shear force, and bending moment diagrams for the following data: $E = 210 \times 10^6$ kN/m^2, $I_1 = 80 \times 10^{-6}$ m^4, $A_1 = 0.005$ m^2, $I_2 = 70 \times 10^{-6}$ m^4, $A_2 = 0.004$ m^2, $I_3 = 90 \times 10^{-6}$ m^4, and $A_3 = 0.006$ m^2. The treatment of the hinged support at node 1 is similar to that followed in example 1.

Answer:

$u_1 = 0$, $w_1 = 0$, $\theta_1 = +0.00796$ rad, $u_2 = -0.00003$ m, $w_2 = -0.00053$ m, θ_2 (for the rigid part of joint 2) $= -0.00493$ rad.

$u_3 = 0$, $w_3 = 0$, $\theta_3 = 0$, $u_4 = 0$, $w_4 = 0$, $\theta_4 = 0$

$R_{X1} = +4.92$ kN, $R_{Z1} = +52.02$ kN, $R_{M1} = 0$,

$R_{X3} = -25.35$ kN, $R_{Z3} = +89.79$ kN, $R_{M3} = -38.88$ kNm,

$R_{X4} = +4.43$ kN, $R_{Z4} = +48.18$ kN, $R_{M4} = +105.47$ kNm.

$$\text{Member 1: } \bar{F}_r^1 = \begin{bmatrix} (\bar{X}_1^1)_r \\ (\bar{Z}_1^1)_r \\ (\bar{M}_1^1)_r \\ (\bar{X}_2^1)_r \\ (\bar{Z}_2^1)_r \\ (\bar{M}_2^1)_r \end{bmatrix} = \begin{bmatrix} +4.92 \\ +52.02 \\ 0 \\ -4.92 \\ +67.98 \\ +47.86 \end{bmatrix},$$

$$\text{Member 2: } \bar{F}_r^2 = \begin{bmatrix} (\bar{X}_2^2)_r \\ (\bar{Z}_2^2)_r \\ (\bar{M}_2^2)_r \\ (\bar{X}_3^2)_r \\ (\bar{Z}_3^2)_r \\ (\bar{M}_3^2)_r \end{bmatrix} = \begin{bmatrix} +89.79 \\ +9.35 \\ -47.86 \\ -89.79 \\ -25.35 \\ -38.88 \end{bmatrix},$$

$$\text{Member 3: } \bar{F}_r^3 = \begin{bmatrix} (\bar{X}_2^3)_r \\ (\bar{Z}_2^3)_r \\ (\bar{M}_2^3) \\ (\bar{X}_4^3)_r \\ (\bar{Z}_4^3)_r \\ (\bar{M}_4^3)_r \end{bmatrix} = \begin{bmatrix} -4.43 \\ +21.82 \\ 0 \\ +4.43 \\ +48.18 \\ +105.47 \end{bmatrix}$$

Chapter 6

Arches

These are some of the most efficient forms of structures particularly for long spans. The horizontal components of the reactions at the supporting foundations produce bending moments that act in the opposite direction to the simple span moments. A theoretical case arises when the arch is in the shape of a parabola and the applied load is uniformly distributed along the whole span then the arch will be in pure compression with no bending moment at all sections. This makes the arch ideal when using a brittle material such as brickwork that is strong in compression but weak in tension. When the arch lies in the xz plane its treatment is similar to that of rigidly connected plane frame. In this chapter, only circular arches will be considered since the treatment of parabolic and elliptic arches leads to very involved expressions due to the additional parameters which define their geometry.

In the analysis of circular arches, the stiffness matrix for an element is derived relative to its local coordinates first and the overall structure stiffness matrix is then assembled relative to global coordinates in the usual way. Obviously the various elements of the arch have different orientations thus the stiffness matrices relative to local coordinates have to be transformed into global coordinates.

Analysis of Structures by Matrix Methods
Fathi Al-Shawi

ISBN 978-981-4968-19-5 (Hardcover), 978-1-003-29130-5 (eBook)
www.jennystanford.com

6.1 Derivation of Stiffness Matrix

Consider an element of the arch subtending an angle β with the centre and its local axis, $\bar{x}$-axis, is defined by the line joining the end points, i and j, of the element with the $\bar{z}$-axis being at right angles to it as shown in Fig. 6.1.

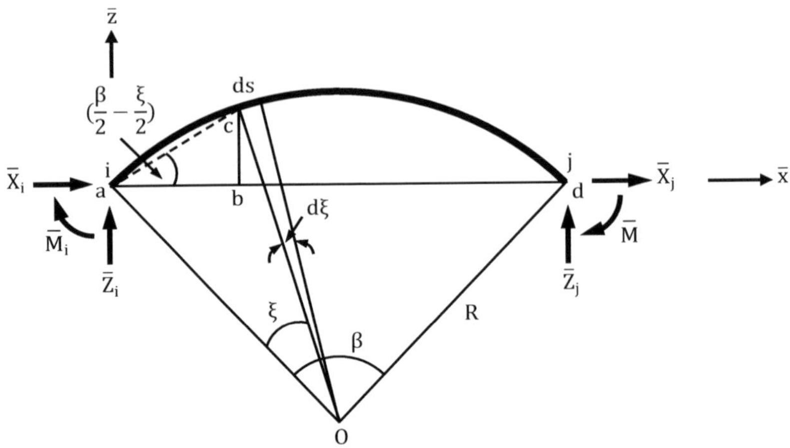

Figure 6.1 A circular arch element.

The derivation of the stiffness matrix is in two parts; in the first part, end i is given displacements $\bar{u}_i$, $\bar{w}_i$, and $\bar{\theta}_i$ while end j is fixed. Application of Castigliano's theorem will give the actions $\bar{X}_i'$, $\bar{Z}_i'$, and $\bar{M}_i'$ and from the equilibrium of the whole element the actions at end j $\bar{X}_j'$, $\bar{Z}_j'$, and $\bar{M}_j'$ can be found. In the second part, end i is fixed and end j is given displacements $\bar{u}_j$, $\bar{w}_j$, and $\bar{\theta}_j$ resulting in the actions $\bar{X}_j''$, $\bar{Z}_j''$, and $\bar{M}_j''$ and the actions at end i $\bar{X}_i''$, $\bar{Z}_i''$, and $\bar{M}_i''$ are found from equilibrium of the whole element. The final actions at ends i and j are determined from the algebraic sum of the two sets.

Consider first the case where end i is given displacements $\bar{u}_i$, $\bar{w}_i$, and $\bar{\theta}_i$ while end j is fixed.

In Fig. 6.2, the internal actions, P', V', and M' at the cut section which is making an angle ξ are drawn in their respective positive directions since they are acting at the right end of the left part of the member.

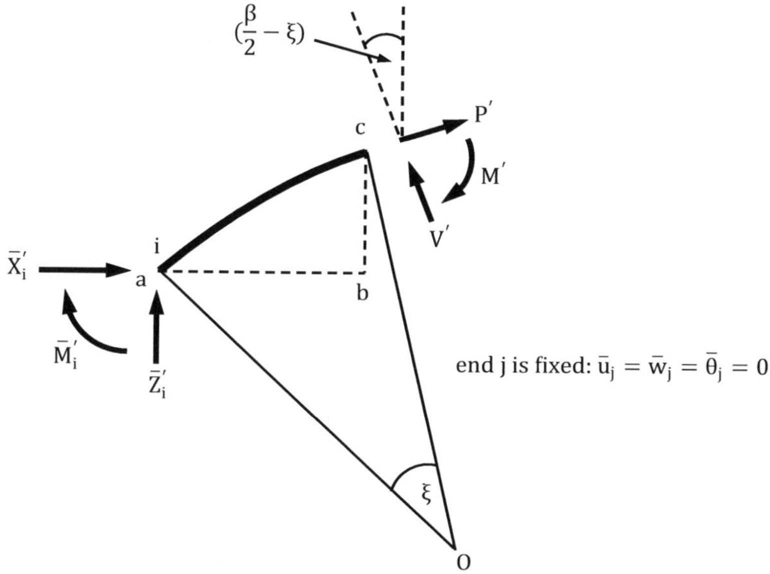

Figure 6.2

Summation of the forces in the $\bar{x}$-direction is zero:

$$\bar{X}_i' + P'\cos\left(\frac{\beta}{2}-\xi\right) - V'\sin\left(\frac{\beta}{2}-\xi\right) = 0 \tag{6.1}$$

Summation of the forces in the $\bar{z}$-direction is zero:

$$\bar{Z}_i' + P'\sin\left(\frac{\beta}{2}-\xi\right) + V'\cos\left(\frac{\beta}{2}-\xi\right) = 0 \tag{6.2}$$

Summation of the moments about the cut section, c, is zero:

$$\bar{M}_i' - \bar{X}_i'(bc) + \bar{Z}_i'(ab) + M' = 0 \tag{6.3}$$

where (from Fig. 6.1)

$$(ab) = (ac)\cos\left(\frac{\beta}{2}-\frac{\xi}{2}\right), (bc) = (ac)\sin\left(\frac{\beta}{2}-\frac{\xi}{2}\right), \text{ and } (ac) = 2R\sin\left(\frac{\xi}{2}\right)$$

Solving (6.1) and (6.2) simultaneously to get

$$P' = -\bar{X}_i'\cos\left(\frac{\beta}{2}-\xi\right) - \bar{Z}_i'\sin\left(\frac{\beta}{2}-\xi\right) \tag{6.4}$$

$$V' = \bar{X}_i'\sin\left(\frac{\beta}{2}-\xi\right) - \bar{Z}_i'\cos\left(\frac{\beta}{2}-\xi\right) \tag{6.5}$$

From (6.3) and with the substitution of the values of (ab) and (bc) we get:

$$M' = -\bar{M}_i' + 2\bar{X}_i' R\sin\left(\frac{\xi}{2}\right)\sin\left(\frac{\beta}{2}-\frac{\xi}{2}\right) - 2\bar{Z}_i' R\sin\left(\frac{\xi}{2}\right)\cos\left(\frac{\beta}{2}-\frac{\xi}{2}\right) \quad (6.6)$$

Neglect the effect of shear force on deformations and apply Castigliano's theorem as follows:

$$\text{The strain energy, } U' = \int \frac{M'^2 ds}{2EI} + \int \frac{P'^2 ds}{2EA}$$

where the length of a small arc, ds = Rdξ and integrating from 0 to β to get

$$U' = \frac{R}{2EI}\int_0^\beta M'^2 d\xi + \frac{R}{2EA}\int_0^\beta P'^2 d\xi \quad (6.7)$$

$$\bar{u}_i = \frac{\partial U'}{\partial \bar{X}_i'} = \frac{\partial U'}{\partial M'}\frac{\partial M'}{\partial \bar{X}_i'} + \frac{\partial U'}{\partial P'}\frac{\partial P'}{\partial \bar{X}_i'} \quad (6.8)$$

$$\bar{w}_i = \frac{\partial U'}{\partial \bar{Z}_i'} = \frac{\partial U'}{\partial M'}\frac{\partial M'}{\partial \bar{Z}_i'} + \frac{\partial U'}{\partial P'}\frac{\partial P'}{\partial \bar{Z}_i'} \quad (6.9)$$

$$\bar{\theta}_i = \frac{\partial U'}{\partial \bar{M}_i'} = \frac{\partial U'}{\partial M'}\frac{\partial M'}{\partial \bar{M}_i'} + \frac{\partial U'}{\partial P'}\frac{\partial P'}{\partial \bar{M}_i'} \quad (6.10)$$

From (6.7) we get

$$\frac{\partial U'}{\partial M'} = \frac{R}{EI}\int_0^\beta M' d\xi \quad \text{and} \quad \frac{\partial U'}{\partial P'} = \frac{R}{EA}\int_0^\beta P' d\xi$$

From equation (6.4) we get:

$$\frac{\partial P'}{\partial \bar{X}_i'} = -\cos\left(\frac{\beta}{2}-\xi\right), \quad \frac{\partial P'}{\partial \bar{Z}_i'} = -\sin\left(\frac{\beta}{2}-\xi\right), \quad \text{and} \quad \frac{\partial P'}{\partial \bar{M}_i'} = 0$$

From equation (6.6) we get:

$$\frac{\partial M'}{\partial \bar{X}_i'} = 2R\sin\left(\frac{\xi}{2}\right)\sin\left(\frac{\beta}{2}-\frac{\xi}{2}\right), \quad \frac{\partial M'}{\partial \bar{Z}_i'} = -2R\sin\left(\frac{\xi}{2}\right)\cos\left(\frac{\beta}{2}-\frac{\xi}{2}\right), \quad \text{and} \quad \frac{\partial M'}{\partial \bar{M}_i'} = -1$$

Substituting (6.4), (6.6), and the relevant derivatives, as appropriate, from above into (6.8) will give

$$\bar{u}_i = \frac{R}{EI}\int_0^{\beta}\left[-\bar{M}_i' + 2\bar{X}_i' R\sin\left(\frac{\xi}{2}\right)\sin\left(\frac{\beta}{2}-\frac{\xi}{2}\right) - 2\bar{Z}_i' R\sin\left(\frac{\xi}{2}\right)\cos\left(\frac{\beta}{2}-\frac{\xi}{2}\right)\right]2R\sin\left(\frac{\xi}{2}\right)$$

$$\sin\left(\frac{\beta}{2}-\frac{\xi}{2}\right)d\xi + \frac{R}{EA}\int_0^{\beta}\left[-\bar{X}_i'\cos\left(\frac{\beta}{2}-\xi\right) - \bar{Z}_i'\sin\left(\frac{\beta}{2}-\xi\right)\right]\left[-\cos\left(\frac{\beta}{2}-\xi\right)\right]d\xi$$

Introducing the parameter $\alpha = I/AR^2$ and integrating from $\xi = 0$ to $\xi = \beta$ to get

$$\bar{u}_i = \frac{R^2}{EI}\Big\{0.5\left[\beta\cos\beta + (\alpha-3)\sin\beta + \beta(\alpha+2)\right]R\bar{X}_i' + \left[\cos\beta + 0.5\beta\sin\beta - 1\right]$$
$$R\bar{Z}_i' + \left[\beta\cos(\beta/2) - 2\sin(\beta/2)\right]\bar{M}_i'\Big\} \tag{6.11}$$

Similarly (6.9) is simplified to give

$$\bar{w}_i = \frac{R^2}{EI}\Big\{\left[\cos\beta + 0.5\beta\sin\beta - 1\right]R\bar{X}_i' - 0.5\left[\beta\cos\beta + (\alpha+1)\sin\beta - \beta(\alpha+2)\right]$$
$$R\bar{Z}_i' + \left[\beta\sin(\beta/2)\right]\bar{M}_i'\Big\} \tag{6.12}$$

And (6.10) is simplified to

$$\bar{\theta}_i = \frac{R}{EI}\left\{\left[\beta\cos(\beta/2) - 2\sin(\beta/2)\right]R\bar{X}_i' + \beta\sin(\beta/2)R\bar{Z}_i' + \beta\bar{M}_i'\right\} \tag{6.13}$$

Solve equations (6.11), (6.12), and (6.13) simultaneously for the unknowns $\bar{X}_i'$, $\bar{Z}_i'$, and $\bar{M}_i'$ to get:

$$\bar{X}_i' = \frac{EI}{R^3}\left(C_1\bar{u}_i + C_2R\bar{\theta}_i\right) \tag{6.14}$$

$$\bar{Z}_i' = \frac{EI}{R^3}\left(C_3\bar{w}_i - C_4R\bar{\theta}_i\right) \tag{6.15}$$

$$\bar{M}_i' = \frac{EI}{R^3}\left(C_2R\bar{u}_i - C_4R\bar{w}_i + C_5R^2\bar{\theta}_i\right) \tag{6.16}$$

Equations (6.14), (6.15), and (6.16) represent $\bar{k}_{ii}$ of the stiffness matrix.

From the overall equilibrium of the arch element the following equations are obtained:

Summation of the forces in the $\bar{x}$-direction:

$$\bar{X}_i' + \bar{X}_j' = 0, \qquad \bar{X}_j' = -\bar{X}_i', \text{ hence}$$

$$\bar{X}_j' = \frac{EI}{R^3}\left(-C_1\bar{u}_i - C_2R\bar{\theta}_i\right) \tag{6.17}$$

Summation of the forces in the $\bar{z}$-direction:

$$\bar{Z}_i' + \bar{Z}_j' = 0, \quad \bar{Z}_j' = -\bar{Z}_i', \text{ hence}$$

$$\bar{Z}_j' = \frac{EI}{R^3}\left(-C_3\bar{w}_i + C_4R\bar{\theta}_i\right) \tag{6.18}$$

Summation of the moments about end j:

$\bar{M}_i' + \bar{Z}_i'L + \bar{M}_j' = 0$ [The span of the arch element, L = 2Rsin(β/2)]

$$\bar{M}_j' = -\bar{M}_i' - 2R\bar{Z}_i'\sin(\beta/2)$$

Substitute for $\bar{Z}_i'$ and $\bar{M}_i'$ from (6.12) and (6.13) respectively we get

$$\bar{M}_j' = \frac{EI}{R^3}\left(-C_2R\bar{u}_i - C_4R\bar{w}_i - C_6R^2\bar{\theta}_i\right) \tag{6.19}$$

Equations (6.17), (6.18), and (6.19) represent $\bar{k}_{ji}$ of the stiffness matrix.

The above process is repeated with end i fixed and end j is given displacements $\bar{u}_j$, $\bar{w}_j$, and $\bar{\theta}_j$.

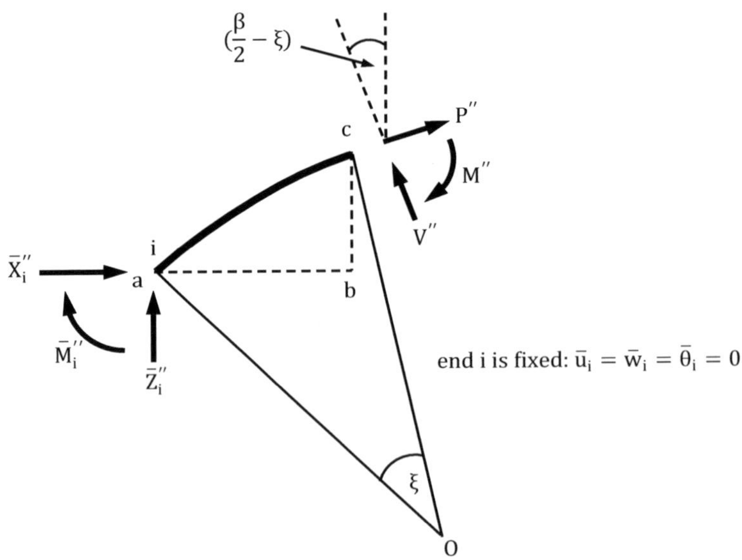

Figure 6.3

The equations for the axial force and moment at section c in this case are similar to (6.4) and (6.6) but the single primes are replaced by double primes as shown in Fig. 6.3, thus

$$P'' = -\bar{X}_i'' \cos\left(\frac{\beta}{2}-\xi\right) - \bar{Z}_i'' \sin\left(\frac{\beta}{2}-\xi\right) \tag{6.20}$$

$$M'' = -\bar{M}_i'' + 2\bar{X}_i'' R\sin\left(\frac{\xi}{2}\right)\sin\left(\frac{\beta}{2}-\frac{\xi}{2}\right) - 2\bar{Z}_i'' R\sin\left(\frac{\xi}{2}\right)\cos\left(\frac{\beta}{2}-\frac{\xi}{2}\right) \tag{6.21}$$

Since we want to find expressions for the displacements $\bar{u}_j$, $\bar{w}_j$, and $\bar{\theta}_j$ the above two equations are written in terms of $\bar{X}_j''$, $\bar{Z}_j''$, and $\bar{M}_j''$ whose derivatives will give the respective displacements. To achieve this, the equilibrium of the whole arch is considered.

Summation of the forces in the $\bar{x}$-direction is zero:

$$\bar{X}_i'' + \bar{X}_j'' = 0, \quad \bar{X}_i'' = -\bar{X}_j''$$

Summation of the forces in the $\bar{z}$-direction is zero:

$$\bar{Z}_i'' + \bar{Z}_j'' = 0, \quad \bar{Z}_i'' = -\bar{Z}_j''$$

Summation of the moments about node i is zero:

$$\bar{M}_i'' + \bar{M}_j'' - \bar{Z}_j'' L = 0, \text{ where } L = 2R\sin(\beta/2)$$

$$\bar{M}_i'' = -\bar{M}_j'' + 2\bar{Z}_j'' R\sin(\beta/2)$$

Substitute the above values of $\bar{X}_i''$, $\bar{Z}_i''$, and $\bar{M}_i''$ in (6.17) and (6.18) respectively to get

$$P'' = \bar{X}_j'' \cos\left(\frac{\beta}{2}-\xi\right) + \bar{Z}_j'' \sin\left(\frac{\beta}{2}-\xi\right) \tag{6.22}$$

$$M'' = \bar{M}_j'' - 2\bar{X}_j'' R\sin\left(\frac{\xi}{2}\right)\sin\left(\frac{\beta}{2}-\frac{\xi}{2}\right) - 2\bar{Z}_j'' R\left[\sin\left(\frac{\beta}{2}\right) - \sin\left(\frac{\xi}{2}\right)\cos\left(\frac{\beta}{2}-\frac{\xi}{2}\right)\right] \tag{6.23}$$

$$U'' = \frac{R}{2EI}\int_0^\beta M''^2 d\xi + \frac{R}{2EA}\int_0^\beta P''^2 d\xi \tag{6.24}$$

$$\bar{u}_j = \frac{\partial U''}{\partial \bar{X}_j''} = \frac{\partial U''}{\partial M''}\frac{\partial M''}{\partial \bar{X}_j''} + \frac{\partial U''}{\partial P''}\frac{\partial P''}{\partial \bar{X}_j''} \tag{6.25}$$

$$\bar{w}_j = \frac{\partial U''}{\partial \bar{Z}_j''} = \frac{\partial U''}{\partial M''}\frac{\partial M''}{\partial \bar{Z}_j''} + \frac{\partial U''}{\partial P''}\frac{\partial P''}{\partial \bar{Z}_j''} \tag{6.26}$$

$$\bar{\theta}_j = \frac{\partial U''}{\partial \bar{M}_j''} = \frac{\partial U''}{\partial M''}\frac{\partial M''}{\partial \bar{M}_j''} + \frac{\partial U''}{\partial P''}\frac{\partial P''}{\partial \bar{M}_j''} \tag{6.27}$$

Equations (6.25), (6.26), and (6.27) are simplified and integrated from $\xi = 0$ to $\xi = \beta$ to give:

$$\bar{u}_j = \frac{R^2}{EI}\Big\{0.5R\big[\beta\cos\beta + (\alpha-3)\sin\beta + \beta(\alpha+2)\big]\bar{X}_j'' + R\big[1-\cos\beta-0.5\beta\sin\beta\big]$$
$$\bar{Z}_j'' + \big[\beta\cos(\beta/2) - 2\sin(\beta/2)\big]\bar{M}_j''\Big\} \tag{6.28}$$

$$\bar{w}_j = \frac{R^2}{EI}\Big\{R\big[1-\cos\beta-0.5\beta\sin\beta\big]\bar{X}_j'' - 0.5R\big[\beta\cos\beta + (\alpha+1)\sin\beta - \beta(\alpha+2)\big]$$
$$\bar{Z}_j'' - \big[\beta\sin(\beta/2)\big]\bar{M}_j''\Big\} \tag{6.29}$$

$$\bar{\theta}_j = \frac{R}{EI}\Big\{R\big[\beta\cos(\beta/2) - 2\sin(\beta/2)\big]\bar{X}_j'' - \big[R\beta\sin(\beta/2)\big]\bar{Z}_j'' + \beta\bar{M}_j''\Big\} \tag{6.30}$$

Solve equations (6.28), (6.29), and (6.30) simultaneously for the unknowns $\bar{X}_j''$, $\bar{Z}_j''$, and $\bar{M}_j''$ to get:

$$\bar{X}_j'' = \frac{EI}{R^3}\left(C_1\bar{u}_j + C_2R\bar{\theta}_j\right) \tag{6.31}$$

$$\bar{Z}_j'' = \frac{EI}{R^3}\left(C_3\bar{w}_j + C_4R\bar{\theta}_j\right) \tag{6.32}$$

$$M_j'' = \frac{EI}{R^3}\left(C_2R\bar{u}_j + C_4R\bar{w}_j + C_5R^2\bar{\theta}_j\right) \tag{6.33}$$

Equations (6.31), (6.32), and (6.33) represent $\bar{k}_{jj}$ of the stiffness matrix.

From the overall equilibrium of the arch

$$\bar{X}_i'' + \bar{X}_j'' = 0, \quad \bar{X}_i'' = -\bar{X}_j''$$

$$\bar{X}_i'' = \frac{EI}{R^3}\left(-C_1\bar{u}_j - C_2R\bar{\theta}_j\right) \tag{6.34}$$

$$\bar{Z}_i'' + \bar{Z}_j'' = 0, \quad \bar{Z}_i'' = -\bar{Z}_j''$$

$$\bar{Z}_i'' = \frac{EI}{R^3}\left(-C_3\bar{w}_j - C_4R\bar{\theta}_j\right) \tag{6.35}$$

Taking moments about end i of the arch

$$\bar{M}_i'' + \bar{M}_j'' - \bar{Z}_j''L = 0, \quad \bar{M}_i'' = -\bar{M}_j'' + 2R\bar{Z}_j''\sin(\beta/2)$$

Substitute for $\bar{Z}_j''$ and $\bar{M}_j''$ from (6.27) and (6.28) respectively to get

$$M_i'' = \frac{EI}{R^3}\left(-C_2R\bar{u}_j + C_4R\bar{w}_j - C_6R^2\bar{\theta}_j\right) \tag{6.36}$$

Equations (6.34), (6.35), and (6.36) represent $\bar{k}_{ij}$ of the stiffness matrix.

Finally, adding quantities with single prime to the corresponding quantities with double primes to get the resultant values of the end forces in terms of the end displacements as shown below.

From (6.14) and (6.34): $\bar{X}_i = \bar{X}_i' + \bar{X}_i''$

From (6.15) and (6.35): $\bar{Z}_i = \bar{Z}_i' + \bar{Z}_i''$

From (6.16) and (6.36): $\bar{M}_i = \bar{M}_i' + \bar{M}_i''$

From (6.17) and (6.31): $\bar{X}_j = \bar{X}_j' + \bar{X}_j''$

From (6.18) and (6.32): $\bar{Z}_j = \bar{Z}_j' + \bar{Z}_j''$

From (6.19) and (6.33): $\bar{M}_j = \bar{M}_j' + \bar{M}_j''$

The above six equations are written in matrix form to give the general stiffness matrix of a circular arch element relative to local coordinates as:

$$\begin{bmatrix} \bar{X}_i \\ \bar{Z}_i \\ \bar{M}_i \\ \bar{X}_j \\ \bar{Z}_j \\ \bar{M}_j \end{bmatrix} = \frac{EI}{R^3}\begin{bmatrix} C_1 & 0 & C_2R & -C_1 & 0 & -C_2R \\ 0 & C_3 & -C_4R & 0 & -C_3 & -C_4R \\ C_2R & -C_4R & C_5R^2 & -C_2R & C_4R & -C_6R^2 \\ -C_1 & 0 & -C_2R & C_1 & 0 & C_2R \\ 0 & -C_3 & C_4R & 0 & C_3 & C_4R \\ -C_2R & -C_4R & C_6R^2 & C_2R & C_4R & C_5R^2 \end{bmatrix}\begin{bmatrix} \bar{u}_i \\ \bar{w}_i \\ \bar{\theta}_i \\ \bar{u}_j \\ \bar{w}_j \\ \bar{\theta}_j \end{bmatrix} \tag{6.37}$$

where $C_1, C_2, \ldots\ldots\ldots, C_6$ are functions of α and β given by

$$C_1 = \frac{a_1}{a_2\alpha+1}, \quad C_2 = \frac{a_3}{a_2\alpha+1}, \quad C_3 = \frac{a_4}{\alpha+1}, \quad C_4 = \frac{a_5}{\alpha+1},$$

$$C_5 = \frac{a_6\alpha^2 + a_7\alpha + a_8}{(\alpha+1)(a_2\alpha+1)}, \quad C_6 = \frac{a_6\alpha^2 - a_9\alpha + a_{10}}{(\alpha+1)(a_2\alpha+1)}, \quad \text{and} \quad \alpha = \frac{I}{AR^2}$$

The above expressions are derived from detailed calculations for the derivation of the stiffness matrix and the values of a_1 to a_{10} are functions of β which have been calculated for different values of β and are given in Table 6.1.

Table 6.1

β	a_1	a_2	a_3	a_4	a_5	a_6	a_7	a_8	a_9	a_{10}
15⁰	587139	152837	3348	671	87.6	583797	2331211	34.4	1163570	11.5
30⁰	18475	9456	419	84.8	21.9	18059	71752	17.1	35611	5.75
45⁰	2463	1838	125	25.6	9.78	2340	9224	11.3	4528	3.84
60⁰	595	569	52.9	11.0	5.52	543	2119	8.41	1021	2.90
75⁰	199	227	27.2	5.83	3.55	173	667	6.65	312	2.33
90⁰	82.3	106	15.9	3.50	2.48	67.3	256	5.46	114	1.96

The stiffness matrix in (6.37) is relative to the local coordinates and for the assembly of the overall stiffness matrix it is required to be written relative to global coordinates. The transformation of the arch element is similar to that of the rigidly connected plane frames discussed in the previous chapter.

6.2 Transformation of Coordinates

Consider an element of the arch whose local $\bar{x}$-axis lies initially along the global x-axis and then it is rotated about the $\bar{y}$-axis by an angle $\varphi_{\bar{y}}$ as shown in Fig. 6.4.

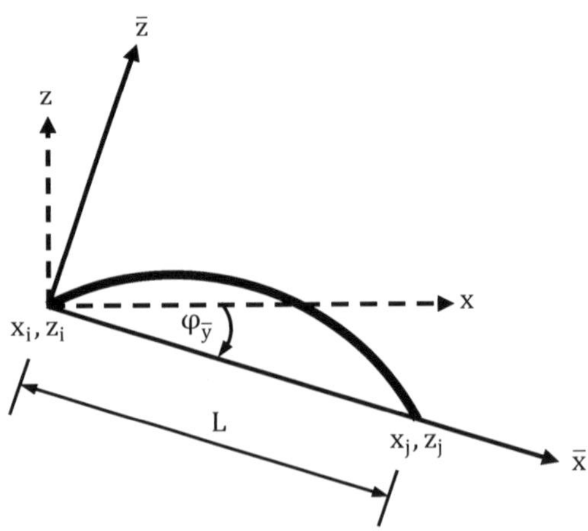

Figure 6.4

Since the element lies in the xz plane, its transformation matrix, r, is the same as that for a straight member in a rigidly connected frame which was derived in chapter five as given by equation (5.6) but with the length L being the straight distance between nodes i and j, thus:

$$r = \begin{bmatrix} x_{ij}/L & z_{ij}/L & 0 & 0 & 0 & 0 \\ -z_{ij}/L & x_{ij}/L & 0 & 0 & 0 & 0 \\ 0 & 0 & 1 & 0 & 0 & 0 \\ 0 & 0 & 0 & x_{ij}/L & z_{ij}/L & 0 \\ 0 & 0 & 0 & -z_{ij}/L & x_{ij}/L & 0 \\ 0 & 0 & 0 & 0 & 0 & 1 \end{bmatrix} \qquad (6.38)$$

where $x_{ij} = x_j - x_i$, $z_{ij} = z_j - z_i$, and $L = \sqrt{x_{ij}^2 + z_{ij}^2}$

and the stiffness matrix relative to global coordinates is $k = r^T \bar{k} r$.

6.3 Calculation of Actions Developed in the Elements

For an arch element lying in the xz plane, the actions that are of interest are the axial force P, shear force V, and the bending moment m rather than the actions relative to the local coordinates as was explained in the previous chapters. A convenient way is to find $\bar{X}$, $\bar{Z}$, and $\bar{M}$ relative to local coordinates and then transformed them into P, V, and m as shown in Fig. 6.5.

With reference to Fig. 6.5 and by resolving the actions that are relative to local coordinates into their components we get

$$P_i = \bar{X}_i \cos(\beta/2) + \bar{Z}_i \sin(\beta/2)$$

$$V_i = -\bar{X}_i \sin(\beta/2) + \bar{Z}_i \cos(\beta/2)$$

The moment, $m_i = \bar{M}_i$ (since both moments are about the same y-axis)

Similarly

$$P_j = \bar{X}_j \cos(\beta/2) - \bar{Z}_j \sin(\beta/2)$$

$$V_j = \bar{X}_j \sin(\beta/2) + \bar{Z}_j \cos(\beta/2)$$

$$m_j = \bar{M}_j$$

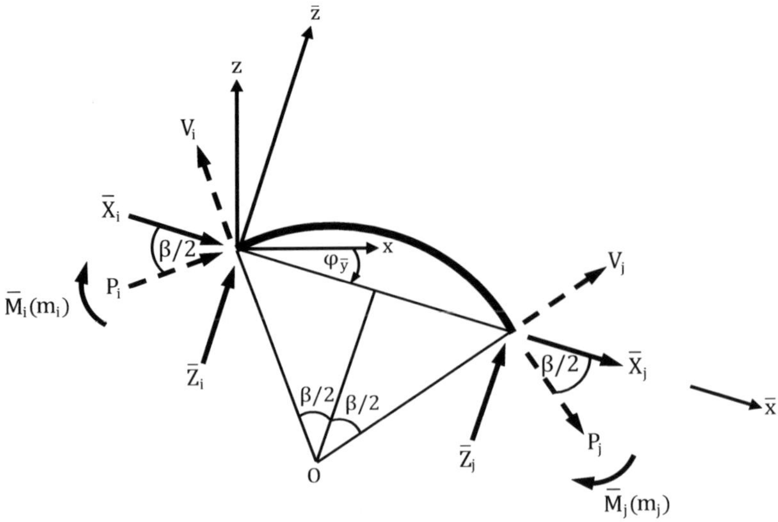

Figure 6.5

In matrix form

$$\begin{bmatrix} P_i \\ V_i \\ m_i \\ P_j \\ V_j \\ m_j \end{bmatrix} = \begin{bmatrix} \cos(\beta/2) & \sin(\beta/2) & 0 & 0 & 0 & 0 \\ -\sin(\beta/2) & \cos(\beta/2) & 0 & 0 & 0 & 0 \\ 0 & 0 & 1 & 0 & 0 & 0 \\ 0 & 0 & 0 & \cos(\beta/2) & -\sin(\beta/2) & 0 \\ 0 & 0 & 0 & \sin(\beta/2) & \cos(\beta/2) & 0 \\ 0 & 0 & 0 & 0 & 0 & 1 \end{bmatrix} \begin{bmatrix} \bar{X}_i \\ \bar{Z}_i \\ \bar{M}_i \\ \bar{X}_j \\ \bar{Z}_j \\ \bar{M}_j \end{bmatrix}$$

$$f = r_\beta \bar{F}$$

where

$$r_\beta = \begin{bmatrix} \cos(\beta/2) & \sin(\beta/2) & 0 & 0 & 0 & 0 \\ -\sin(\beta/2) & \cos(\beta/2) & 0 & 0 & 0 & 0 \\ 0 & 0 & 1 & 0 & 0 & 0 \\ 0 & 0 & 0 & \cos(\beta/2) & -\sin(\beta/2) & 0 \\ 0 & 0 & 0 & \sin(\beta/2) & \cos(\beta/2) & 0 \\ 0 & 0 & 0 & 0 & 0 & 1 \end{bmatrix} \qquad (6.39)$$

$$f = \begin{bmatrix} P_i \\ V_i \\ m_i \\ P_j \\ V_j \\ m_j \end{bmatrix} \quad \text{and} \quad \overline{F} = \begin{bmatrix} \overline{X}_i \\ \overline{Z}_i \\ \overline{M}_i \\ \overline{X}_j \\ \overline{Z}_j \\ \overline{M}_j \end{bmatrix}$$

But, $\overline{F} = \overline{k}\overline{\delta}$ and $\overline{\delta} = r\delta$, therefore

$$f = r_\beta \overline{k} r \delta \tag{6.40}$$

Example

Analyse the circular arch whose geometry and the forces acting on it as shown in Fig. 6.6 using the following data:

$E = 210 \times 10^6$ kN/m^2, $I = 0.00048$ m^4, $A = 0.0125$ m^2, and $R = 16$ m.

$$\alpha = \frac{I}{AR^2} = \frac{0.00048}{0.0125 \times 16^2} = 0.00015$$

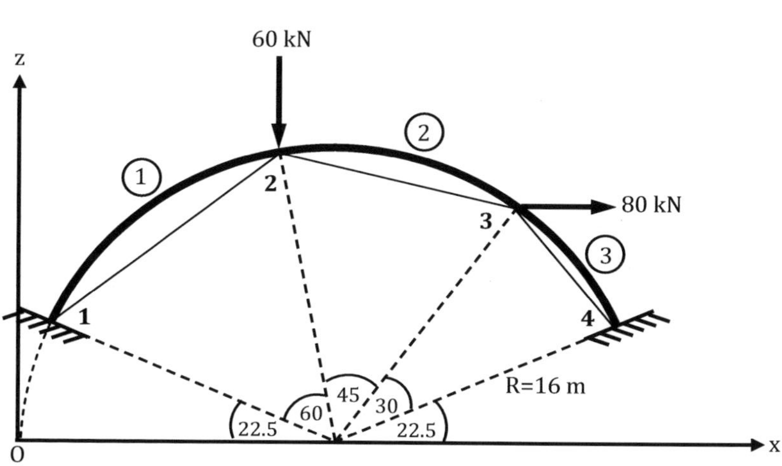

Figure 6.6

Coordinates of nodes (Taking point O as the origin)

Node 1:

$x_1 = R - R\cos(22.5°) = 16 - 16\cos(22.5°) = 1.22$ m

$z_1 = R\sin(22.5°) = 16\sin(22.5°) = 6.12$ m

Node 2:

$x_2 = R - R\cos(82.5°) = 16 - 16\cos(82.5°) = 13.91$ m

$z_2 = R\sin(82.5°) = 16\sin(82.5°) = 15.86$ m

Node 3:

$x_3 = R + R\cos(52.5°) = 16 + 16\cos(52.5°) = 25.74$ m

$z_3 = R\sin(52.5°) = 16\sin(52.5°) = 12.69$ m

Node 4:

$x_4 = R + R\cos(22.5°) = 16 + 16\cos(22.5°) = 30.78$ m

$z_4 = R\sin(22.5°) = 16\sin(22.5°) = 6.12$ m

Element 1 ($\beta = 60°$), from Table 6.1

$a_1 = 595$, $a_2 = 569$, $a_3 = 52.9$, $a_4 = 11.0$, $a_5 = 5.5$,

$a_6 = 543$, $a_7 = 2119$, $a_8 = 8.41$, $a_9 = 1021$, $a_{10} = 2.90$

$$C_1 = \frac{a_1}{a_2\alpha + 1} = \frac{595}{569 \times 0.00015 + 1} = 548, \; C_2 = \frac{a_3}{a_2\alpha + 1} = \frac{52.9}{569 \times 0.00015 + 1} = 48.7$$

$$C_3 = \frac{a_4}{\alpha + 1} = \frac{11.0}{0.00015 + 1} = 11.0, \; C_4 = \frac{a_5}{\alpha + 1} = \frac{5.52}{0.00015 + 1} = 5.52$$

$$C_5 = \frac{a_6\alpha^2 + a_7\alpha + a_8}{(\alpha+1)(a_2\alpha+1)} = \frac{543 \times 0.00015^2 + 2119 \times 0.00015 + 8.41}{(0.00015+1)(569 \times 0.00015 + 1)} = 8.04$$

$$C_6 = \frac{a_6\alpha^2 - a_9\alpha + a_{10}}{(\alpha+1)(a_2\alpha+1)} = \frac{543 \times 0.00015^2 - 1021 \times 0.00015 + 2.90}{(0.00015+1)(569 \times 0.00015 + 1)} = 2.53$$

From (6.37)

$$\bar{k}^1 = \frac{210 \times 10^6 \times 0.00048}{16^3} \begin{bmatrix} 548 & 0 & 48.7\times 16 & -548 & 0 & -48.7\times 16 \\ 0 & 11.0 & -5.52\times 16 & 0 & -11.0 & -5.52\times 16 \\ 48.7\times 16 & -5.52\times 16 & 8.04\times 16^2 & -48.7\times 16 & 5.52\times 16 & -2.53\times 16^2 \\ -548 & 0 & -48.7\times 16 & 548 & 0 & 48.7\times 16 \\ 0 & -11.0 & 5.52\times 16 & 0 & 11.0 & 5.52\times 16 \\ -48.7\times 16 & -5.52\times 16 & -2.53\times 16^2 & 48.7\times 16 & 5.52\times 16 & 8.04\times 16^2 \end{bmatrix}$$

$$\bar{k}^1 = \begin{bmatrix} 13486 & 0 & 19176 & -13486 & 0 & -19176 \\ 0 & 271 & -2174 & 0 & -271 & -2174 \\ 19176 & -2174 & 50652 & -19176 & 2174 & -15939 \\ -13486 & 0 & -19176 & 13486 & 0 & 19176 \\ 0 & -271 & 2174 & 0 & 271 & 2174 \\ -19176 & -2174 & -15939 & 19176 & 2174 & 50652 \end{bmatrix} \tag{6.41}$$

$x_i = x_1 = 1.22$ m, $x_j = x_2 = 13.91$ m,
$x_{ij} = x_j - x_i = 13.91 - 1.22 = 12.69$ m
$z_i = z_1 = 6.12$ m, $z_j = z_2 = 15.86$ m,
$z_{ij} = z_j - z_i = 15.86 - 6.12 = 9.74$ m

$$L = \sqrt{x_{ij}^2 + z_{ij}^2} = \sqrt{12.69^2 + 9.74^2} = 16.00 \text{ m}$$

$$\frac{x_{ij}}{L} = \frac{12.69}{16.00} = 0.793, \quad \frac{z_{ij}}{L} = \frac{9.74}{16.00} = 0.609$$

From (6.38)

$$r^1 = \begin{bmatrix} 0.793 & 0.609 & 0 & 0 & 0 & 0 \\ -0.609 & 0.793 & 0 & 0 & 0 & 0 \\ 0 & 0 & 1 & 0 & 0 & 0 \\ 0 & 0 & 0 & 0.793 & 0.609 & 0 \\ 0 & 0 & 0 & -0.609 & 0.793 & 0 \\ 0 & 0 & 0 & 0 & 0 & 1 \end{bmatrix} \tag{6.42}$$

$$k^1 = (r^1)^T \bar{k}^1 r^1$$

$$k^1 = \begin{bmatrix} 8581 & 6382 & 16531 & -8581 & -6382 & -13883 \\ 6382 & 5172 & 9954 & -6382 & -5172 & -13402 \\ 16531 & 9954 & 50652 & -16531 & -9954 & -15939 \\ -8581 & -6382 & -16531 & 8581 & 6382 & 13883 \\ -6382 & -5172 & -9954 & 6382 & 5172 & 13402 \\ -13883 & -13402 & -15939 & 13883 & 13402 & 50652 \end{bmatrix} \tag{6.43}$$

Element 2 ($\beta = 45°$), from Table 6.1

$a_1 = 2463$, $a_2 = 1838$, $a_3 = 125$, $a_4 = 25.6$, $a_5 = 9.78$,
$a_6 = 2340$, $a_7 = 9224$, $a_8 = 11.3$, $a_9 = 4528$, $a_{10} = 3.84$

$$C_1 = \frac{a_1}{a_2\alpha + 1} = \frac{2463}{1838 \times 0.00015 + 1} = 1931,$$

$$C_2 = \frac{a_3}{a_2\alpha + 1} = \frac{125}{1838 \times 0.00015 + 1} = 98.0$$

$$C_3 = \frac{a_4}{\alpha + 1} = \frac{25.6}{0.00015 + 1} = 25.6, \quad C_4 = \frac{a_5}{\alpha + 1} = \frac{9.78}{0.00015 + 1} = 9.78$$

$$C_5 = \frac{a_6\alpha^2 + a_7\alpha + a_8}{(\alpha + 1)(a_2\alpha + 1)} = \frac{2340 \times 0.00015^2 + 9224 \times 0.00015 + 11.3}{(0.00015 + 1)(1838 \times 0.00015 + 1)} = 9.94$$

$$C_6 = \frac{a_6\alpha^2 - a_9\alpha + a_{10}}{(\alpha + 1)(a_2\alpha + 1)} = \frac{2340 \times 0.00015^2 - 4528 \times 0.00015 + 3.84}{(0.00015 + 1)(1838 \times 0.00015 + 1)} = 2.48$$

$$\bar{k}^2 = \begin{bmatrix} 47521 & 0 & 38588 & -47521 & 0 & -38588 \\ 0 & 630 & -3851 & 0 & -630 & -3851 \\ 38588 & -3851 & 62622 & -38588 & 3851 & -15624 \\ -47521 & 0 & -38588 & 47521 & 0 & 38588 \\ 0 & -630 & 3851 & 0 & 630 & 3851 \\ -38588 & -3851 & -15624 & 38588 & 3851 & 62622 \end{bmatrix} \tag{6.44}$$

$x_i = x_2 = 13.91$ m, $x_j = x_3 = 25.74$ m,
$x_{ij} = x_j - x_i = 25.74 - 13.91 = 11.83$ m
$z_i = z_2 = 15.86$ m, $z_j = z_3 = 12.69$ m,
$z_{ij} = z_j - z_i = 12.69 - 15.86 = -3.17$ m

$$L = \sqrt{x_{ij}^2 + z_{ij}^2} = \sqrt{11.83^2 + (-3.17)^2} = 12.25\ \text{m}$$

$$\frac{x_{ij}}{L} = \frac{11.83}{12.25} = 0.966, \quad \frac{z_{ij}}{L} = \frac{-3.17}{12.25} = -0.259$$

$$r^2 = \begin{bmatrix} 0.966 & -0.259 & 0 & 0 & 0 & 0 \\ 0.259 & 0.966 & 0 & 0 & 0 & 0 \\ 0 & 0 & 1 & 0 & 0 & 0 \\ 0 & 0 & 0 & 0.966 & -0.259 & 0 \\ 0 & 0 & 0 & 0.259 & 0.966 & 0 \\ 0 & 0 & 0 & 0 & 0 & 1 \end{bmatrix} \quad (6.45)$$

$$k^2 = (r^2)^T \bar{k}^2 r^2$$

$$k^2 = \begin{bmatrix} 44387 & -11732 & 36279 & -44387 & 11732 & -38273 \\ -11732 & 3776 & -13714 & 11732 & -3776 & 6274 \\ 36279 & -13714 & 62622 & -36279 & 13714 & -15624 \\ -44387 & 11732 & -36279 & 44387 & -11732 & 38273 \\ 11732 & -3776 & 13714 & -11732 & 3776 & -6274 \\ -38273 & 6274 & -15624 & 38273 & -6274 & 62622 \end{bmatrix} \quad (6.46)$$

Element 3 ($\beta = 30°$), from Table 6.1

$a_1 = 18475$, $a_2 = 9456$, $a_3 = 419$, $a_4 = 84.8$, $a_5 = 21.9$,
$a_6 = 18059$, $a_7 = 71752$, $a_8 = 17.1$, $a_9 = 35611$, $a_{10} = 5.75$

$$C_1 = \frac{a_1}{a_2\alpha+1} = \frac{18475}{9456\times0.00015+1} = 7639,$$

$$C_2 = \frac{a_3}{a_2\alpha+1} = \frac{419}{9456\times0.00015+1} = 173$$

$$C_3 = \frac{a_4}{\alpha+1} = \frac{84.8}{0.00015+1} = 84.8, \quad C_4 = \frac{a_5}{\alpha+1} = \frac{21.9}{0.00015+1} = 21.9$$

$$C_5 = \frac{a_6\alpha^2+a_7\alpha+a_8}{(\alpha+1)(a_2\alpha+1)} = \frac{18059\times0.00015^2+71752\times0.00015+17.1}{(0.00015+1)(9456\times0.00015+1)} = 11.52$$

$$C_6 = \frac{a_6\alpha^2-a_9\alpha+a_{10}}{(\alpha+1)(a_2\alpha+1)} = \frac{18059\times0.00015^2-35611\times0.00015+5.75}{(0.00015+1)(9456\times0.00015+1)} = 0.169$$

$$\bar{k}^3 = \begin{bmatrix} 187991 & 0 & 68119 & -187991 & 0 & -68119 \\ 0 & 2087 & -8623 & 0 & -2087 & -8623 \\ 68119 & -8623 & 72576 & -68119 & 8623 & -1065 \\ -187991 & 0 & -68119 & 187991 & 0 & 68119 \\ 0 & -2087 & 8623 & 0 & 2087 & 8623 \\ -68119 & -8623 & -1065 & 68119 & 8623 & 72576 \end{bmatrix} \quad (6.47)$$

$x_i = x_3 = 25.74$ m, $x_j = x_4 = 30.78$ m,
$x_{ij} = x_j - x_i = 30.78 - 25.74 = 5.04$ m
$z_i = z_3 = 12.69$ m, $z_j = z_4 = 6.12$ m,
$z_{ij} = z_j - z_i = 6.12 - 12.69 = -6.57$ m

$$L = \sqrt{x_{ij}^2 + z_{ij}^2} = \sqrt{5.04^2 + (-6.57)^2} = 8.28\,\text{m}$$

$$\frac{x_{ij}}{L} = \frac{5.04}{8.28} = 0.609, \quad \frac{z_{ij}}{L} = \frac{-6.57}{8.28} = -0.794$$

$$r^3 = \begin{bmatrix} 0.609 & -0.794 & 0 & 0 & 0 & 0 \\ 0.794 & 0.609 & 0 & 0 & 0 & 0 \\ 0 & 0 & 1 & 0 & 0 & 0 \\ 0 & 0 & 0 & 0.609 & -0.794 & 0 \\ 0 & 0 & 0 & 0.794 & 0.609 & 0 \\ 0 & 0 & 0 & 0 & 0 & 1 \end{bmatrix} \quad (6.48)$$

$$k^3 = (r^3)^T \bar{k}^3 r^3$$

$$k^3 = \begin{bmatrix} 71038 & -89893 & 34638 & -71038 & 89893 & -48331 \\ -89893 & 119290 & -59338 & 89893 & -119290 & 48835 \\ 34638 & -59338 & 72576 & -34638 & 59338 & -1065 \\ -71038 & 89893 & -34638 & 71038 & -89893 & 48331 \\ 89893 & -119290 & 59338 & -89893 & 119290 & -48835 \\ -48331 & 48835 & -1065 & 48331 & -48835 & 72576 \end{bmatrix} \quad (6.49)$$

Assembly of the overall structure stiffness matrix, K:

K =

$K_{11} = k^1_{ii}$	$K_{12} = k^1_{ij}$	0	0
$K_{21} = k^1_{ji}$	$K_{22} = k^1_{jj} + k^2_{ii}$	$K_{23} = k^2_{ij}$	0
0	$K_{32} = k^2_{ji}$	$K_{33} = k^2_{jj} + k^3_{ii}$	$K_{34} = k^3_{ij}$
0	0	$K_{43} = k^3_{ji}$	$K_{44} = k^3_{jj}$

where k^1, k^2, and k^3 are given in (6.43), (6.46), and (6.49), respectively.

Load vector

At node 1 the reactions on the structure are: the force in the x-direction R_{X1}, the force in the z-direction R_{Z1}, and the moment R_{M1}.

At node 2 the external force of –60 kN in the z-direction and at node 3 the external force of 80 kN in the x-direction.

At node 4 the reactions on the structure are: the force in the x-direction R_{X4}, the force in the z-direction R_{Z4}, and the moment R_{M4}.

$$F = \begin{bmatrix} F_1 \\ F_2 \\ F_3 \\ F_4 \end{bmatrix} = \begin{bmatrix} X_1 \\ Z_1 \\ M_1 \\ X_2 \\ Z_2 \\ M_2 \\ X_3 \\ Z_3 \\ M_3 \\ X_4 \\ Z_4 \\ M_4 \end{bmatrix} = \begin{bmatrix} R_{X1} \\ R_{Z1} \\ R_{M1} \\ 0 \\ -60 \\ 0 \\ +80 \\ 0 \\ 0 \\ R_{X4} \\ R_{Z4} \\ R_{M4} \end{bmatrix}$$

8581	6382	16531	−8581	−6382	−13883	0	0	0	0	0	0	u_1		R_{X1}
6382	5172	9954	−6382	−5172	−13402	0	0	0	0	0	0	w_1		R_{Z1}
16531	9954	50652	−16531	−9954	−15939	0	0	0	0	0	0	θ_1		R_{M1}
−8581	−6382	−16531	8581 +44387	6382 −11732	13883 +36279	−44387	11732	−38273	0	0	0	u_2		0
−6382	−5172	−9954	6382 −11732	5172 +3776	13402 −13714	11732	−3779	6274	0	0	0	w_2		-60
−13883	−13402	−15939	13883 +36279	13402 −13714	50652 +62622	−36279	13714	−15624	0	0	0	θ_2	=	0
0	0	0	−44387	11732	−36279	44387 +71038	−11732 −89893	38273 +34638	−71038	89893	−48331	u_3		80
0	0	0	11732	−3776	13714	−11732 −89893	3776 +119290	−6274 −59338	−48331	−119290	48835	w_3		0
0	0	0	−39273	6274	−15624	38273 +34638	−6274 −59338	62622 +72576	−34638	59338	−1065	θ_3		0
0	0	0	0	0	0	−71038	89893	−34638	71038	−89893	48331	u_4		R_{X4}
0	0	0	0	0	0	89893	−119290	59338	−89893	119290	−48835	w_4		R_{Z4}
0	0	0	0	0	0	−48331	48835	−1065	48331	−48835	72576	θ_4		R_{M4}

(6.50)

The reduced matrix is obtained by applying the boundary conditions as follows:

At node 1

$u_1 = 0$, $w_1 = 0$, and $\theta_1 = 0$, therefore delete rows and columns 1, 2, and 3.

At node 4

$U_4 = 0$, $w_4 = 0$, and $\theta_4 = 0$, hence delete rows and columns 10, 11, and 12.

Thus the reduced matrix is

$$\begin{bmatrix} 52968 & -5350 & 50162 & -44387 & 11732 & -38273 \\ -5350 & 8948 & -312 & 11732 & -3776 & 6274 \\ 50162 & -312 & 113274 & -36279 & 13714 & -15624 \\ -44387 & 11732 & -36279 & 115425 & -101625 & 72911 \\ 11732 & -3776 & 13714 & -101625 & 123066 & -65612 \\ -38273 & 6274 & -15624 & 72911 & -65612 & 135198 \end{bmatrix} \begin{bmatrix} u_2 \\ w_2 \\ \theta_2 \\ u_3 \\ w_3 \\ \theta_3 \end{bmatrix} = \begin{bmatrix} 0 \\ -60 \\ 0 \\ 80 \\ 0 \\ 0 \end{bmatrix}$$

The solution of the above simultaneous equations is:

$u_2 = 0.018931$ m, $w_2 = -0.023316$ m, $\theta_2 = -0.002081$ rad

$u_3 = 0.027037$ m, $w_3 = 0.021005$ m, $\theta_3 = 0.001814$ rad

So the full displacement vector is:

$$\delta = \begin{bmatrix} \delta_1 \\ \delta_2 \\ \delta_3 \\ \delta_4 \end{bmatrix} = \begin{bmatrix} u_1 \\ w_1 \\ \theta_1 \\ u_2 \\ w_2 \\ \theta_2 \\ u_3 \\ w_3 \\ \theta_3 \\ u_4 \\ w_4 \\ \theta_4 \end{bmatrix} \begin{bmatrix} 0 \\ 0 \\ 0 \\ +0.018931 \\ -0.023316 \\ -0.002081 \\ +0.027037 \\ +0.021005 \\ +0.001814 \\ 0 \\ 0 \\ 0 \end{bmatrix}$$

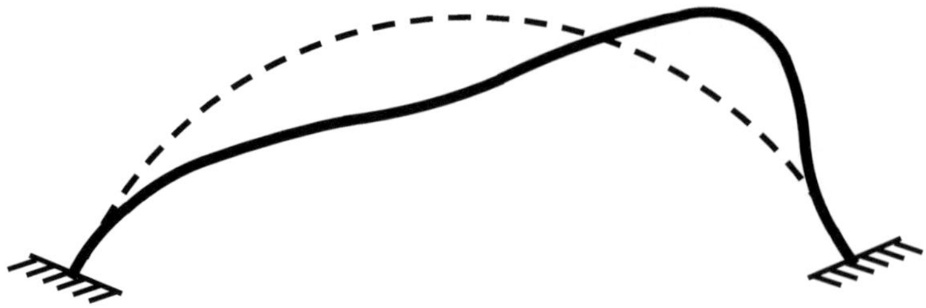

Figure 6.7 Deflected shape of the arch.

Calculation of the reactions at the supports

These are usually calculated relative to the global coordinates and may be found from the full overall stiffness matrix (6.50) as shown below.

For node 1:

From row 1

$$8581u_1 + 6382w_1 + 16531\theta_1 - 8581u_2 - 6382w_2 - 13883\theta_2 = R_{X1}$$

$$8581 \times 0 + 6382 \times 0 + 16531 \times 0 - 8581 \times 0.018931 - 6382 \times (-0.023316) - 13883 \times (-0.002081) = R_{X1},$$

$R_{X1} = +15.25$ kN

From row 2

$$6382u_1 + 5172w_1 + 9954\theta_1 - 6382u_2 - 5172w_2 - 13402\theta_2 = R_{Z1}$$

$$6382 \times 0 + 5172 \times 0 + 9954 \times 0 - 6382 \times 0.018931 - 5172 \times (-0.023316) - 13402 \times (-0.002081) = R_{Z1},$$

$R_{Z1} = +27.66$ kN

From row 3

$$16531u_1 + 9954w_1 + 50652\theta_1 - 16531u_2 - 9954w_2 - 15939\theta_2 = R_{M1}$$

$$16531 \times 0 + 9954 \times 0 + 50652 \times 0 - 16531 \times 0.018931 - 9954 \times (-0.023316) - 15939 \times (-0.002081) = R_{M1},$$

$R_{M1} = -47.69$ kNm

Similarly for node 4:

From row 10, $R_{X4} = -95.29$ kN

From row 11, $R_{Z4} = +32.39$ kN

From row 12, $R_{M4} = -282.88$ kNm

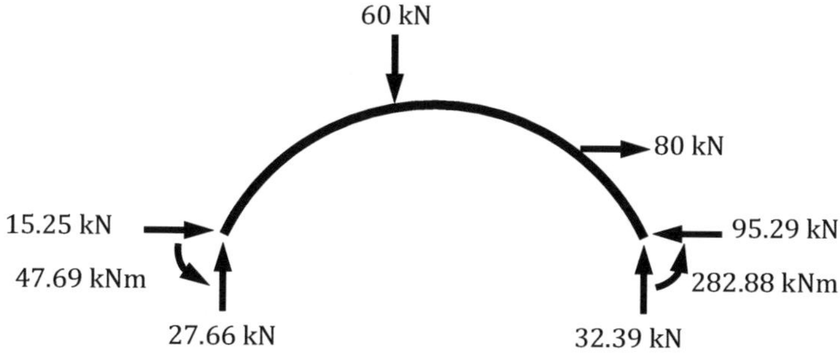

Figure 6.8

Calculation of actions on the elements

Element 1

From (39) with $\beta = 60°$

$$r_\beta^1 = \begin{bmatrix} 0.866 & 0.500 & 0 & 0 & 0 & 0 \\ -0.500 & 0.866 & 0 & 0 & 0 & 0 \\ 0 & 0 & 1 & 0 & 0 & 0 \\ 0 & 0 & 0 & 0.866 & -0.500 & 0 \\ 0 & 0 & 0 & 0.500 & 0.866 & 0 \\ 0 & 0 & 0 & 0 & 0 & 1 \end{bmatrix}$$

$\bar{k}^1$ and r^1 are given in (6.41) and (6.42) respectively.

$$\delta^1 = \begin{bmatrix} \delta_i^1 \\ \delta_j^1 \end{bmatrix} = \begin{bmatrix} \delta_1 \\ \delta_2 \end{bmatrix} = \begin{bmatrix} u_1 \\ w_1 \\ \theta_1 \\ u_2 \\ w_2 \\ \theta_2 \end{bmatrix} = \begin{bmatrix} 0 \\ 0 \\ 0 \\ +0.018931 \\ -0.023316 \\ -0.002081 \end{bmatrix}$$

From (6.40)

$$f^1 = \begin{bmatrix} f_i^1 \\ f_j^1 \end{bmatrix} = \begin{bmatrix} f_1 \\ f_2 \end{bmatrix} = \begin{bmatrix} P_1 \\ V_1 \\ m_1 \\ P_2 \\ V_2 \\ m_2 \end{bmatrix} = r_\beta^1 \bar{k}^1 r^1 \delta^1 = \begin{bmatrix} +31.39 \\ -3.51 \\ -47.68 \\ -18.74 \\ -25.43 \\ -155.08 \end{bmatrix}$$

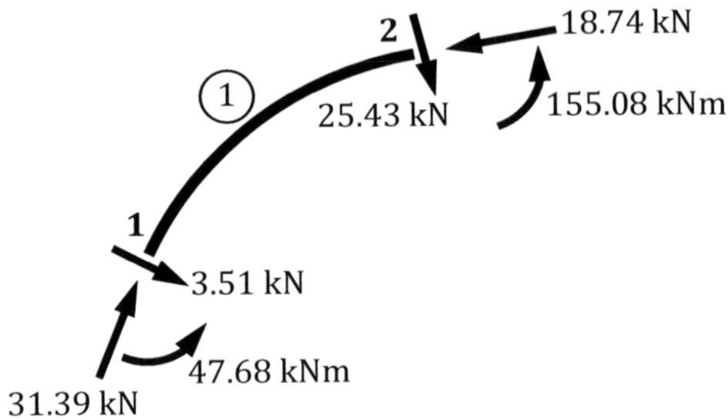

Figure 6.9

Element 2

From (39) with $\beta = 45°$

$$r_\beta^2 = \begin{bmatrix} 0.924 & 0.383 & 0 & 0 & 0 & 0 \\ -0.383 & 0.924 & 0 & 0 & 0 & 0 \\ 0 & 0 & 1 & 0 & 0 & 0 \\ 0 & 0 & 0 & 0.924 & -0.383 & 0 \\ 0 & 0 & 0 & 0.383 & 0.924 & 0 \\ 0 & 0 & 0 & 0 & 0 & 1 \end{bmatrix}$$

$\bar{k}^2$ and r^2 are given in (6.44) and (6.45) respectively.

$$\delta^2 = \begin{bmatrix} \delta_i^2 \\ \delta_j^2 \end{bmatrix} = \begin{bmatrix} \delta_2 \\ \delta_3 \end{bmatrix} = \begin{bmatrix} u_2 \\ w_2 \\ \theta_2 \\ u_3 \\ w_3 \\ \theta_3 \end{bmatrix} = \begin{bmatrix} +0.018931 \\ -0.023316 \\ -0.002081 \\ +0.027037 \\ +0.021005 \\ +0.001814 \end{bmatrix}$$

From (6.40)

$$f^2 = \begin{bmatrix} f_i^2 \\ f_j^2 \end{bmatrix} = \begin{bmatrix} f_2 \\ f_3 \end{bmatrix} = \begin{bmatrix} P_2 \\ V_2 \\ m_2 \\ P_3 \\ V_3 \\ m_3 \end{bmatrix} = r_\beta^2 \bar{k}^2 r^2 \delta^2 = \begin{bmatrix} +10.89 \\ -34.04 \\ +155.10 \\ -31.78 \\ +16.35 \\ +178.27 \end{bmatrix}$$

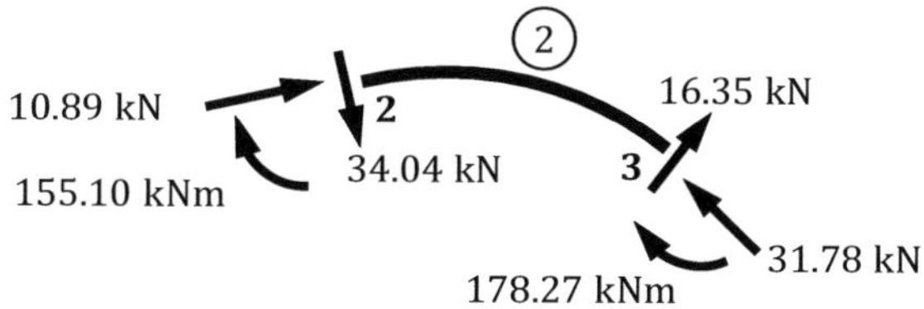

Figure 6.10

Element 3

From (39) with $\beta = 30°$

$$r_\beta^3 = \begin{bmatrix} 0.966 & 0.259 & 0 & 0 & 0 & 0 \\ -0.259 & 0.966 & 0 & 0 & 0 & 0 \\ 0 & 0 & 1 & 0 & 0 & 0 \\ 0 & 0 & 0 & 0.966 & -0.259 & 0 \\ 0 & 0 & 0 & 0.259 & 0.966 & 0 \\ 0 & 0 & 0 & 0 & 0 & 1 \end{bmatrix}$$

$\bar{k}^3$ and r^3 are given in (6.47) and (6.48) respectively.

$$\delta^3 = \begin{bmatrix} \delta_i^3 \\ \delta_j^3 \end{bmatrix} = \begin{bmatrix} \delta_3 \\ \delta_4 \end{bmatrix} = \begin{bmatrix} u_3 \\ w_3 \\ \theta_3 \\ u_4 \\ w_4 \\ \theta_4 \end{bmatrix} = \begin{bmatrix} +0.027037 \\ +0.021005 \\ +0.001814 \\ 0 \\ 0 \\ 0 \end{bmatrix}$$

From (6.40)

$$f^3 = \begin{bmatrix} f_i^3 \\ f_j^3 \end{bmatrix} = \begin{bmatrix} f_3 \\ f_4 \end{bmatrix} = \begin{bmatrix} P_3 \\ V_3 \\ m_3 \\ P_4 \\ V_4 \\ m_4 \end{bmatrix} = r_\beta^3 \bar{k}^3 r^3 \delta^3 = \begin{bmatrix} +95.26 \\ +32.30 \\ -178.23 \\ -66.32 \\ -75.62 \\ -282.88 \end{bmatrix}$$

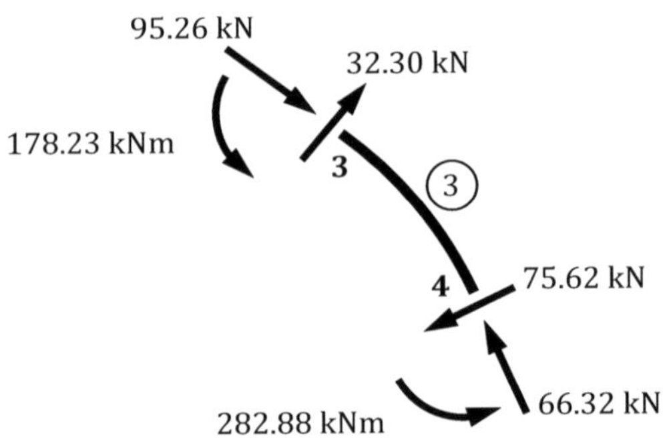

Figure 6.11

Notice the discontinuity in the values of P_2, V_2, P_3, and V_3 which is due to the presence of the 60 kN in the negative z-direction at node 2 and the 80 kN in the positive x-direction at node 3.

Problems

P6.1. Analyse the circular arch whose geometry and the forces acting on it as shown in Fig. P6.1 using the following data: $A = 0.0036\ m^2$, $I = 0.00004\ m^4$, $E = 210 \times 10^6\ kN/m^2$.

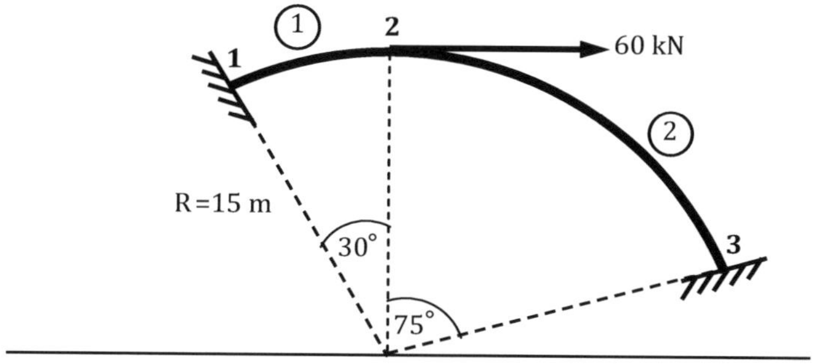

Figure P6.1

Answer:

$u_1 = 0$, $w_1 = 0$, $\theta_1 = 0$, $u_2 = 0.009889$ m, $w_2 = 1.027113$ m,

$\theta_2 = -0.002584$ rad, $u_3 = 0$, $w_3 = 0$, $\theta_3 = 0$

$R_{X1} = -52.51$ kN, $R_{Z1} = -5.60$ kN, $R_{M1} = -46.61$ kNm

$R_{X3} = -7.49$ kN, $R_{Z3} = +5.60$ kN, $R_{M3} = -19.12$ kNm

$$\text{Element 1: } \begin{bmatrix} P_1 \\ V_1 \\ M_1 \\ P_2 \\ V_2 \\ M_2 \end{bmatrix} = \begin{bmatrix} -48.27 \\ +21.43 \\ -46.61 \\ +52.52 \\ +5.57 \\ -16.97 \end{bmatrix}, \quad \text{Element 2: } \begin{bmatrix} P_2 \\ V_2 \\ M_2 \\ P_3 \\ V_3 \\ M_3 \end{bmatrix} = \begin{bmatrix} +7.49 \\ -5.59 \\ +16.97 \\ -7.34 \\ -5.78 \\ -19.12 \end{bmatrix}$$

P6.2. Analyse the circular arch whose geometry and the forces acting on it as shown in Fig. P6.2 using the following data: $A = 0.08\ m^2$, $I = 0.0012\ m^4$, $E = 25 \times 10^6\ kN/m^2$.

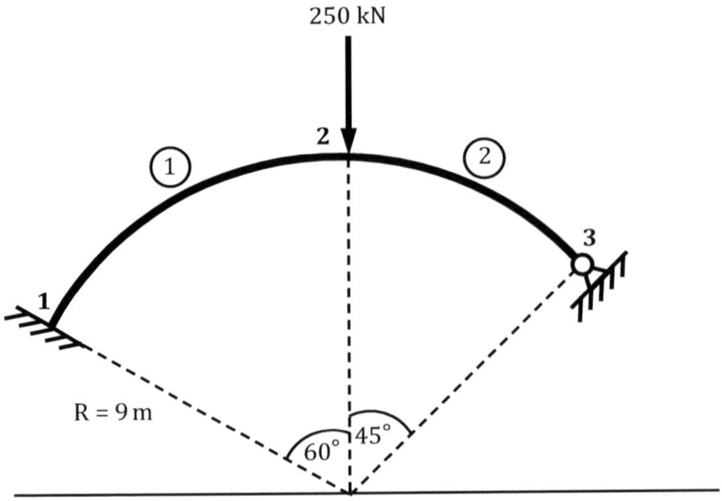

Figure P6.2

Answer:

$u_1 = 0$, $w_1 = 0$, $\theta_1 = 0$, $u_2 = -0.003555$ m, $w_2 = -0.018162$ m,

$\theta_2 = -0.001149$ rad, $u_3 = 0$, $w_3 = 0$, $\theta_3 = 0.002123$ rad

$R_{X1} = 213.71$ kN, $R_{Z1} = 128.00$ kN, $R_{M1} = 177.19$ kNm

$R_{X3} = -213.71$ kN, $R_{Z3} = 122.000$ kN, $R_{M3} = 0$

$$\text{Element 1: } \begin{bmatrix} P_1 \\ V_1 \\ M_1 \\ P_2 \\ V_2 \\ M_2 \end{bmatrix} = \begin{bmatrix} +217.62 \\ -121.21 \\ +177.18 \\ -213.61 \\ -128.14 \\ -213.10 \end{bmatrix}, \text{ Element 2: } \begin{bmatrix} P_2 \\ V_2 \\ M_2 \\ P_3 \\ V_3 \\ M_3 \end{bmatrix} = \begin{bmatrix} +213.62 \\ -122.15 \\ +213.10 \\ -237.43 \\ -64.64 \\ 0 \end{bmatrix}$$

P6.3. Analyse the circular arch whose geometry and the forces acting on it as shown in Fig. P6.3 using the following data: $A = 0.40\ \text{m}^2$, $I = 0.025\ \text{m}^4$, $E = 30 \times 10^6\ \text{kN/m}^2$.

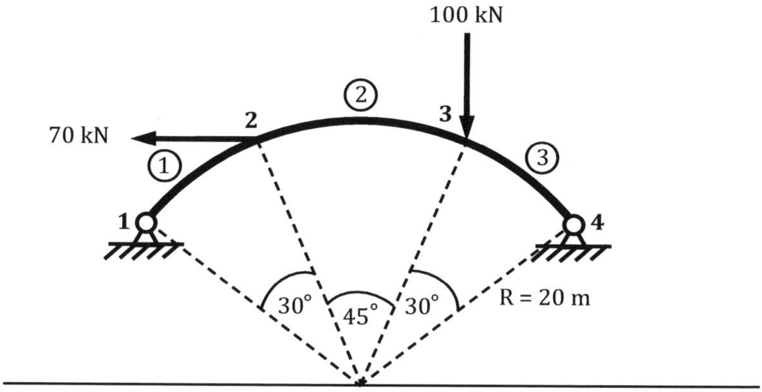

Figure P6.3

Answer:

$u_1 = 0$, $w_1 = 0$, $\theta_1 = -0.002195$ rad,
$u_2 = -0.008941$ m, $w_2 = 0.009568$ m, $\theta_2 = 0.000311$ rad,
$u_3 = -0.008941$, $w_3 = -0.010202$, $\theta_3 = 0.000103$ rad,
$u_4 = 0$, $w_4 = 0$, $\theta_4 = -0.001913$ rad
$R_{X1} = 94.64$ kN, $R_{Z1} = 39.78$ kN, $R_{M1} = 0$
$R_{X4} = -24.64$ kN, $R_{Z4} = 60.22$ kN, $R_{M4} = 0$

$$\text{Element 1:} \begin{bmatrix} P_1 \\ V_1 \\ M_1 \\ P_2 \\ V_2 \\ M_2 \end{bmatrix} = \begin{bmatrix} +89.17 \\ -50.87 \\ 0 \\ -102.66 \\ -0.51 \\ +269.76 \end{bmatrix}, \quad \text{Element 2:} \begin{bmatrix} P_2 \\ V_2 \\ M_2 \\ P_3 \\ V_3 \\ M_3 \end{bmatrix} = \begin{bmatrix} +37.98 \\ +27.34 \\ -269.76 \\ -7.56 \\ -46.18 \\ -339.24 \end{bmatrix},$$

$$\text{Element 3:} \begin{bmatrix} P_3 \\ V_3 \\ M_3 \\ P_4 \\ V_4 \\ M_4 \end{bmatrix} = \begin{bmatrix} +45.82 \\ -46.19 \\ +339.24 \\ -62.77 \\ +17.10 \\ 0 \end{bmatrix}$$

P6.4. Analyse the circular arch whose geometry and the forces acting on it as shown in Fig. P6.4 using the following data: $A = 0.004\ m^2$, $I = 0.00006\ m^4$, $E = 210 \times 10^6\ kN/m^2$.

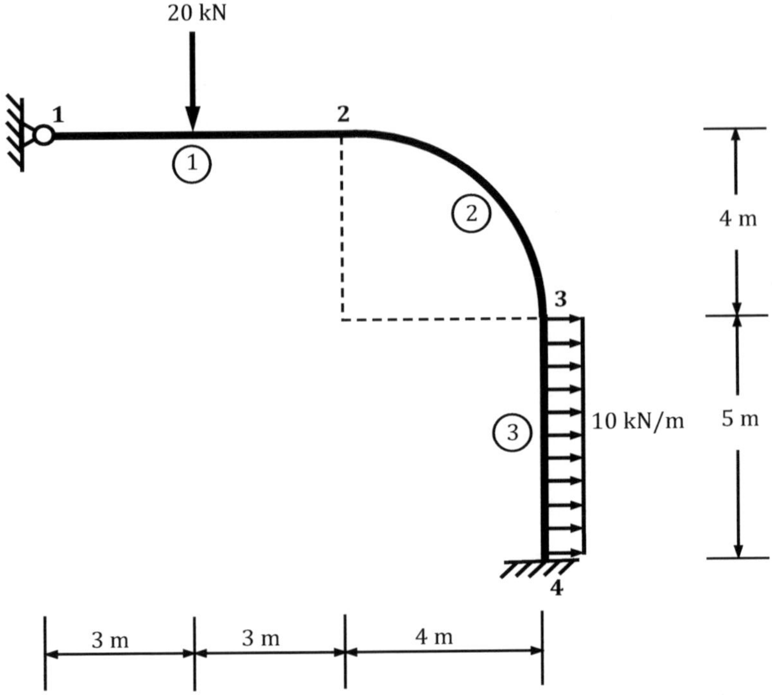

Figure P6.4

Answer:

$u_1 = 0$, $w_1 = 0$, $\theta_1 = 0.008167$ rad,

$u_2 = 0.000038$ m, $w_2 = -0.018513$ m, $\theta_2 = -0.003507$ rad,

$u_3 = 0.015093$ m, $w_3 = -0.000041$ m, $\theta_3 = -0.000471$ rad,

$u_4 = 0$, $w_4 = 0$, $\theta_4 = 0$

$R_{X1} = -5.32$ kN, $R_{Z1} = 13.17$ kN, $R_{M1} = 0$

$R_{X4} = -44.68$ kN, $R_{Z4} = 6.83$ kN, $R_{M4} = -68.85$ kNm

$$\text{Element 1: } \begin{bmatrix} \bar{X}_1 \\ \bar{Z}_1 \\ \bar{M}_1 \\ \bar{X}_2 \\ \bar{Z}_2 \\ \bar{M}_2 \end{bmatrix} = \begin{bmatrix} -5.32 \\ +13.17 \\ 0 \\ +5.32 \\ +6.83 \\ -19.03 \end{bmatrix}, \quad \text{Element 2: } \begin{bmatrix} P_2 \\ V_2 \\ M_2 \\ P_3 \\ V_3 \\ M_3 \end{bmatrix} = \begin{bmatrix} -5.33 \\ -6.82 \\ +19.03 \\ -6.84 \\ +5.30 \\ +29.56 \end{bmatrix},$$

$$\text{Element 3: } \begin{bmatrix} \bar{X}_3 \\ \bar{Z}_3 \\ \bar{M}_3 \\ \bar{X}_4 \\ \bar{Z}_4 \\ \bar{M}_4 \end{bmatrix} = \begin{bmatrix} +6.83 \\ -5.32 \\ -29.56 \\ -6.83 \\ -44.68 \\ -68.85 \end{bmatrix}$$

Chapter 7

Grillage Analysis

A typical example of this type of frame is a bridge deck carrying gravity loads resulting from road or rail traffic. The bridge deck transfers the loads to beams running in two directions which are usually at right angles to each other as shown in Fig. 7.1. There are three degrees of freedom at each node, which are the translation w in the z-direction and the rotations Φ and θ about the x- and y-axes respectively as shown in Fig. 7.2. The beams are connected rigidly at their intersections, thus they develop bending moment (and the associated shear force) as well as torsion as shown in Fig. 7.3. In order to derive the stiffness matrix for a grillage member, the bending and torsion are combined together to give the full stiffness matrix. The stiffness matrix for the bending of beams was derived in Chapter 4 and the stiffness matrix for the torsion of bars will be derived in the next section.

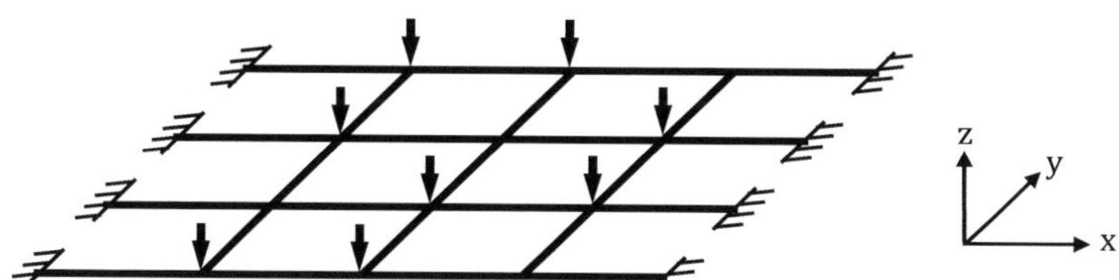

Figure 7.1 Grillage structure.

Analysis of Structures by Matrix Methods
Fathi Al-Shawi

ISBN 978-981-4968-19-5 (Hardcover), 978-1-003-29130-5 (eBook)
www.jennystanford.com

Figure 7.2 Displacement at the ends of a typical member.

Figure 7.3 Actions at the ends of a typical member.

7.1 Derivation of Stiffness Matrix

Consider first a bar subjected to torques $\bar{T}_i$ and $\bar{T}_j$ at its ends i and j with the corresponding twists $\bar{\Phi}_i$ and $\bar{\Phi}_j$ about the $\bar{x}$-axis as shown in Fig. 7.4.

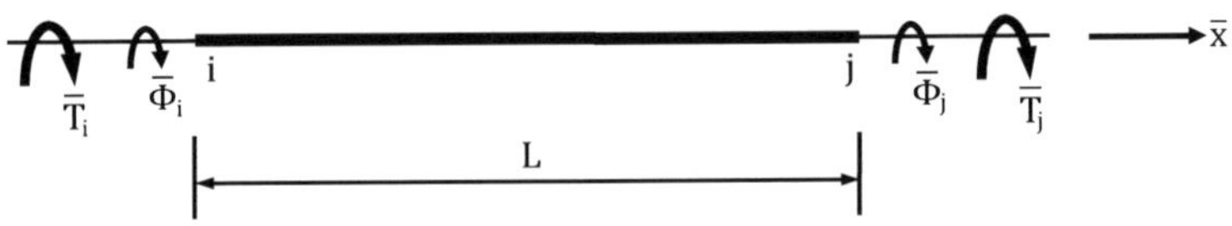

Figure 7.4 Bar subjected to torsion.

Assume that end j is fixed as shown in Fig. 7.5, i.e. $\bar{\Phi}_j = 0$:

$$\bar{T}_i' = \frac{GJ}{L}\bar{\Phi}_i \tag{a}$$

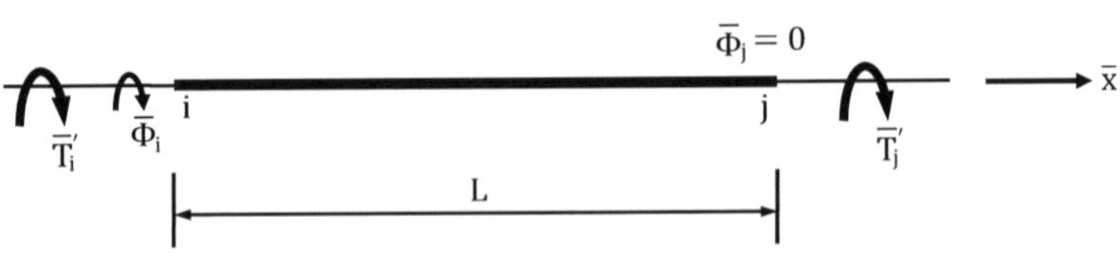

Figure 7.5

From equilibrium, $\overline{T}_i^{'} + \overline{T}_j^{'} = 0, \quad \overline{T}_j^{'} = -\overline{T}_i^{'}$

Thus, $$\overline{T}_j^{'} = -\frac{GJ}{L}\overline{\Phi}_i \qquad \text{(b)}$$

Assume that end i is fixed as shown in Fig. 7.6, i.e. $\overline{\Phi}_i = 0$:

$$\overline{T}_j^{''} = \frac{GJ}{L}\overline{\Phi}_j \qquad \text{(c)}$$

From equilibrium, $\overline{T}_i^{''} + \overline{T}_j^{''} = 0, \quad \overline{T}_i^{''} = -\overline{T}_j^{''}$

So, $$\overline{T}_i^{''} = -\frac{GJ}{L}\overline{\Phi}_j \qquad \text{(d)}$$

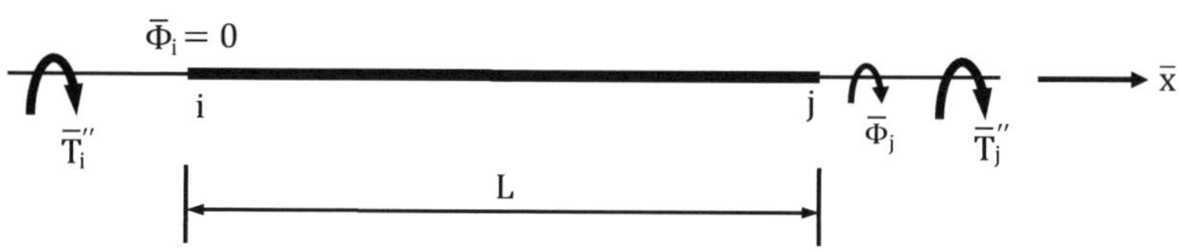

Figure 7.6

The total twisting moment at each end is the sum of the above two cases, thus from (a) and (d) we get

$$\overline{T}_i = \overline{T}_i^{'} + \overline{T}_i^{''} = +\frac{GJ}{L}\overline{\Phi}_i - \frac{GJ}{L}\overline{\Phi}_j$$

and from (b) and (c)

$$\overline{T}_j = \overline{T}_j^{'} + \overline{T}_j^{''} = -\frac{GJ}{L}\overline{\Phi}_i + \frac{GJ}{L}\overline{\Phi}_j$$

In matrix form

$$\begin{bmatrix} \overline{T}_i \\ \overline{T}_j \end{bmatrix} = \begin{bmatrix} \frac{GJ}{L} & -\frac{GJ}{L} \\ -\frac{GJ}{L} & \frac{GJ}{L} \end{bmatrix} \begin{bmatrix} \overline{\Phi}_i \\ \overline{\Phi}_j \end{bmatrix} \qquad (7.1)$$

The above relationships can alternatively be derived by a finite element approach using the so-called interpolation polynomial which defines the displacement along the element as explained in Appendix 3.

From Chapter 4 about bending of beams we had the relation

(4.17) as:

$$\begin{bmatrix} \bar{Z}_i \\ \bar{M}_i \\ \bar{Z}_j \\ \bar{M}_j \end{bmatrix} = \begin{bmatrix} \frac{12EI}{L^3} & -\frac{6EI}{L^2} & -\frac{12EI}{L^3} & -\frac{6EI}{L^2} \\ -\frac{6EI}{L^2} & \frac{4EI}{L} & \frac{6EI}{L^2} & \frac{2EI}{L} \\ -\frac{12EI}{L^3} & \frac{6EI}{L^2} & \frac{12EI}{L^3} & \frac{6EI}{L^2} \\ -\frac{6EI}{L^2} & \frac{2EI}{L} & \frac{6EI}{L^2} & \frac{4EI}{L} \end{bmatrix} \begin{bmatrix} \bar{w}_i \\ \bar{\theta}_i \\ \bar{w}_j \\ \bar{\theta}_j \end{bmatrix}$$

Combining the above relation with (7.1) we get:

$$\begin{bmatrix} \bar{Z}_i \\ \bar{T}_i \\ \bar{M}_i \\ \bar{Z}_j \\ \bar{T}_j \\ \bar{M}_j \end{bmatrix} = \begin{bmatrix} \frac{12EI}{L^3} & 0 & -\frac{6EI}{L^2} & -\frac{12EI}{L^3} & 0 & -\frac{6EI}{L^2} \\ 0 & \frac{GJ}{L} & 0 & 0 & -\frac{GJ}{L} & 0 \\ -\frac{6EI}{L^2} & 0 & \frac{4EI}{L} & \frac{6EI}{L^2} & 0 & \frac{2EI}{L} \\ -\frac{12EI}{L^3} & 0 & \frac{6EI}{L^2} & \frac{12EI}{L^3} & 0 & \frac{6EI}{L^2} \\ 0 & -\frac{GJ}{L} & 0 & 0 & \frac{GJ}{L} & 0 \\ -\frac{6EI}{L^2} & 0 & \frac{2EI}{L} & \frac{6EI}{L^2} & 0 & \frac{4EI}{L} \end{bmatrix} \begin{bmatrix} \bar{w}_i \\ \bar{\Phi}_i \\ \bar{\theta}_i \\ \bar{w}_j \\ \bar{\Phi}_j \\ \bar{\theta}_j \end{bmatrix} \qquad (7.2)$$

i.e. $\bar{F} = \bar{k}\bar{\delta}$, where

$$\bar{k} = \begin{bmatrix} \frac{12EI}{L^3} & 0 & -\frac{6EI}{L^2} & -\frac{12EI}{L^3} & 0 & -\frac{6EI}{L^2} \\ 0 & \frac{GJ}{L} & 0 & 0 & -\frac{GJ}{L} & 0 \\ -\frac{6EI}{L^2} & 0 & \frac{4EI}{L} & \frac{6EI}{L^2} & 0 & \frac{2EI}{L} \\ -\frac{12EI}{L^3} & 0 & \frac{6EI}{L^2} & \frac{12EI}{L^3} & 0 & \frac{6EI}{L^2} \\ 0 & -\frac{GJ}{L} & 0 & 0 & \frac{GJ}{L} & 0 \\ -\frac{6EI}{L^2} & 0 & \frac{2EI}{L} & \frac{6EI}{L^2} & 0 & \frac{4EI}{L} \end{bmatrix} \qquad (7.3)$$

$$\bar{\delta} = \begin{bmatrix} \bar{\delta}_i \\ \bar{\delta}_j \end{bmatrix} = \begin{bmatrix} \bar{w}_i \\ \bar{\Phi}_i \\ \bar{\theta}_i \\ \bar{w}_j \\ \bar{\Phi}_j \\ \bar{\theta}_j \end{bmatrix}, \quad \bar{F} = \begin{bmatrix} \bar{F}_i \\ \bar{F}_j \end{bmatrix} = \begin{bmatrix} \bar{Z}_i \\ \bar{T}_i \\ \bar{M}_i \\ \bar{Z}_j \\ \bar{T}_j \\ \bar{M}_j \end{bmatrix},$$

E is the modulus of elasticity,

G is the modulus of rigidity $= \dfrac{E}{2(1+\mu)}$,

μ is Poisson's ratio = 0.15 for concrete and 0.30 for steel,

I is the second moment of area of the cross section about its neutral axis,

J is the torsion constant. For a circular cross section, $J = \pi d^4/32$ where d is the diameter and for a rectangular cross section, $J = chb^3$ with b as the short side, h is the long side and c is a constant given in *Roark's Formulas for Stress and Strain* and simplified to the following equation

$$c = \frac{\left(\frac{h}{b}\right)^5 - 0.630\left(\frac{h}{b}\right)^4 + 0.053}{3\left(\frac{h}{b}\right)^5} \qquad (7.4)$$

For open sections consisting of rectangular parts, for example an 'I' section,

$J = \sum_{i=1}^{i=n} c_i h_i b_i^3$, where n is the number of parts that make the section.

7.2 Transformation from Local to Global Coordinates

The above relationships are relative to the local coordinates and they need to be transformed to the global coordinates system if the member $\bar{x}$-axis does not lie along the global x-axis. Since the grillage is in the xy plane the orientation of any member is defined by the rotation about the $\bar{z}$ -axis as shown in Fig. 7.7.

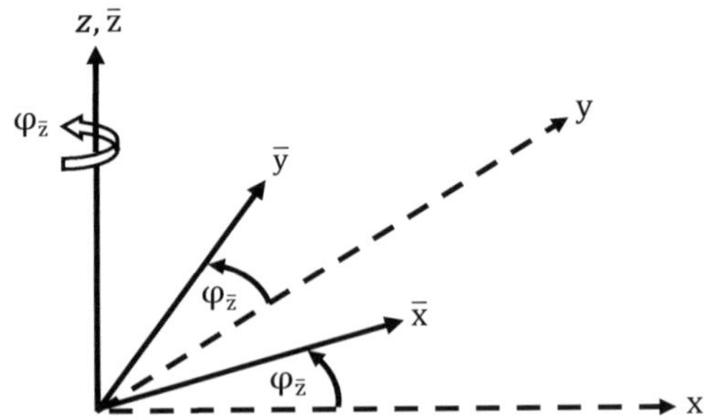

Figure 7.7

Consider the member shown in Fig. 7.8 which has taken its final location by a rotation about the $\bar{z}$-axis by an angle $\varphi_{\bar{z}}$ in the clockwise direction.

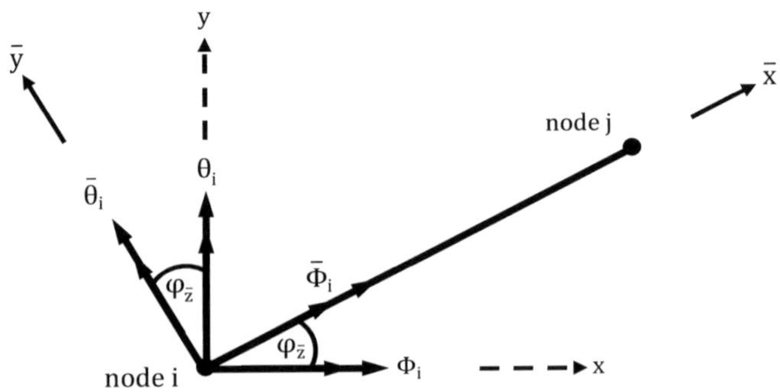

Figure 7.8

Since the member has taken up its final position by rotating the $\bar{z}$-axis only and this means that the local $\bar{z}$-axis is coincident with the global z-axis then the translational displacement relative to the local coordinates is not changed thus:

$$\bar{w}_i = w_i$$

The rotational displacements are vectors, so they are resolved into components along the relevant axes in the same way as the translational displacements vectors as shown in Fig. 7.8.

The rotational displacement $\overline{\Phi}_i$ is equal to the algebraic sum of the components of the rotational displacements Φ_i and θ_i along the $\overline{x}$ -axis and is given by:

$$\overline{\Phi}_i = \Phi_i \cos\varphi_{\overline{z}} + \theta_i \sin\varphi_{\overline{z}}$$

Similarly, the rotational displacement $\overline{\theta}_i$ is equal to the algebraic sum of the components of the rotational displacements Φ_i and θ_i along the $\overline{y}$ -axis and is given by:

$$\overline{\theta}_i = -\Phi_i \sin\varphi_{\overline{z}} + \theta_i \cos\varphi_{\overline{z}}$$

The above transformations are written in matrix form as

$$\begin{bmatrix} \overline{w}_i \\ \overline{\Phi}_i \\ \overline{\theta}_i \end{bmatrix} = \begin{bmatrix} 1 & 0 & 0 \\ 0 & \cos\varphi_{\overline{z}} & \sin\varphi_{\overline{z}} \\ 0 & -\sin\varphi_{\overline{z}} & \cos\varphi_{\overline{z}} \end{bmatrix} \begin{bmatrix} w_i \\ \Phi_i \\ \theta_i \end{bmatrix}$$

or $\overline{\delta}_i = \rho_{\overline{z}} \delta_i$

where

$\rho_{\overline{z}} = \begin{bmatrix} 1 & 0 & 0 \\ 0 & \cos\varphi_{\overline{z}} & \sin\varphi_{\overline{z}} \\ 0 & -\sin\varphi_{\overline{z}} & \cos\varphi_{\overline{z}} \end{bmatrix}$ is the transformation matrix, r_i, for node i.

Similarly the transformation matrix for node j, $r_j = \rho_{\overline{z}}$.

The transformation matrix for nodes i and j is

$r = \begin{bmatrix} r_i & O \\ O & r_j \end{bmatrix} = \begin{bmatrix} \rho_{\overline{z}} & O \\ O & \rho_{\overline{z}} \end{bmatrix}$, where $O = \begin{bmatrix} 0 & 0 \\ 0 & 0 \end{bmatrix}$, hence

$$r = \begin{bmatrix} 1 & 0 & 0 & 0 & 0 & 0 \\ 0 & \cos\varphi_{\overline{z}} & \sin\varphi_{\overline{z}} & 0 & 0 & 0 \\ 0 & -\sin\varphi_{\overline{z}} & \cos\varphi_{\overline{z}} & 0 & 0 & 0 \\ 0 & 0 & 0 & 1 & 0 & 0 \\ 0 & 0 & 0 & 0 & \cos\varphi_{\overline{z}} & \sin\varphi_{\overline{z}} \\ 0 & 0 & 0 & 0 & -\sin\varphi_{\overline{z}} & \cos\varphi_{\overline{z}} \end{bmatrix}$$

The transformation matrix r can be written in a more convenient form by expressing $\sin\varphi_{\bar{z}}$ and $\cos\varphi_{\bar{z}}$ in terms of the coordinates at the ends of the member as shown in Fig. 7.9

where $\sin\varphi_{\bar{z}} = \dfrac{y_j - y_i}{L} = \dfrac{y_{ij}}{L}$, $\cos\varphi_{\bar{z}} = \dfrac{x_j - x_i}{L} = \dfrac{x_{ij}}{L}$, and

$$L = \sqrt{(x_j - x_i)^2 + (y_j - y_i)^2} = \sqrt{x_{ij}^2 + y_{ij}^2}$$

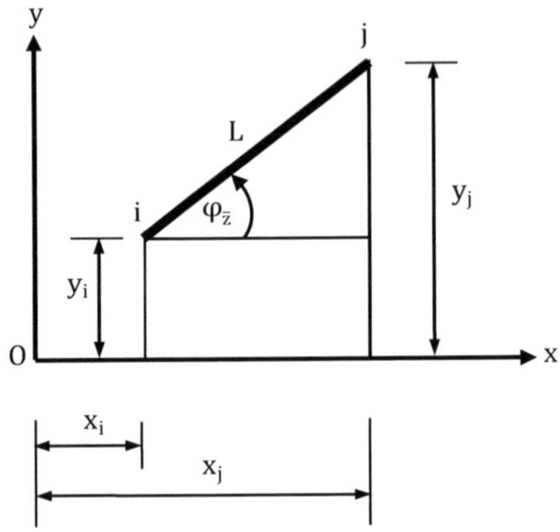

Figure 7.9

Substitute in (7.4) to get:

$$r = \begin{bmatrix} 1 & 0 & 0 & 0 & 0 & 0 \\ 0 & x_{ij}/L & y_{ij}/L & 0 & 0 & 0 \\ 0 & -y_{ij}/L & x_{ij}/L & 0 & 0 & 0 \\ 0 & 0 & 0 & 1 & 0 & 0 \\ 0 & 0 & 0 & 0 & x_{ij}/L & y_{ij}/L \\ 0 & 0 & 0 & 0 & -y_{ij}/L & x_{ij}/L \end{bmatrix} \quad (7.5)$$

The stiffness matrix relative to global coordinates is $k = r^T \bar{k} r$ with $\bar{k}$ from (7.3) and r from (7.5), thus

$$k = \left[\begin{array}{ccc|ccc}
\frac{12EI}{L^3} & \frac{6EIy_{ij}}{L^3} & -\frac{6EIx_{ij}}{L^3} & -\frac{12EI}{L^3} & \frac{6EIy_{ij}}{L^3} & -\frac{6EIx_{ij}}{L^3} \\
\frac{6EIy_{ij}}{L^3} & \frac{4EIy_{ij}^2}{L^3}+\frac{GJx_{ij}^2}{L^3} & -\frac{4EIx_{ij}y_{ij}}{L^3}+\frac{GJx_{ij}y_{ij}}{L^3} & -\frac{6EIy_{ij}}{L^3} & \frac{2EIy_{ij}^2}{L^3}-\frac{GJx_{ij}^2}{L^3} & -\frac{2EIx_{ij}y_{ij}}{L^3}-\frac{GJx_{ij}y_{ij}}{L^3} \\
-\frac{6EIx_{ij}}{L^3} & -\frac{4EIx_{ij}y_{ij}}{L^3}+\frac{GJx_{ij}y_{ij}}{L^3} & \frac{4EIx_{ij}^2}{L^3}+\frac{GJy_{ij}^2}{L^3} & \frac{6EIx_{ij}}{L^3} & -\frac{2EIx_{ij}y_{ij}}{L^3}-\frac{GJx_{ij}y_{ij}}{L^3} & \frac{2EIx_{ij}^2}{L^3}-\frac{GJy_{ij}^2}{L^3} \\
\hline
-\frac{12EI}{L^3} & -\frac{6EIy_{ij}}{L^3} & \frac{6EIx_{ij}}{L^3} & \frac{12EI}{L^3} & -\frac{6EIy_{ij}}{L^3} & \frac{6EIx_{ij}}{L^3} \\
\frac{6EIy_{ij}}{L^3} & \frac{2EIy_{ij}^2}{L^3}-\frac{GJx_{ij}^2}{L^3} & -\frac{2EIx_{ij}y_{ij}}{L^3}-\frac{GJx_{ij}y_{ij}}{L^3} & -\frac{6EIy_{ij}}{L^3} & \frac{4EIy_{ij}^2}{L^3}+\frac{GJx_{ij}^2}{L^3} & -\frac{4EIx_{ij}y_{ij}}{L^3}+\frac{GJx_{ij}y_{ij}}{L^3} \\
-\frac{6EIx_{ij}}{L^3} & -\frac{2EIx_{ij}y_{ij}}{L^3}-\frac{GJx_{ij}y_{ij}}{L^3} & \frac{2EIx_{ij}^2}{L^3}-\frac{GJy_{ij}^2}{L^3} & \frac{6EIx_{ij}}{L^3} & -\frac{4EIx_{ij}y_{ij}}{L^3}+\frac{GJx_{ij}y_{ij}}{L^3} & \frac{4EIx_{ij}^2}{L^3}+\frac{GJy_{ij}^2}{L^3}
\end{array}\right] \tag{7.6}$$

Example

Analyse the grillage shown in Fig. 7.10 given that the properties of all members are: $E = 210 \times 10^6$ kN/m², $\mu = 0.30$, $I = 0.00059$ m⁴, $J = 0.00092$ m⁴.

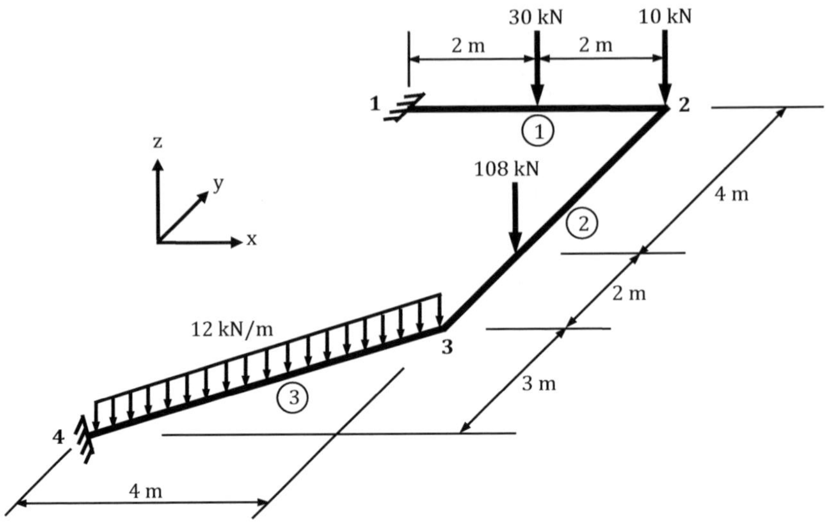

Figure 7.10

$$G = \frac{E}{2(1+\mu)} = \frac{210 \times 10^6}{2(1+0.30)} = 80\ \text{kN/m}^2$$

Stiffness matrices from (7.6)

Member 1

$x_i = 0, x_j = 4$ m, $x_{ij} = x_j - x_i = 4 - 0 = 4$ m
$y_i = 0, y_j = 0$ m, $y_{ij} = y_j - y_i = 0 - 0 = 0$

$$L = \sqrt{x_{ij}^2 + y_{ij}^2} = \sqrt{4^2 + 0^2} = 4\ \text{m}$$

$$
k^1 = \overset{\delta^1}{\overbrace{\begin{array}{cccccc} \overset{\delta_i=\delta_1}{\overbrace{w_i \quad \Phi_i \quad \theta_i}} & & & \overset{\delta_j=\delta_2}{\overbrace{w_j \quad \Phi_j \quad \theta_j}} & & \\ w_1 & \Phi_1 & \theta_1 & w_2 & \Phi_2 & \theta_2 \end{array}}}
\begin{bmatrix} 23231 & 0 & -46462 & -23231 & 0 & -46462 \\ 0 & 18400 & 0 & 0 & -18400 & 0 \\ -46462 & 0 & 123900 & 46462 & 0 & 61950 \\ -23231 & 0 & 46462 & 23231 & 0 & 46462 \\ 0 & -18400 & 0 & 0 & 18400 & 0 \\ -46462 & 0 & 61950 & 46462 & 0 & 123900 \end{bmatrix}
\begin{matrix} w_1 \\ \Phi_1 \\ \theta_1 \\ w_2 \\ \Phi_2 \\ \theta_2 \end{matrix}
$$

Member 2

$x_i = 4$ m, $x_j = 4$ m, $x_{ij} = x_j - x_i = 4 - 4 = 0$
$y_i = 0$, $y_j = -6$ m, $y_{ij} = y_j - y_i = -6 - 0 = -6$ m

$$L = \sqrt{x_{ij}^2 + y_{ij}^2} = \sqrt{0^2 + (-6)^2} = 6 \text{ m}$$

$$
k^2 = \overset{\delta^2}{\overbrace{\begin{array}{cccccc} \overset{\delta_i=\delta_2}{\overbrace{w_i \quad \Phi_i \quad \theta_i}} & & & \overset{\delta_j=\delta_3}{\overbrace{u_j \quad w_j \quad \theta_j}} & & \\ w_2 & \Phi_2 & \theta_2 & u_3 & w_3 & \theta_3 \end{array}}}
\begin{bmatrix} 6883 & -20650 & 0 & -6883 & -20650 & 0 \\ -20650 & 82600 & 0 & 20650 & 41300 & 0 \\ 0 & 0 & 12267 & 0 & 0 & -12267 \\ -6883 & 20650 & 0 & 6883 & 20650 & 0 \\ -20650 & 41300 & 0 & 20650 & 82600 & 0 \\ 0 & 0 & -12267 & 0 & 0 & 12267 \end{bmatrix}
\begin{matrix} w_2 \\ \Phi_2 \\ \theta_2 \\ w_3 \\ \Phi_3 \\ \theta_3 \end{matrix}
$$

Member 3

$x_i = 4$ m, $x_j = 0$, $x_{ij} = x_j - x_i = 0 - 4 = -4$ m

$y_i = -6$ m, $y_j = -9$ m, $y_{ij} = y_j - y_i = -9 - (-6) = -3$ m

$$L = \sqrt{x_{ij}^2 + y_{ij}^2} = \sqrt{(-4)^2 + (-3)^2} = 5\,\text{m}$$

$$k^3 = \overbrace{\begin{matrix} \overbrace{\begin{matrix} w_i & \Phi_i & \theta_i \\ w_3 & \Phi_3 & \theta_3 \end{matrix}}^{\delta_i = \delta_3} & \overbrace{\begin{matrix} w_j & \Phi_j & \theta_j \\ w_4 & \Phi_4 & \theta_4 \end{matrix}}^{\delta_j = \delta_4} \end{matrix}}^{\delta^3}$$

$$k^3 = \begin{bmatrix} 11894 & -17842 & 23789 & -11894 & -17842 & 23789 \\ -17842 & 45104 & -40512 & 17842 & 8421 & -30854 \\ 23789 & -40512 & 68736 & -23789 & -30854 & 26419 \\ -11894 & 17842 & -23789 & 11894 & 17842 & -23789 \\ -17842 & 8421 & -30854 & 17842 & 45104 & -40512 \\ 23789 & -30854 & 26419 & -23789 & -40512 & 68736 \end{bmatrix} \begin{matrix} w_3 \\ \Phi_3 \\ \theta_3 \\ w_4 \\ \Phi_4 \\ \theta_4 \end{matrix}$$

K=

	δ_1	δ_2	δ_3	δ_4	
	$K_{11} = k_{ii}^1$	$K_{12} = k_{ij}^1$	0	0	δ_1
	$K_{21} = k_{ji}^1$	$K_{22} = k_{jj}^1 + k_{ii}^2$	$K_{23} = k_{ij}^2$	0	δ_2
	0	$K_{32} = k_{ji}^2$	$K_{33} = k_{jj}^2 + k_{ii}^3$	$K_{34} = k_{ij}^3$	δ_3
	0	0	$K_{43} = k_{ji}^3$	$K_{44} = k_{jj}^3$	δ_4

The overall stiffness matrix of the structure is:

K =

23231	0	−46462	−23231	0	−46462	0	0	0	0	0	0
0	18400	0	0	−18400	0	0	0	0	0	0	0
−46462	0	123900	46462	0	61950	0	0	0	0	0	0
−23231	0	46462	30114	−20650	46462	−6883	−20650	0	0	0	0
0	−18400	0	−20650	101000	0	20650	41300	0	0	0	0
−46462	0	61950	46462	0	136167	0	0	−12267	0	0	0
0	0	0	−6883	20650	0	18777	2808	23789	−11894	−17842	23789
0	0	0	−20650	41300	0	2808	127704	−40512	17842	8421	−30854
0	0	0	0	0	−12267	23789	−40512	81003	−23789	−30854	26419
0	0	0	0	0	0	−11894	17842	−23789	11894	17842	−23789
0	0	0	0	0	0	−17842	8421	−30854	17842	45104	−40512
0	0	0	0	0	0	23789	−30854	26419	−23789	−40512	68736

(7.7)

Load vector

Member 1

Actions on member 1, W = –30 kN

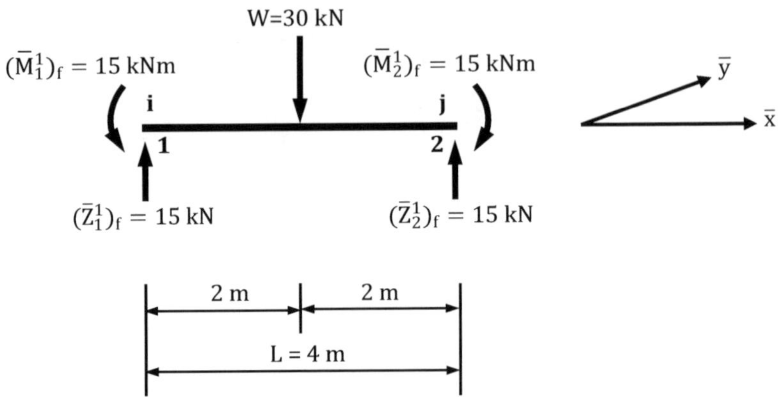

$$(\bar{Z}_1^1)_f = -\frac{W}{2} = -\frac{(-30)}{2} = +15\text{ kN}$$

$$(\bar{Z}_2^1)_f = -\frac{W}{2} = -\frac{(-30)}{2} = +15\text{ kN}$$

$$(\bar{M}_1^1)_f = +\frac{WL}{8} = +\frac{(-30)\times 4}{8} = -15\text{ kN}$$

$$(\bar{M}_2^1)_f = -\frac{WL}{8} = -\frac{(-30)\times 4}{8} = +15\text{ kN}$$

$$\bar{F}_f^1 = \begin{bmatrix} (\bar{Z}_1^1)_f \\ (\bar{T}_1^1)_f \\ (\bar{M}_1^1)_f \\ (\bar{Z}_2^1)_f \\ (\bar{T}_2^1)_f \\ (\bar{M}_2^1)_f \end{bmatrix} = \begin{bmatrix} +15 \\ 0 \\ -15 \\ +15 \\ 0 \\ +15 \end{bmatrix} \tag{7.8}$$

The load vector on joints 1 and 2 is given by

$$F_S^1 = -(r^1)^T \bar{F}_f^1$$

where r^1 is the transformation matrix which is given by (7.5) and since this member coincides with the positive global x-axis, then r^1 = I (the unit matrix), thus

$$F_S^1 = \begin{bmatrix} (Z_1^1)_S \\ (T_1^1)_S \\ (M_1^1)_S \\ (Z_2^1)_S \\ (T_2^1)_S \\ (M_2^1)_S \end{bmatrix} = -\bar{F}_f^1 = \begin{bmatrix} -15 \\ 0 \\ +15 \\ -15 \\ 0 \\ -15 \end{bmatrix} \tag{7.9}$$

Member 2

Actions on member 2, W = –108 kN

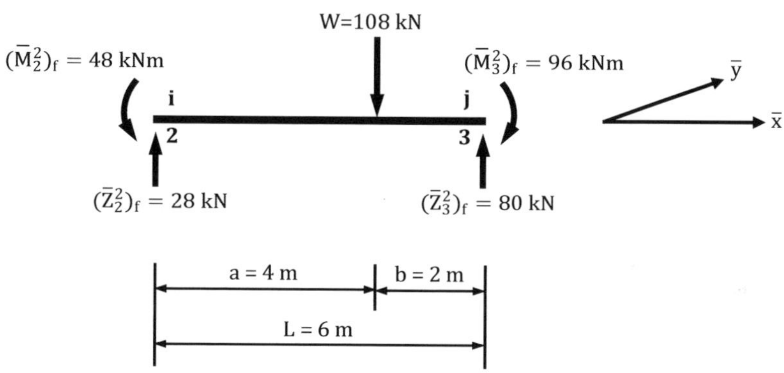

$$(\bar{Z}_2^2)_f = -\frac{Wb}{L^3}\left(L^2 + ab - a^2\right) = -\frac{(-108)\times 2}{6^3}\left(6^2 + 4\times 2 - 4^2\right) = +28\text{ kN}$$

$$(\bar{Z}_3^2)_f = -\frac{Wa}{L^3}\left(L^2 + ab - b^2\right) = -\frac{(-108)\times 4}{6^3}\left(6^2 + 4\times 2 - 2^2\right) = +80\text{ kN}$$

$$(\bar{M}_2^2)_f = +\frac{Wab^2}{L^2} = +\frac{(-108)\times 4\times 2^2}{6^2} = -48\text{ kNm}$$

$$(\bar{M}_3^2)_f = -\frac{Wa^2b}{L^2} = -\frac{(-108)\times 4^2 \times 2}{6^2} = +96 \text{ kNm}$$

The action vector for member 2 is:

$$\bar{F}_f^2 = \begin{bmatrix} (\bar{Z}_2^2)_f \\ (\bar{T}_2^2)_f \\ (\bar{M}_2^2)_f \\ (\bar{Z}_3^2)_f \\ (\bar{T}_3^2)_f \\ (\bar{M}_3^2)_f \end{bmatrix} = \begin{bmatrix} +28 \\ 0 \\ -48 \\ +80 \\ 0 \\ +96 \end{bmatrix} \quad (7.10)$$

The load vector on joints 2 and 3 is given by

$$F_S^2 = -(r^2)^T \bar{F}_f^2$$

where r^2 is the transformation matrix which is given by (7.5). Thus for member 2 where $x_{ij} = 0$, $y_{ij} = -6$ m and $L = 6$ m, it is given by

$$r^2 = \begin{bmatrix} 1 & 0 & 0 & 0 & 0 & 0 \\ 0 & 0 & -1 & 0 & 0 & 0 \\ 0 & 1 & 0 & 0 & 0 & 0 \\ 0 & 0 & 0 & 1 & 0 & 0 \\ 0 & 0 & 0 & 0 & 0 & -1 \\ 0 & 0 & 0 & 0 & 1 & 0 \end{bmatrix}$$

$$F_S^2 = \begin{bmatrix} (Z_2^2)_S \\ (T_2^2)_S \\ (M_2^2)_S \\ (Z_3^2)_S \\ (T_3^2)_S \\ (M_3^2)_S \end{bmatrix} = -\begin{bmatrix} 1 & 0 & 0 & 0 & 0 & 0 \\ 0 & 0 & 1 & 0 & 0 & 0 \\ 0 & -1 & 0 & 0 & 0 & 0 \\ 0 & 0 & 0 & 1 & 0 & 0 \\ 0 & 0 & 0 & 0 & 0 & 1 \\ 0 & 0 & 0 & 0 & -1 & 0 \end{bmatrix} \begin{bmatrix} +28 \\ 0 \\ -48 \\ +80 \\ 0 \\ +96 \end{bmatrix} = \begin{bmatrix} -28 \\ +48 \\ 0 \\ -80 \\ -96 \\ 0 \end{bmatrix} \quad (7.11)$$

Member 3

Actions on member 3, $n = -12$ kN/m

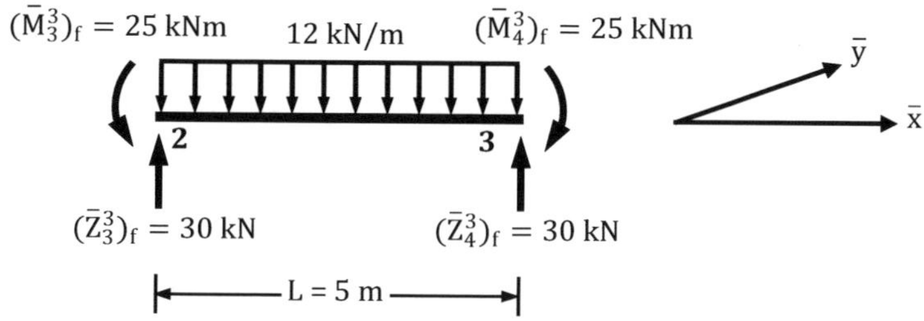

$$(\bar{Z}_3^3)_f = -\frac{nL}{2} = -\frac{(-12)\times 5}{2} = +30\text{ kN}$$

$$(\bar{Z}_4^3)_f = -\frac{nL}{2} = -\frac{(-12)\times 5}{2} = +30\text{ kN}$$

$$(\bar{M}_3^3)_f = +\frac{nL^2}{12} = +\frac{(-12)\times 5^2}{12} = -25\text{ kNm}$$

$$(\bar{M}_4^3)_f = -\frac{nL^2}{12} = -\frac{(-12)\times 5^2}{12} = +25\text{ kNm}$$

It follows that the action vector on the member assumed fixed at its ends is given by:

$$\bar{F}_f^3 = \begin{bmatrix} (\bar{Z}_3^3)_f \\ (\bar{T}_3^3)_f \\ (\bar{M}_3^3)_f \\ (\bar{Z}_4^3)_f \\ (\bar{T}_4^3)_f \\ (\bar{M}_4^3)_f \end{bmatrix} = \begin{bmatrix} +30 \\ 0 \\ -25 \\ +30 \\ 0 \\ +25 \end{bmatrix} \tag{7.12}$$

The load vector on joints 2 and 3 is given by

$$F_S^3 = -(r^3)^T \bar{F}_f^3$$

where r^3 is the transformation matrix which is given by (7.5). Thus for this member, where $x_{ij} = -4$ m, $y_{ij} = -3$ m and L = 5 m, it is given by

$$r^3 = \begin{bmatrix} 1.0 & 0 & 0 & 0 & 0 & 0 \\ 0 & -0.8 & -0.6 & 0 & 0 & 0 \\ 0 & 0.6 & -0.8 & 0 & 0 & 0 \\ 0 & 0 & 0 & 1.0 & 0 & 0 \\ 0 & 0 & 0 & 0 & -0.8 & -0.6 \\ 0 & 0 & 0 & 0 & 0.6 & -0.8 \end{bmatrix}$$

$$F_S^3 = \begin{bmatrix} (Z_3^3)_S \\ (T_3^3)_S \\ (M_3^3)_S \\ (Z_4^3)_S \\ (T_4^3)_S \\ (M_4^3)_S \end{bmatrix} = -\begin{bmatrix} 1.0 & 0 & 0 & 0 & 0 & 0 \\ 0 & -0.8 & 0.6 & 0 & 0 & 0 \\ 0 & -0.6 & -0.8 & 0 & 0 & 0 \\ 0 & 0 & 0 & 1.0 & 0 & 0 \\ 0 & 0 & 0 & 0 & -0.8 & 0.6 \\ 0 & 0 & 0 & 0 & -0.6 & -0.8 \end{bmatrix} \begin{bmatrix} +30 \\ 0 \\ -25 \\ +30 \\ 0 \\ +25 \end{bmatrix} = \begin{bmatrix} -30 \\ +15 \\ -20 \\ -30 \\ -15 \\ +20 \end{bmatrix} \quad (7.13)$$

The load vector, F_S, due to the forces acting on the members is

$$F_S = \begin{bmatrix} (F_1)_S \\ (F_2)_S \\ (F_3)_S \\ (F_4)_S \end{bmatrix} = F_S^1 + F_S^2 + F_S^3$$

From (7.9), (7.11), and (7.13)

$$F_S = \begin{bmatrix} (F_1)_S \\ (F_2)_S \\ (F_3)_S \\ (F_4)_S \end{bmatrix} = \begin{bmatrix} (Z_1)_S \\ (T_1)_S \\ (M_1)_S \\ (Z_2)_S \\ (T_2)_S \\ (M_2)_S \\ (Z_3)_S \\ (T_3)_S \\ (M_3)_S \\ (Z_4)_S \\ (T_4)_S \\ (M_4)_S \end{bmatrix} = \begin{bmatrix} -15 \\ 0 \\ +15 \\ -15 \\ 0 \\ -15 \\ 0 \\ 0 \\ 0 \\ 0 \\ 0 \\ 0 \end{bmatrix} + \begin{bmatrix} 0 \\ 0 \\ 0 \\ -28 \\ +48 \\ 0 \\ -80 \\ -96 \\ 0 \\ 0 \\ 0 \\ 0 \end{bmatrix} + \begin{bmatrix} 0 \\ 0 \\ 0 \\ 0 \\ 0 \\ 0 \\ -30 \\ +15 \\ -20 \\ -30 \\ -15 \\ +20 \end{bmatrix} = \begin{bmatrix} -15 \\ 0 \\ +15 \\ -43 \\ +48 \\ -15 \\ -110 \\ -81 \\ -20 \\ -30 \\ -15 \\ +20 \end{bmatrix} \quad (7.14a)$$

The load vector, F_N, due to external forces directly applied at the nodes

A load of −10 kN applied at node 2 in the z-direction, thus

$$F_N = \begin{bmatrix} (F_1)_N \\ (F_2)_N \\ (F_3)_N \\ (F_4)_N \end{bmatrix} = \begin{bmatrix} (Z_1)_N \\ (T_1)_N \\ (M_1)_N \\ (Z_2)_N \\ (T_2)_N \\ (M_2)_N \\ (Z_3)_N \\ (T_3)_N \\ (M_3)_N \\ (Z_4)_N \\ (T_4)_N \\ (M_4)_N \end{bmatrix} = \begin{bmatrix} 0 \\ 0 \\ 0 \\ -10 \\ 0 \\ 0 \\ 0 \\ 0 \\ 0 \\ 0 \\ 0 \\ 0 \end{bmatrix} \tag{7.14b}$$

The load vector, F_C, due to the reactions at the supports

At node 1 the reactions on the structure are: the force in the z-direction R_{Z1}, the moment about the x-axis R_{T1}, and the moment about the y-axis R_{M1}. Similarly, at node 4 the reactions are R_{Z4}, R_{T4}, and R_{M4}.

$$F_C = \begin{bmatrix} (F_1)_C \\ (F_2)_C \\ (F_3)_C \\ (F_4)_C \end{bmatrix} = \begin{bmatrix} (Z_1)_C \\ (T_1)_C \\ (M_1)_C \\ (Z_2)_C \\ (T_2)_C \\ (M_2)_C \\ (Z_3)_C \\ (T_3)_C \\ (M_3)_C \\ (Z_4)_C \\ (T_4)_C \\ (M_4)_C \end{bmatrix} = \begin{bmatrix} R_{Z1} \\ R_{T1} \\ R_{M1} \\ 0 \\ 0 \\ 0 \\ 0 \\ 0 \\ 0 \\ R_{Z4} \\ R_{T4} \\ R_{M4} \end{bmatrix} \tag{7.15}$$

The total load vector on the joints of the structure is obtained from the algebraic addition of (7.14a), (7.14b), and (7.15) as:

$$F = F_S + F_N + F_C =$$

$$\begin{bmatrix} F_1 \\ F_2 \\ F_3 \\ F_4 \end{bmatrix} = \begin{bmatrix} Z_1 \\ T_1 \\ M_1 \\ Z_2 \\ T_2 \\ M_2 \\ Z_3 \\ T_3 \\ M_3 \\ Z_4 \\ T_4 \\ M_4 \end{bmatrix} = \begin{bmatrix} -15 \\ 0 \\ +15 \\ -43 \\ +48 \\ -15 \\ -110 \\ -81 \\ -20 \\ -30 \\ -15 \\ +20 \end{bmatrix} + \begin{bmatrix} 0 \\ 0 \\ 0 \\ -10 \\ 0 \\ 0 \\ 0 \\ 0 \\ 0 \\ 0 \\ 0 \\ 0 \end{bmatrix} + \begin{bmatrix} R_{Z1} \\ R_{T1} \\ R_{M1} \\ 0 \\ 0 \\ 0 \\ 0 \\ 0 \\ 0 \\ R_{Z4} \\ R_{T4} \\ R_{M4} \end{bmatrix} = \begin{bmatrix} -15+R_{Z1} \\ R_{T1} \\ +15+R_{M1} \\ -53 \\ +48 \\ -15 \\ -110 \\ -81 \\ -20 \\ -30+R_{Z4} \\ -15+R_{T4} \\ +20+R_{M4} \end{bmatrix} \tag{7.16}$$

Substitute (7.7) and (7.16) in the general relationship $K\delta = F$ to get (see equation 7.17 on next page).

The boundary conditions are w_1, Φ_1, θ_1, w_4, Φ_4, and θ_4, thus delete rows and columns 1, 2, 3, 10, 11, and 12 to get:

$$\begin{bmatrix} 30114 & -20650 & 46462 & -6883 & -20650 & 0 \\ -20650 & 101000 & 0 & 20650 & 41300 & 0 \\ 46462 & 0 & 136167 & 0 & 0 & -12267 \\ -6883 & 20650 & 0 & 18777 & 2808 & 23789 \\ -20650 & 41300 & 0 & 2808 & 127704 & -40512 \\ 0 & 0 & -12267 & 23789 & -40512 & 81003 \end{bmatrix} \begin{bmatrix} w_2 \\ \Phi_2 \\ \theta_2 \\ w_3 \\ \Phi_3 \\ \theta_3 \end{bmatrix} = \begin{bmatrix} -53 \\ +48 \\ -15 \\ -110 \\ -81 \\ -20 \end{bmatrix}$$

The solution of the above set is:

$w_2 = -0.013445$ m, $\Phi_2 = 0.002555$ rad, $\theta_2 = 0.005028$ rad,

$w_3 = -0.021157$ m, $\Phi_3 = -0.001230$ rad, $\theta_3 = 0.006113$ rad,

23231	0	−46462	−23231	0	−46462	0	0	0	0	0	0	w_1		$-15 + R_{Z1}$
0	18400	0	0	−18400	0	0	0	0	0	0	0	Φ_1		R_{T1}
−46462	0	123900	46462	0	61950	0	0	0	0	0	0	θ_1		$+15 + R_{M1}$
−23231	0	46462	30114	−20650	46462	−6883	−20650	0	0	0	0	w_2		−53
0	−18400	0	−20650	101000	0	20650	41300	0	0	0	0	Φ_2		+48
−46462	0	61950	46462	0	136167	0	0	−12267	0	0	0	θ_2	=	−15
0	0	0	−6883	20650	0	18777	2808	23789	−11894	−17842	23789	w_3		−110
0	0	0	−20650	41300	0	2808	127704	−40512	17842	8421	−30854	Φ_3		−81
0	0	0	0	0	−12267	23789	−40512	81003	−23789	−30854	26419	θ_3		−20
0	0	0	0	0	0	−11894	17842	−23789	11894	17842	−23789	w_4		$-30 + R_{Z4}$
0	0	0	0	0	0	−17842	8421	−30854	17842	45104	−40512	Φ_4		$-15 + R_{T4}$
0	0	0	0	0	0	23789	−30854	26419	−23789	−40512	68736	θ_4		$+20 + R_{M4}$

(7.17)

The full displacement vector is

$$\delta = \begin{bmatrix} w_1 \\ \Phi_1 \\ \theta_1 \\ w_2 \\ \Phi_2 \\ \theta_2 \\ w_3 \\ \Phi_3 \\ \theta_3 \\ w_4 \\ \Phi_4 \\ \theta_4 \end{bmatrix} = \begin{bmatrix} 0 \\ 0 \\ 0 \\ -0.013445 \\ +0.002555 \\ +0.005028 \\ -0.021157 \\ -0.001230 \\ +0.006113 \\ 0 \\ 0 \\ 0 \end{bmatrix}$$

Calculation of reactions at the supports from (7.17)

The first row

$-23231\ w_2 - 464632 = -15 + R_{Z1}$

$-23231 \times (-0.013445) - 46462 \times 0.005028 = -15 + R_{Z1},$

$R_{Z1} = +93.73$ kN

The second row

$-18400 \times 0.002555 = R_{T1}$, $R_{T1} = -47.01$ kNm

And the third row

$46462 w_2 + 61950\theta_2 = +15 + R_{M1}$

$46462 \times (-0.013445) + 61950 \times 0.005028 = +15 + R_{M1},$

$R_{M1} = -328.20$ kNm

Similarly, from rows 10, 11, and 12, $R_{Z4} = +114.27$ kN, $R_{T4} = +193.52$ kNm, and $R_{M4} = -323.85$ kNm, respectively.

Calculation of actions on the members

The resultant actions on any member of the frame is given by

$$\bar{F}_r = \bar{F}_d + \bar{F}_f \tag{7.18}$$

where

$\bar{F}_d = \bar{k}\bar{\delta}$ is the actions on the member due to displacements at the ends of the member.

$\overline{F}_f$ is the column vector of actions on the member due to the applied loads assuming that the member is fixed at its ends.

$\overline{k}$ is the stiffness matrix of the member relative to local coordinates given by (7.3).

$\overline{\delta} = r\delta$ is the column vector of the displacements at the ends of the member relative to local coordinates.

r is the transformation matrix of the member given by (7.5).

δ is the column vector of the displacements relative to global coordinates at the ends of the member. Thus equation (7.18) becomes

$$\overline{F}_r = \overline{k}\overline{\delta} + \overline{F}_f \tag{7.19}$$

Member 1

From (7.3)

$$\overline{k}^1 = \begin{bmatrix} 23231 & 0 & -46462 & -23231 & 0 & -46462 \\ 0 & 18400 & 0 & 0 & -18400 & 0 \\ -46462 & 0 & 123900 & 46462 & 0 & 61950 \\ -23231 & 0 & 46462 & 23231 & 0 & 46462 \\ 0 & -18400 & 0 & 0 & 18400 & 0 \\ -46462 & 0 & 61950 & 46462 & 0 & 123900 \end{bmatrix}$$

$$x_{ij} = 4 \text{ m}, y_{ij} = 0, L = 4 \text{ m}$$

where r^1 is the transformation matrix which is given by (7.5) and since this member coincides with the positive global x-axis, then $r^1 = I$ (the unit matrix), thus

$$r^1 = \begin{bmatrix} 1 & 0 & 0 & 0 & 0 & 0 \\ 0 & 1 & 0 & 0 & 0 & 0 \\ 0 & 0 & 1 & 0 & 0 & 0 \\ 0 & 0 & 0 & 1 & 0 & 0 \\ 0 & 0 & 0 & 0 & 1 & 0 \\ 0 & 0 & 0 & 0 & 0 & 1 \end{bmatrix}, \quad \delta^1 = \begin{bmatrix} \delta_i^1 \\ \delta_j^1 \end{bmatrix} = \begin{bmatrix} \delta_1 \\ \delta_2 \end{bmatrix} = \begin{bmatrix} w_1 \\ \Phi_1 \\ \theta_1 \\ w_2 \\ \Phi_2 \\ \theta_2 \end{bmatrix} = \begin{bmatrix} 0 \\ 0 \\ 0 \\ -0.013445 \\ +0.002555 \\ +0.005028 \end{bmatrix}$$

$$\bar{\delta}^1 = r^1\delta^1 = \begin{bmatrix} 1 & 0 & 0 & 0 & 0 & 0 \\ 0 & 1 & 0 & 0 & 0 & 0 \\ 0 & 0 & 1 & 0 & 0 & 0 \\ 0 & 0 & 0 & 1 & 0 & 0 \\ 0 & 0 & 0 & 0 & 1 & 0 \\ 0 & 0 & 0 & 0 & 0 & 1 \end{bmatrix} \begin{bmatrix} 0 \\ 0 \\ 0 \\ -0.013445 \\ +0.002555 \\ +0.005028 \end{bmatrix} = \begin{bmatrix} 0 \\ 0 \\ 0 \\ -0.013445 \\ +0.002555 \\ +0.005028 \end{bmatrix}$$

$$\bar{F}_d^1 = \bar{k}^1\bar{\delta}^1$$

$$\bar{F}_d^1 = \begin{bmatrix} 23231 & 0 & -46462 & -23231 & 0 & -46462 \\ 0 & 18400 & 0 & 0 & -18400 & 0 \\ -46462 & 0 & 123900 & 46462 & 0 & 61950 \\ -23231 & 0 & 46462 & 23231 & 0 & 46462 \\ 0 & -18400 & 0 & 0 & 18400 & 0 \\ -46462 & 0 & 61950 & 46462 & 0 & 123900 \end{bmatrix} \begin{bmatrix} 0 \\ 0 \\ 0 \\ -0.013445 \\ +0.002555 \\ +0.005028 \end{bmatrix}$$

$$\bar{F}_d^1 = \begin{bmatrix} +78.71 \\ -47.01 \\ -313.19 \\ -78.71 \\ +47.01 \\ -1.71 \end{bmatrix} \text{ and from (7.12) } \bar{F}_f^1 = \begin{bmatrix} +15 \\ 0 \\ -15 \\ +15 \\ 0 \\ +15 \end{bmatrix}$$

$$\begin{bmatrix} (\bar{Z}_1^1)_r \\ (\bar{T}_1^1)_r \\ (\bar{M}_1^1)_r \\ (\bar{Z}_2^1)_r \\ (\bar{T}_2^1)_r \\ (\bar{M}_2^1)_r \end{bmatrix} = \bar{F}_r^1 = \bar{F}_d^1 + \bar{F}_f^1 = \begin{bmatrix} +78.71 \\ -47.01 \\ -313.19 \\ -78.71 \\ +47.01 \\ -1.71 \end{bmatrix} + \begin{bmatrix} +15 \\ 0 \\ -15 \\ +15 \\ 0 \\ +15 \end{bmatrix} = \begin{bmatrix} +93.71 \\ -47.01 \\ -328.19 \\ -63.71 \\ +47.01 \\ +13.29 \end{bmatrix}$$

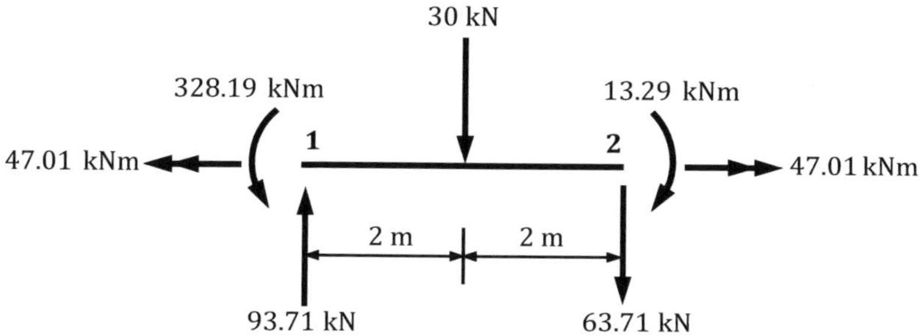

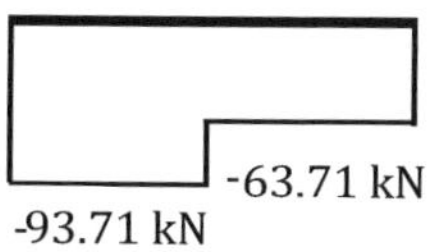

Shear force diagram, $\overline{V}$

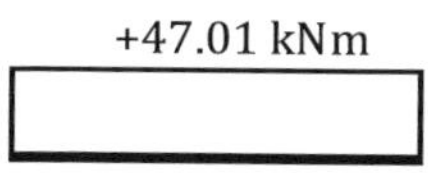

Torque diagram, $\overline{T}$

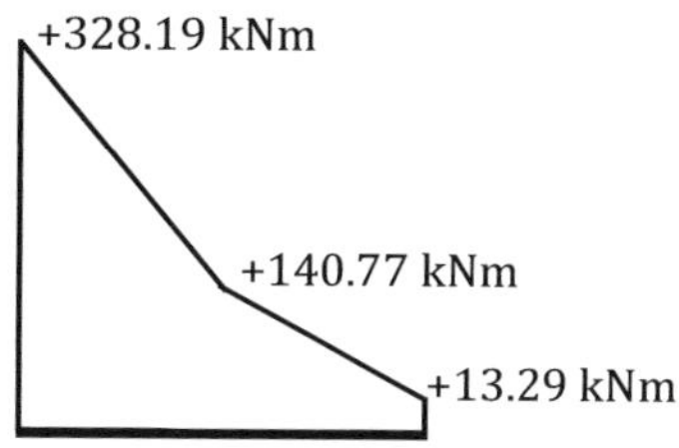

Bending moment diagram, $\overline{M}$

Member 2

$x_{ij} = 0$, $y_{ij} = -6$ m, L = 6 m

$$\overline{k}^2 = \begin{bmatrix} 6883 & 0 & -20650 & -6883 & 0 & -20650 \\ 0 & 12267 & 0 & 0 & -12267 & 0 \\ -20650 & 0 & 82600 & 20650 & 0 & 41300 \\ -6883 & 0 & 20650 & 6883 & 0 & 20650 \\ 0 & -12267 & 0 & 0 & 12267 & 0 \\ -20650 & 0 & 41300 & 20650 & 0 & 82600 \end{bmatrix}$$

$$r^2 = \begin{bmatrix} 1 & 0 & 0 & 0 & 0 & 0 \\ 0 & 0 & -1 & 0 & 0 & 0 \\ 0 & 1 & 0 & 0 & 0 & 0 \\ 0 & 0 & 0 & 1 & 0 & 0 \\ 0 & 0 & 0 & 0 & 0 & -1 \\ 0 & 0 & 0 & 0 & 1 & 0 \end{bmatrix}, \; \delta^2 = \begin{bmatrix} \delta_i^2 \\ \delta_j^2 \end{bmatrix} = \begin{bmatrix} \delta_2 \\ \delta_3 \end{bmatrix} = \begin{bmatrix} w_2 \\ \Phi_2 \\ \theta_2 \\ w_3 \\ \Phi_3 \\ \theta_3 \end{bmatrix} = \begin{bmatrix} -0.013445 \\ +0.002555 \\ +0.005028 \\ -0.021157 \\ -0.001230 \\ +0.006113 \end{bmatrix}$$

$$\bar{\delta}^2 = r^2\delta^2 = \begin{bmatrix} 1 & 0 & 0 & 0 & 0 & 0 \\ 0 & 0 & -1 & 0 & 0 & 0 \\ 0 & 1 & 0 & 0 & 0 & 0 \\ 0 & 0 & 0 & 1 & 0 & 0 \\ 0 & 0 & 0 & 0 & 0 & -1 \\ 0 & 0 & 0 & 0 & 1 & 0 \end{bmatrix} \begin{bmatrix} -0.013445 \\ +0.002555 \\ +0.005028 \\ -0.021157 \\ -0.001230 \\ +0.006113 \end{bmatrix} = \begin{bmatrix} -0.013445 \\ -0.005028 \\ 0.002555 \\ -0.021157 \\ -0.006113 \\ -0.001230 \end{bmatrix}$$

$$\bar{F}_d^2 = \bar{k}^2\bar{\delta}^2$$

$$\bar{F}_d^2 = \begin{bmatrix} 6883 & 0 & -20650 & -6883 & 0 & -20650 \\ 0 & 12267 & 0 & 0 & -12267 & 0 \\ -20650 & 0 & 82600 & 20650 & 0 & 41300 \\ -6883 & 0 & 20650 & 6883 & 0 & 20650 \\ 0 & -12267 & 0 & 0 & 12267 & 0 \\ -20650 & 0 & 41300 & 20650 & 0 & 82600 \end{bmatrix} \begin{bmatrix} -0.013445 \\ -0.005028 \\ 0.002555 \\ -0.021157 \\ -0.006113 \\ -0.001230 \end{bmatrix}$$

$$\bar{F}_d^2 = \begin{bmatrix} +25.72 \\ +13.31 \\ +0.99 \\ -25.72 \\ -13.31 \\ -155.33 \end{bmatrix} \text{ and from (7.10), } \bar{F}_f^2 = \begin{bmatrix} +28 \\ 0 \\ -48 \\ +80 \\ 0 \\ +96 \end{bmatrix}$$

$$\begin{bmatrix} (\bar{Z}_2^2)_r \\ (\bar{T}_2^2)_r \\ (\bar{M}_2^2)_r \\ (\bar{Z}_3^2)_r \\ (\bar{T}_3^2)_r \\ (\bar{M}_3^2)_r \end{bmatrix} = \bar{F}_r^2 = \bar{F}_d^2 + \bar{F}_f^2 = \begin{bmatrix} +25.72 \\ +13.31 \\ +0.99 \\ -25.72 \\ -13.31 \\ -155.33 \end{bmatrix} + \begin{bmatrix} +28 \\ 0 \\ -48 \\ +80 \\ 0 \\ +96 \end{bmatrix} = \begin{bmatrix} +53.72 \\ +13.31 \\ -47.01 \\ +54.28 \\ -13.31 \\ -59.33 \end{bmatrix}$$

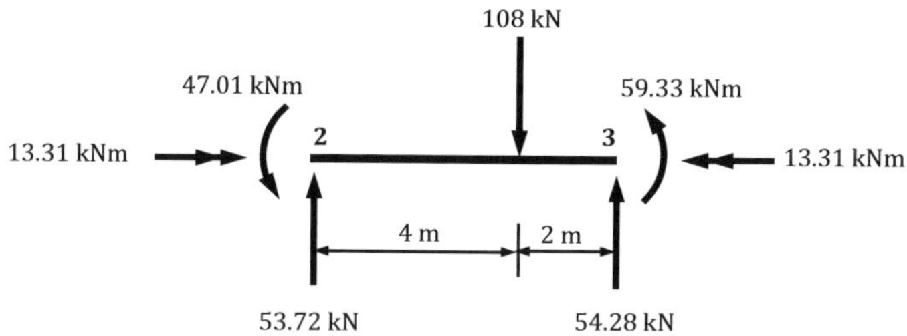

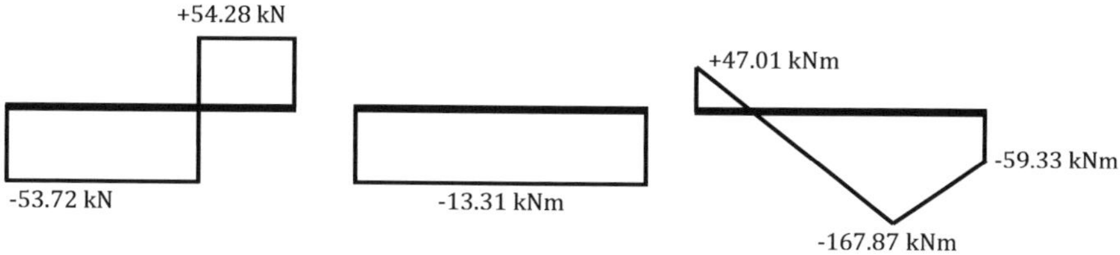

Shear force diagram, $\overline{V}$ Torque diagram, $\overline{T}$ Bending moment diagram, $\overline{M}$

Member 3

$x_{ij} = -4$ m, $y_{ij} = -3$ m, L = 5 m

$$\overline{k}^3 = \begin{bmatrix} 11894 & 0 & -29736 & -11894 & 0 & -29736 \\ 0 & 14720 & 0 & 0 & -14720 & 0 \\ -29736 & 0 & 99120 & 29736 & 0 & 49560 \\ -11894 & 0 & 29736 & 11894 & 0 & 29736 \\ 0 & -14720 & 0 & 0 & 14720 & 0 \\ -29736 & 0 & 49560 & 29736 & 0 & 99120 \end{bmatrix}$$

$$r^3 = \begin{bmatrix} 1.0 & 0 & 0 & 0 & 0 & 0 \\ 0 & -0.8 & -0.6 & 0 & 0 & 0 \\ 0 & 0.6 & -0.8 & 0 & 0 & 0 \\ 0 & 0 & 0 & 1.0 & 0 & 0 \\ 0 & 0 & 0 & 0 & -0.8 & -0.6 \\ 0 & 0 & 0 & 0 & 0.6 & -0.8 \end{bmatrix}$$

$$\delta^3 = \begin{bmatrix} \delta_i^3 \\ \delta_j^3 \end{bmatrix} = \begin{bmatrix} \delta_3 \\ \delta_4 \end{bmatrix} = \begin{bmatrix} w_3 \\ \Phi_3 \\ \theta_3 \\ w_4 \\ \Phi_4 \\ \theta_4 \end{bmatrix} = \begin{bmatrix} -0.021157 \\ -0.001230 \\ +0.006113 \\ 0 \\ 0 \\ 0 \end{bmatrix}$$

$$\bar{\delta}^3 = r^3\delta^3 = \begin{bmatrix} 1.0 & 0 & 0 & 0 & 0 & 0 \\ 0 & -0.8 & -0.6 & 0 & 0 & 0 \\ 0 & 0.6 & -0.8 & 0 & 0 & 0 \\ 0 & 0 & 0 & 1.0 & 0 & 0 \\ 0 & 0 & 0 & 0 & -0.8 & -0.6 \\ 0 & 0 & 0 & 0 & 0.6 & -0.8 \end{bmatrix} \begin{bmatrix} -0.021157 \\ -0.001230 \\ +0.006113 \\ 0 \\ 0 \\ 0 \end{bmatrix} = \begin{bmatrix} -0.021157 \\ -0.002684 \\ -0.005628 \\ 0 \\ 0 \\ 0 \end{bmatrix}$$

$$\bar{F}_d^3 = \bar{k}^3\bar{\delta}^3 = \begin{bmatrix} 11894 & 0 & -29736 & -11894 & 0 & -29736 \\ 0 & 14720 & 0 & 0 & -14720 & 0 \\ -29736 & 0 & 99120 & 29736 & 0 & 49560 \\ -11894 & 0 & 29736 & 11894 & 0 & 29736 \\ 0 & -14720 & 0 & 0 & 14720 & 0 \\ -29736 & 0 & 49560 & 29736 & 0 & 99120 \end{bmatrix} \begin{bmatrix} -0.021157 \\ -0.002684 \\ -0.005628 \\ 0 \\ 0 \\ 0 \end{bmatrix}$$

$$\bar{F}_d^3 = \begin{bmatrix} -84.30 \\ -39.51 \\ +71.28 \\ +84.30 \\ +39.51 \\ +350.20 \end{bmatrix} \text{ and from (7.12), } \bar{F}_f^3 = \begin{bmatrix} +30 \\ 0 \\ -25 \\ +30 \\ 0 \\ +25 \end{bmatrix}$$

$$\begin{bmatrix} (\bar{Z}_3^3)_r \\ (\bar{T}_3^3)_r \\ (\bar{M}_3^3)_r \\ (\bar{Z}_4^3)_r \\ (\bar{T}_4^3)_r \\ (\bar{M}_4^3)_r \end{bmatrix} = \bar{F}_r^3 = \bar{F}_d^3 + \bar{F}_f^3 = \begin{bmatrix} -84.30 \\ -39.51 \\ +71.28 \\ +84.30 \\ +39.51 \\ +350.20 \end{bmatrix} + \begin{bmatrix} +30 \\ 0 \\ -25 \\ +30 \\ 0 \\ +25 \end{bmatrix} = \begin{bmatrix} -54.30 \\ -39.51 \\ +46.28 \\ +114.30 \\ +39.51 \\ +375.20 \end{bmatrix}$$

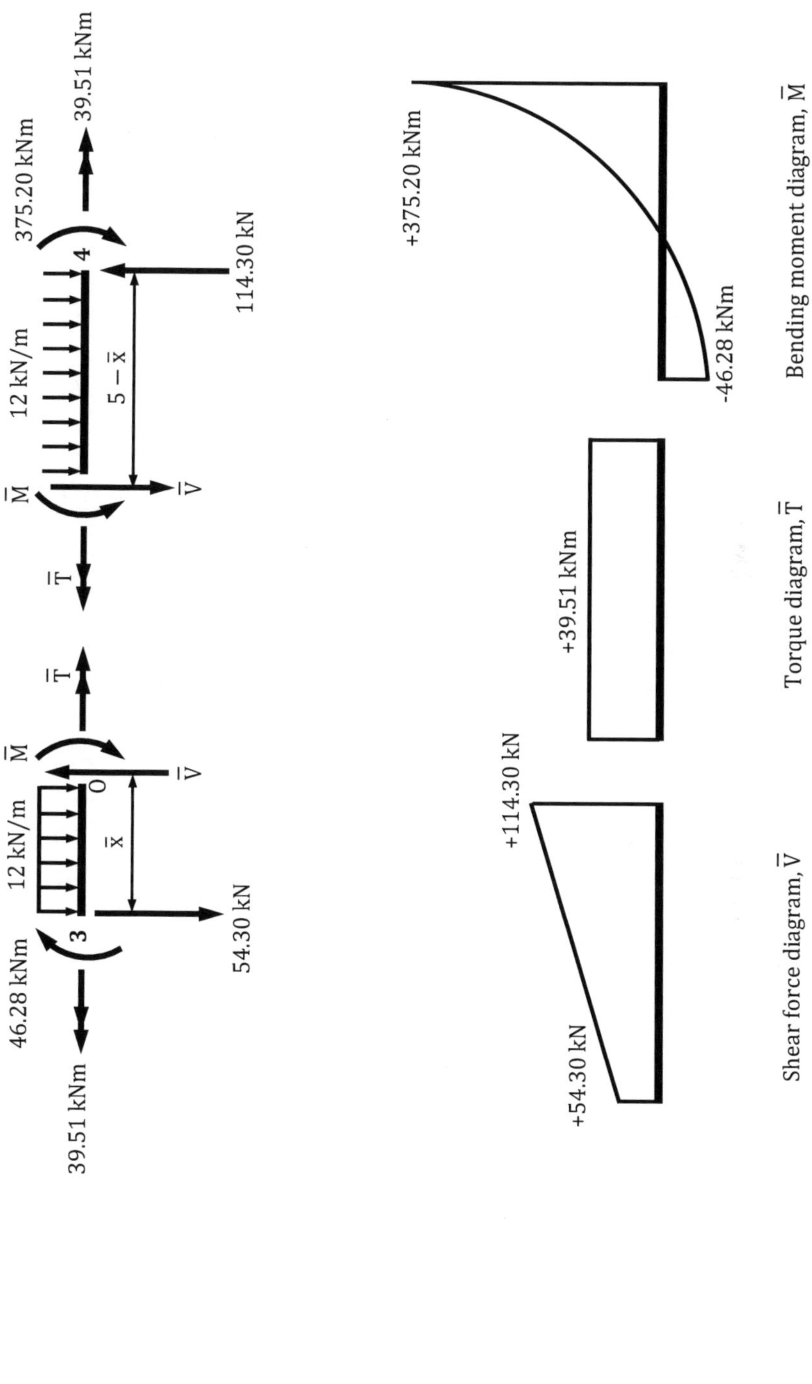
39.51 kNm
46.28 kNm
3
12 kN/m
x̄
O
M̄
V̄
T̄
54.30 kN
T̄
M̄
12 kN/m
5 − x̄
V̄
4
375.20 kNm
39.51 kNm
114.30 kN
+54.30 kN
+114.30 kN
Shear force diagram, V̄
+39.51 kNm
Torque diagram, T̄
+375.20 kNm
-46.28 kNm
Bending moment diagram, M̄

Calculation of internal actions for member 3

Considering the left part of the member, the summation of the forces in the $\bar{z}$-direction is zero:

$$-54.30-12\bar{x}+\bar{V}=0,\quad \bar{V}=54.30+12\bar{x}\ \text{kN}$$

Summation of the moments about the $\bar{x}$-axis is zero:

$$-39.51+\bar{T}=0,\quad \bar{T}=+39.51\ \text{kNm}$$

Summation of the moments about the $\bar{y}$-axis about point O is zero:

$$+46.28-54.30\bar{x}-\frac{12\bar{x}^2}{2}+\bar{M}=0$$

$$\bar{M}=-46.28+54.30\bar{x}+6\bar{x}^2\ \text{kNm}$$

Problems

P7.1. The frame shown in Fig. P7.1 is fixed at supports 1 and 3 and carries a point load of 50 kN at node 2. Analyse the frame and draw the shear force, bending moment and twisting moment diagrams for the following data:

$E = 32 \times 10^6$ kN/m^2, $G = 14 \times 10^6$ kN/m^2 and all members have the same cross section with $I = 0.003$ m^4 and $J = 0.002$ m^4.

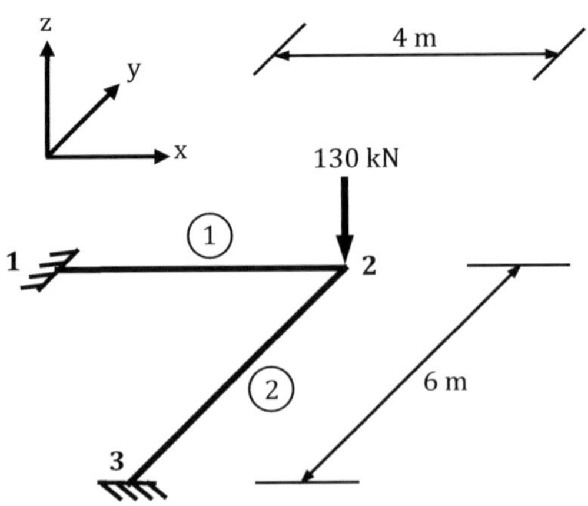

Figure P7.1

Answer:

$w_2 = -0.01897$ m, $\Phi_2 = -0.00428$ rad, $\theta_2 = +1.00678$ rad

$w_3 = 0$, $\Phi_3 = 0$, $\theta_3 = 0$

$R_{Z1} = +97.23$ kN, $R_{T1} = +29.92$ kNm, $R_{M1} = -357.26$ kNm

$R_{Z3} = +32.77$ kN, $R_{T3} = +166.71$ kNm, $R_{M3} = -31.66$ kNm

$$\text{Member 1:} \begin{bmatrix} (\bar{Z}_1^1)_r \\ (\bar{T}_1^1)_r \\ (\bar{M}_1^1)_r \\ (\bar{Z}_2^1)_r \\ (\bar{T}_2^1)_r \\ (\bar{M}_2^1)_r \end{bmatrix} = \begin{bmatrix} +97.23 \\ +29.92 \\ -357.26 \\ -97.23 \\ -29.92 \\ -31.66 \end{bmatrix}, \text{Member 2:} \begin{bmatrix} (\bar{Z}_2^2)_r \\ (\bar{T}_2^2)_r \\ (\bar{M}_2^2)_r \\ (\bar{Z}_3^2)_r \\ (\bar{T}_3^2)_r \\ (\bar{M}_3^2)_r \end{bmatrix} = \begin{bmatrix} -32.77 \\ -31.66 \\ +29.92 \\ +32.77 \\ +31.66 \\ +166.71 \end{bmatrix}$$

P7.2. The frame shown in Fig. P7.2 is fixed at support 1 and is pinned at support 3, so that the external moment exerted on the frame about the global x-axis at that support is zero, and. Analyse the frame and draw the shear force, bending moment, and twisting moment diagrams for the following data:

$E = 210 \times 10^6$ kN/m^2, $G = 80 \times 10^6$ kN/m^2, $I_1 = 0.000190$ m^4, $J_1 = 0.000230$ m^4, $I_2 = 0.000120$ m^4, $J_2 = 0.000170$ m^4.

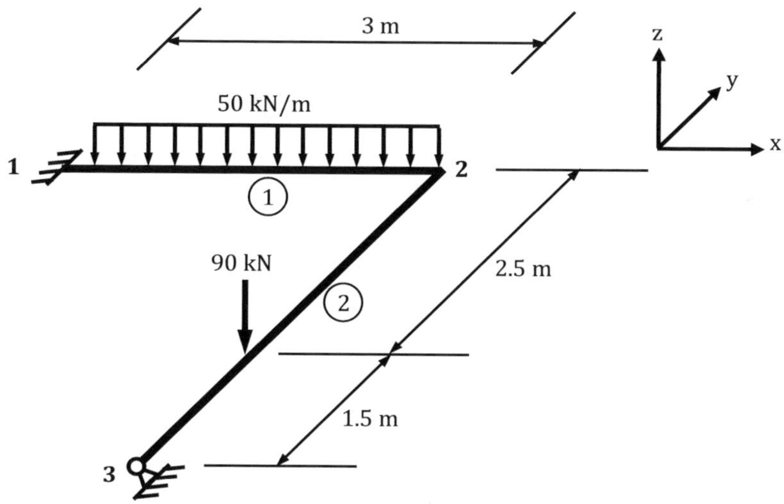

Figure P7.2

Answer:

$w_1 = 0, \Phi_1 - 0, \theta_1 = 0$

$w_2 = -0.01715$ m, $\Phi_2 = -0.00092$ rad, $\theta_2 = +0.00740$ rad

$w_3 = 0, \Phi_3 = -0.00806$ rad, $\theta_3 = 0$

$R_{Z1} = +182.34$ kN, $R_{T1} = +5.64$ kNm, $R_{M1} = -296.88$ kNm

$R_{Z3} = +57.66$ kN, $R_{T3} = 0$ kNm, $R_{M3} = -25.15$ kNm

$$\text{Member 1:} \begin{bmatrix} (\bar{Z}_1^1)_r \\ (\bar{T}_1^1)_r \\ (\bar{M}_1^1)_r \\ (\bar{Z}_2^1)_r \\ (\bar{T}_2^1)_r \\ (\bar{M}_2^1)_r \end{bmatrix} = \begin{bmatrix} +182.34 \\ +5.64 \\ -296.88 \\ -32.34 \\ -5.64 \\ -25.15 \end{bmatrix}, \quad \text{Member 2:} \begin{bmatrix} (\bar{Z}_2^2)_r \\ (\bar{T}_2^2)_r \\ (\bar{M}_2^2)_r \\ (\bar{Z}_3^2)_r \\ (\bar{T}_3^2)_r \\ (\bar{M}_3^2)_r \end{bmatrix} = \begin{bmatrix} +32.34 \\ -25.15 \\ +5.64 \\ +57.66 \\ +25.15 \\ 0 \end{bmatrix}$$

P7.3. The frame shown in Fig. P7.3 is fixed at the supports 3 and 4. Analyse the frame and draw the shear force, bending moment, and twisting moment diagrams for the following data:

$I_1 = I_2 = I_3 = 0.00022$ m^4, $J_1 = J_2 = J_3 = 0.00014$ m^4, $I_4 = 0.00030$ m^4, $J_4 = 0.00017$ m^4,

$E = 210 \times 10^6$ kN/m^2, and $G = 80 \times 10^6$ kN/m^2.

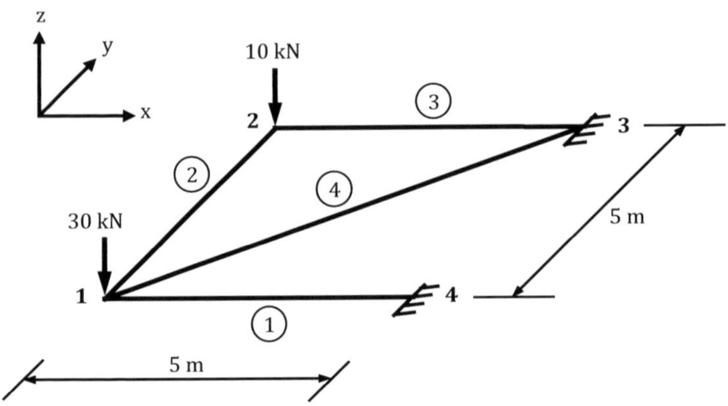

Figure P7.3

Answer:

$w_1 = -0.01615$ m, $\Phi_1 = +0.00081$ rad, $\theta_1 = -0.00447$ rad

$w_2 = -0.01118$ m, $\Phi_2 = +0.00102$ rad, $\theta_2 = -0.00342$ rad

$w_3 = 0, \Phi_3 = 0, \theta_3 = 0$
$w_4 = 0, \Phi_4 = 0, \theta_4 = 0$
$R_{Z3} = +17.97$ kN, $R_{T3} = -38.01$ kNm, $R_{M3} = +103.58$ kNm
$R_{Z4} = +22.03$ kN, $R_{T4} = -1.82$ kNm, $R_{M4} = +96.42$ kNm

$$\text{Member 1: } \begin{bmatrix} (\bar{Z}_1^1)_r \\ (\bar{T}_1^1)_r \\ (\bar{M}_1^1)_r \\ (\bar{Z}_4^1)_r \\ (\bar{T}_4^1)_r \\ (\bar{M}_4^1)_r \end{bmatrix} = \begin{bmatrix} -22.03 \\ +1.82 \\ +13.75 \\ +22.03 \\ -1.82 \\ +96.42 \end{bmatrix}, \quad \text{Member 2: } \begin{bmatrix} (\bar{Z}_1^2)_r \\ (\bar{T}_1^2)_r \\ (\bar{M}_1^2)_r \\ (\bar{Z}_2^2)_r \\ (\bar{T}_2^2)_r \\ (\bar{M}_2^2)_r \end{bmatrix} = \begin{bmatrix} -1.69 \\ -2.36 \\ +6.16 \\ +1.69 \\ +2.36 \\ +2.29 \end{bmatrix},$$

$$\text{Member 3: } \begin{bmatrix} (\bar{Z}_2^3)_r \\ (\bar{T}_2^3)_r \\ (\bar{M}_2^3)_r \\ (\bar{Z}_3^3)_r \\ (\bar{T}_3^3)_r \\ (\bar{M}_3^3)_r \end{bmatrix} = \begin{bmatrix} -11.69 \\ +2.29 \\ -2.36 \\ +11.69 \\ -2.29 \\ +60.82 \end{bmatrix}, \quad \text{Member 4: } \begin{bmatrix} (\bar{Z}_1^4)_r \\ (\bar{T}_1^4)_r \\ (\bar{M}_1^4)_r \\ (\bar{Z}_3^4)_r \\ (\bar{T}_3^4)_r \\ (\bar{M}_3^4)_r \end{bmatrix} = \begin{bmatrix} -6.28 \\ -4.98 \\ -11.12 \\ +6.28 \\ +4.98 \\ +55.50 \end{bmatrix}$$

P7.4. The frame shown in Fig. P7.4 is fixed at supports 3 and 4 with a rigid joint at node 2 and is pinned to the support at node 1 so that it cannot develop moment about the global

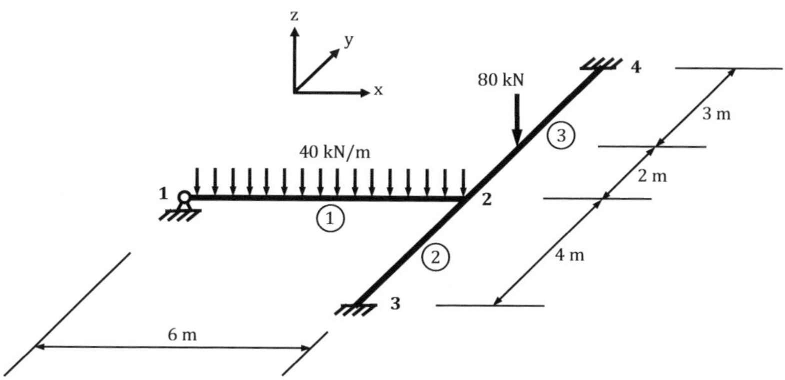

Figure P7.4

y-axis. Analyse the frame and draw the shear force, bending moment, and twisting moment diagrams for the following data:

$E = 27 \times 10^6$ kN/m^2, $G = 12 \times 10^6$ kN/m^2, $I = 0.0015$ m^4, and $J = 0.0012$ m^4.

Answer:

$w_1 = 0, \Phi_1 = 0, \theta_1 = +0.01086$ rad

$w_2 = -0.01634$ m, $\Phi_2 = -0.00195$ rad, $\theta_2 = -0.00467$ rad

$w_3 = 0, \Phi_3 = 0, \theta_3 = 0, w_4 = 0, \Phi_4 = 0, \theta_4 = 0$

$R_{Z1} = +114.96$ kN, $R_{T1} = +4.68$ kNm, $R_{M1} = 0$

$R_{Z3} = +94.41$ kN, $R_{T3} = +208.58$ kNm, $R_{M3} = +16.82$ kNm

$R_{Z4} = +110.63$ kN, $R_{T4} = -228.78$ kNm, $R_{M4} = +13.45$ kNm

$$\text{Member1:}\begin{bmatrix}(\bar{Z}_1^1)_r\\(\bar{T}_1^1)_r\\(\bar{M}_1^1)_r\\(\bar{Z}_2^1)_r\\(\bar{T}_2^1)_r\\(\bar{M}_2^1)_r\end{bmatrix}=\begin{bmatrix}+114.96\\+4.68\\0\\+125.05\\-4.68\\+30.27\end{bmatrix},\text{Member2:}\begin{bmatrix}(\bar{Z}_2^2)_r\\(\bar{T}_2^2)_r\\(\bar{M}_2^2)_r\\(\bar{Z}_3^2)_r\\(\bar{T}_3^2)_r\\(\bar{M}_3^2)_r\end{bmatrix}=\begin{bmatrix}-94.41\\+16.82\\+169.07\\+94.41\\-16.82\\+208.58\end{bmatrix},$$

$$\text{Member 3:}\begin{bmatrix}(\bar{Z}_2^3)_r\\(\bar{T}_2^3)_r\\(\bar{M}_2^3)_r\\(\bar{Z}_4^3)_r\\(\bar{T}_4^3)_r\\(\bar{M}_4^3)_r\end{bmatrix}=\begin{bmatrix}-30.63\\-13.45\\+164.39\\+110.63\\+13.45\\+228.78\end{bmatrix}$$

Chapter 8

Beams Curved in Plan

This type of beams occurs in curved bridges and buildings where the plan is of a curved shape as shown in Fig. 8.1. In most common case the shape of the curve is circular and this chapter deals specifically with that type of curved beams. The behaviour of these beams is characterised by torsion that develops due to their curvature in plan. So, a beam curved in plan subjected to gravity loads will develop torsion in addition to the bending moment, and the shear force that occur in straight beams.

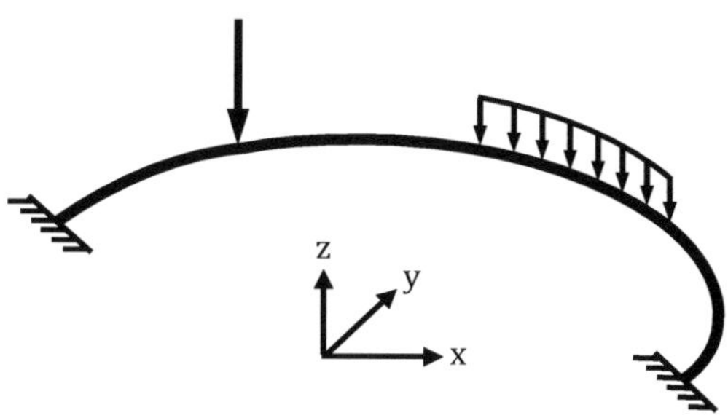

Figure 8.1 Beam curved in plan.

Analysis of Structures by Matrix Methods
Fathi Al-Shawi

ISBN 978-981-4968-19-5 (Hardcover), 978-1-003-29130-5 (eBook)
www.jennystanford.com

8.1 Derivation of Stiffness Matrix

Consider an element of a beam curved in plan whose local $\bar{x}$-axis is defined by the line joining its two ends and is coincident with the global x-axis as shown by the line (ae) in Fig. 8.2. At node i the element is subjected to a shear force $\bar{Z}_i$ in the $\bar{z}$-direction and moments $\bar{T}_i$ and $\bar{M}_i$ about the local $\bar{x}$- and $\bar{y}$-axes respectively. The corresponding displacements are translation $\bar{w}_i$ in the $\bar{z}$-direction and rotations $\bar{\Phi}_i$ and $\bar{\theta}_i$ about the $\bar{x}$- and $\bar{y}$-axes respectively. Similar actions and displacements occur at the other end of the element but with the subscript j.

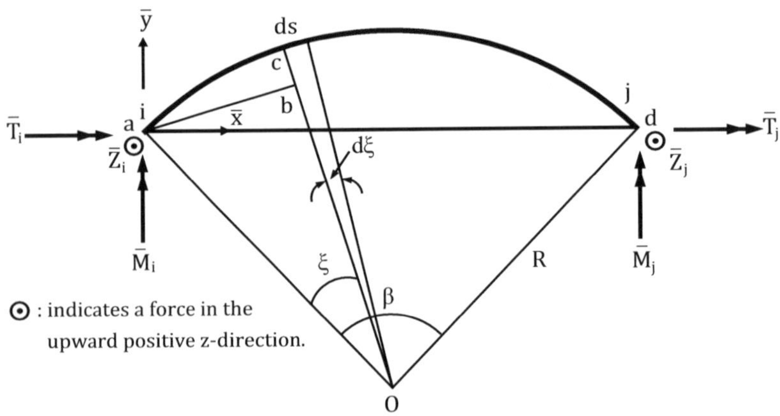

Figure 8.2 A circular beam curved in plan element.

First consider end i and assume that it is displaced by a translational displacement $\bar{w}_i$ and rotational displacements $\bar{\Phi}_i$ and $\bar{\theta}_i$ about the $\bar{x}$- and $\bar{y}$-axes respectively, while end j is fixed, i.e. $\bar{w}_j = 0$, $\bar{\Phi}_j = 0$, and $\bar{\theta}_j = 0$. For this case, the forces and moments developed at the ends of the element are calculated in terms of $\bar{w}_i$, $\bar{\Phi}_i$, and $\bar{\theta}_i$ and they are superscripted by a single prime. The second part of the derivation assumes that end j is given displacements $\bar{w}_j$, $\bar{\Phi}_j$, and $\bar{\theta}_j$ while end i is fixed, i.e. $\bar{w}_i = 0$, $\bar{\Phi}_i = 0$, and $\bar{\theta}_i = 0$. This will lead to forces and moments developed at the ends of the element in terms of $\bar{w}_j$, $\bar{\Phi}_j$, and $\bar{\theta}_j$, these are superscripted by double primes. The final forces and moments are obtained by adding the quantities from the two steps.

With reference to the circular beam element of radius R shown in Fig. 8.2, consider section c which makes an angle ξ with the line oa, thus

$$bc = R - R\cos\xi = R(1 - \cos\xi) \text{ and } ab = R\sin\xi$$

Consider the equilibrium of part ac shown in Fig. 8.3.

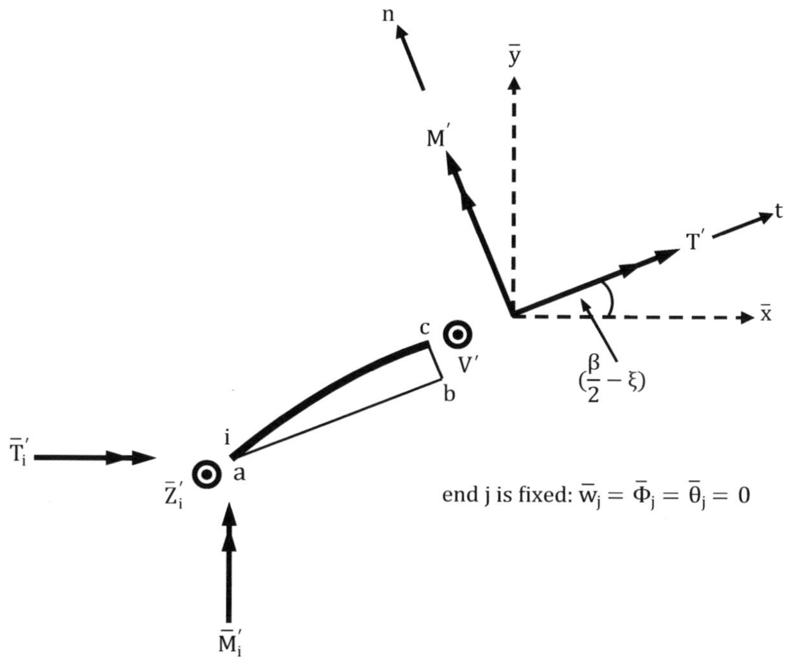

Figure 8.3

Summation of the moments about the tangential t-axis:

$$\bar{T}_i' \cos\left(\frac{\beta}{2} - \xi\right) + \bar{M}_i' \sin\left(\frac{\beta}{2} - \xi\right) - \bar{Z}_i'(bc) + T' = 0$$

$$T' = -\bar{T}_i' \cos\left(\frac{\beta}{2} - \xi\right) - \bar{M}_i' \sin\left(\frac{\beta}{2} - \xi\right) + \bar{Z}_i' R(1 - \cos\xi) \qquad (8.1)$$

Summation of the moments about the normal n-axis:

$$-\bar{T}_i' \sin\left(\frac{\beta}{2} - \xi\right) + \bar{M}_i' \cos\left(\frac{\beta}{2} - \xi\right) + \bar{Z}_i'(ab) + M' = 0$$

$$M' = \bar{T}_i' \sin\left(\frac{\beta}{2} - \xi\right) - \bar{M}_i' \cos\left(\frac{\beta}{2} - \xi\right) - \bar{Z}_i' R \sin\xi \tag{8.2}$$

where T′ and M′ are the internal moments about the tangential and normal axes respectively at point c.

Neglecting the effect of shearing forces on the deformation the strain energy is given by:

$$U' = \int \frac{M'^2 ds}{2EI} + \int \frac{T'^2 ds}{2GJ}, \quad \text{where} \quad ds = R d\xi$$

$$U' = \frac{R}{2EI}\int_0^{\beta} M'^2 d\xi + \frac{R}{2GJ}\int_0^{\beta} T'^2 d\xi \tag{8.3}$$

Applying Castigliano's theorem leads to the following equations:

$$\bar{w}_i = \frac{\partial U'}{\partial \bar{Z}_i'} = \frac{\partial U'}{\partial M'}\frac{\partial M'}{\partial \bar{Z}_i'} + \frac{\partial U'}{\partial T'}\frac{\partial T'}{\partial \bar{Z}_i'} \tag{8.4}$$

$$\bar{\Phi}_i = \frac{\partial U'}{\partial \bar{T}_i'} = \frac{\partial U'}{\partial M'}\frac{\partial M'}{\partial \bar{T}_i'} + \frac{\partial U'}{\partial T'}\frac{\partial T'}{\partial \bar{T}_i'} \tag{8.5}$$

$$\bar{\theta}_i = \frac{\partial U'}{\partial \bar{M}_i'} = \frac{\partial U'}{\partial M'}\frac{\partial M'}{\partial \bar{M}_i'} + \frac{\partial U'}{\partial T'}\frac{\partial T'}{\partial \bar{M}_i'} \tag{8.6}$$

From (8.3) we get

$$\frac{\partial U'}{\partial M'} = \frac{R}{EI}\int_0^{\beta} M d\xi \quad \text{and} \quad \frac{\partial U'}{\partial T'} = \frac{R}{GJ}\int_0^{\beta} T' d\xi$$

From (8.1)

$$\frac{\partial T'}{\partial \bar{Z}_i'} = R(1-\cos\xi), \quad \frac{\partial T'}{\partial \bar{T}_i'} = -\cos\left(\frac{\beta}{2} - \xi\right) \text{and} \quad \frac{\partial T'}{\partial \bar{M}_i'} = -\sin\left(\frac{\beta}{2} - \xi\right)$$

From (8.2)

$$\frac{\partial M'}{\partial \bar{Z}_i'} = -R\sin\xi, \quad \frac{\partial M'}{\partial \bar{T}_i'} = \sin\left(\frac{\beta}{2} - \xi\right) \quad \text{and} \quad \frac{\partial M'}{\partial \bar{M}_i'} = -\cos\left(\frac{\beta}{2} - \xi\right)$$

Substitution of (8.1), (8.2), and the relevant derivatives, as appropriate, from above into (8.4) will give

$$\overline{w}_i = \frac{R}{EI}\int_0^{\beta}\left(\overline{T}_i' \sin(\beta/2-\xi) - \overline{M}_i' \cos(\beta/2-\xi) - \overline{Z}_i' R\sin\xi\right)(-R\sin\xi)d\xi \; +$$

$$\frac{R}{GJ}\int_0^{\beta}(-\overline{T}_i' \cos(\beta/2-\xi) - \overline{M}_i' \sin(\beta/2-\xi) + \overline{Z}_i' R(1-\cos\xi))R(1-\cos\xi)d\xi$$

Introducing the parameter $\alpha = EI/GJ$ and integrating from $\xi = 0$ to $\xi = \beta$ to get

$$\overline{w}_i = (R^2/EI)\Big\{\big[0.5\beta(3\alpha+1) - 2\alpha\sin\beta + 0.5(\alpha-1)\sin\beta\cos\beta\big]R\overline{Z}_i' +$$

$$\big[0.25(\alpha-1)\sin(3\beta/2) + 0.5\beta(\alpha+1)\cos(\beta/2) - 0.25(7\alpha+1)\sin(\beta/2)\big]\overline{T}_i' +$$

$$\big[0.25(\alpha-1)\cos(3\beta/2) + 0.5\beta(\alpha+1)\sin(\beta/2) - 0.25(\alpha-1)\cos(\beta/2)\big]\overline{M}_i'\Big\} \tag{8.7}$$

Similarly (8.5) is simplified to give

$$\overline{\Phi}_i = (R/EI)\big\{[0.25(\alpha-1)\sin(3\beta/2) + 0.5\beta(\alpha+1)\cos(\beta/2) -$$

$$0.25(7\alpha+1)\sin(\beta/2)]R\overline{Z}_i' + \big[0.5\beta(\alpha+1) + 0.5(\alpha-1)\sin\beta\big]\overline{T}_i'\big\} \tag{8.8}$$

And (8.6) is simplified to

$$\overline{\theta}_i = (R/EI)\Big\{[0.25(\alpha-1)\cos(3\beta/2) + 0.5\beta(\alpha+1)\sin(\beta/2) \; -$$

$$0.25(\alpha-1)\cos(\beta/2)]R\overline{Z}_i' + \big[0.5\beta(\alpha+1) - 0.5(\alpha-1)\sin\beta\big]\overline{M}_i'\Big\} \tag{8.9}$$

Solve equations (8.7), (8.8), and (8.9) simultaneously for the unknowns $\overline{Z}_i'$, $\overline{T}_i'$, and $\overline{M}_i'$ to get:

$$\overline{Z}_i' = \frac{EI}{R^3}\left(C_1\overline{w}_i + C_2R\overline{\Phi}_i - C_3R\overline{\theta}_i\right) \tag{8.10}$$

$$\overline{T}_i' = \frac{EI}{R^3}\left(C_2R\overline{w}_i + C_4R^2\overline{\Phi}_i - C_5R^2\overline{\theta}_i\right) \tag{8.11}$$

$$\overline{M}_i' = \frac{EI}{R^3}\left(-C_3R\overline{w}_i - C_5R^2\overline{\Phi}_i + C_6R^2\overline{\theta}_i\right) \tag{8.12}$$

Equations (8.10), (8.11), and (8.12) represent $\overline{k}_{ii}$ of the stiffness matrix.

From the overall equilibrium of the element the following equations are obtained:

$\sum Z = 0: \bar{Z}_i' + \bar{Z}_j' = 0,\ \bar{Z}_j' = -\bar{Z}_i'$ and from (8.10) we get

$$\bar{Z}_j' = \frac{EI}{R^3}\left(-C_1\bar{w}_i - C_2R\bar{\Phi}_i + C_3R\bar{\theta}_i\right) \tag{8.13}$$

$\sum T = 0: \bar{T}_i' + \bar{T}_j' = 0,\ \bar{T}_j' = -\bar{T}_i'$ and from (8.11) we get

$$\bar{T}_j' = \frac{EI}{R^3}\left(-C_2R\bar{w}_i - C_4R^2\bar{\Phi}_i + C_5R^2\bar{\theta}_i\right) \tag{8.14}$$

$\sum M = 0: \bar{M}_i' + \bar{Z}_i'L + \bar{M}_j' = 0,\ \bar{M}_j' = -\bar{M}_i' - \bar{Z}_i'L$, where L is the length of the straight line joining ends i and j of the element and is equal to $2R\sin(\beta/2)$.

Substitute $\bar{M}_i'$ and $\bar{Z}_i'$ as given in (8.12) and (8.13) respectively in the above equation leads to:

$$\bar{M}_j' = \frac{EI}{R^3}(-C_3R\bar{w}_i - C_5R^2\bar{\Phi}_i + C_7R^2\bar{\theta}_i) \tag{8.15}$$

Equations (8.13), (8.14), and (8.15) represent $\bar{k}_{ji}$ of the stiffness matrix.

And $C_1, C_2, \ldots\ldots\ldots, C_7$ are functions of α and β.

The above process is repeated with end i fixed and end j is given displacements

$$\bar{w}_j,\ \bar{\Phi}_j, \text{ and } \bar{\theta}_j.$$

The equations for the torque and moment at section c in this case are similar to (8.1) and (8.2) but the single primes are replaced by double primes as shown in Fig. 8.4, thus

$$T'' = -\bar{T}_i''\cos\left(\frac{\beta}{2} - \xi\right) - \bar{M}_i''\sin\left(\frac{\beta}{2} - \xi\right) + \bar{Z}_i''R(1 - \cos\xi) \tag{8.16}$$

$$M'' = \bar{T}_i''\sin\left(\frac{\beta}{2} - \xi\right) - \bar{M}_i''\cos\left(\frac{\beta}{2} - \xi\right) - \bar{Z}_i''R\sin\xi \tag{8.17}$$

Since we want to find expressions for the displacements $\bar{w}_j$, $\bar{\Phi}_j$, and $\bar{\theta}_j$ the above two equations are written in terms of $\bar{Z}_j''$, $\bar{T}_j''$, and $\bar{M}_j''$ whose derivatives will give the respective displacements. To achieve this, the equilibrium of the whole beam is considered.

Summation of the forces in the $\bar{z}$-direction is zero:

$$\bar{Z}_i'' + \bar{Z}_j'' = 0,\ \ \bar{Z}_i'' = -\bar{Z}_j''$$

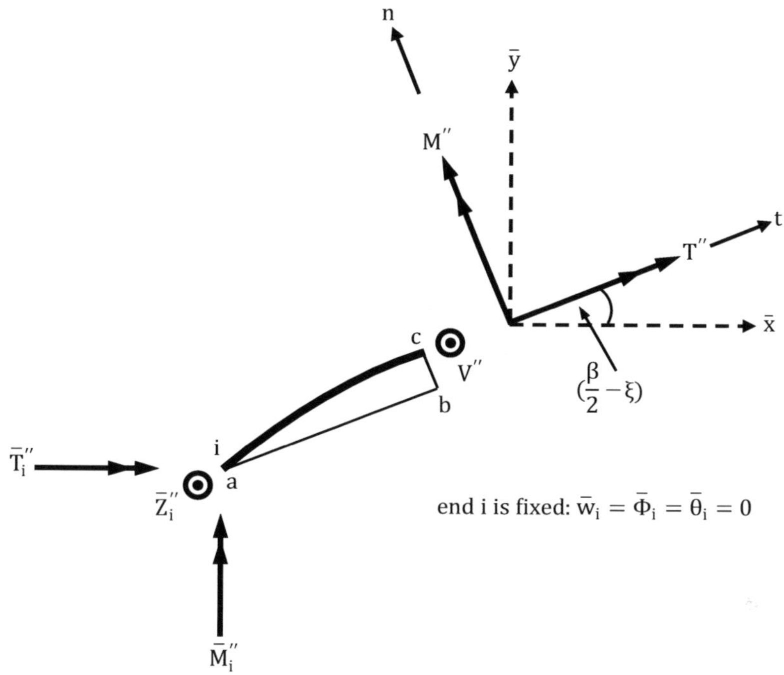

Figure 8.4

Summation of moments about the $\bar{x}$-axis is zero:

$$\bar{T}_i'' + \bar{T}_j'' = 0, \quad \bar{T}_i'' = -\bar{T}_j''$$

Summation of moments about node I is zero:

$$\bar{M}_i'' - \bar{Z}_j'' L + \bar{M}_j'' = 0, \quad \bar{M}_i'' = 2R\bar{Z}_j'' \sin\left(\frac{\beta}{2}\right) - \bar{M}_j'' = 0,$$

Substitute the above values of $\bar{Z}_i''$, $\bar{T}_i''$, and $\bar{M}_i''$ in (8.16) and (8.17) respectively and simplify to get

$$T'' = \bar{T}_j'' \cos\left(\frac{\beta}{2} - \xi\right) + \bar{M}_j'' \sin\left(\frac{\beta}{2} - \xi\right) - \bar{Z}_j'' R\big(1 - \cos(\beta - \xi)\big) \quad (8.18)$$

$$M'' = -\bar{T}_j'' \sin\left(\frac{\beta}{2} - \xi\right) + \bar{M}_j'' \cos\left(\frac{\beta}{2} - \xi\right) - \bar{Z}_j'' R \sin(\beta - \xi) \quad (8.19)$$

$$U'' = \frac{R}{2EI}\int_0^\beta M''^2 d\xi + \frac{R}{2GJ}\int_0^\beta T''^2 d\xi \quad (8.20)$$

$$\bar{w}_j = \frac{\partial U''}{\partial \bar{Z}_j''} = \frac{\partial U''}{\partial M''}\frac{\partial M''}{\partial \bar{Z}_j''} + \frac{\partial U''}{\partial T''}\frac{\partial T''}{\partial \bar{Z}_j''} \tag{8.21}$$

$$\bar{\Phi}_j = \frac{\partial U''}{\partial \bar{T}_j''} = \frac{\partial U''}{\partial M''}\frac{\partial M''}{\partial \bar{T}_j''} + \frac{\partial U''}{\partial T''}\frac{\partial T''}{\partial \bar{T}_j''} \tag{8.22}$$

$$\bar{\theta}_j = \frac{\partial U''}{\partial \bar{M}_j''} = \frac{\partial U''}{\partial M''}\frac{\partial M''}{\partial \bar{M}_j''} + \frac{\partial U''}{\partial T''}\frac{\partial T''}{\partial \bar{M}_j''} \tag{8.23}$$

Equations (8.21), (8.22), and (8.23) are respectively simplified and integrated from $\xi = 0$ to $\xi = \beta$ to give:

$$\begin{aligned} w_j = \left(R^2/EI\right)\Big\{&\left[0.25(\alpha-1)\sin 2\beta - 2\alpha\sin\beta + 0.5\beta(3\alpha+1)\right]RZ_j'' \\ &+[0.25(\alpha-1)\sin(3\beta/2) + 0.5\beta(\alpha+1)\cos(\beta/2) - 0.25(7\alpha+1)\sin(\beta/2)]T_j'' \\ &+[0.25(1-\alpha)\cos(3\beta/2) + 0.25(\alpha-1)\cos(\beta/2) - 0.5\beta(\alpha+1)\sin(\beta/2)]M_j''\Big\} \end{aligned} \tag{8.24}$$

$$\begin{aligned} \Phi_j = (R/EI)\Big\{&[0.25(\alpha-1)\sin(3\beta/2) + 0.5\beta(\alpha+1)\cos(\beta/2) \\ &-0.25(7\alpha+1)\sin(\beta/2)]RZ_j'' + [0.5(\alpha-1)\sin\beta + 0.5\beta(\alpha+1)]T_j''\Big\} \end{aligned} \tag{8.25}$$

$$\begin{aligned} \theta_j = (R/EI)\Big\{&[0.25(1-\alpha)\cos(3\beta/2) + 0.25(\alpha-1)\cos(\beta/2) \\ &-0.5\beta(\alpha+1)\sin(\beta/2)]RZ_j'' + [0.5(1-\alpha)\sin\beta + 0.5\beta(\alpha+1)]M_j''\Big\} \end{aligned} \tag{8.26}$$

Solve equations (8.24), (8.25), and (8.26) simultaneously for the unknowns $\bar{Z}_j''$, $\bar{T}_j''$, and $\bar{M}_j''$ to get:

$$\bar{Z}_j'' = \frac{EI}{R^3}\left(C_1\bar{w}_j + C_2R\bar{\Phi}_j + C_3R\bar{\theta}_j\right) \tag{8.27}$$

$$\bar{T}_j'' = \frac{EI}{R^3}\left(C_2R\bar{w}_j + C_4R^2\bar{\Phi}_j + C_5R^2\bar{\theta}_j\right) \tag{8.28}$$

$$\bar{M}_j'' = \frac{EI}{R^3}\left(C_3R\bar{w}_j + C_5R^2\bar{\Phi}_j + C_6R^2\bar{\theta}_j\right) \tag{8.29}$$

Equations (8.27), (8.28), and (8.29) represent $\bar{k}_{jj}$ of the stiffness matrix.

From the overall equilibrium of the element the following equations are obtained:

$\sum Z = 0:\ \bar{Z}_i'' + \bar{Z}_j'' = 0,\ \bar{Z}_i' = -\bar{Z}_j'$ and from (8.27) we get

$$\bar{Z}_i'' = \frac{EI}{R^3}\left(-C_1\bar{w}_j - C_2R\bar{\Phi}_j - C_3R\bar{\theta}_j\right) \tag{8.30}$$

$\sum T = 0: \ \bar{T}_i'' + \bar{T}_j'' = 0, \bar{T}_i'' = -\bar{T}_j''$ and from (8.28) we get

$$\bar{T}_i'' = \frac{EI}{R^3}\left(-C_2R\bar{w}_j - C_4R^2\bar{\Phi}_j - C_5R^2\bar{\theta}_j\right) \tag{8.31}$$

$\sum M = 0: \ \bar{M}_i'' - \bar{Z}_j'L + \bar{M}_j'' = 0, \bar{M}_i'' = -\bar{M}_j'' + \bar{Z}_j''L$,

where $L = 2R\sin(\beta/2)$.

Substitute $\bar{Z}_j''$ and $\bar{M}_j''$ as given in (8.27) and (8.29) respectively in the above equation leads to:

$$\bar{M}_i'' = \frac{EI}{R^3}\left(C_3R\bar{w}_j + C_5R^2\bar{\Phi}_j + C_7R^2\bar{\theta}_j\right) \tag{8.32}$$

Equations (8.30), (8.31), and (8.32) represent $\bar{k}_{ij}$ of the stiffness matrix.

Finally

from (8.10) and (8.30): $\bar{Z}_i = \bar{Z}_i' + \bar{Z}_i''$

from (8.11) and (8.31): $\bar{T}_i = \bar{T}_i' + \bar{T}_i''$

from (8.12) and (8.32): $\bar{M}_i = \bar{M}_i' + \bar{M}_i''$

from (8.13) and (8.27): $\bar{Z}_j = \bar{Z}_j' + \bar{Z}_j''$

from (8.14) and (8.28): $\bar{T}_j = \bar{T}_j' + \bar{T}_j''$

from (8.15) and (8.29): $\bar{M}_j = \bar{M}_j' + \bar{M}_j''$

The above six equations are written in matrix form to give the general stiffness matrix of a beam curved in plan element relative to local coordinates as:

$$\begin{bmatrix} \bar{Z}_i \\ \bar{T}_i \\ \bar{M}_i \\ \bar{Z}_j \\ \bar{T}_j \\ \bar{M}_j \end{bmatrix} = \frac{EI}{R^3}\begin{bmatrix} C_1 & C_2R & -C_3R & -C_1 & -C_2R & -C_3R \\ C_2R & C_4R^2 & -C_5R^2 & -C_2R & -C_4R^2 & -C_5R^2 \\ -C_3R & -C_5R^2 & C_6R^2 & C_3R & C_5R^2 & C_7R^2 \\ -C_1 & -C_2R & C_3R & C_1 & C_2R & C_3R \\ -C_2R & -C_4R^2 & C_5R^2 & C_2R & C_4R^2 & C_5R^2 \\ -C_3R & -C_5R^2 & C_7R^2 & C_3R & C_5R^2 & C_6R^2 \end{bmatrix}\begin{bmatrix} \bar{w}_i \\ \bar{\Phi}_i \\ \bar{\theta}_i \\ \bar{w}_j \\ \bar{\Phi}_j \\ \bar{\theta}_j \end{bmatrix}$$

(8.33)

$$\bar{k} = \frac{EI}{R^3}\begin{bmatrix} C_1 & C_2R & -C_3R & -C_1 & -C_2R & -C_3R \\ C_2R & C_4R^2 & -C_5R^2 & -C_2R & -C_4R^2 & -C_5R^2 \\ -C_3R & -C_5R^2 & C_6R^2 & C_3R & C_5R^2 & C_7R^2 \\ -C_1 & -C_2R & C_3R & C_1 & C_2R & C_3R \\ -C_2R & -C_4R^2 & C_5R^2 & C_2R & C_4R^2 & C_5R^2 \\ -C_3R & -C_5R^2 & C_7R^2 & C_3R & C_5R^2 & C_6R^2 \end{bmatrix} \quad (8.34)$$

Expressions for calculating the values of C_1, C_2, ... , C_7 are as given below.

$$C_1 = \frac{a_1\alpha + a_2}{\alpha(a_3\alpha+1)}, C_2 = \frac{a_4\alpha - a_5}{\alpha(a_3\alpha+1)}, C_3 = \frac{a_6\alpha + a_7}{\alpha(a_3\alpha+1)}, C_4 = \frac{a_8\alpha + a_9}{\alpha(a_3\alpha+1)},$$

$$C_5 = \frac{a_{10}\alpha - a_{11}}{\alpha(a_3\alpha+1)}, C_6 = \frac{a_{12}\alpha^2 + a_{13}\alpha + a_{14}}{\alpha(a_3\alpha+1)(a_{15}\alpha+1)}, C_7 = \frac{a_{16}\alpha^2 + a_{17}\alpha + a_{14}}{\alpha(a_3\alpha+1)(a_{15}\alpha+1)}$$

where $\alpha = EI/GJ$ and the values of a_1, a_2, ... , a_{17} for various values of β are shown in Table 8.1.

8.2 Transformation from Local to Global Coordinates

Since the resulting stiffness matrices for the elements are derived relative to local coordinates, transformation is necessary to make them relative to the global coordinates system. The direction of the secant, AB, is assumed to represent the local $\bar{x}$-axis. For the element shown in Fig. 8.5, the secant has rotated about the $\bar{z}$-axis by an angle φ_z.

The stiffness matrix relative to global coordinates is given by:

$k = r^T\bar{k}r$, where the transformation matrix, r is the same as that for a grillage member which was derived in Chapter 7 in (7.5) as

$$r = \begin{bmatrix} 1 & 0 & 0 & 0 & 0 & 0 \\ 0 & x_{ij}/L & y_{ij}/L & 0 & 0 & 0 \\ 0 & -y_{ij}/L & x_{ij}/L & 0 & 0 & 0 \\ 0 & 0 & 0 & 1 & 0 & 0 \\ 0 & 0 & 0 & 0 & x_{ij}/L & y_{ij}/L \\ 0 & 0 & 0 & 0 & -y_{ij}/L & x_{ij}/L \end{bmatrix} \quad (8.35)$$

Table 8.1

β	a_1	a_2	a_3	a_4	a_5	a_6	a_7	a_8	a_9	a_{10}	a_{11}	a_{12}	a_{13}	a_{14}	a_{15}	a_{16}	a_{17}
15°	667.285	3.820	0.001	7.616	3.787	87.166	0.499	0.090	3.754	0.993	0.494	0.069	15.191	0.065	0.006	0.061	7.518
30°	82.803	1.909	0.005	3.769	1.845	21.427	0.494	0.180	1.781	0.977	0.478	0.137	7.510	0.128	0.023	0.119	3.600
45°	24.252	1.273	0.010	2.471	1.176	9.284	0.487	0.265	1.087	0.944	0.451	0.201	4.918	0.187	0.054	0.173	2.231
60°	10.085	0.955	0.019	1.808	0.827	5.048	0.478	0.344	0.716	0.905	0.414	0.258	3.592	0.239	0.095	0.220	1.500
75°	5.064	0.764	0.029	1.403	0.606	3.082	0.465	0.415	0.481	0.855	0.369	0.309	2.798	0.283	0.151	0.257	1.040
90°	2.869	0.637	0.043	1.127	0.450	2.027	0.450	0.476	0.318	0.796	0.318	0.351	2.280	0.318	0.222	0.285	0.725

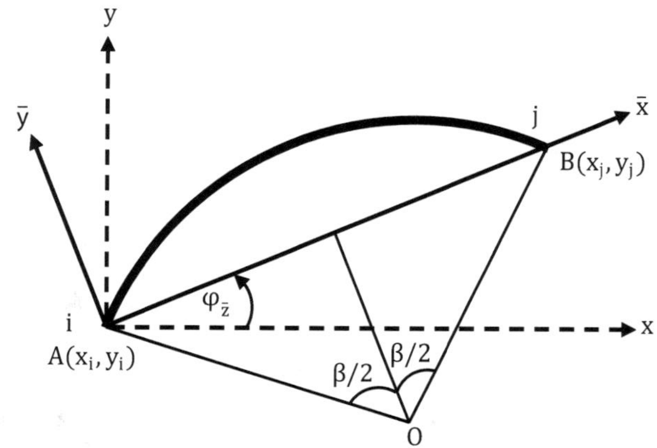

Figure 8.5

where $x_{ij} = x_j - x_i$, $y_{ij} = y_j - y_i$, and $L = \sqrt{x_{ij}^2 + y_{ij}^2}$.

8.3 Calculation of Actions Developed in the Elements

The actions on an element in a curved beam are calculated relative to the tangent and normal to the curve at the ends of the element. So the twisting moment, t, is the resultant moment about the tangent to the curve and the bending moment, m, is the resultant moment about the normal to the curve. Expressions for t and m are derived from the moments $\overline{T}$ and $\overline{M}$ about the local $\overline{x}$- and $\overline{y}$-axes respectively as shown in Fig. 8.6. The shear force, V, acting on the element in the $\overline{z}$-direction is the same as the force, $\overline{Z}$.

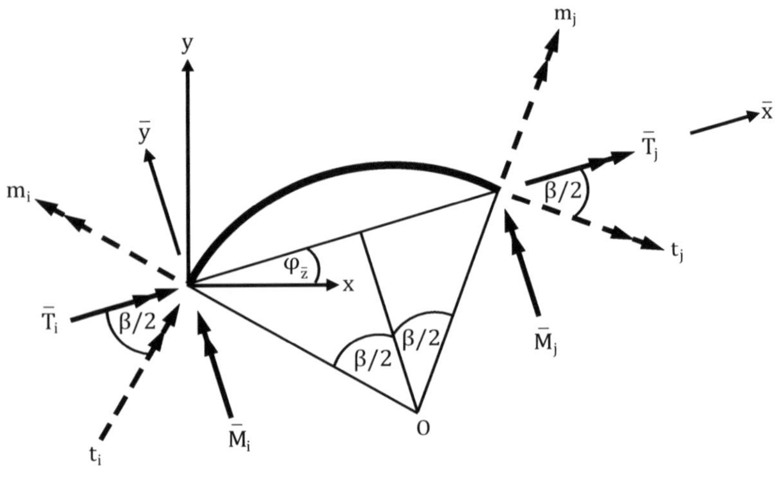

Figure 8.6

It follows that:

$$V_i = \overline{Z}_i$$

$$t_i = \overline{T}_i \cos(\beta/2) + \overline{M}_i \sin(\beta/2)$$

$$m_i = -\overline{T}_i \sin(\beta/2) + \overline{M}_i \cos(\beta/2)$$

$$V_j = \overline{Z}_j$$

$$t_j = \bar{T}_j \cos(\beta/2) - \bar{M}_j \sin(\beta/2)$$

$$m_j = \bar{T}_j \sin(\beta/2) + \bar{M}_j \cos(\beta/2)$$

In matrix form

$$\begin{bmatrix} V_i \\ t_i \\ m_i \\ V_j \\ t_j \\ m_j \end{bmatrix} = \begin{bmatrix} 1 & 0 & 0 & 0 & 0 & 0 \\ 0 & \cos\beta/2 & \sin\beta/2 & 0 & 0 & 0 \\ 0 & -\sin\beta/2 & \cos\beta/2 & 0 & 0 & 0 \\ 0 & 0 & 0 & 1 & 0 & 0 \\ 0 & 0 & 0 & 0 & \cos\beta/2 & -\sin\beta/2 \\ 0 & 0 & 0 & 0 & \sin\beta/2 & \cos\beta/2 \end{bmatrix} \begin{bmatrix} \bar{Z}_i \\ \bar{T}_i \\ \bar{M}_i \\ \bar{Z}_j \\ \bar{T}_j \\ \bar{M}_j \end{bmatrix}$$

or $f = r_\beta \bar{F}$, where

$$r_\beta = \begin{bmatrix} 1 & 0 & 0 & 0 & 0 & 0 \\ 0 & \cos\beta/2 & \sin\beta/2 & 0 & 0 & 0 \\ 0 & -\sin\beta/2 & \cos\beta/2 & 0 & 0 & 0 \\ 0 & 0 & 0 & 1 & 0 & 0 \\ 0 & 0 & 0 & 0 & \cos\beta/2 & -\sin\beta/2 \\ 0 & 0 & 0 & 0 & \sin\beta/2 & \cos\beta/2 \end{bmatrix} \tag{8.36}$$

and $\bar{F} = \bar{k}\bar{\delta} = \bar{k}(r\delta)$, thus

$$\begin{bmatrix} V_i \\ t_i \\ m_i \\ V_j \\ t_j \\ m_j \end{bmatrix} = f = r_\beta \bar{k}(r\delta) \tag{8.37}$$

Example

The beam shown in Fig. 8.7 is circular in plan and has a radius R = 4 m and a rectangular cross section with width, b = 0.3 m and depth, h = 0.6 m. The beam is made of concrete with modulus of elasticity, E = 25 × 10^6 kN/m^2 and Poisson's ratio, μ = 0.15. Calculate the displacement, shear force, twisting moment, and bending moment at the nodes 1 to 4.

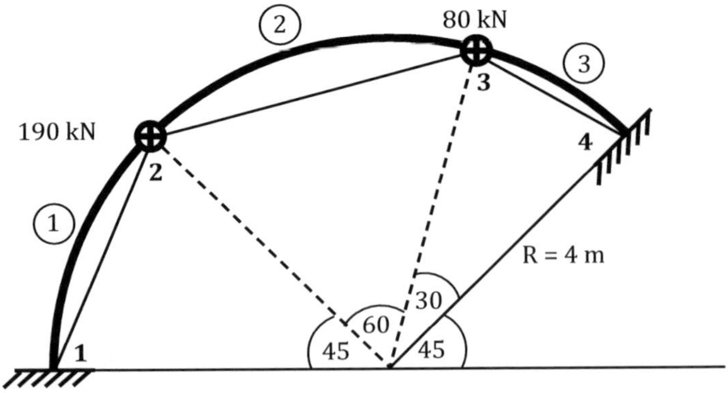

Figure 8.7

$$I = \frac{bh^3}{12} = \frac{0.3 \times 0.6^3}{12} = 0.0054\ \mathrm{m}^4$$

$$G = \frac{E}{2(1+\mu)} = \frac{25 \times 10^6}{2(1+0.15)} = 10.9 \times 10^6\ \mathrm{kN/m^2}$$

From Chapter 7, equation 7.4, $c = \dfrac{\left(\frac{h}{b}\right)^5 - 0.630\left(\frac{h}{b}\right)^4 + 0.053}{3\left(\frac{h}{b}\right)^5}$

$$c = \frac{\left(\frac{0.6}{0.3}\right)^5 - 0.630\left(\frac{0.6}{0.3}\right)^4 + 0.053}{3\left(\frac{0.6}{0.3}\right)^5} = 0.229$$

$$J = cb^3h = 0.229 \times 0.3^3 \times 0.6 = 0.0037\ \mathrm{m}^4$$

$$\alpha = \frac{EI}{GJ} = \frac{25 \times 10^6 \times 0.0054}{10.9 \times 10^6 \times 0.0037} = 3.35$$

Coordinates of nodes

Node 1:

$$x_1 = 0, \quad y_1 = 0$$

Node 2:

$$x_2 = 4 - 4\cos45 = 1.172 \text{ m}, \quad y_2 = 4\sin45 = 2.828 \text{ m}$$

Node 3:

$$X_3 = 4 + 4\cos75 = 5.035 \text{ m}, \quad y_3 = 4\sin75 = 3.864 \text{ m}$$

Node 4:

$$X_4 = 4 + 4\cos45 = 6.828 \text{ m}, \quad y_4 = 4\sin45 = 2.828 \text{ m}$$

Stiffness matrices

Element 1

$$\alpha = 3.35$$

From Table 8.1, for $\beta = 45^0$:

$a_1 = 24.252, a_2 = 1.273, a_3 = 0.010, a_4 = 2.471, a_5 = 1.176, a_6 = 9.284,$

$a_7 = 0.487, a_8 = 0.265, a_9 = 1.087, a_{10} = 0.944, a_{11} = 0.451, a_{12} = 0.201,$

$a_{13} = 4.918, a_{14} = 0.187, a_{15} = 0.054, a_{16} = 0.173, a_{17} = 2.231$

$$C_1 = \frac{a_1\alpha + a_2}{\alpha(a_3\alpha+1)} = \frac{24.252\times3.35+1.273}{3.35(0.010\times3.35+1)} = 23.83$$

$$C_2 = \frac{a_4\alpha - a_5}{\alpha(a_3\alpha+1)} = \frac{2.471\times3.35-1.176}{3.35(0.010\times3.35+1)} = 2.05$$

$$C_3 = \frac{a_6\alpha + a_7}{\alpha(a_3\alpha+1)} = \frac{9.284\times3.35+0.487}{3.35(0.010\times3.35+1)} = 9.12$$

$$C_4 = \frac{a_8\alpha + a_9}{\alpha(a_3\alpha+1)} = \frac{0.265\times3.35+1.087}{3.35(0.010\times3.35+1)} = 0.57$$

$$C_5 = \frac{a_{10}\alpha - a_{11}}{\alpha(a_3\alpha+1)} = \frac{0.944\times3.35-0.451}{3.35(0.010\times3.35+1)} = 0.78$$

$$C_6 = \frac{a_{12}\alpha^2 + a_{13}\alpha + a_{14}}{\alpha(a_3\alpha+1)(a_{15}\alpha+1)}$$

$$= \frac{0.201\times3.35^2 + 4.918\times3.35+0.187}{3.35(0.010\times3.35+1)(0.054\times3.35+1)} = 4.63$$

$$C_7 = \frac{a_{16}\alpha^2 + a_{17}\alpha + a_{14}}{\alpha(a_3\alpha + 1)(a_{15}\alpha + 1)}$$

$$= \frac{0.173 \times 3.35^2 + 2.231 \times 3.35 + 0.187}{3.35(0.010 \times 3.35 + 1)(0.054 \times 3.35 + 1)} = 2.35$$

From (8.34)

$$\bar{k}^1 = \frac{25 \times 10^6 \times 0.0054}{4^3}$$

$$\begin{bmatrix} 23.83 & 2.05\times4 & -9.12\times4 & -23.83 & -2.05\times4 & -9.12\times4 \\ 2.05\times4 & 0.57\times4^2 & -0.78\times4^2 & -2.05\times4 & -0.57\times4^2 & -0.78\times4^2 \\ -9.12\times4 & -0.78\times4^2 & 4.63\times4^2 & 9.12\times4 & 0.78\times4^2 & 2.35\times4^2 \\ -23.83 & -2.05\times4 & 9.12\times4 & 23.83 & 2.05\times4 & 9.12\times4 \\ -2.05\times4 & -0.57\times4^2 & 0.78\times4^2 & 2.05\times4 & 0.57\times4^2 & 0.78\times4^2 \\ -9.12\times4 & -0.78\times4^2 & 2.35\times4^2 & 9.12\times4 & 0.78\times4^2 & 4.63\times4^2 \end{bmatrix}$$

$$\bar{k}^1 = \begin{bmatrix} 50266 & 17297 & -76950 & -50266 & -17297 & -76950 \\ 17297 & 19238 & -26325 & -17297 & -19238 & -26325 \\ -76950 & -26325 & 156263 & 76950 & 26325 & 79313 \\ -50266 & -17297 & 76950 & 50266 & 17297 & 76950 \\ -17297 & -19238 & 26325 & 17297 & 19238 & 26325 \\ -76950 & -26325 & 79313 & 76950 & 26325 & 156263 \end{bmatrix} \tag{8.38}$$

$$x_i = x_1 = 0,\ x_j = x_2 = 1.172\text{ m},\ x_{ij} = x_j - x_i = 1.172 - 0 = 1.172\text{ m}$$
$$y_i = y_1 = 0,\ y_j = y_2 = 2.828\text{ m},\ y_{ij} = y_j - y_i = 2.828 - 0 = 2.828\text{ m}$$

$$L = \sqrt{x_{ij}^2 + y_{ij}^2} = \sqrt{1.172^2 + 2.828^2} = 3.061\text{ m}$$

From (8.35)

$$r^1 = \begin{bmatrix} 1 & 0 & 0 & 0 & 0 & 0 \\ 0 & 0.383 & 0.924 & 0 & 0 & 0 \\ 0 & -0.924 & 0.383 & 0 & 0 & 0 \\ 0 & 0 & 0 & 1 & 0 & 0 \\ 0 & 0 & 0 & 0 & 0.383 & 0.924 \\ 0 & 0 & 0 & 0 & -0.924 & 0.383 \end{bmatrix} \tag{8.39}$$

$$k^1 = (r^1)^T \bar{k}^1 r^1$$

$$k^1 = \begin{bmatrix} 50266 & 77709 & -13481 & -50266 & 64465 & -45440 \\ 77709 & 154799 & -29856 & -77709 & 64868 & -61180 \\ -13481 & -29856 & 20701 & 13481 & -8530 & -4793 \\ -50266 & -77709 & 13481 & 50266 & -64465 & 45440 \\ 64465 & 64868 & -8530 & -64465 & 117557 & -67071 \\ -45440 & -61180 & -4793 & 45440 & -67071 & 57943 \end{bmatrix} \tag{8.40}$$

Similarly for **element 2** with $\beta = 60^0$

$$\bar{k}^2 = \begin{bmatrix} 20588 & 12403 & -41175 & -20588 & -12403 & -41175 \\ 12403 & 17888 & -24975 & -12403 & -17888 & -24975 \\ -41175 & -24975 & 109013 & 41175 & 24975 & 55688 \\ -20588 & -12403 & 41175 & 20588 & 12403 & 41175 \\ -12403 & -17888 & 24975 & 12403 & 17888 & 24975 \\ -41175 & -24975 & 55688 & 41175 & 24975 & 109013 \end{bmatrix} \tag{8.41}$$

$x_i = x_2 = 1.172\,\text{m},\ x_j = x_3 = 5.035\,\text{m},\ x_{ij} = x_j - x_i = 5.035 - 1.172 = 3.863\,\text{m}$

$y_i = y_2 = 2.828,\ y_j = y_3 = 3.864\,\text{m},\ y_{ij} = y_j - y_i = 3.864 - 2.828 = 1.036\,\text{m}$

$$L = \sqrt{x_{ij}^2 + y_{ij}^2} = \sqrt{3.863^2 + 1.036^2} = 4.000\,\text{m}$$

$$r^2 = \begin{bmatrix} 1 & 0 & 0 & 0 & 0 & 0 \\ 0 & 0.966 & 0.259 & 0 & 0 & 0 \\ 0 & -0.259 & 0.966 & 0 & 0 & 0 \\ 0 & 0 & 0 & 1 & 0 & 0 \\ 0 & 0 & 0 & 0 & 0.966 & 0.259 \\ 0 & 0 & 0 & 0 & -0.259 & 0.966 \end{bmatrix} \tag{8.42}$$

$$k^2 = (r^2)^T \bar{k}^2 r^2$$

$$k^2 = \begin{bmatrix} 20588 & 22645 & -36557 & -20588 & -1314 & -42983 \\ 22645 & 36499 & -44422 & -22645 & -12951 & -43383 \\ -36557 & -44422 & 90401 & 36557 & 6567 & 50751 \\ -20588 & -22645 & 36557 & 20588 & 1314 & 42983 \\ -1314 & -12951 & 6567 & 1314 & 11505 & -1175 \\ -42983 & -43383 & 50751 & 42983 & -1175 & 115395 \end{bmatrix} \tag{8.43}$$

And for **element 3** with $\beta = 30^0$

$$\bar{k}^3 = \begin{bmatrix} 173285 & 26747 & -179381 & -173285 & -26747 & -179381 \\ 26747 & 23625 & -27675 & -26747 & -23625 & -27675 \\ -179381 & -27675 & 247050 & 179381 & 27675 & 124538 \\ -173285 & -26747 & 179381 & 173285 & 26747 & 179381 \\ -26747 & -23625 & 27675 & 26747 & 23625 & 27675 \\ -179381 & -27675 & 124538 & 179381 & 27675 & 247050 \end{bmatrix} \tag{8.44}$$

$$x_i = x_3 = 5.035\text{ m},\ x_j = x_4 = 6.828\text{ m},\ x_{ij} = x_j - x_i = 6.828 - 5.035 = 1.793\text{ m}$$

$$y_i = y_3 = 3.864,\ y_j = y_4 = 2.828\text{ m},\ y_{ij} = y_j - y_i = 2.828 - 3.864 = -1.036\text{ m}$$

$$L = \sqrt{x_{ij}^2 + y_{ij}^2} = \sqrt{1.793^2 + (-1.036)^2} = 2.071\text{ m}$$

$$r^3 = \begin{bmatrix} 1 & 0 & 0 & 0 & 0 & 0 \\ 0 & 0.866 & -0.500 & 0 & 0 & 0 \\ 0 & 0.500 & 0.866 & 0 & 0 & 0 \\ 0 & 0 & 0 & 1 & 0 & 0 \\ 0 & 0 & 0 & 0 & 0.866 & -0.500 \\ 0 & 0 & 0 & 0 & 0.500 & 0.866 \end{bmatrix} \tag{8.45}$$

$$k^3 = (r^3)^T \bar{k}^3 r^3$$

$$k^3 = \begin{bmatrix} 173285 & -66584 & -168700 & -173285 & -112902 & -141937 \\ -66584 & 55570 & 82963 & 66584 & 13459 & 36506 \\ -168700 & 82963 & 215105 & 168700 & 91856 & 87453 \\ -173285 & 66584 & 168700 & 173285 & 112902 & 141937 \\ -112902 & 13459 & 91856 & 112902 & 103524 & 110605 \\ -141937 & 36506 & 87453 & 141937 & 110605 & 167152 \end{bmatrix} \tag{8.46}$$

K=

$K_{11}=k_{ii}^1$	$K_{12}=k_{ij}^1$	0	0
$K_{21}=k_{ji}^1$	$K_{22}=k_{jj}^1+k_{ii}^2$	$K_{23}=k_{ij}^2$	0
0	$K_{32}=k_{ji}^2$	$K_{33}=k_{jj}^2+k_{ii}^3$	$K_{34}=k_{ij}^3$
0	0	$K_{43}=k_{ji}^3$	$K_{44}=k_{jj}^3$

Substitute k^1, k^2, and k^3 from (8.40), (8.43), and (8.46) respectively to get the overall structure matrix.

Load vector

At node 1 the reactions on the structure are: the force in the z-direction R_{Z1}, the moment about the x-axis R_{T1}, and the moment about the y-axis R_{M1}.

At node 2 the external force of −190 kN in the z-direction and at node 3 the external force of −80 kN in the z-direction.

At node 4 the reactions on the structure are: the force in the z-direction R_{Z4}, the moment about the x-axis R_{T4}, and the moment about the y-axis R_{M4}.

$$F=\begin{bmatrix}F_1\\F_2\\F_3\\F_4\end{bmatrix}=\begin{bmatrix}Z_1\\T_1\\M_1\\Z_2\\T_2\\M_2\\Z_3\\T_3\\M_3\\Z_4\\T_4\\M_4\end{bmatrix}=\begin{bmatrix}R_{Z1}\\R_{T1}\\R_{M1}\\-190\\0\\0\\-80\\0\\0\\R_{Z4}\\R_{T4}\\R_{M4}\end{bmatrix}$$

50266	77709	−13481	−50266	64465	−45440	0	0	0	0	0	0	w_1		R_{Z1}
77709	154799	−29856	−77709	64868	−61180	0	0	0	0	0	0	Φ_1		R_{T1}
−13481	−29856	20701	13481	−8530	−4793	0	0	0	0	0	0	θ_1		R_{M1}
−50266	−77709	13481	50266 +20588	−64465 +22645	45440 −36557	−20588	−1314	−42983	0	0	0	w_2		−190
64465	64868	−8530	−64465 +22645	117557 +36499	−67071 −44422	−22645	−12951	−43383	0	0	0	Φ_2		0
−45440	−61180	−4793	45440 −36557	−67071 −44422	57943 +90401	36557	6567	50751	0	0	0	θ_2		0
0	0	0	−20588	−22645	36557	20588 +173285	1314 −66584	42983 −168700	−173285	−112902	−141937	w_3	=	−80
0	0	0	−1314	−12951	6567	1314 −66584	11505 +55570	−1175 +82963	66584	13459	36506	Φ_3		0
0	0	0	−42983	−43383	50751	42983 −168700	−1175 +82963	115395 +215105	168700	91856	87453	θ_3		0
0	0	0	0	0	0	−173285	66584	168700	173285	112902	141937	w_4		R_{Z4}
0	0	0	0	0	0	−112902	13459	91856	112902	103524	110605	Φ_4		R_{T4}
0	0	0	0	0	0	−141937	36506	87453	141937	110605	167152	θ_4		R_{M4}

(8.47)

For the boundary conditions that node 1 is fixed, i.e. $w_1 = 0$, $\Phi_1 = 0$, and $\theta_1 = 0$, therefore delete rows 1, 2, and 3 and columns 1, 2, and 3.

The boundary conditions that node 4 is fixed, i.e. $w_4 = 0$, $\Phi_4 = 0$, and $\theta_4 = 0$, therefore delete rows 10, 11, and 12 and columns 10, 11, and 12. The resulting reduced matrix is as shown below.

$$\begin{bmatrix} 70854 & -41820 & 8883 & -20588 & -1314 & -42983 \\ -41820 & 154056 & -111493 & -22645 & -12951 & -43383 \\ 8883 & -111493 & 148344 & 36557 & 6567 & 50751 \\ -20588 & -22645 & 36557 & 193873 & -65270 & -125717 \\ -1314 & -12951 & 6567 & -65270 & 67075 & 81788 \\ -42983 & -43383 & 50751 & -125717 & 81788 & 330500 \end{bmatrix} \begin{bmatrix} w_2 \\ \Phi_2 \\ \theta_2 \\ w_3 \\ \Phi_3 \\ \theta_3 \end{bmatrix} = \begin{bmatrix} -190 \\ 0 \\ 0 \\ -80 \\ 0 \\ 0 \end{bmatrix}$$

The solution of the above set of simultaneous equations is:

$w_2 = -0.00716$ m, $\Phi_2 = -0.00367$ rad, $\theta_2 = -0.00065$ rad,

$w_3 = -0.00340$ m, $\Phi_3 = -0.00131$ rad, and $\theta_3 = -0.00228$ rad.

The full displacement vector is

$$\begin{bmatrix} 0 \\ 0 \\ 0 \\ -0.00716 \\ -0.00367 \\ -0.00065 \\ -0.00340 \\ -0.00131 \\ -0.00228 \\ 0 \\ 0 \\ 0 \end{bmatrix}$$

Calculation of reactions at the supports

These are calculated relative to global coordinates from (8.47) as follows:

From row 1

$$50266w_1 + 77709\Phi_1 - 13481\theta_1 - 50266w_2 + 64465\Phi_2 - 45440\theta_2 = R_{Z1}$$

$$\begin{aligned} R_{Z1} &= -50266(-0.00716) + 64465(-0.00367) - 45440(-0.00065) \\ &= +152.85\,\text{kN} \end{aligned}$$

From row 2

$$77709w_1 + 154799\Phi_1 - 29856\theta_1 - 77709w_2 + 64868\Phi_2 - 61180\theta_2 = R_{T1}$$

$$\begin{aligned} R_{T1} &= -77709(-0.00716) + 64868(-0.00367) - 61180(-0.00065) \\ &= +358.10\,\text{kNm} \end{aligned}$$

From row 3

$$-13481w_1 - 29856\Phi_1 + 20701\theta_1 + 13481w_2 - 8530\Phi_2 - 4793\theta_2 = R_{M1}$$

$$\begin{aligned} R_{M1} &= 13481(-0.00716) - 8530(-0.00367) - 4793(-0.00065) \\ &= -62.10\,\text{kNm} \end{aligned}$$

Similarly, rows 10, 11, and 12 respectively give:

$R_{Z4} = +117.31\,\text{kN}$, $R_{T4} = +156.80\,\text{kNm}$, and $R_{M4} = +235.37\,\text{kNm}$.

Calculation of actions on the elements

These are calculated from (8.37) as shown below:

Element 1

$$\begin{bmatrix} V_i^1 \\ t_i^1 \\ m_i^1 \\ V_j^1 \\ t_j^1 \\ m_j^1 \end{bmatrix} = f^1 = r_\beta^1 \bar{k}^1 r^1 \delta^1, \text{ where } \delta^1 = \begin{bmatrix} \delta_i^1 \\ \delta_j^1 \end{bmatrix} = \begin{bmatrix} \delta_1 \\ \delta_2 \end{bmatrix} = \begin{bmatrix} w_1 \\ \Phi_1 \\ \theta_1 \\ w_2 \\ \Phi_2 \\ \theta_2 \end{bmatrix} = \begin{bmatrix} 0 \\ 0 \\ 0 \\ -0.00716 \\ -0.00367 \\ -0.00065 \end{bmatrix},$$

from (8.36) with $\beta = 45^0$,

$$r_\beta^1 = \begin{bmatrix} 1 & 0 & 0 & 0 & 0 & 0 \\ 0 & 0.924 & 0.383 & 0 & 0 & 0 \\ 0 & -0.383 & 0.924 & 0 & 0 & 0 \\ 0 & 0 & 0 & 1 & 0 & 0 \\ 0 & 0 & 0 & 0 & 0.924 & -0.383 \\ 0 & 0 & 0 & 0 & 0.383 & 0.924 \end{bmatrix}$$

and $\bar{k}^1$ and r^1 from (8.38) and (8.39) respectively, thus

$$f^1 = \begin{bmatrix} V_1^1 \\ t_1^1 \\ m_1^1 \\ V_2^1 \\ t_2^1 \\ m_2^1 \end{bmatrix} = \begin{bmatrix} 1 & 0 & 0 & 0 & 0 & 0 \\ 0 & 0.924 & 0.383 & 0 & 0 & 0 \\ 0 & -0.383 & 0.924 & 0 & 0 & 0 \\ 0 & 0 & 0 & 1 & 0 & 0 \\ 0 & 0 & 0 & 0 & 0.924 & -0.383 \\ 0 & 0 & 0 & 0 & 0.383 & 0.924 \end{bmatrix}.$$

$$\begin{bmatrix} 50266 & 17297 & -76950 & -50266 & -17297 & -76950 \\ 17297 & 19238 & -26325 & -17297 & -19238 & -26325 \\ -76950 & -26325 & 156263 & 76950 & 26325 & 79313 \\ -50266 & -17297 & 76950 & 50266 & 17297 & 76950 \\ -17297 & -19238 & 26325 & 17297 & 19238 & 26325 \\ -76950 & -26325 & 79313 & 76950 & 26325 & 156263 \end{bmatrix}.$$

$$\begin{bmatrix} 1 & 0 & 0 & 0 & 0 & 0 \\ 0 & 0.383 & 0.924 & 0 & 0 & 0 \\ 0 & -0.924 & 0.383 & 0 & 0 & 0 \\ 0 & 0 & 0 & 1 & 0 & 0 \\ 0 & 0 & 0 & 0 & 0.383 & 0.924 \\ 0 & 0 & 0 & 0 & -0.924 & 0.383 \end{bmatrix} \begin{bmatrix} 0 \\ 0 \\ 0 \\ -0.00716 \\ -0.00367 \\ -0.00065 \end{bmatrix} = \begin{bmatrix} +153.11 \\ -62.06 \\ -358.40 \\ -153.11 \\ -30.37 \\ -135.27 \end{bmatrix}$$

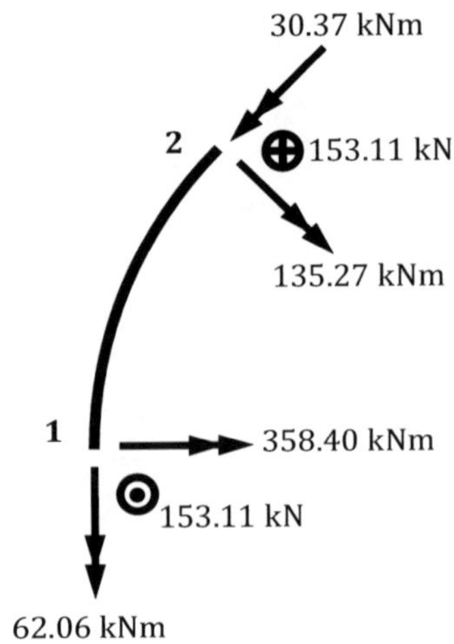

Element 2

$$\begin{bmatrix} V_i^2 \\ t_i^2 \\ m_i^2 \\ V_j^2 \\ t_j^2 \\ m_j^2 \end{bmatrix} = f^2 = r_\beta^2 \bar{k}^2 r^2 \delta^2 \text{, where } \delta^2 = \begin{bmatrix} \delta_i^2 \\ \delta_j^2 \end{bmatrix} = \begin{bmatrix} \delta_2 \\ \delta_3 \end{bmatrix} = \begin{bmatrix} w_2 \\ \Phi_2 \\ \theta_2 \\ w_3 \\ \Phi_3 \\ \theta_3 \end{bmatrix} = \begin{bmatrix} -0.00716 \\ -0.00367 \\ -0.00065 \\ -0.00340 \\ -0.00131 \\ -0.00228 \end{bmatrix},$$

from (8.36) with $\beta = 60^0$,

$$r_\beta^2 = \begin{bmatrix} 1 & 0 & 0 & 0 & 0 & 0 \\ 0 & 0.866 & 0.500 & 0 & 0 & 0 \\ 0 & -0.500 & 0.866 & 0 & 0 & 0 \\ 0 & 0 & 0 & 1 & 0 & 0 \\ 0 & 0 & 0 & 0 & 0.866 & -0.500 \\ 0 & 0 & 0 & 0 & 0.500 & 0.866 \end{bmatrix}$$

and $\bar{k}^2$ and r^2 from (8.41) and (8.42) respectively, thus

$$f^2 = \begin{bmatrix} V_2^2 \\ t_2^2 \\ m_2^2 \\ V_3^2 \\ t_3^2 \\ m_3^2 \end{bmatrix} = \begin{bmatrix} 1 & 0 & 0 & 0 & 0 & 0 \\ 0 & 0.866 & 0.500 & 0 & 0 & 0 \\ 0 & -0.500 & 0.866 & 0 & 0 & 0 \\ 0 & 0 & 0 & 1 & 0 & 0 \\ 0 & 0 & 0 & 0 & 0.866 & -0.500 \\ 0 & 0 & 0 & 0 & 0.500 & 0.866 \end{bmatrix}.$$

$$\begin{bmatrix} 20588 & 12403 & -41175 & -20588 & -12403 & -41175 \\ 12403 & 17888 & -24975 & -12403 & -17888 & -24975 \\ -41175 & -24975 & 109013 & 41175 & 24975 & 55688 \\ -20588 & -12403 & 41175 & 20588 & 12403 & 41175 \\ -12403 & -17888 & 24975 & 12403 & 17888 & 24975 \\ -41175 & -24975 & 55688 & 41175 & 24975 & 109013 \end{bmatrix}.$$

$$\begin{bmatrix} 1 & 0 & 0 & 0 & 0 & 0 \\ 0 & 0.966 & 0.259 & 0 & 0 & 0 \\ 0 & -0.259 & 0.966 & 0 & 0 & 0 \\ 0 & 0 & 0 & 1 & 0 & 0 \\ 0 & 0 & 0 & 0 & 0.966 & 0.259 \\ 0 & 0 & 0 & 0 & -0.259 & 0.966 \end{bmatrix} \begin{bmatrix} -0.00716 \\ -0.00367 \\ -0.00065 \\ -0.00340 \\ -0.00131 \\ -0.00228 \end{bmatrix} = \begin{bmatrix} -36.90 \\ +30.42 \\ +135.26 \\ +36.90 \\ +27.82 \\ +34.38 \end{bmatrix}$$

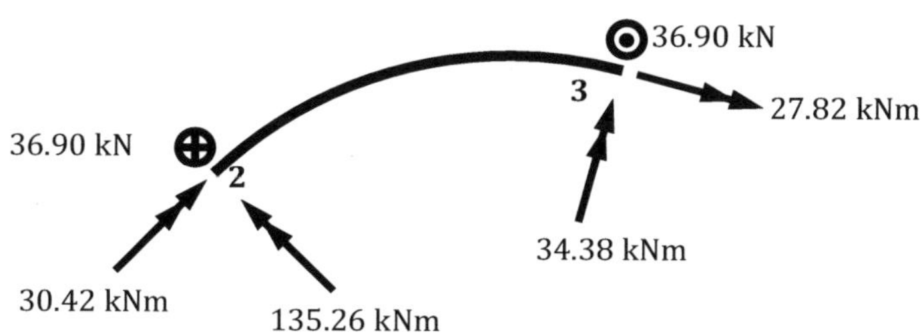

Element 3

$$\begin{bmatrix} V_i^3 \\ t_i^3 \\ m_i^3 \\ V_j^3 \\ t_j^3 \\ m_j^3 \end{bmatrix} = f^3 = r_\beta^3 \bar{k}^3 r^3 \delta^3, \text{ where } \delta^3 = \begin{bmatrix} \delta_i^3 \\ \delta_j^3 \end{bmatrix} = \begin{bmatrix} \delta_3 \\ \delta_4 \end{bmatrix} = \begin{bmatrix} w_3 \\ \Phi_3 \\ \theta_3 \\ w_4 \\ \Phi_4 \\ \theta_4 \end{bmatrix} = \begin{bmatrix} -0.00340 \\ -0.00131 \\ -0.00228 \\ 0 \\ 0 \\ 0 \end{bmatrix}$$

from (8.36) with $\beta = 30^0$,

$$r_\beta^3 = \begin{bmatrix} 1 & 0 & 0 & 0 & 0 & 0 \\ 0 & 0.966 & 0.259 & 0 & 0 & 0 \\ 0 & -0.259 & 0.966 & 0 & 0 & 0 \\ 0 & 0 & 0 & 1 & 0 & 0 \\ 0 & 0 & 0 & 0 & 0.966 & -0.259 \\ 0 & 0 & 0 & 0 & 0.259 & 0.966 \end{bmatrix}$$

and $\bar{k}^3$ and r^3 from (8.44) and (8.45) respectively, thus

$$f^3 = \begin{bmatrix} V_3^3 \\ t_3^3 \\ m_3^3 \\ V_4^3 \\ t_4^3 \\ m_4^3 \end{bmatrix} = \begin{bmatrix} 1 & 0 & 0 & 0 & 0 & 0 \\ 0 & 0.966 & 0.259 & 0 & 0 & 0 \\ 0 & -0.259 & 0.966 & 0 & 0 & 0 \\ 0 & 0 & 0 & 1 & 0 & 0 \\ 0 & 0 & 0 & 0 & 0.966 & -0.259 \\ 0 & 0 & 0 & 0 & 0.259 & 0.966 \end{bmatrix}.$$

$$\begin{bmatrix} 173285 & 26747 & -179381 & -173285 & -26747 & -179381 \\ 26747 & 23625 & -27675 & -26747 & -23625 & -27675 \\ -179381 & -27675 & 247050 & 179381 & 27675 & 124538 \\ -173285 & -26747 & 179381 & 173285 & 26747 & 179381 \\ -26747 & -23625 & 27675 & 26747 & 23625 & 27675 \\ -179381 & -27675 & 124538 & 179381 & 27675 & 247050 \end{bmatrix}$$

$$\begin{bmatrix} 1 & 0 & 0 & 0 & 0 & 0 \\ 0 & 0.866 & -0.500 & 0 & 0 & 0 \\ 0 & 0.500 & 0.866 & 0 & 0 & 0 \\ 0 & 0 & 0 & 1 & 0 & 0 \\ 0 & 0 & 0 & 0 & 0.866 & -0.500 \\ 0 & 0 & 0 & 0 & 0.500 & 0.866 \end{bmatrix} \begin{bmatrix} -0.00340 \\ -0.00131 \\ -0.00228 \\ 0 \\ 0 \\ 0 \end{bmatrix} = \begin{bmatrix} -116.90 \\ -27.81 \\ -34.39 \\ +116.90 \\ -55.61 \\ +276.95 \end{bmatrix}$$

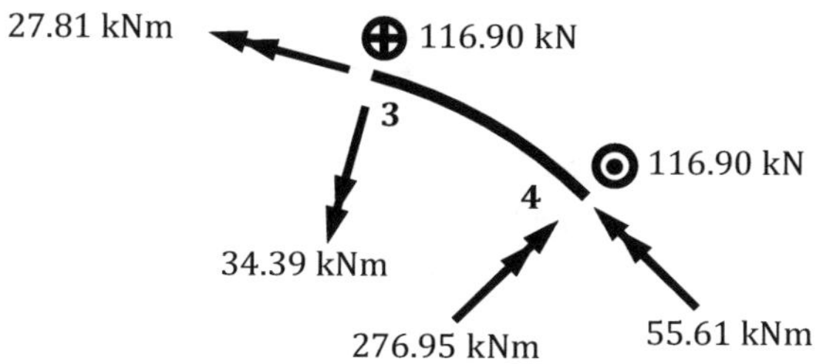

Problems

Analyse problems P8.1 to P8.4 below for the circular beams which are curved in plan for the data and loading shown.

P8.1. An aluminium beam with E = 70 × 10^6 kN/m^2, G = 26 × 10^6 kN/m^2, I = 0.000012 m^4, and J = 0.000019 m^4.

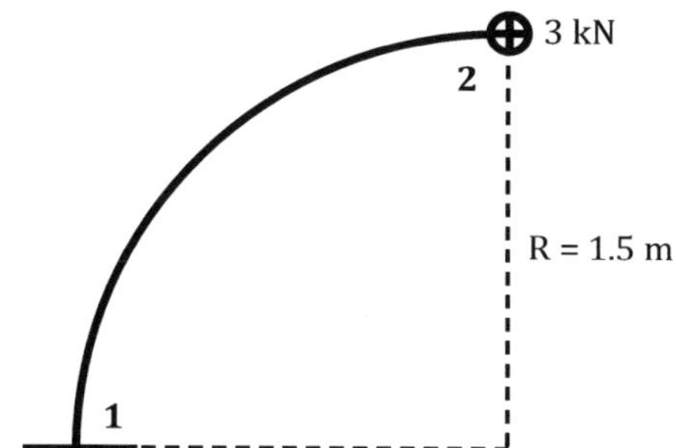

Figure P8.1

Answer:

$w_1 = 0, \Phi_1 = 0, \theta_1 = 0,$

$w_2 = -0.01677 \text{ m}, \Phi_2 = -0.00338 \text{ rad}, \theta_2 = +0.01087 \text{ rad},$

$R_{Z1} = +3.00\text{ kN}, R_{T1} = +4.50\text{ kNm}, R_{M1} = -4.50\text{ kNm}$

$$\begin{bmatrix} V_1 \\ t_1 \\ m_1 \\ V_2 \\ t_2 \\ m_2 \end{bmatrix} = \begin{bmatrix} +3.00 \\ -4.50 \\ -4.50 \\ -3.00 \\ 0 \\ 0 \end{bmatrix}$$

P8.2. A timber beam with E = 8 × 10^6 kN/m^2, G = 0.5 × 10^6 kN/m^2, I = 0.000020 m^4, and J = 0.000015 m^4.

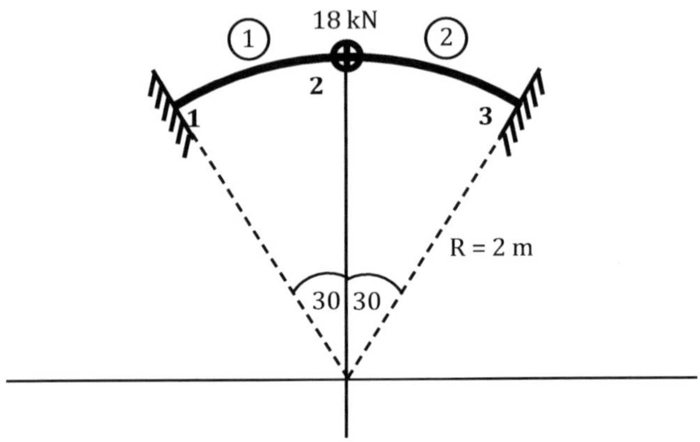

Figure P8.2

Answer:

$w_1 = 0, \Phi_1 = 0, \theta_1 = 0,$

$w_2 = -0.00749\text{ m}, \Phi_2 = -0.03185\text{ rad}, \theta_2 = 0,$

$w_3 = 0, \Phi_3 = 0, \theta_3 = 0,$

$R_{Z1} = +9.00\text{ kN}, R_{T1} = +2.41\text{ kNm}, R_{M1} = -5.07\text{ kNm},$

$R_{Z3} = +9.00\text{ kN}, R_{T3} = +2.41\text{ kNm}, R_{M3} = +5.07\text{ kNm}$

$$\text{Element 1}: \begin{bmatrix} V_1^1 \\ t_1^1 \\ m_1^1 \\ V_2^1 \\ t_2^1 \\ m_2^1 \end{bmatrix} = \begin{bmatrix} +9.00 \\ -0.45 \\ -5.60 \\ -9.00 \\ - \\ -3.93 \end{bmatrix}, \text{ Element 2}: \begin{bmatrix} V_2^2 \\ t_2^2 \\ m_2^2 \\ V_3^2 \\ t_3^2 \\ m_3^2 \end{bmatrix} = \begin{bmatrix} -9.00 \\ 0 \\ +3.93 \\ +9.00 \\ -0.45 \\ +5.60 \end{bmatrix}$$

P8.3. A concrete beam with E = 30 × 10^6 kN/m^2, G = 13 × 10^6 kN/m^2, I = 0.00099 m^4, and J = 0.00038 m^4.

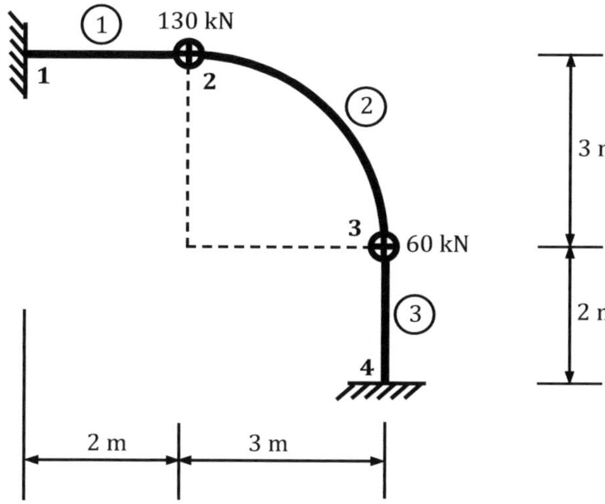

Figure P8.3

Answer:

$$w_1 = 0, \Phi_1 = 0, \theta_1 = 0,$$
$$w_2 = -0.00915\text{ m}, \Phi_2 = -0.00468\text{ rad}, \theta_2 = +0.00640\text{ rad},$$
$$w_3 = -0.00680\text{ m}, \Phi_3 = -0.00528\text{ rad}, \theta_3 = +0.00201\text{ rad},$$
$$w_4 = 0, \Phi_4 = 0, \theta_4 = 0,$$
$$R_{Z1} = +122.50\text{ kN}, R_{T1} = +11.55\text{ kNm}, R_{M1} = -217.56\text{ kNm},$$
$$R_{Z4} = +67.50\text{ kN}, R_{T4} = +145.93\text{ kNm}, R_{M4} = -4.96\text{ kNm}$$

$$\text{Element 1:} \begin{bmatrix} V_1^1 \\ t_1^1 \\ m_1^1 \\ V_2^1 \\ t_2^1 \\ m_2^1 \end{bmatrix} = \begin{bmatrix} +122.50 \\ +11.55 \\ -217.56 \\ -122.50 \\ -11.55 \\ -27.45 \end{bmatrix},$$

$$\text{Element 2:}\begin{bmatrix} V_2^2 \\ t_2^2 \\ m_2^2 \\ V_3^2 \\ t_3^2 \\ m_3^2 \end{bmatrix} = \begin{bmatrix} -7.50 \\ +11.55 \\ +27.45 \\ +7.50 \\ +4.96 \\ +10.94 \end{bmatrix},$$

$$\text{Element 3:}\begin{bmatrix} V_3^3 \\ t_3^3 \\ m_3^3 \\ V_4^3 \\ t_4^3 \\ m_4^3 \end{bmatrix} = \begin{bmatrix} -67.50 \\ -4.96 \\ -10.94 \\ +67.50 \\ +4.96 \\ +145.93 \end{bmatrix}$$

P8.4. A steel beam with $E = 210 \times 10^6$ kN/m², $G = 80 \times 10^6$ kN/m², $I = 0.00022$ m⁴, and $J = 0.00014$ m⁴.

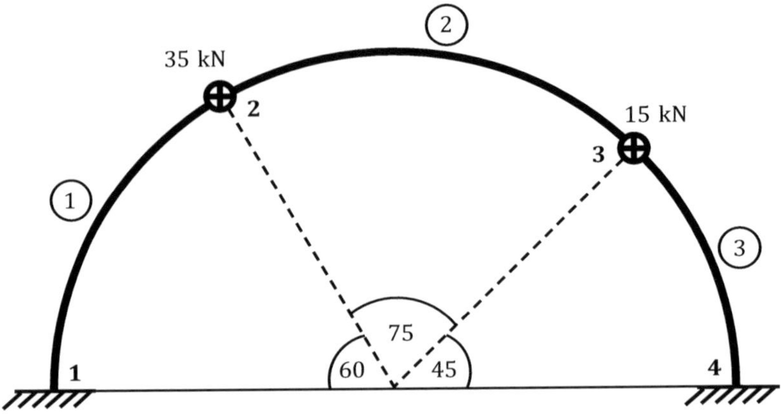

Figure P8.4

Answer:

$$w_1 = 0, \Phi_1 = 0, \theta_1 = 0,$$
$$w_2 = -0.02264\ \text{m}, \Phi_2 = -0.00706\ \text{rad}, \theta_2 = +0.00102\ \text{rad},$$

$w_3 = -0.01267\ \mathrm{m}, \Phi_3 = -0.00433\ \mathrm{rad}, \theta_3 = -0.00293\ \mathrm{rad},$
$w_4 = 0, \Phi_4 = 0, \theta_4 = 0,$
$R_{Z1} = +28.93\ \mathrm{kN}, R_{T1} = +116.75\ \mathrm{kNm}, R_{M1} = -34.55\ \mathrm{kNm},$
$R_{Z4} = +21.07\ \mathrm{kN}, R_{T4} = +87.84\ \mathrm{kNm}, R_{M4} = +29.76\ \mathrm{kNm}$

$$\text{Element 1}: \begin{bmatrix} V_1^1 \\ t_1^1 \\ m_1^1 \\ V_2^1 \\ t_2^1 \\ m_2^1 \end{bmatrix} = \begin{bmatrix} +28.93 \\ +34.55 \\ -116.75 \\ -28.93 \\ -11.52 \\ -36.96 \end{bmatrix},$$

$$\text{Element 2}: \begin{bmatrix} V_2^2 \\ t_2^2 \\ m_2^2 \\ V_3^2 \\ t_3^2 \\ m_3^2 \end{bmatrix} = \begin{bmatrix} -6.07 \\ +11.52 \\ +36.96 \\ +6.07 \\ +10.21 \\ +8.65 \end{bmatrix},$$

$$\text{Element 3}: \begin{bmatrix} V_3^3 \\ t_3^3 \\ m_3^3 \\ V_4^3 \\ t_4^3 \\ m_4^3 \end{bmatrix} = \begin{bmatrix} -21.07 \\ -10.21 \\ -8.65 \\ +21.07 \\ -29.76 \\ +87.84 \end{bmatrix}$$

Chapter 9

Pin-Connected Space Frames

These are three-dimensional structures that consist of pin-connected members in which no moments are transferred through the pin joints. The analysis of such frames is similar to pin-connected plane frames as explained in Chapter 3 where the members develop axial forces only.

9.1 Derivation of the Stiffness Matrix

It was shown in Chapter 2 that the stiffness matrix relative to local coordinates for a member subjected to axial forces is given by (2.4) as:

$$\begin{bmatrix} \bar{X}_i \\ \bar{X}_j \end{bmatrix} = \begin{bmatrix} \frac{EA}{L} & -\frac{EA}{L} \\ -\frac{EA}{L} & \frac{EA}{L} \end{bmatrix} \begin{bmatrix} \bar{u}_i \\ \bar{u}_j \end{bmatrix}$$

The above equation is for a bar lying along the x-axis and in order to write it in a general form, the displacements and forces in the y- and z-axes are introduced as shown in the relationship below.

Analysis of Structures by Matrix Methods
Fathi Al-Shawi

ISBN 978-981-4968-19-5 (Hardcover), 978-1-003-29130-5 (eBook)
www.jennystanford.com

$$\begin{bmatrix} \bar{X}_i \\ \bar{Y}_i \\ \bar{Z}_i \\ \bar{X}_j \\ \bar{Y}_j \\ \bar{Z}_j \end{bmatrix} = \begin{bmatrix} EA/L & 0 & 0 & -EA/L & 0 & 0 \\ 0 & 0 & 0 & 0 & 0 & 0 \\ 0 & 0 & 0 & 0 & 0 & 0 \\ -EA/L & 0 & 0 & EA/L & 0 & 0 \\ 0 & 0 & 0 & 0 & 0 & 0 \\ 0 & 0 & 0 & 0 & 0 & 0 \end{bmatrix} \begin{bmatrix} \bar{u}_i \\ \bar{v}_i \\ \bar{w}_i \\ \bar{u}_j \\ \bar{v}_j \\ \bar{w}_j \end{bmatrix}, \tag{9.1}$$

or $\bar{F} = \bar{k}\bar{\delta}$.

9.2 Transformation of Coordinates

Assume that the member local $\bar{x}$-axis lies initially along the global x-axis and its direction is OA as shown in Fig. 9.1a. The final position of the member is OA'' which is achieved by two rotations. The first is a rotation about the $\bar{y}$-axis by an angle $\varphi_{\bar{y}}$ to get to the position along the line OA' where the $\bar{x}$-axis has moved to $\bar{x}'$ and the $\bar{z}$-axis to $\bar{z}'$. The second stage is a rotation about the $\bar{z}'$-axis by an angle $\varphi_{\bar{z}}$ to the position of OA'' where $\bar{x}'$-axis has moved to $\bar{x}''$ and $\bar{y}$-axis to $\bar{y}'$. So, the final directions of the local coordinates $\overline{xyz}$ are now defined by $\bar{x}''\bar{y}'\bar{z}'$.

With reference to Fig. 9.1b the angles of rotation can be defined by the coordinates of the ends of the member as:

$$\sin\varphi_{\bar{y}} = \frac{z_i - z_j}{L} = -\frac{z_j - z_i}{s} = -\frac{z_{ij}}{s}.$$

Notice that for positive rotation $\varphi_{\bar{y}}$, $z_j < z_i$ and hence z_{ij} is negative.

$$\cos\varphi_{\bar{y}} = \frac{x_j - x_i}{s} = \frac{x_{ij}}{s}, \quad \sin\varphi_{\bar{z}} = \frac{y_j - y_i}{L} = \frac{y_{ij}}{L}, \quad \cos\varphi_{\bar{z}} = \frac{s}{L}$$

where $s = \sqrt{x_{ij}^2 + z_{ij}^2}, L = \sqrt{s^2 + y_{ij}^2} = \sqrt{x_{ij}^2 + y_{ij}^2 + z_{ij}^2}$.

One way of deriving the transformation matrix is to consider the effect of rotating the $\bar{y}$- and $\bar{z}$-axes separately and then combining the two effects as shown below.

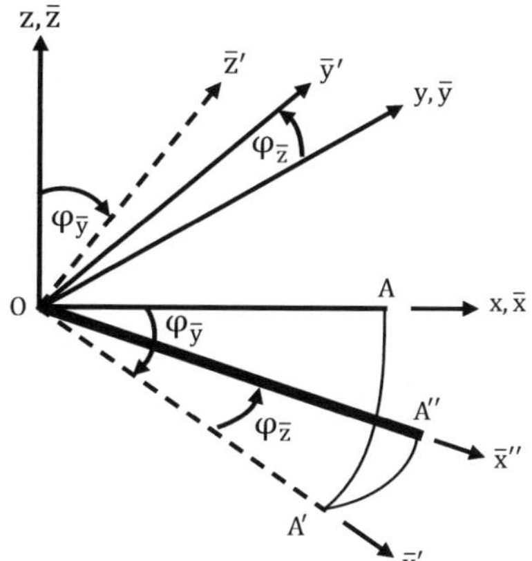
z, z̄
z̄′
ȳ′
y, ȳ
φz̄
φȳ
O
A
x, x̄
φȳ
A″
φz̄
x̄″
A′
x̄′

(a)

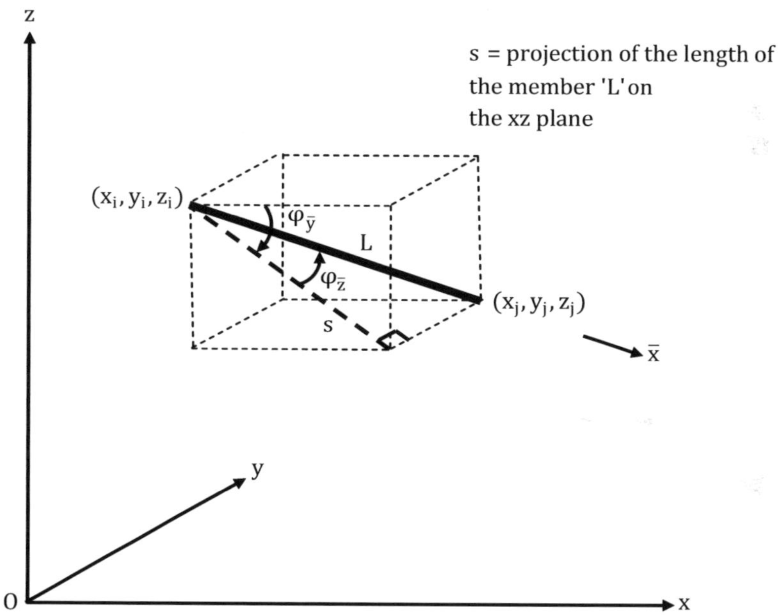
z
s = projection of the length of the member 'L' on the xz plane
(xi, yi, zi)
φȳ
L
φz̄
(xj, yj, zj)
s
x̄
y
O
x

(b)

Figure 9.1

9.2.1 Rotation about the $\bar{y}$-axis by an Angle $\varphi_{\bar{y}}$

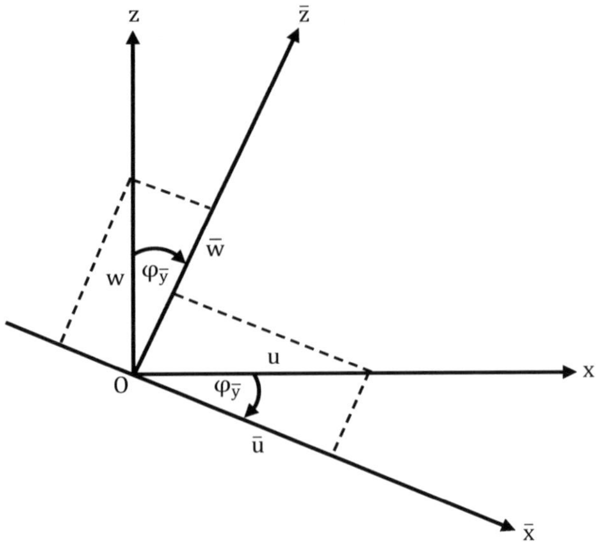

Figure 9.2

Although the $\bar{y}$-axis is rotated axially it remains pointing in the same direction as the y-axis which means that displacements and forces relative to the rotated $\bar{y}$-axis are the same as those relative to the y-axis, thus

$$\bar{v} = v$$

The displacement along the $\bar{x}$-axis is equal to the algebraic sum of the components of the displacements along the x- and z-axes respectively, hence

$$\bar{u} = u\cos\varphi_{\bar{y}} - w\sin\varphi_{\bar{y}}.$$

The displacement along the $\bar{z}$-axis is equal to the algebraic sum of the components of the displacements along the x- and z-axes respectively, thus

$$\bar{w} = u\sin\varphi_{\bar{y}} + w\cos\varphi_{\bar{y}}$$

and in matrix form

$$\begin{bmatrix} \bar{u} \\ \bar{v} \\ \bar{w} \end{bmatrix} = \begin{bmatrix} \cos\varphi_{\bar{y}} & 0 & -\sin\varphi_{\bar{y}} \\ 0 & 1 & 0 \\ \sin\varphi_{\bar{y}} & 0 & \cos\varphi_{\bar{y}} \end{bmatrix} \begin{bmatrix} u \\ v \\ w \end{bmatrix} \quad \text{or } \bar{\delta} = \rho_{\bar{y}}\delta$$

where $\bar{\delta} = \begin{bmatrix} \bar{u} \\ \bar{v} \\ \bar{w} \end{bmatrix}, \delta = \begin{bmatrix} u \\ v \\ w \end{bmatrix}$ and

$$\rho_{\bar{y}} = \begin{bmatrix} \cos\varphi_{\bar{y}} & 0 & -\sin\varphi_{\bar{y}} \\ 0 & 1 & 0 \\ \sin\varphi_{\bar{y}} & 0 & \cos\varphi_{\bar{y}} \end{bmatrix} \tag{9.2}$$

9.2.2 Rotation about the $\bar{z}$-axis by an Angle $\varphi_{\bar{z}}$

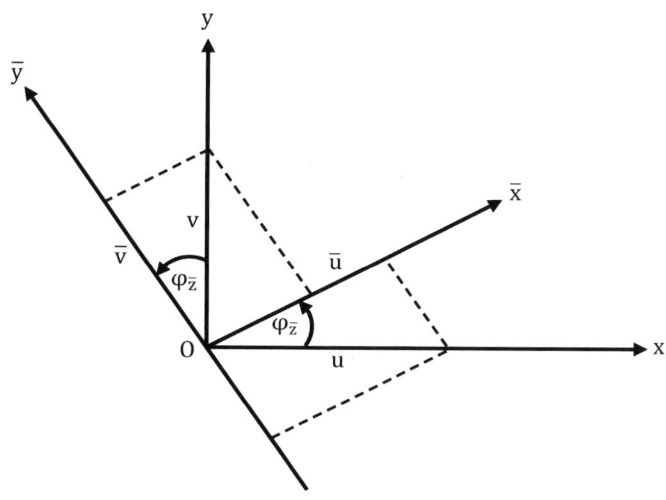

Figure 9.3

Although the $\bar{z}$-axis is rotated axially it remains pointing in the same direction as the z-axis which means that displacements and forces relative to the rotated $\bar{z}$-axis are the same as those relative to the z-axis, thus

$$\bar{w} = w.$$

The displacement along the $\bar{x}$-axis is equal to the algebraic sum of the components of the displacements along the x- and y-axes respectively, hence

$$\bar{u} = u\cos\varphi_{\bar{z}} + v\sin\varphi_{\bar{z}}.$$

The displacement along the $\bar{y}$-axis is equal to the algebraic sum of the components of the displacements along the x- and y-axes respectively, hence

$$\bar{v} = -u\sin\varphi_{\bar{z}} + v\cos\varphi_{\bar{z}}$$

and in matrix form

$$\begin{bmatrix} \bar{u} \\ \bar{v} \\ \bar{w} \end{bmatrix} = \begin{bmatrix} \cos\varphi_{\bar{z}} & \sin\varphi_{\bar{z}} & 0 \\ -\sin\varphi_{\bar{z}} & \cos\varphi_{\bar{z}} & 0 \\ 0 & 0 & 1 \end{bmatrix} \begin{bmatrix} u \\ v \\ w \end{bmatrix} \qquad \text{or} \quad \bar{\delta} = \rho_{\bar{z}}\delta$$

where $\bar{\delta} = \begin{bmatrix} \bar{u} \\ \bar{v} \\ \bar{w} \end{bmatrix}$, $\delta = \begin{bmatrix} u \\ v \\ w \end{bmatrix}$, and

$$\rho_{\bar{z}} = \begin{bmatrix} \cos\varphi_{\bar{z}} & \sin\varphi_{\bar{z}} & 0 \\ -\sin\varphi_{\bar{z}} & \cos\varphi_{\bar{z}} & 0 \\ 0 & 0 & 1 \end{bmatrix}. \tag{9.3}$$

The combined effect of rotations about the $\bar{y}$- and $\bar{z}$-axes respectively is achieved by pre-multiplying (9.2) by (9.3) to give

$$\bar{\delta} = \rho_{\bar{z}}\rho_{\bar{y}}\delta \quad \text{or} \quad \bar{\delta} = \rho\delta$$

where $\rho = \rho_{\bar{z}}\rho_{\bar{y}} = \begin{bmatrix} \cos\varphi_{\bar{z}} & \sin\varphi_{\bar{z}} & 0 \\ -\sin\varphi_{\bar{z}} & \cos\varphi_{\bar{z}} & 0 \\ 0 & 0 & 1 \end{bmatrix} \begin{bmatrix} \cos\varphi_{\bar{y}} & 0 & -\sin\varphi_{\bar{y}} \\ 0 & 1 & 0 \\ \sin\varphi_{\bar{y}} & 0 & \cos\varphi_{\bar{y}} \end{bmatrix}$

$$\rho = \begin{bmatrix} \cos\varphi_{\bar{z}}\cos\varphi_{\bar{y}} & \sin\varphi_{\bar{z}} & -\cos\varphi_{\bar{z}}\sin\varphi_{\bar{y}} \\ -\sin\varphi_{\bar{z}}\cos\varphi_{\bar{y}} & \cos\varphi_{\bar{z}} & \sin\varphi_{\bar{z}}\sin\varphi_{\bar{y}} \\ \sin\varphi_{\bar{y}} & 0 & \cos\varphi_{\bar{y}} \end{bmatrix}. \tag{9.4}$$

The above treatment may be considered to apply to one end of the member, and if the transformation is carried out for the two ends of the member, then

$$r = \begin{bmatrix} \rho & 0 \\ 0 & \rho \end{bmatrix} \quad \text{and} \quad 0 = \begin{bmatrix} 0 & 0 & 0 \\ 0 & 0 & 0 \\ 0 & 0 & 0 \end{bmatrix}$$

$$r = \begin{bmatrix} \cos\varphi_{\bar{z}}\cos\varphi_{\bar{y}} & \sin\varphi_{\bar{z}} & -\cos\varphi_{\bar{z}}\sin\varphi_{\bar{y}} & 0 & 0 & 0 \\ -\sin\varphi_{\bar{z}}\cos\varphi_{\bar{y}} & \cos\varphi_{\bar{z}} & \sin\varphi_{\bar{z}}\sin\varphi_{\bar{y}} & 0 & 0 & 0 \\ \sin\varphi_{\bar{y}} & 0 & \cos\varphi_{\bar{y}} & 0 & 0 & 0 \\ 0 & 0 & 0 & \cos\varphi_{\bar{z}}\cos\varphi_{\bar{y}} & \sin\varphi_{\bar{z}} & -\cos\varphi_{\bar{z}}\sin\varphi_{\bar{y}} \\ 0 & 0 & 0 & -\sin\varphi_{\bar{z}}\cos\varphi_{\bar{y}} & \cos\varphi_{\bar{z}} & \sin\varphi_{\bar{z}}\sin\varphi_{\bar{y}} \\ 0 & 0 & 0 & \sin\varphi_{\bar{y}} & 0 & \cos\varphi_{\bar{y}} \end{bmatrix}$$

Substitute the values of $\sin\varphi_{\bar{y}}$, $\cos\varphi_{\bar{y}}$, $\sin\varphi_{\bar{z}}$, and $\cos\varphi_{\bar{z}}$ as derived previously to get:

$$r = \begin{bmatrix} x_{ij}/L & y_{ij}/L & z_{ij}/L & 0 & 0 & 0 \\ -y_{ij}x_{ij}/Ls & s/L & -y_{ij}z_{ij}/Ls & 0 & 0 & 0 \\ -z_{ij}/s & 0 & x_{ij}/s & 0 & 0 & 0 \\ 0 & 0 & 0 & x_{ij}/L & y_{ij}/L & z_{ij}/L \\ 0 & 0 & 0 & -y_{ij}x_{ij}/Ls & s/L & -y_{ij}z_{ij}/Ls \\ 0 & 0 & 0 & -z_{ij}/s & 0 & x_{ij}/s \end{bmatrix}. \tag{9.5}$$

The above transformation matrix has been derived for the displacements and it applies equally well for the forces since both are vectors in the respective directions.

A special case arises when the local $\bar{x}$-axis of the member is coincident with the global y-axis where $x_{ij} = 0$ and $z_{ij} = 0$. And since $s = \sqrt{x_{ij}^2 + z_{ij}^2}$ then $x_{ij}/s = 0/0$ and $z_{ij}/s = 0/0$ which are indeterminate quantities. To overcome this situation we revert back to the original transformation matrix in (9.4) and substitute the rotation about the $\bar{y}$-axis, $\varphi_{\bar{y}} = 0$ and the rotation about the $\bar{z}$-axis, $\varphi_{\bar{z}} = 90$ degrees to get:

$$\rho = \begin{bmatrix} 0 & 1 & 0 \\ -1 & 0 & 0 \\ 0 & 0 & 1 \end{bmatrix}.$$

It should be noted that rotation of the $\bar{x}$-axis was not considered in the above transformation since only axial forces will develop in the members of a pin-connected frame and these are not changed if the member is rotated about its own axis.

The stiffness matrix $\bar{k}$ in (9.1) is transformed into global coordinates by applying the following equation:

$k = r^T \bar{k} r$ which leads to:

$$k = \frac{EA}{L^3} \begin{bmatrix} x_{ij}^2 & x_{ij}y_{ij} & x_{ij}z_{ij} & -x_{ij}^2 & -x_{ij}y_{ij} & -x_{ij}z_{ij} \\ x_{ij}y_{ij} & y_{ij}^2 & y_{ij}z_{ij} & -x_{ij}y_{ij} & -y_{ij}^2 & -y_{ij}z_{ij} \\ x_{ij}z_{ij} & y_{ij}z_{ij} & z_{ij}^2 & -y_{ij}z_{ij} & -y_{ij}z_{ij} & -z_{ij}^2 \\ -x_{ij}^2 & -x_{ij}y_{ij} & -x_{ij}z_{ij} & x_{ij}^2 & x_{ij}y_{ij} & x_{ij}z_{ij} \\ -x_{ij}y_{ij} & -y_{ij}^2 & -y_{ij}z_{ij} & x_{ij}y_{ij} & y_{ij}^2 & y_{ij}z_{ij} \\ -x_{ij}z_{ij} & -y_{ij}z_{ij} & -z_{ij}^2 & x_{ij}z_{ij} & y_{ij}z_{ij} & z_{ij}^2 \end{bmatrix}. \quad (9.6)$$

Example 1:

Determine the displacements at the nodes and the forces developed in the members of the ball-connected space frame shown in Fig. 9.4. Given that the cross-sectional area, A, of all the members is 700×10^{-6} m^2 and the modulus of elasticity, $E = 210 \times 10^6$ kN/m^2. The forces acting on the structure are as follows:

At node 4: $X_4 = +30$ kN and at node 5: $Z_5 = -80$ kN.

The coordinates of the joints of the frame are given in the table below.

Node number	x (m)	y (m)	z (m)
1	0	0	0
2	6	0	0
3	3	4	0
4	4	2	6
5	9	5	8

The member and structure addresses are shown below.

Member number	Node i	Node j
1	1	4
2	2	4
3	3	4
4	2	5
5	3	5
6	4	5

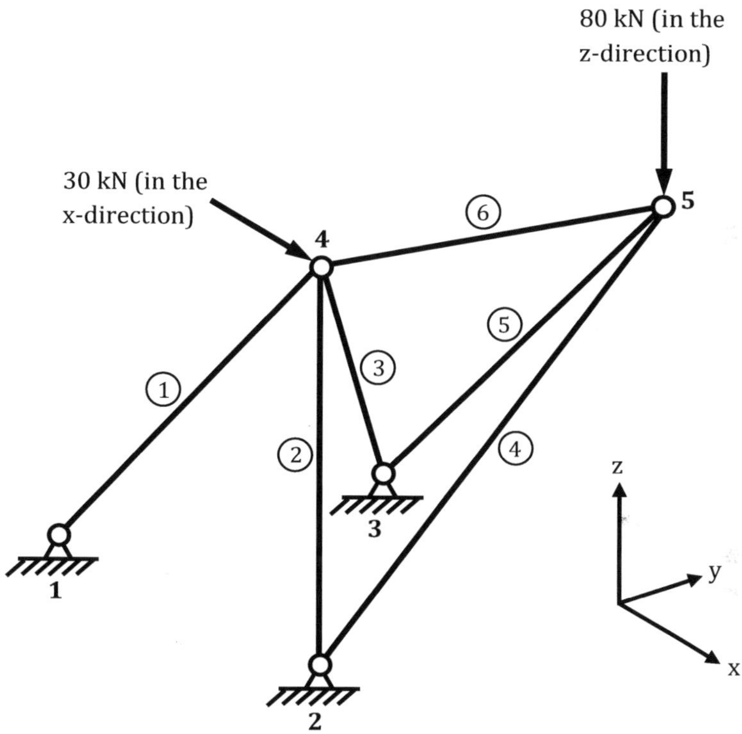

Figure 9.4

Calculation of the stiffness matrices of the members

$E = 210 \times 10^6\ \text{kN/m}^2$, $A = 700 \times 10^{-6}\ \text{m}^2$ for all the members.

Member 1

Node i of the member is node 1 in the structure and node j of the member is node 4 in the structure.

$x_i = 0, x_j = 4$ m, $x_{ij} = x_j - x_i = 4 - 0 = 4$ m

$y_i = 0, y_j = 2$ m, $y_{ij} = y_j - y_i = 2 - 0 = 2$ m

$z_i = 0, z_j = 6$ m, $z_{ij} = z_j - z_i = 6 - 0 = 6$ m

$$s = \sqrt{x_{ij}^2 + z_{ij}^2} = \sqrt{4^2 + 6^2} = 7.211 \text{ m}$$

$$L = \sqrt{x_{ij}^2 + y_{ij}^2 + z_{ij}^2} = \sqrt{4^2 + 2^2 + 6^2} = 7.483 \text{ m}$$

From (9.6)

$$\begin{array}{c} \overbrace{\qquad\qquad\qquad\qquad\qquad\qquad\qquad\qquad\qquad\qquad}^{\delta^1} \\ \overbrace{\qquad\qquad\qquad\qquad}^{\delta_i=\delta_1} \quad \overbrace{\qquad\qquad\qquad\qquad}^{\delta_j=\delta_4} \\ \begin{array}{cccccc} u_i & v_i & w_i & u_j & v_j & w_j \\ u_1 & v_1 & w_1 & u_4 & v_4 & w_4 \end{array} \end{array}$$

$$k^1 = 10^3 \begin{bmatrix} 5.61 & 2.81 & 8.42 & -5.61 & -2.81 & -8.42 \\ 2.81 & 1.40 & 4.21 & -2.81 & -1.40 & -4.21 \\ 8.42 & 4.21 & 12.63 & -8.42 & -4.21 & -12.63 \\ -5.61 & -2.81 & -8.42 & 5.61 & 2.81 & 8.42 \\ -2.81 & -1.40 & -4.21 & 2.81 & 1.40 & 4.21 \\ -8.42 & -4.21 & -12.63 & 8.42 & 4.21 & 12.63 \end{bmatrix} \begin{matrix} u_i = u_1 \\ v_i = v_1 \\ w_i = w_1 \\ u_j = u_4 \\ v_j = v_4 \\ w_j = w_4 \end{matrix}$$

Member 2

Node i of the member is node 2 in the structure and node j of the member is node 4 in the structure.

$x_i = 6$ m, $x_j = 4$ m, $x_{ij} = x_j - x_i = 4 - 6 = -2$ m

$y_i = 0, y_j = 2$ m, $y_{ij} = y_j - y_i = 2 - 0 = 2$ m

$z_i = 0, z_j = 6$ m, $z_{ij} = z_j - z_i = 6 - 0 = 6$ m

$$s = \sqrt{x_{ij}^2 + z_{ij}^2} = \sqrt{(-2)^2 + 6^2} = 6.325 \text{ m}$$

$$L = \sqrt{x_{ij}^2 + y_{ij}^2 + z_{ij}^2} = \sqrt{(-2)^2 + 2^2 + 6^2} = 6.633 \text{ m}$$

$$k^2 = 10^3 \begin{bmatrix} 2.02 & -2.02 & -6.04 & -2.02 & 2.02 & 6.04 \\ -2.02 & 2.02 & 6.04 & 2.02 & -2.02 & -6.04 \\ -6.04 & 6.04 & 18.13 & 6.04 & -6.04 & -18.13 \\ -2.02 & 2.02 & 6.04 & 2.02 & -2.02 & -6.04 \\ 2.02 & -2.02 & -6.04 & -2.02 & 2.02 & 6.04 \\ 6.04 & -6.04 & -18.13 & -6.04 & 6.04 & 18.13 \end{bmatrix} \begin{matrix} u_i = u_2 \\ v_i = v_2 \\ w_i = w_2 \\ u_j = u_4 \\ v_j = v_4 \\ w_j = w_4 \end{matrix}$$

Columns: δ^2: $\delta_i = \delta_2$ ($u_i = u_2$, $v_i = v_2$, $w_i = w_2$), $\delta_j = \delta_4$ ($u_j = u_4$, $v_j = v_4$, $w_j = w_4$)

Member 3

Node i of the member is node 3 in the structure and node j of the member is node 4 in the structure.

$x_i = 3$ m, $x_j = 4$ m, $x_{ij} = x_j - x_i = 4 - 3 = 1$ m

$y_i = 4$ m, $y_j = 2$ m, $y_{ij} = y_j - y_i = 2 - 4 = -2$ m

$z_i = 0$, $z_j = 6$ m, $z_{ij} = z_j - z_i = 6 - 0 = 6$ m

$$s = \sqrt{x_{ij}^2 + z_{ij}^2} = \sqrt{1^2 + 6^2} = 6.083 \text{ m}$$

$$L = \sqrt{x_{ij}^2 + y_{ij}^2 + z_{ij}^2} = \sqrt{1^2 + (-2)^2 + 6^2} = 6.403 \text{ m}$$

Columns: δ^3: $\delta_i = \delta_3$ ($u_i = u_3$, $v_i = v_3$, $w_i = w_3$), $\delta_j = \delta_4$ ($u_j = u_4$, $v_j = v_4$, $w_j = w_4$)

$$k^3 = 10^3 \begin{bmatrix} 0.56 & -1.12 & 3.36 & -0.56 & 1.12 & -3.36 \\ -1.12 & 2.24 & -6.72 & 1.12 & -2.24 & 6.72 \\ 3.36 & -6.72 & 20.16 & -3.36 & 6.72 & -20.16 \\ -0.56 & 1.12 & -3.36 & 0.56 & -1.12 & 3.36 \\ 1.12 & -2.24 & 6.72 & -1.12 & 2.24 & -6.72 \\ -3.36 & 6.72 & -20.16 & 3.36 & -6.72 & 20.16 \end{bmatrix} \begin{matrix} u_i = u_3 \\ v_i = v_3 \\ w_i = w_3 \\ u_j = u_4 \\ v_j = v_4 \\ w_j = w_4 \end{matrix}$$

Member 4

Node i of the member is node 2 in the structure and node j of the member is node 5 in the structure.

$x_i = 6$ m, $x_j = 9$ m, $x_{ij} = x_j - x_i = 9 - 6 = 3$ m

$y_i = 0$, $y_j = 5$ m, $y_{ij} = y_j - y_i = 5 - 0 = 5$ m

$z_i = 0$, $z_j = 8$ m, $z_{ij} = z_j - z_i = 8 - 0 = 8$ m

$$s = \sqrt{x_{ij}^2 + z_{ij}^2} = \sqrt{3^2 + 8^2} = 8.544 \text{ m}$$

$$L = \sqrt{x_{ij}^2 + y_{ij}^2 + z_{ij}^2} = \sqrt{3^2 + 5^2 + 8^2} = 9.900 \text{ m}$$

$$\overbrace{\overbrace{\begin{matrix} u_i & v_i & w_i \\ u_2 & v_2 & w_2 \end{matrix}}^{\delta_i=\delta_2} \quad \overbrace{\begin{matrix} u_j & v_j & w_j \\ u_5 & v_5 & w_5 \end{matrix}}^{\delta_j=\delta_5}}^{\delta^4}$$

$$k^4 = 10^3 \begin{bmatrix} 1.36 & 2.27 & 3.64 & -1.36 & -2.27 & -3.64 \\ 2.27 & 3.79 & 6.06 & -2.27 & -3.79 & -6.06 \\ 3.64 & 6.06 & 9.70 & -3.64 & -6.06 & -9.70 \\ -1.36 & -2.27 & -3.64 & 1.36 & 2.27 & 3.64 \\ -2.27 & -3.79 & -6.06 & 2.27 & 3.79 & 6.06 \\ -3.64 & -6.06 & -9.70 & 3.64 & 6.06 & 9.70 \end{bmatrix} \begin{matrix} u_2 \\ v_2 \\ w_2 \\ u_5 \\ v_5 \\ w_5 \end{matrix}$$

Member 5

Node i of the member is node 3 in the structure and node j of the member is node 5 in the structure.

$x_i = 3$ m, $x_j = 9$ m, $x_{ij} = x_j - x_i = 9 - 3 = 6$ m

$y_i = 4$ m, $y_j = 5$ m, $y_{ij} = y_j - y_i = 5 - 4 = 1$ m

$z_i = 0$, $z_j = 8$ m, $z_{ij} = z_j - z_i = 8 - 0 = 8$ m

$$s = \sqrt{x_{ij}^2 + z_{ij}^2} = \sqrt{6^2 + 8^2} = 10.000 \text{ m}$$

$$L = \sqrt{x_{ij}^2 + y_{ij}^2 + z_{ij}^2} = \sqrt{6^2 + 1^2 + 8^2} = 10.050 \text{ m}$$

$$\overbrace{\overbrace{\begin{matrix} u_i & v_i & w_i \\ u_3 & v_3 & w_3 \end{matrix}}^{\delta_i=\delta_3} \quad \overbrace{\begin{matrix} u_j & v_j & w_j \\ u_5 & v_5 & w_5 \end{matrix}}^{\delta_j=\delta_5}}^{\delta^5}$$

$$k^5 = 10^3 \begin{bmatrix} 5.21 & 0.87 & 6.95 & -5.21 & -0.87 & -6.95 \\ 0.87 & 0.15 & 1.16 & -0.87 & -0.15 & -1.16 \\ 6.95 & 1.16 & 9.27 & -6.95 & -1.16 & -9.27 \\ -5.21 & -0.87 & -6.95 & 5.21 & 0.87 & 6.95 \\ -0.87 & -0.15 & -1.16 & 0.87 & 0.15 & 1.16 \\ -6.95 & -1.16 & -9.27 & 6.95 & 1.16 & 9.27 \end{bmatrix} \begin{matrix} u_3 \\ v_3 \\ w_3 \\ u_5 \\ v_5 \\ w_5 \end{matrix}$$

Member 6

Node i of the member is node 4 in the structure and node j of the member is node 5 in the structure.

$x_i = 4$ m, $x_j = 9$ m, $x_{ij} = x_j - x_i = 9 - 4 = 5$ m

$y_i = 2$ m, $y_j = 5$ m, $y_{ij} = y_j - y_i = 5 - 2 = 3$ m

$z_i = 6$ m, $z_j = 8$ m, $z_{ij} = z_j - z_i = 8 - 6 = 2$ m

$$s = \sqrt{x_{ij}^2 + z_{ij}^2} = \sqrt{5^2 + 2^2} = 5.385 \text{ m}$$

$$L = \sqrt{x_{ij}^2 + y_{ij}^2 + z_{ij}^2} = \sqrt{5^2 + 3^2 + 2^2} = 6.164 \text{ m}$$

$$\overbrace{\overbrace{\begin{matrix} u_i & v_i & w_i \\ u_4 & v_4 & w_4 \end{matrix}}^{\delta_i=\delta_4} \quad \overbrace{\begin{matrix} u_j & v_j & w_j \\ u_5 & v_5 & w_5 \end{matrix}}^{\delta_j=\delta_5}}^{\delta^6}$$

$$k^6 = 10^3 \begin{bmatrix} 15.69 & 9.41 & 6.28 & -15.69 & -9.41 & -6.28 \\ 9.41 & 5.65 & 3.77 & -9.41 & -5.65 & -3.77 \\ 6.28 & 3.77 & 2.51 & -6.28 & -3.77 & -2.51 \\ -15.69 & -9.41 & -6.28 & 15.69 & 9.41 & 6.28 \\ -9.41 & -5.65 & -3.77 & 9.41 & 5.65 & 3.77 \\ -6.28 & -3.77 & -2.51 & 6.28 & 3.77 & 2.51 \end{bmatrix} \begin{matrix} u_4 \\ v_4 \\ w_4 \\ u_5 \\ v_5 \\ w_5 \end{matrix}$$

Assembly of the overall structure stiffness matrix

$K\delta = F$

$$\begin{bmatrix} K_{11} & K_{12} & K_{13} & K_{14} & K_{15} \\ K_{21} & K_{22} & K_{23} & K_{24} & K_{25} \\ K_{31} & K_{32} & K_{33} & K_{34} & K_{35} \\ K_{41} & K_{42} & K_{43} & K_{44} & K_{45} \\ K_{51} & K_{52} & K_{53} & K_{54} & K_{55} \end{bmatrix} \begin{bmatrix} \delta_1 \\ \delta_2 \\ \delta_3 \\ \delta_4 \\ \delta_5 \end{bmatrix} = \begin{bmatrix} F_1 \\ F_2 \\ F_3 \\ F_4 \\ F_5 \end{bmatrix}$$

$K_{11} = k_{ii}^1$	0	0	$K_{14} = k_{ij}^1$	0	u_1 v_1 w_1		X_1 Y_1 Z_1
0	$K_{22} = k_{ii}^2 + k_{ii}^4$	0	$K_{24} = k_{ij}^2$	$K_{25} = k_{ij}^4$	u_2 v_2 w_2		X_2 Y_2 Z_2
0	0	$K_{33} = k_{ii}^3 + k_{ii}^5$	$K_{34} = k_{ij}^3$	$K_{35} = k_{ij}^5$	u_3 v_3 w_3	=	X_3 Y_3 Z_3
$K_{41} = k_{ji}^1$	$K_{42} = k_{ji}^2$	$K_{43} = k_{ji}^3$	$K_{44} = k_{jj}^1 + k_{jj}^2 + k_{jj}^3 + k_{ii}^6$	$K_{45} = k_{ij}^6$	u_4 v_4 w_4		X_4 Y_4 Z_4
0	$K_{52} = k_{ji}^4$	$K_{53} = k_{ji}^5$	$K_{54} = k_{ji}^6$	$K_{55} = k_{jj}^4 + k_{jj}^5 + k_{jj}^6$	u_5 v_5 w_5		X_5 Y_5 Z_5

$$10^3\left[\begin{array}{ccc|ccc|ccc|ccc|ccc}
5.61 & 2.81 & 8.42 & 0 & 0 & 0 & 0 & 0 & 0 & -5.61 & -2.81 & -8.42 & 0 & 0 & 0 \\
2.81 & 1.40 & 4.21 & 0 & 0 & 0 & 0 & 0 & 0 & -2.81 & -1.40 & -4.21 & 0 & 0 & 0 \\
8.42 & 4.21 & 12.63 & 0 & 0 & 0 & 0 & 0 & 0 & -8.42 & -4.21 & -12.63 & 0 & 0 & 0 \\
\hline
0 & 0 & 0 & 3.38 & 0.26 & -2.41 & 0 & 0 & 0 & -2.02 & 2.02 & 6.04 & -1.36 & -2.27 & -3.64 \\
0 & 0 & 0 & 0.26 & 5.80 & 12.11 & 0 & 0 & 0 & 2.02 & -2.02 & -6.04 & -2.27 & -3.79 & -6.06 \\
0 & 0 & 0 & -2.41 & 12.11 & 27.83 & 0 & 0 & 0 & 6.04 & -6.04 & -18.13 & -3.64 & -6.06 & -9.70 \\
\hline
0 & 0 & 0 & 0 & 0 & 0 & 5.77 & -0.25 & 10.31 & -0.56 & 1.12 & -3.36 & -5.21 & -0.87 & -6.95 \\
0 & 0 & 0 & 0 & 0 & 0 & -0.25 & 2.39 & -5.56 & 1.12 & -2.24 & 6.72 & -0.87 & -0.15 & -1.16 \\
0 & 0 & 0 & 0 & 0 & 0 & 10.31 & -5.56 & 29.43 & -3.36 & 6.72 & -20.16 & -6.95 & -1.16 & -9.27 \\
\hline
-5.61 & -2.81 & -8.42 & -2.02 & 2.02 & 6.04 & -0.56 & 1.12 & -3.36 & 23.88 & 9.08 & 12.01 & -15.70 & -9.41 & -6.28 \\
-2.81 & -1.40 & -4.21 & 2.02 & -2.02 & -6.04 & 1.12 & -2.24 & 6.72 & 9.08 & 11.31 & 7.30 & -9.41 & -5.65 & -3.77 \\
-8.42 & -4.21 & -12.63 & 6.04 & -6.04 & -18.13 & -3.36 & 6.72 & -20.16 & 12.01 & 7.30 & 53.43 & -6.28 & -3.77 & -2.51 \\
\hline
0 & 0 & 0 & -1.36 & -2.27 & -3.64 & -5.21 & -0.87 & -6.95 & -15.69 & -9.41 & -6.28 & 22.27 & 12.56 & 16.86 \\
0 & 0 & 0 & -2.27 & -3.79 & -6.06 & -0.87 & -0.15 & -1.16 & -9.41 & -5.65 & -3.77 & 12.56 & 9.58 & 10.99 \\
0 & 0 & 0 & -3.64 & -6.06 & -9.70 & -6.95 & -1.16 & -9.27 & -6.28 & -3.77 & -2.51 & 16.86 & 10.99 & 21.48
\end{array}\right]
\left[\begin{array}{c}
u_1 \\ v_1 \\ w_1 \\ \hline u_2 \\ v_2 \\ w_2 \\ \hline u_3 \\ v_3 \\ w_3 \\ \hline u_4 \\ v_4 \\ w_4 \\ \hline u_5 \\ v_5 \\ w_5
\end{array}\right]
=
\left[\begin{array}{c}
R_{X1} \\ R_{Y1} \\ R_{Z1} \\ \hline R_{X2} \\ R_{Y2} \\ R_{Z2} \\ \hline R_{X3} \\ R_{Y3} \\ R_{Z3} \\ \hline +30 \\ 0 \\ 0 \\ \hline 0 \\ 0 \\ -80
\end{array}\right] \quad (9.6)$$

The boundary conditions are: $u_1 = 0$, $v_1 = 0$, $w_1 = 0$, $u_2 = 0$, $v_2 = 0$, $w_2 = 0$, $u_3 = 0$, $v_3 = 0$, and $w_3 = 0$, so delete rows and columns 1 to 9, respectively. The resulting set of simultaneous equations will then be:

$$10^3\begin{bmatrix} 23.88 & 9.08 & 12.01 & -15.69 & -9.41 & -6.28 \\ 9.08 & 11.31 & 7.30 & -9.41 & -5.65 & -3.77 \\ 12.01 & 7.30 & 53.43 & -6.28 & -3.77 & -2.51 \\ -15.69 & -9.41 & -6.28 & 22.27 & 12.56 & 16.86 \\ -9.41 & -5.65 & -3.77 & 12.56 & 9.58 & 10.99 \\ -6.28 & -3.77 & -2.51 & 16.86 & 10.99 & 21.48 \end{bmatrix}\begin{bmatrix} u_4 \\ v_4 \\ w_4 \\ u_5 \\ v_5 \\ w_5 \end{bmatrix} = \begin{bmatrix} +30 \\ 0 \\ 0 \\ 0 \\ 0 \\ -80 \end{bmatrix}$$

The solution to the above equations is:

$u_4 = 0.01237$ m, $v_4 = 0.00802$ m, $w_4 = -0.00147$ m,
$u_5 = 0.01919$ m, $v_5 = 0.01725$ m, $w_5 = -0.02276$ m,

Calculation of reactions

The set of equations (9.6) will give the external reactions on the structure by substituting the values of the above displacements.

From row 1 of (9.6)

$$10^3(5.61u_1 + 2.81v_1 + 8.42w_1 - 5.61u_4 - 2.81v_4 - 8.42w_4 = R_{X1}$$

$$10^3(5.61\times 0 + 2.81\times 0 + 8.42\times 0 - 5.61\times 0.01237 - 2.81\times 0.00802$$
$$-8.42\times(-0.00147) = R_{X1},\quad R_{X1} = -79.55\text{ kN}$$

From row 2

$$10^3(2.81u_1 + 1.40v_1 + 4.21w_1 - 2.81u_4 - 1.40v_4 - 4.21w_4 = R_{Y1}$$

$$10^3(2.81\times 0 + 1.40\times 0 + 4.21\times 0 - 2.81\times 0.01237 - 1.40\times 0.00802$$
$$-4.21\times(-0.00147) = R_{Y1},\quad R_{Y1} = -39.80\text{ kN}$$

From row 3

$$10^3(8.42u_1 + 4.21v_1 + 12.63w_1 - 8.42u_4 - 4.21v_4 - 12.63w_4 = R_{Z1}$$

$$10^3(8.42\times 0 + 4.21\times 0 + 12.63\times 0 - 8.42\times 0.01237 - 4.21\times 0.00802$$
$$-12.63\times(-0.00147) = R_{Z1},\quad R_{Z1} = -119.35\text{ kN}$$

Similarly rows 4 to 9 respectively give the values of the reactions at nodes 2 and 3 as:
$R_{X2} = -0.08$ kN, $R_{Y2} = 46.65$ kN, $R_{Z2} = 99.31$ kN, $R_{X3} = 50.19$ kN, $R_{Y3} = -6.87$ kN, and $R_{Z3} = 99.57$ kN.

Calculation of the forces in the members

The forces in the members are calculated relative to local coordinates from the relation $\bar{F} = \bar{k}\bar{\delta}$ with $\bar{k}$ from (9.2) and $\bar{\delta} = r\delta$ with the transformation matrix r as given in (9.5).

Member 1

From (9.1)

$x_{ij} = 4$ m, $y_{ij} = 2$ m, $z_{ij} = 6$ m, s = 7.211 m, L = 7.483 m

$$\frac{EA}{L} = \frac{210 \times 10^6 \times 700 \times 10^{-6}}{7.483} = 19.65 \times 10^3 \text{ kN/m}$$

$$\bar{k}^1 = 10^3 \begin{bmatrix} 19.65 & 0 & 0 & -19.65 & 0 & 0 \\ 0 & 0 & 0 & 0 & 0 & 0 \\ 0 & 0 & 0 & 0 & 0 & 0 \\ -19.65 & 0 & 0 & 19.65 & 0 & 0 \\ 0 & 0 & 0 & 0 & 0 & 0 \\ 0 & 0 & 0 & 0 & 0 & 0 \end{bmatrix}$$

From (9.5)

$$r^1 \begin{bmatrix} 0.53 & 0.27 & 0.80 & 0 & 0 & 0 \\ -0.15 & 0.96 & -0.22 & 0 & 0 & 0 \\ -0.83 & 0 & 0.55 & 0 & 0 & 0 \\ 0 & 0 & 0 & 0.53 & 0.27 & 0.80 \\ 0 & 0 & 0 & -0.15 & 0.96 & -0.22 \\ 0 & 0 & 0 & -0.83 & 0 & 0.55 \end{bmatrix}$$

$$\delta^1 = \begin{bmatrix} \delta_i^1 \\ \delta_j^1 \end{bmatrix} = \begin{bmatrix} \delta_1 \\ \delta_4 \end{bmatrix} = \begin{bmatrix} u_1 \\ v_1 \\ w_1 \\ u_4 \\ v_4 \\ w_4 \end{bmatrix} = \begin{bmatrix} 0 \\ 0 \\ 0 \\ +0.01237 \\ +0.00802 \\ -0.00147 \end{bmatrix}$$

$$\bar{\delta}^1 = r^1\delta^1 = \begin{bmatrix} 0.53 & 0.27 & 0.80 & 0 & 0 & 0 \\ -0.15 & 0.96 & -0.22 & 0 & 0 & 0 \\ -0.83 & 0 & 0.55 & 0 & 0 & 0 \\ 0 & 0 & 0 & 0.53 & 0.27 & 0.80 \\ 0 & 0 & 0 & -0.15 & 0.96 & -0.22 \\ 0 & 0 & 0 & -0.83 & 0 & 0.55 \end{bmatrix} \begin{bmatrix} 0 \\ 0 \\ 0 \\ +0.01237 \\ +0.00802 \\ -0.00147 \end{bmatrix} = \begin{bmatrix} 0 \\ 0 \\ 0 \\ +0.00755 \\ +0.00617 \\ -0.01108 \end{bmatrix}$$

$$\bar{F}^1 = \bar{k}^1\bar{\delta}^1 = 10^3 \begin{bmatrix} 19.65 & 0 & 0 & -19.65 & 0 & 0 \\ 0 & 0 & 0 & 0 & 0 & 0 \\ 0 & 0 & 0 & 0 & 0 & 0 \\ -19.65 & 0 & 0 & 19.65 & 0 & 0 \\ 0 & 0 & 0 & 0 & 0 & 0 \\ 0 & 0 & 0 & 0 & 0 & 0 \end{bmatrix} \begin{bmatrix} 0 \\ 0 \\ 0 \\ +0.00755 \\ +0.00617 \\ -0.01108 \end{bmatrix} = \begin{bmatrix} -148.36 \\ 0 \\ 0 \\ +148.36 \\ 0 \\ 0 \end{bmatrix}$$

$$\bar{F}^1 = \begin{bmatrix} \bar{X}_1^1 \\ \bar{Y}_1^1 \\ \bar{Z}_1^1 \\ \bar{X}_4^1 \\ \bar{Y}_4^1 \\ \bar{Z}_4^1 \end{bmatrix} = \begin{bmatrix} -148.36 \\ 0 \\ 0 \\ +148.36 \\ 0 \\ 0 \end{bmatrix} \text{(i.e., the member is in tension)}$$

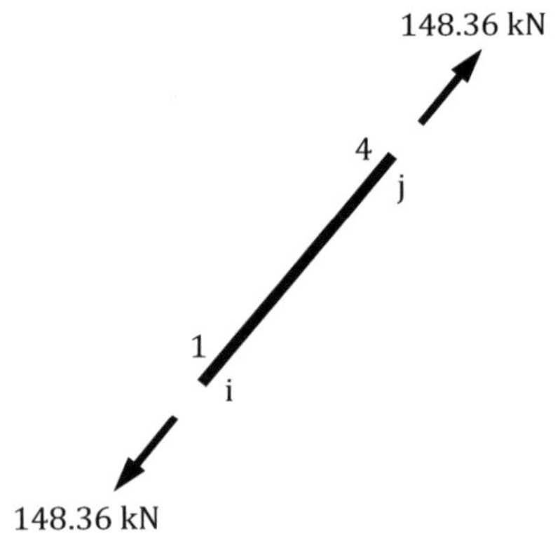

Member 3

$$x_{ij} = 1\text{ m},\ y_{ij} = -2\text{ m},\ z_{ij} = 6\text{ m},\ s = 6.083\text{ m},\ L = 6.403\text{ m}$$

$$\frac{EA}{L} = \frac{210\times10^{6}\times700\times10^{-6}}{6.403} = 22.96\times10^{3}\text{ kN/m}$$

$$\bar{k}^3 = 10^3\begin{bmatrix} 22.96 & 0 & 0 & -22.96 & 0 & 0 \\ 0 & 0 & 0 & 0 & 0 & 0 \\ 0 & 0 & 0 & 0 & 0 & 0 \\ -22.96 & 0 & 0 & 22.96 & 0 & 0 \\ 0 & 0 & 0 & 0 & 0 & 0 \\ 0 & 0 & 0 & 0 & 0 & 0 \end{bmatrix}$$

$$r^3 = \begin{bmatrix} 0.16 & -0.31 & 0.94 & 0 & 0 & 0 \\ 0.05 & 0.95 & 0.31 & 0 & 0 & 0 \\ -0.99 & 0 & 0.16 & 0 & 0 & 0 \\ 0 & 0 & 0 & 0.16 & -0.31 & 0.94 \\ 0 & 0 & 0 & 0.05 & 0.95 & 0.31 \\ 0 & 0 & 0 & -0.99 & & 0.16 \end{bmatrix}$$

$$\delta^3 = \begin{bmatrix} \delta_i^3 \\ \delta_j^3 \end{bmatrix} = \begin{bmatrix} \delta_3 \\ \delta_4 \end{bmatrix} = \begin{bmatrix} u_3 \\ v_3 \\ w_3 \\ u_4 \\ v_4 \\ w_4 \end{bmatrix} = \begin{bmatrix} 0 \\ 0 \\ 0 \\ +0.01237 \\ +0.00802 \\ -0.00147 \end{bmatrix}$$

$$\bar{\delta}^3 = r^3\delta^3 = \begin{bmatrix} 0.16 & -0.31 & 0.94 & 0 & 0 & 0 \\ 0.05 & 0.95 & 0.31 & 0 & 0 & 0 \\ -0.99 & 0 & 0.16 & 0 & 0 & 0 \\ 0 & 0 & 0 & 0.16 & -0.31 & 0.94 \\ 0 & 0 & 0 & 0.05 & 0.95 & 0.31 \\ 0 & 0 & 0 & -0.99 & 0 & 0.16 \end{bmatrix}\begin{bmatrix} 0 \\ 0 \\ 0 \\ +0.01237 \\ +0.00802 \\ -0.00147 \end{bmatrix} = \begin{bmatrix} 0 \\ 0 \\ 0 \\ -0.00189 \\ +0.00778 \\ -0.01248 \end{bmatrix}$$

$$\bar{F}^3 = \bar{k}^3\bar{\delta}^3 = 10^3\begin{bmatrix} 22.96 & 0 & 0 & -22.96 & 0 & 0 \\ 0 & 0 & 0 & 0 & 0 & 0 \\ 0 & 0 & 0 & 0 & 0 & 0 \\ -22.96 & 0 & 0 & 22.96 & 0 & 0 \\ 0 & 0 & 0 & 0 & 0 & 0 \\ 0 & 0 & 0 & 0 & 0 & 0 \end{bmatrix}\begin{bmatrix} 0 \\ 0 \\ 0 \\ -0.00189 \\ +0.00778 \\ -0.01248 \end{bmatrix} = \begin{bmatrix} +43.39 \\ 0 \\ 0 \\ -43.39 \\ 0 \\ 0 \end{bmatrix}$$

$$\bar{F}^3 = \begin{bmatrix} \bar{X}_3^3 \\ \bar{Y}_3^3 \\ \bar{Z}_3^3 \\ \bar{X}_4^3 \\ \bar{Y}_4^3 \\ \bar{Z}_4^3 \end{bmatrix} = \begin{bmatrix} +43.39 \\ 0 \\ 0 \\ -43.39 \\ 0 \\ 0 \end{bmatrix} \text{(i.e., the member is in compression)}$$

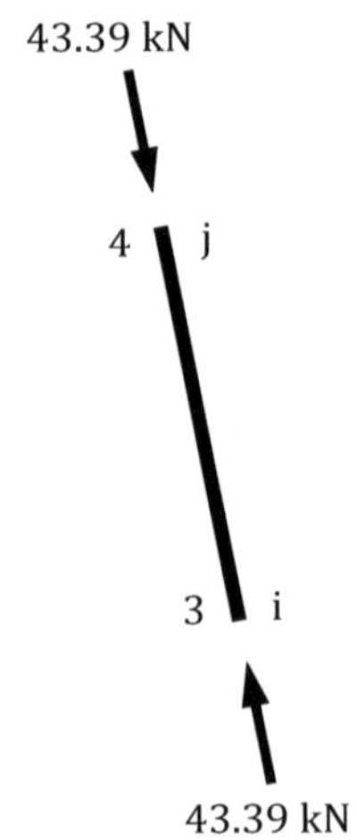

Member 6

$x_{ij} = 5$ m, $y_{ij} = 3$ m, $z_{ij} = 2$ m, s = 5.385 m, L = 6.164 m

$$\frac{EA}{L} = \frac{210\times10^6\times700\times10^{-6}}{6.164} = 23.85\times10^3 \text{ kN/m}$$

$$\bar{k}^6 = 10^3 \begin{bmatrix} 23.85 & 0 & 0 & -23.85 & 0 & 0 \\ 0 & 0 & 0 & 0 & 0 & 0 \\ 0 & 0 & 0 & 0 & 0 & 0 \\ -23.85 & 0 & 0 & 23.85 & 0 & 0 \\ 0 & 0 & 0 & 0 & 0 & 0 \\ 0 & 0 & 0 & 0 & 0 & 0 \end{bmatrix}$$

$$r^6 = \begin{bmatrix} 0.81 & 0.49 & 0.32 & 0 & 0 & 0 \\ -0.45 & 0.87 & -0.18 & 0 & 0 & 0 \\ -0.37 & 0 & 0.93 & 0 & 0 & 0 \\ 0 & 0 & 0 & 0.81 & 0.49 & 0.32 \\ 0 & 0 & 0 & -0.45 & 0.87 & -0.18 \\ 0 & 0 & 0 & -0.37 & & 0.93 \end{bmatrix}$$

$$\delta^6 = \begin{bmatrix} \delta_i^6 \\ \delta_j^6 \end{bmatrix} = \begin{bmatrix} \delta_4 \\ \delta_5 \end{bmatrix} = \begin{bmatrix} u_4 \\ v_4 \\ w_4 \\ u_5 \\ v_5 \\ w_5 \end{bmatrix} = \begin{bmatrix} +0.01237 \\ +0.00802 \\ -0.00147 \\ +0.01919 \\ +0.01725 \\ -0.02276 \end{bmatrix}$$

$$\bar{\delta}^6 = r^6\delta^6 = \begin{bmatrix} 0.81 & 0.49 & 0.32 & 0 & 0 & 0 \\ -0.45 & 0.87 & -0.18 & 0 & 0 & 0 \\ -0.37 & 0 & 0.93 & 0 & 0 & 0 \\ 0 & 0 & 0 & 0.81 & 0.49 & 0.32 \\ 0 & 0 & 0 & -0.45 & 0.87 & -0.18 \\ 0 & 0 & 0 & -0.37 & 0 & 0.93 \end{bmatrix} \begin{bmatrix} +0.01237 \\ +0.00802 \\ -0.00147 \\ +0.01919 \\ +0.01725 \\ -0.02276 \end{bmatrix} = \begin{bmatrix} +0.01348 \\ +0.00168 \\ -0.00594 \\ +0.01671 \\ +0.01047 \\ -0.02827 \end{bmatrix}$$

$$\bar{F}^6 = \bar{k}^6\bar{\delta}^6 = 10^3 \begin{bmatrix} 23.85 & 0 & 0 & -23.85 & 0 & 0 \\ 0 & 0 & 0 & 0 & 0 & 0 \\ 0 & 0 & 0 & 0 & 0 & 0 \\ -23.85 & 0 & 0 & 23.85 & 0 & 0 \\ 0 & 0 & 0 & 0 & 0 & 0 \\ 0 & 0 & 0 & 0 & 0 & 0 \end{bmatrix} \begin{bmatrix} +0.01348 \\ +0.00168 \\ -0.00594 \\ +0.01671 \\ +0.01047 \\ -0.02827 \end{bmatrix} = \begin{bmatrix} -77.04 \\ 0 \\ 0 \\ +77.04 \\ 0 \\ 0 \end{bmatrix}$$

$$\bar{F}^6 = \begin{bmatrix} \bar{X}_4^6 \\ \bar{Y}_4^6 \\ \bar{Z}_4^6 \\ \bar{X}_5^6 \\ \bar{Y}_5^6 \\ \bar{Z}_5^6 \end{bmatrix} = \begin{bmatrix} -77.04 \\ 0 \\ 0 \\ +77.04 \\ 0 \\ 0 \end{bmatrix} \text{ (i.e., the member is in tension)}$$

The forces in members 2, 4, and 5 are calculated in a similar manner to be as follows:

$$\bar{F}^2 = \begin{bmatrix} \bar{X}_2^2 \\ \bar{Y}_2^2 \\ \bar{Z}_2^2 \\ \bar{X}_4^2 \\ \bar{Y}_4^2 \\ \bar{Z}_4^2 \end{bmatrix} = \begin{bmatrix} +58.55 \\ 0 \\ 0 \\ -58.55 \\ 0 \\ 0 \end{bmatrix} \text{ (Compression),}$$

$$\bar{F}^4 = \begin{bmatrix} \bar{X}_2^4 \\ \bar{Y}_2^4 \\ \bar{Z}_2^4 \\ \bar{X}_5^4 \\ \bar{Y}_5^4 \\ \bar{Z}_5^4 \end{bmatrix} = \begin{bmatrix} +57.62 \\ 0 \\ 0 \\ -57.62 \\ 0 \\ 0 \end{bmatrix} \text{ (Compression),}$$

$$\overline{F}^5 = \begin{bmatrix} \overline{X}_3^5 \\ \overline{Y}_3^5 \\ \overline{Z}_3^5 \\ \overline{X}_5^5 \\ \overline{Y}_5^5 \\ \overline{Z}_5^5 \end{bmatrix} = \begin{bmatrix} +72.71 \\ 0 \\ 0 \\ -72.71 \\ 0 \\ 0 \end{bmatrix} \text{(Compression).}$$

Problems

Analyse the pin-connected space frames for the loading and data shown in problems P9.1 to P9.3.

P9.1. $E = 210 \times 10^6$ kN/m^2, $A_1 = 0.0024$ m^2, $A_2 = 0.0036$ m^2, $A_3 = 0.0028$ m^2, $X_4 = +100$ kN, and $Z_4 = -140$ kN.

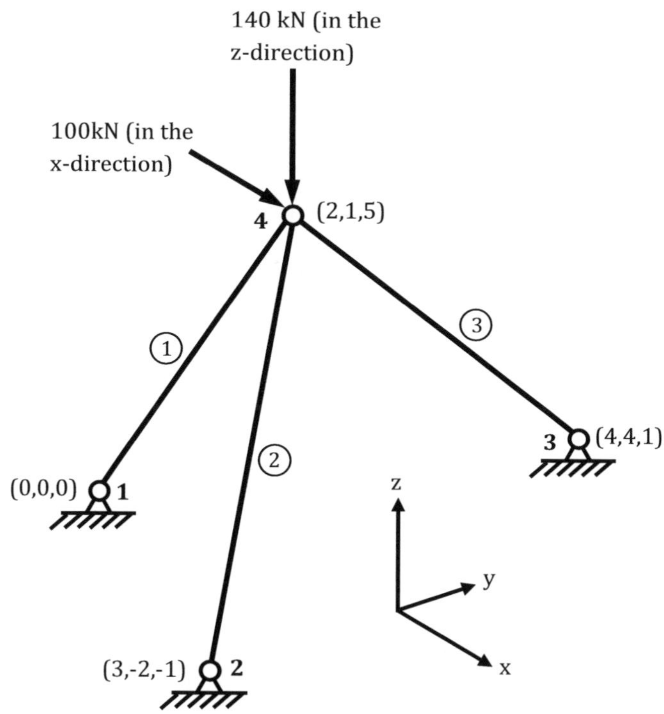

Figure P9.1

Answer:

$$u_1 = 0, v_1 = 0, w_1 = 0, u_2 = 0, v_2 = 0, w_2 = 0, u_3 = 0, v_3 = 0, w_3 = 0,$$

$$u_4 = 0.00434\ \text{m}, v_4 = -0.00152\ \text{m}, w_4 = -0.00026\ \text{m}$$

$$R_{X1} = -35.88\ \text{kN}, R_{Y1} = -17.94\ \text{kN},\ R_{Z1} = -89.69\ \text{kN},$$

$$R_{X2} = -25.36\ \text{kN},\ R_{Y2} = 76.08\ \text{kN},\ R_{Z2} = 152.17\ \text{kN},$$

$$R_{X3} = -38.76\ \text{kN},\ R_{Y3} = -58.14\ \text{kN},\ R_{Z3} = 77.53\ \text{kN}\ .$$

$$\bar{F}^1 = \begin{bmatrix} \bar{X}_1^1 \\ \bar{Y}_1^1 \\ \bar{Z}_1^1 \\ \bar{X}_4^1 \\ \bar{Y}_4^1 \\ \bar{Z}_4^1 \end{bmatrix} = \begin{bmatrix} -98.25 \\ 0 \\ 0 \\ +98.25 \\ 0 \\ 0 \end{bmatrix} \text{(Tension)},$$

$$\bar{F}^2 = \begin{bmatrix} \bar{X}_2^2 \\ \bar{Y}_2^2 \\ \bar{Z}_2^2 \\ \bar{X}_4^2 \\ \bar{Y}_4^2 \\ \bar{Z}_4^2 \end{bmatrix} = \begin{bmatrix} +172.01 \\ 0 \\ 0 \\ -172.01 \\ 0 \\ 0 \end{bmatrix} \text{(Compression)},$$

$$\bar{F}^3 = \begin{bmatrix} \bar{X}_3^3 \\ \bar{Y}_3^3 \\ \bar{Z}_3^3 \\ \bar{X}_4^3 \\ \bar{Y}_4^3 \\ \bar{Z}_4^3 \end{bmatrix} = \begin{bmatrix} +104.37 \\ 0 \\ 0 \\ -104.37 \\ 0 \\ 0 \end{bmatrix} \text{(Compression)}.$$

P9.2. For all members: E = 70 × 10^6 kN/m^2 and A = 0.0019 m^2, X_4 = +20 kN, Z_4 = −60 kN, and Y_5 = −40 kN.

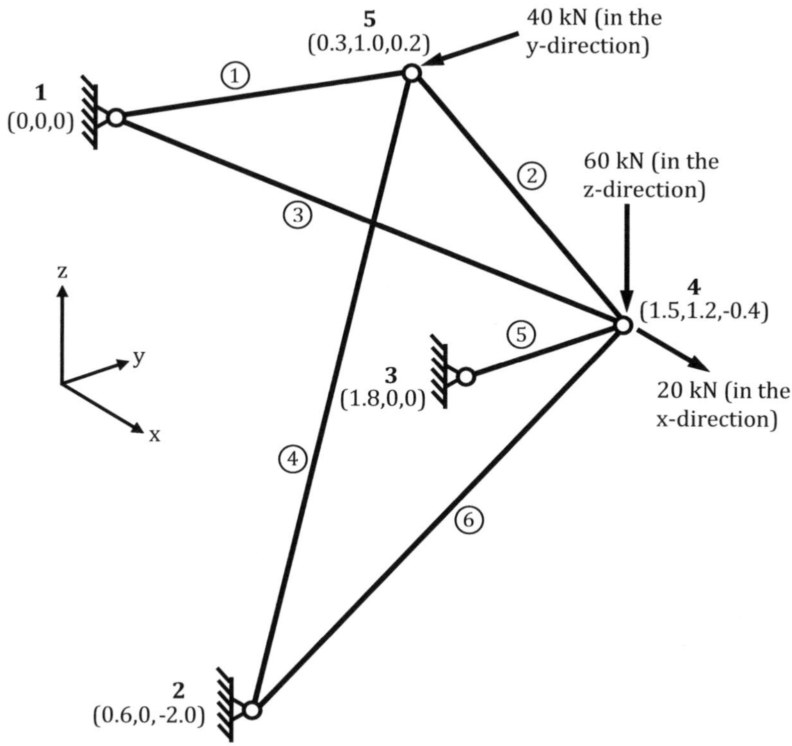

Figure P9.2

Answer:

$$u_1 = 0, v_1 = 0, w_1 = 0, u_2 = 0, v_2 = 0, w_2 = 0, u_3 = 0, v_3 = 0, w_3 = 0,$$

$$u_4 = 0.00147\,\text{m},\ v_4 = -0.00046\,\text{m},\ w_4 = -0.00217\,\text{m},$$

$$u_5 = 0.00402\,\text{m},\ v_5 = -0.00206\,\text{m},\ w_5 = 0.00192\,\text{m},$$

$$R_{X1} = -50.97\,\text{kN},\ R_{Y1} = -1.33\,\text{kN},\ R_{Z1} = 28.12\text{kN},$$

$$R_{X2} = 33.30\,\text{kN},\ R_{Y2} = 32.00\,\text{kN},\ R_{Z2} = 34.99\,\text{kN},$$

$$R_{X3} = -2.33\,\text{kN},\ R_{Y3} = 9.33\,\text{kN},\ R_{Z3} = -3.11\,\text{kN}.$$

$$\bar{F}^1 = \begin{bmatrix} \bar{X}_1^1 \\ \bar{Y}_1^1 \\ \bar{Z}_1^1 \\ \bar{X}_5^1 \\ \bar{Y}_5^1 \\ \bar{Z}_5^1 \end{bmatrix} = \begin{bmatrix} +55.17 \\ 0 \\ 0 \\ -55.17 \\ 0 \\ 0 \end{bmatrix} \text{(Compression)},$$

$$\bar{F}^2 = \begin{bmatrix} \bar{X}_5^2 \\ \bar{Y}_5^2 \\ \bar{Z}_5^2 \\ \bar{X}_4^2 \\ \bar{Y}_4^2 \\ \bar{Z}_4^2 \end{bmatrix} = \begin{bmatrix} +20.61 \\ 0 \\ 0 \\ -20.61 \\ 0 \\ 0 \end{bmatrix} \text{(Compression)},$$

$$\bar{F}^3 = \begin{bmatrix} \bar{X}_1^3 \\ \bar{Y}_1^3 \\ \bar{Z}_1^3 \\ \bar{X}_4^3 \\ \bar{Y}_4^3 \\ \bar{Z}_4^3 \end{bmatrix} = \begin{bmatrix} -87.04 \\ 0 \\ 0 \\ +87.04 \\ 0 \\ 0 \end{bmatrix} \text{(Tension)}, \ \bar{F}^4 = \begin{bmatrix} \bar{X}_2^4 \\ \bar{Y}_2^4 \\ \bar{Z}_2^4 \\ \bar{X}_5^4 \\ \bar{Y}_5^4 \\ \bar{Z}_5^4 \end{bmatrix} = \begin{bmatrix} -21.58 \\ 0 \\ 0 \\ +21.58 \\ 0 \\ 0 \end{bmatrix} \text{(Tension)},$$

$$\bar{F}^5 = \begin{bmatrix} \bar{X}_3^5 \\ \bar{Y}_3^5 \\ \bar{Z}_3^5 \\ \bar{X}_4^5 \\ \bar{Y}_4^5 \\ \bar{Z}_4^5 \end{bmatrix} = \begin{bmatrix} +10.11 \\ 0 \\ 0 \\ -10.11 \\ 0 \\ 0 \end{bmatrix} \text{(Compression)},$$

$$\overline{F}^6 = \begin{bmatrix} \overline{X}_2^6 \\ \overline{Y}_2^6 \\ \overline{Z}_2^6 \\ \overline{X}_4^6 \\ \overline{Y}_4^6 \\ \overline{Z}_4^6 \end{bmatrix} = \begin{bmatrix} +74.68 \\ 0 \\ 0 \\ -74.68 \\ 0 \\ 0 \end{bmatrix} \text{(Compression).}$$

P9.3. For all members: $E = 210 \times 10^6\ \text{kN/m}^2$, $A = 0.0012\ \text{m}^2$.
$X_4 = +70$ kN, $Y_5 = +30$ kN, and $Z_5 = -50$ kN.

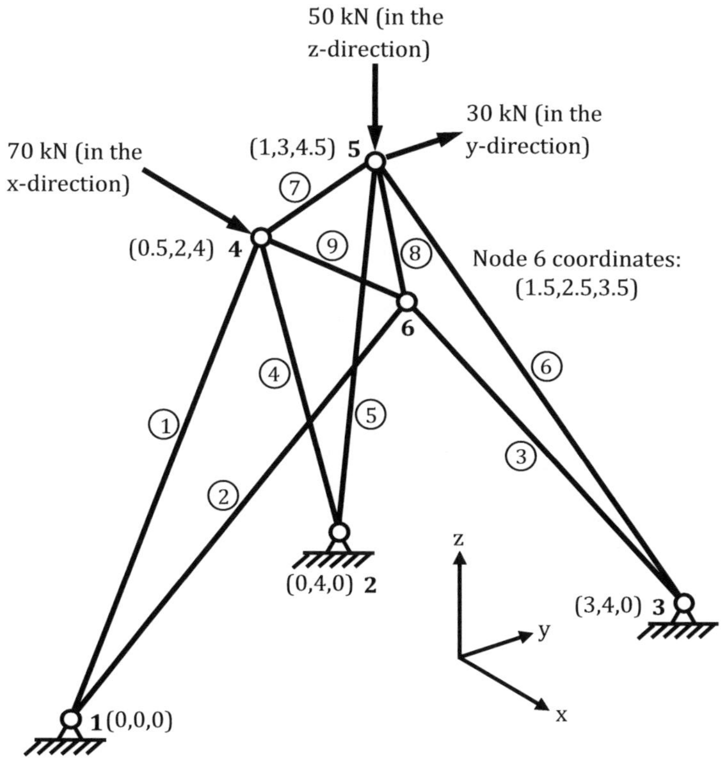

Figure P9.3

Answer:

$$u_1 = 0, v_1 = 0, w_1 = 0, u_2 = 0, v_2 = 0, w_2 = 0, u_3 = 0, v_3 = 0, w_3 = 0,$$

$$u_4 = +0.00498\,\text{m},\ v_4 = -0.00193\,\text{m},\ w_4 = -0.00035\,\text{m},$$

$$u_5 = +0.00222\,\text{m},\ v_5 = -0.00037\,\text{m},\ w_5 = -0.00091\,\text{m},$$

$$u_6 = +0.00478\,\text{m},\ v_6 = -0.00147\,\text{m},\ w_6 = +0.00059\,\text{m},$$

$$R_{X1} = -18.43\,\text{kN},\ R_{Y1} = -21.76\,\text{kN},\ R_{Z1} = -21.25\ \text{kN},$$

$$R_{X2} = -3.28\,\text{kN},\ R_{Y2} = +23.86\,\text{kN},\ R_{Z2} = -38.75\,\text{kN},$$

$$R_{X3} = -48.30\,\text{kN},\ R_{Y3} = -32.10\,\text{kN},\ R_{Z3} = +110.00\ \text{kN}.$$

$$\bar{F}^1 = \begin{bmatrix} \bar{X}_1^1 \\ \bar{Y}_1^1 \\ \bar{Z}_1^1 \\ \bar{X}_4^1 \\ \bar{Y}_4^1 \\ \bar{Z}_4^1 \end{bmatrix} = \begin{bmatrix} +34.53 \\ 0 \\ 0 \\ -34.53 \\ 0 \\ 0 \end{bmatrix},\ \bar{F}^2 = \begin{bmatrix} \bar{X}_1^2 \\ \bar{Y}_1^2 \\ \bar{Z}_1^2 \\ \bar{X}_6^2 \\ \bar{Y}_6^2 \\ \bar{Z}_6^2 \end{bmatrix} = \begin{bmatrix} -67.60 \\ 0 \\ 0 \\ +67.60 \\ 0 \\ 0 \end{bmatrix},\ \bar{F}^3 = \begin{bmatrix} \bar{X}_3^3 \\ \bar{Y}_3^3 \\ \bar{Z}_3^3 \\ \bar{X}_6^3 \\ \bar{Y}_6^3 \\ \bar{Z}_6^3 \end{bmatrix} = \begin{bmatrix} +43.39 \\ 0 \\ 0 \\ -43.39 \\ 0 \\ 0 \end{bmatrix},$$

$$\bar{F}^4 = \begin{bmatrix} \bar{X}_2^4 \\ \bar{Y}_2^4 \\ \bar{Z}_2^4 \\ \bar{X}_4^4 \\ \bar{Y}_4^4 \\ \bar{Z}_4^4 \end{bmatrix} = \begin{bmatrix} -61.74 \\ 0 \\ 0 \\ +61.74 \\ 0 \\ 0 \end{bmatrix},\ \bar{F}^5 = \begin{bmatrix} \bar{X}_2^5 \\ \bar{Y}_2^5 \\ \bar{Z}_2^5 \\ \bar{X}_5^5 \\ \bar{Y}_5^5 \\ \bar{Z}_5^5 \end{bmatrix} = \begin{bmatrix} +16.91 \\ 0 \\ 0 \\ -16.91 \\ 0 \\ 0 \end{bmatrix},\ \bar{F}^6 = \begin{bmatrix} \bar{X}_3^6 \\ \bar{Y}_3^6 \\ \bar{Z}_3^6 \\ \bar{X}_5^6 \\ \bar{Y}_5^6 \\ \bar{Z}_5^6 \end{bmatrix} = \begin{bmatrix} +81.40 \\ 0 \\ 0 \\ -81.40 \\ 0 \\ 0 \end{bmatrix},$$

$$\bar{F}^7 = \begin{bmatrix} \bar{X}_4^7 \\ \bar{Y}_4^7 \\ \bar{Z}_4^7 \\ \bar{X}_5^7 \\ \bar{Y}_5^7 \\ \bar{Z}_5^7 \end{bmatrix} = \begin{bmatrix} +15.19 \\ 0 \\ 0 \\ -15.19 \\ 0 \\ 0 \end{bmatrix},\ \bar{F}^8 = \begin{bmatrix} \bar{X}_5^8 \\ \bar{Y}_5^8 \\ \bar{Z}_5^8 \\ \bar{X}_6^8 \\ \bar{Y}_6^8 \\ \bar{Z}_6^8 \end{bmatrix} = \begin{bmatrix} -55.39 \\ 0 \\ 0 \\ +55.39 \\ 0 \\ 0 \end{bmatrix},\ \bar{F}^9 = \begin{bmatrix} \bar{X}_4^9 \\ \bar{Y}_4^9 \\ \bar{Z}_4^9 \\ \bar{X}_6^9 \\ \bar{Y}_6^9 \\ \bar{Z}_6^9 \end{bmatrix} = \begin{bmatrix} +74.44 \\ 0 \\ 0 \\ -74.44 \\ 0 \\ 0 \end{bmatrix}.$$

Chapter 10

Rigidly Connected Space Frames

These are three-dimensional frames composed of members that are rigidly connected at their joints. In contrast with pin-connected space frames where the members develop axial forces only, rigidly connected space frames develop shear forces and bending moments as well.

10.1 Derivation of Stiffness Matrix

The stiffness matrix is derived from the combination of stiffness matrices for bending about the $\bar{y}$ - and $\bar{z}$ -axes, torsion about the $\bar{x}$ -axis and axial force along the $\bar{x}$ -axis as follows:

(i) For axial force along the $\bar{x}$ -axis the stiffness matrix was derived in Chapter 2 as:

$$\begin{bmatrix} \bar{X}_i \\ \bar{X}_j \end{bmatrix} = \begin{bmatrix} \frac{EA}{L} & -\frac{EA}{L} \\ -\frac{EA}{L} & \frac{EA}{L} \end{bmatrix} \begin{bmatrix} \bar{u}_i \\ \bar{u}_j \end{bmatrix} \tag{10.1}$$

(ii) The stiffness matrix for torsion about the $\bar{x}$ -axis was derived in Chapter 7 as:

Analysis of Structures by Matrix Methods
Fathi Al-Shawi

ISBN 978-981-4968-19-5 (Hardcover), 978-1-003-29130-5 (eBook)
www.jennystanford.com

$$\begin{bmatrix} \bar{T}_i \\ \bar{T}_j \end{bmatrix} = \begin{bmatrix} \frac{GJ}{L} & -\frac{GJ}{L} \\ -\frac{GJ}{L} & \frac{GJ}{L} \end{bmatrix} \begin{bmatrix} \bar{\phi}_i \\ \bar{\phi}_j \end{bmatrix} \tag{10.2}$$

(iii) The stiffness matrix for bending about the $\bar{y}$ -axis was derived in Chapter 4 as:

$$\begin{bmatrix} \bar{Z}_i \\ \bar{M}_i \\ \bar{Z}_j \\ \bar{M}_j \end{bmatrix} = \begin{bmatrix} \frac{12EI_{\bar{y}}}{L^3} & -\frac{6EI_{\bar{y}}}{L^2} & -\frac{12EI_{\bar{y}}}{L^3} & -\frac{6EI_{\bar{y}}}{L^2} \\ -\frac{6EI_{\bar{y}}}{L^2} & \frac{4EI_{\bar{y}}}{L} & \frac{6EI_{\bar{y}}}{L^2} & \frac{2EI_{\bar{y}}}{L} \\ -\frac{12EI_{\bar{y}}}{L^3} & \frac{6EI_{\bar{y}}}{L^2} & \frac{12EI_{\bar{y}}}{L^3} & \frac{6EI_{\bar{y}}}{L^2} \\ -\frac{6EI_{\bar{y}}}{L^2} & \frac{2EI_{\bar{y}}}{L} & \frac{6EI_{\bar{y}}}{L^2} & \frac{4EI_{\bar{y}}}{L} \end{bmatrix} \begin{bmatrix} \bar{w}_i \\ \bar{\theta}_i \\ \bar{w}_j \\ \bar{\theta}_j \end{bmatrix} \tag{10.3}$$

(iv) The stiffness matrix for bending about the $\bar{z}$ -axis is derived in Appendix 2 as:

$$\begin{bmatrix} \bar{Y}_i \\ \bar{N}_i \\ \bar{Y}_j \\ \bar{N}_j \end{bmatrix} = \begin{bmatrix} \frac{12EI_{\bar{z}}}{L^3} & \frac{6EI_{\bar{z}}}{L^2} & -\frac{12EI_{\bar{z}}}{L^3} & \frac{6EI_{\bar{z}}}{L^2} \\ \frac{6EI_{\bar{z}}}{L^2} & \frac{4EI_{\bar{z}}}{L} & -\frac{6EI_{\bar{z}}}{L^2} & \frac{2EI_{\bar{z}}}{L} \\ -\frac{12EI_{\bar{z}}}{L^3} & -\frac{6EI_{\bar{z}}}{L^2} & \frac{12EI_{\bar{z}}}{L^3} & -\frac{6EI_{\bar{z}}}{L^2} \\ \frac{6EI_{\bar{z}}}{L^2} & \frac{2EI_{\bar{z}}}{L} & -\frac{6EI_{\bar{z}}}{L^2} & \frac{4EI_{\bar{z}}}{L} \end{bmatrix} \begin{bmatrix} \bar{v}_i \\ \bar{\Psi}_i \\ \bar{v}_j \\ \bar{\Psi}_j \end{bmatrix} \tag{10.4}$$

Notice the change in sign of some of the coefficients in comparison with the matrix for bending about the $\bar{y}$ -axis.

The resulting stiffness matrix for the general case is obtained by combining cases (i), (ii), (iii), and (iv) as given by (10.1), (10.2), (10.3), and (10.4) to get:

$$\begin{bmatrix} \frac{EA}{L} & 0 & 0 & 0 & 0 & 0 & -\frac{EA}{L} & 0 & 0 & 0 & 0 & 0 \\ 0 & \frac{12EI_{\bar{z}}}{L^3} & 0 & 0 & 0 & \frac{6EI_{\bar{z}}}{L^2} & 0 & -\frac{12EI_{\bar{z}}}{L^3} & 0 & 0 & 0 & \frac{6EI_{\bar{z}}}{L^2} \\ 0 & 0 & \frac{12EI_{\bar{y}}}{L^3} & 0 & -\frac{6EI_{\bar{y}}}{L^2} & 0 & 0 & 0 & -\frac{12EI_{\bar{y}}}{L^3} & 0 & -\frac{6EI_{\bar{y}}}{L^2} & 0 \\ 0 & 0 & 0 & \frac{GJ}{L} & 0 & 0 & 0 & 0 & 0 & -\frac{GJ}{L} & 0 & 0 \\ 0 & 0 & -\frac{6EI_{\bar{y}}}{L^2} & 0 & \frac{4EI_{\bar{y}}}{L} & 0 & 0 & 0 & \frac{6EI_{\bar{y}}}{L^2} & 0 & \frac{2EI_{\bar{y}}}{L} & 0 \\ 0 & \frac{6EI_{\bar{z}}}{L^2} & 0 & 0 & 0 & \frac{4EI_{\bar{z}}}{L} & 0 & -\frac{6EI_{\bar{z}}}{L^2} & 0 & 0 & 0 & \frac{2EI_{\bar{z}}}{L} \\ -\frac{EA}{L} & 0 & 0 & 0 & 0 & 0 & \frac{EA}{L} & 0 & 0 & 0 & 0 & 0 \\ 0 & -\frac{12EI_{\bar{z}}}{L^3} & 0 & 0 & 0 & -\frac{6EI_{\bar{z}}}{L^2} & 0 & \frac{12EI_{\bar{z}}}{L^3} & 0 & 0 & 0 & -\frac{6EI_{\bar{z}}}{L^2} \\ 0 & 0 & -\frac{12EI_{\bar{y}}}{L^3} & 0 & \frac{6EI_{\bar{y}}}{L^2} & 0 & 0 & 0 & \frac{12EI_{\bar{y}}}{L^3} & 0 & \frac{6EI_{\bar{y}}}{L^2} & 0 \\ 0 & 0 & 0 & -\frac{GJ}{L} & 0 & 0 & 0 & 0 & 0 & \frac{GJ}{L} & 0 & 0 \\ 0 & 0 & -\frac{6EI_{\bar{y}}}{L^2} & 0 & \frac{2EI_{\bar{y}}}{L} & 0 & 0 & 0 & \frac{6EI_{\bar{y}}}{L^2} & 0 & \frac{4EI_{\bar{y}}}{L} & 0 \\ 0 & \frac{6EI_{\bar{z}}}{L^2} & 0 & 0 & 0 & \frac{2EI_{\bar{z}}}{L} & 0 & -\frac{6EI_{\bar{z}}}{L^2} & 0 & 0 & 0 & \frac{4EI_{\bar{z}}}{L} \end{bmatrix} \begin{Bmatrix} \bar{u}_i \\ \bar{v}_i \\ \bar{w}_i \\ \bar{\Phi}_i \\ \bar{\theta}_i \\ \bar{\Psi}_i \\ \bar{u}_j \\ \bar{v}_j \\ \bar{w}_j \\ \bar{\Phi}_j \\ \bar{\theta}_j \\ \bar{\Psi}_j \end{Bmatrix} = \begin{Bmatrix} \bar{X}_i \\ \bar{Y}_i \\ \bar{Z}_i \\ \bar{T}_i \\ \bar{M}_i \\ \bar{N}_i \\ \bar{X}_j \\ \bar{Y}_j \\ \bar{Z}_j \\ \bar{T}_j \\ \bar{M}_j \\ \bar{N}_j \end{Bmatrix} \quad (10.5)$$

or simply,

$$\bar{k}\bar{\delta} = \bar{F}. \tag{10.6}$$

10.2 Transformation to Global Coordinates

The quantities in (10.6) are relative to local coordinates and in the assembly of the stiffness matrix for the overall structure they need to be transformed and written relative to the global coordinates.

In Chapter 9 the members of the space frame develop axial forces only since the frame is pin-connected and only rotations of the $\bar{y}$- and $\bar{z}$-axes are considered in the transformation as shown in Fig. 10.1a.

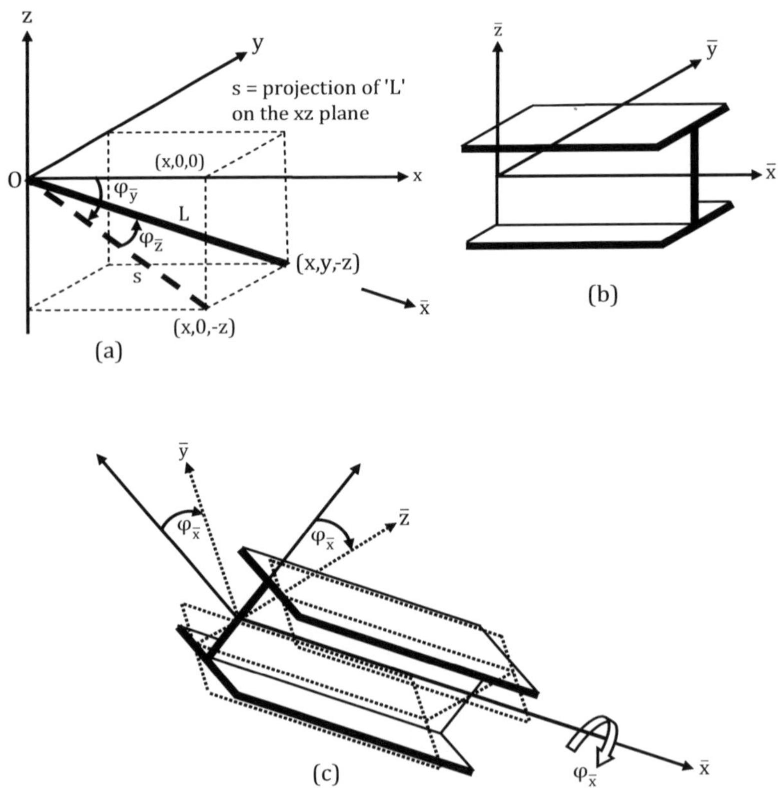

Figure 10.1

In this chapter where the members develop moments (as well as axial forces) due to the rigidity of the joints, the rotation about the $\bar{x}$-axis is also considered since this rotation will change the orientation of the principal axes of the cross section which affects the bending stiffness of the member relative to the global axes. Therefore, transformation from local to global coordinates due to rotations of the $\bar{y}$- and $\bar{z}$- as well as the $\bar{x}$-axes is carried out. The transformation is done in three separate stages which are combined to give the final transformation matrix. In the first stage a rotation $\varphi_{\bar{y}}$ about the $\bar{y}$-axis is made to calculate the transformation matrix $\rho_{\bar{y}}$ and in the second stage a rotation $\varphi_{\bar{z}}$ about the $\bar{z}$-axis is made to calculate the transformation matrix $\rho_{\bar{z}}$. These two transformation matrices have been derived in Chapter 9 for the pin-connected space frame as given in (9.2) and (9.3), respectively, as follows:

For rotation of the $\bar{y}$-axis

$$\rho_{\bar{y}} = \begin{bmatrix} \cos\varphi_{\bar{y}} & 0 & -\sin\varphi_{\bar{y}} \\ 0 & 1 & 0 \\ \sin\varphi_{\bar{y}} & 0 & \cos\varphi_{\bar{y}} \end{bmatrix} \tag{10.7}$$

and for rotation of the $\bar{z}$-axis

$$\rho_{\bar{z}} = \begin{bmatrix} \cos\varphi_{\bar{z}} & \sin\varphi_{\bar{z}} & 0 \\ -\sin\varphi_{\bar{z}} & \cos\varphi_{\bar{z}} & 0 \\ 0 & 0 & 1 \end{bmatrix} \tag{10.8}$$

Sometimes the member is rotated about its longitudinal axis (i.e., its $\bar{x}$-axis) to achieve more efficient use of its cross-sectional geometric properties or make the details of the connections between members of the structure more practical. In such a case a third separate transformation is required for a rotation of the $\bar{x}$-axis. Consider a member that is originally lying along the x-axis and is rotated about its own axis by an angle $\varphi_{\bar{x}}$ in the clockwise direction as shown in Fig. 10.1c and Fig. 10.2.

Although the $\bar{x}$-axis is rotated axially it remains pointing in the same direction which means that displacements and forces relative to (along) the rotated $\bar{x}$-axis are the same as those relative to the x-axis, thus

$$\bar{u} = u.$$

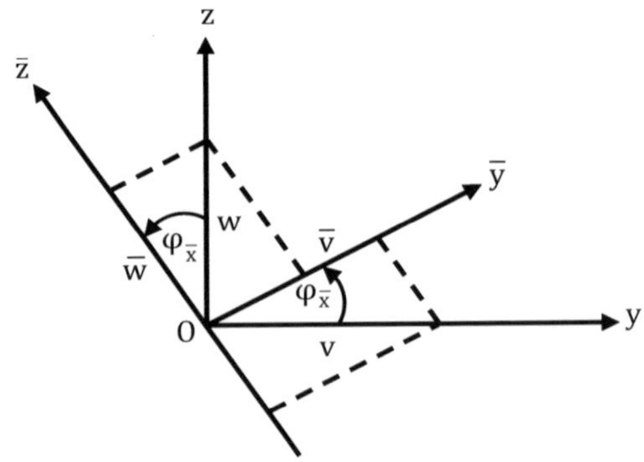

Figure 10.2

The displacement along the $\bar{y}$ -axis is equal to the algebraic sum of the components of the displacements along the y- and z-axes, respectively, thus

$$\bar{v} = v\cos\varphi_{\bar{x}} + w\sin\varphi_{\bar{x}}.$$

The displacement along the $\bar{z}$ -axis is equal to the algebraic sum of the components of the displacements along the y- and z-axes, respectively, hence

$$\bar{w} = -v\sin\varphi_{\bar{x}} + w\cos\varphi_{\bar{x}}.$$

In matrix form

$$\begin{bmatrix} \bar{u} \\ \bar{v} \\ \bar{w} \end{bmatrix} = \begin{bmatrix} 1 & 0 & 0 \\ 0 & \cos\varphi_{\bar{x}} & \sin\varphi_{\bar{x}} \\ 0 & -\sin\varphi_{\bar{x}} & \cos\varphi_{\bar{x}} \end{bmatrix} \begin{bmatrix} u \\ v \\ w \end{bmatrix}$$

$$\begin{bmatrix} \bar{u} \\ \bar{v} \\ \bar{w} \end{bmatrix} = \rho_{\bar{x}} \begin{bmatrix} u \\ v \\ w \end{bmatrix},$$

where

$$\rho_{\bar{x}} = \begin{bmatrix} 1 & 0 & 0 \\ 0 & \cos\varphi_{\bar{x}} & \sin\varphi_{\bar{x}} \\ 0 & -\sin\varphi_{\bar{x}} & \cos\varphi_{\bar{x}} \end{bmatrix} \qquad (10.9)$$

The complete transformation matrix ρ is obtained by multiplying the three transformation matrices (10.7), (10.8), and (10.9) in the order of rotations of the local axes, thus

$$\rho = \rho_{\bar{x}}\rho_{\bar{z}}\rho_{\bar{y}}$$

$$\rho = \begin{bmatrix} 1 & 0 & 0 \\ 0 & \cos\varphi_{\bar{x}} & \sin\varphi_{\bar{x}} \\ 0 & -\sin\varphi_{\bar{x}} & \cos\varphi_{\bar{x}} \end{bmatrix} \begin{bmatrix} \cos\varphi_{\bar{z}} & \sin\varphi_{\bar{z}} & 0 \\ -\sin\varphi_{\bar{z}} & \cos\varphi_{\bar{z}} & 0 \\ 0 & 0 & 1 \end{bmatrix} \begin{bmatrix} \cos\varphi_{\bar{y}} & 0 & -\sin\varphi_{\bar{y}} \\ 0 & 1 & 0 \\ \sin\varphi_{\bar{y}} & 0 & \cos\varphi_{\bar{y}} \end{bmatrix}$$

$$\rho = \begin{bmatrix} \cos\varphi_{\bar{z}}\cos\varphi_{\bar{y}} & \sin\varphi_{\bar{z}} & -\cos\varphi_{\bar{z}}\sin\varphi_{\bar{y}} \\ \sin\varphi_{\bar{x}}\sin\varphi_{\bar{y}} - \sin\varphi_{\bar{z}}\cos\varphi_{\bar{x}}\cos\varphi_{\bar{y}} & \cos\varphi_{\bar{z}}\cos\varphi_{\bar{x}} & \sin\varphi_{\bar{z}}\cos\varphi_{\bar{x}}\sin\varphi_{\bar{y}} + \sin\varphi_{\bar{x}}\cos\varphi_{\bar{y}} \\ \sin\varphi_{\bar{z}}\sin\varphi_{\bar{x}}\cos\varphi_{\bar{y}} + \cos\varphi_{\bar{x}}\sin\varphi_{\bar{y}} & -\cos\varphi_{\bar{z}}\sin\varphi_{\bar{x}} & \cos\varphi_{\bar{x}}\cos\varphi_{\bar{y}} - \sin\varphi_{\bar{z}}\sin\varphi_{\bar{x}}\sin\varphi_{\bar{y}} \end{bmatrix} \tag{10.10}$$

$$\begin{bmatrix} \bar{u} \\ \bar{v} \\ \bar{w} \end{bmatrix} = \rho \begin{bmatrix} u \\ v \\ w \end{bmatrix}$$

The above transformation is for the translational displacements $\bar{u}$, $\bar{v}$, and $\bar{w}$ and since the rotational displacements $\bar{\Phi}$, $\bar{\theta}$, and $\bar{\Psi}$ are vectors about the same axes, then the same transformation matrix, ρ, will apply, thus

$$\begin{bmatrix} \bar{\Phi} \\ \bar{\theta} \\ \bar{\Psi} \end{bmatrix} = \rho \begin{bmatrix} \Phi \\ \theta \\ \Psi \end{bmatrix}$$

For transformation of the displacement vector at node i: $\bar{\delta}_i = r_i\delta_i$, where

$$R_i = \begin{bmatrix} \rho & O \\ O & \rho \end{bmatrix}, \quad O = \begin{bmatrix} 0 & 0 & 0 \\ 0 & 0 & 0 \\ 0 & 0 & 0 \end{bmatrix}, \quad \bar{\delta}_i = \begin{bmatrix} \bar{u}_i \\ \bar{v}_i \\ \bar{w}_i \\ \bar{\Phi}_i \\ \bar{\theta}_i \\ \bar{\Psi}_i \end{bmatrix}, \text{ and } \delta_i = \begin{bmatrix} u_i \\ v_i \\ w_i \\ \Phi_i \\ \theta_i \\ \Psi_i \end{bmatrix},$$

thus
$$\begin{bmatrix}\bar{u}_i\\\bar{v}_i\\\bar{w}_i\\\bar{\Phi}_i\\\bar{\theta}_i\\\bar{\Psi}_i\end{bmatrix}=\begin{bmatrix}\rho & 0\\0 & \rho\end{bmatrix}\begin{bmatrix}u_i\\v_i\\w_i\\\Phi_i\\\theta_i\\\Psi_i\end{bmatrix}$$

Similarly, for the transformation of the displacement vector at node j: $\bar{\delta}_j = r_j\delta_j$

$$\begin{bmatrix}\bar{u}_j\\\bar{v}_j\\\bar{w}_j\\\bar{\Phi}_j\\\bar{\theta}_j\\\bar{\Psi}_j\end{bmatrix}=\begin{bmatrix}\rho & 0\\0 & \rho\end{bmatrix}\begin{bmatrix}u_j\\v_j\\w_j\\\Phi_j\\\theta_j\\\Psi_j\end{bmatrix}$$

and for both nodes, $\bar{\delta} = r\delta$, where $r=\begin{bmatrix}r_i & 0\\0 & r_j\end{bmatrix}$ which results in the final transformation matrix as:

$$r=\begin{bmatrix}\rho & 0 & 0 & 0\\0 & \rho & 0 & 0\\0 & 0 & \rho & 0\\0 & 0 & 0 & \rho\end{bmatrix}\text{ (i.e., a }12\times 12\text{ matrix)}$$

The above transformation matrix is for the displacements and it is the same for the actions since both displacements and actions are vectors relative to the same axes, thus

$$\begin{bmatrix}\bar{X}\\\bar{Y}\\\bar{Z}\end{bmatrix}=\rho\begin{bmatrix}X\\Y\\Z\end{bmatrix}\text{ and }\begin{bmatrix}\bar{T}\\\bar{M}\\\bar{N}\end{bmatrix}=\rho\begin{bmatrix}T\\M\\N\end{bmatrix}$$

or $\bar{F} = rF$.

The stiffness matrix relative to global coordinates is given by:

$$k = r^T\bar{k}r \qquad (10.11)$$

with $\bar{k}$ and r from (10.5) and (10.10), respectively.

The transformation matrix in (10.10) can be simplified by referring to Fig. 10.1 and making the following substitutions

$$\sin\varphi_{\bar{y}} = -\frac{z_{ij}}{s}, \cos\varphi_{\bar{y}} = \frac{x_{ij}}{s}, \sin\varphi_{\bar{z}} = \frac{y_{ij}}{L}, \cos\varphi_{\bar{z}} = \frac{s}{L}, \text{ where } x_{ij} = x_j - x_i,$$

$$y_{ij} = y_j - y_i, z_{ij} = z_j - z_i, s = \sqrt{x_{ij}^2 + z_{ij}^2}, \text{ and } L = \sqrt{x_{ij}^2 + y_{ij}^2 + z_{ij}^2}.$$

$$\rho = \begin{bmatrix} \dfrac{x_{ij}}{L} & \dfrac{y_{ij}}{L} & \dfrac{z_{ij}}{L} \\ -\dfrac{z_{ij}}{s}\sin\varphi_{\bar{x}} - \dfrac{y_{ij}x_{ij}}{Ls}\cos\varphi_{\bar{x}} & \dfrac{s}{L}\cos\varphi_{\bar{x}} & -\dfrac{y_{ij}z_{ij}}{Ls}\cos\varphi_{\bar{x}} + \dfrac{x_{ij}}{s}\sin\varphi_{\bar{x}} \\ \dfrac{y_{ij}x_{ij}}{Ls}\sin\varphi_{\bar{x}} - \dfrac{z_{ij}}{s}\cos\varphi_{\bar{x}} & -\dfrac{s}{L}\sin\varphi_{\bar{x}} & \dfrac{x_{ij}}{s}\cos\varphi_{\bar{x}} + \dfrac{y_{ij}z_{ij}}{Ls}\sin\varphi_{\bar{x}} \end{bmatrix} \tag{10.12}$$

Example 1:

Analyse the rigidly connected space frame shown in Fig. 10.3 given that all the members have the same cross section with the following properties:

$A = 17800 \times 10^{-6}\text{m}^2$, $I_{\bar{y}} = 1117.77 \times 10^{-6}\text{m}^4$, $I_{\bar{z}} = 45.05 \times 10^{-6}\text{m}^4$, $J = 2.16 \times 10^{-6}\text{m}^4$, $E = 210 \times 10^6\text{kN/m}^2$, and $\mu = 0.3$.

$$G = \frac{E}{2(1+\mu)} = \frac{210 \times 10^6}{2(1+0.3)} = 80.77 \times 10^6 \text{kN/m}^2.$$

The coordinates of the joints of the frame are given in the table below.

Node number	x (m)	y (m)	z (m)
1	0	0	0
2	0	0	6
3	0	4	6
4	5	4	6

The member and structure addresses are shown below.

Member number	Node i	Node j
1	1	2
2	2	3
3	3	4

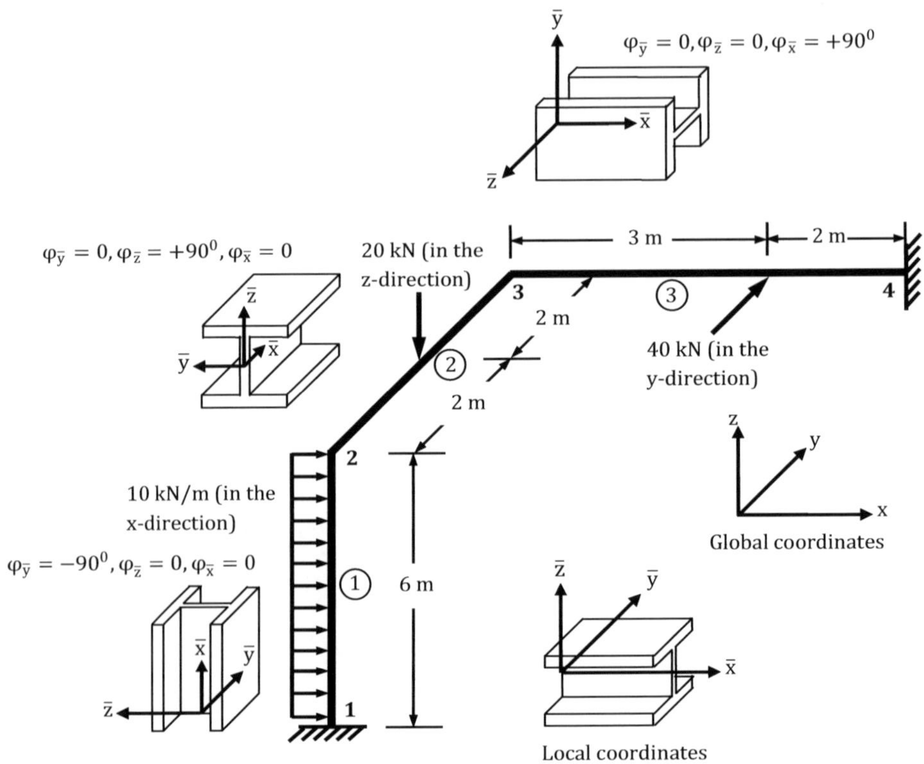

Figure 10.3

Calculation of member stiffness matrices

Member 1, L = 6 m

From (10.5)

$$
\bar{\mathrm{k}}^1 = \left[\begin{array}{cccccc|cccccc}
623000 & 0 & 0 & 0 & 0 & 0 & -623000 & 0 & 0 & 0 & 0 & 0 \\
0 & 526 & 0 & 0 & 0 & 1577 & 0 & -526 & 0 & 0 & 0 & 1577 \\
0 & 0 & 13041 & 0 & -39122 & 0 & 0 & 0 & -13041 & 0 & -39122 & 0 \\
0 & 0 & 0 & 29 & 0 & 0 & 0 & 0 & 0 & -29 & 0 & 0 \\
0 & 0 & -39122 & 0 & 156488 & 0 & 0 & 0 & 39122 & 0 & 78244 & 0 \\
0 & 1577 & 0 & 0 & 0 & 6307 & 0 & -1577 & 0 & 0 & 0 & 3154 \\
\hline
-623000 & 0 & 0 & 0 & 0 & 0 & 623000 & 0 & 0 & 0 & 0 & 0 \\
0 & -526 & 0 & 0 & 0 & -1577 & 0 & 526 & 0 & 0 & 0 & -1577 \\
0 & 0 & -13041 & 0 & 39122 & 0 & 0 & 0 & 13041 & 0 & 39122 & 0 \\
0 & 0 & 0 & -29 & 0 & 0 & 0 & 0 & 0 & 29 & 0 & 0 \\
0 & 0 & -39122 & 0 & 78244 & 0 & 0 & 0 & 39122 & 0 & 156488 & 0 \\
0 & 1577 & 0 & 0 & 0 & 3154 & 0 & -1577 & 0 & 0 & 0 & 6307
\end{array}\right] \tag{10.13}
$$

$x_i = 0,\ x_j = 0,\ x_{ij} = x_j - x_i = 0 - 0 = 0$

$y_i = 0,\ y_j = 0,\ y_{ij} = y_j - y_i = 0 - 0 = 0$

$z_i = 0,\ z_j = 6\text{ m},\ z_{ij} = z_j - z_i = 6 - 0 = 6\text{ m}$

$$s = \sqrt{x_{ij}^2 + z_{ij}^2} = \sqrt{0^2 + 6^2} = 6\text{ m}$$

$$L = \sqrt{x_{ij}^2 + y_{ij}^2 + z_{ij}^2} = \sqrt{0^2 + 0^2 + 6^2} = 6\text{ m}$$

$$\varphi_{\bar{x}} = 0, \sin\varphi_{\bar{x}} = 0, \cos\varphi_{\bar{x}} = 1$$

From (10.12)

$$\rho^1 = \begin{bmatrix} \dfrac{0}{6} & \dfrac{0}{6} & \dfrac{6}{6} \\ -\dfrac{6}{6}\times 0 - \dfrac{0\times 0}{6\times 6}\times 1 & \dfrac{6}{6}\times 1 & -\dfrac{0\times 6}{6\times 6}\times 1 + \dfrac{0}{6}\times 0 \\ \dfrac{0\times 0}{6\times 6}\times 0 - \dfrac{6}{6}\times 1 & -\dfrac{6}{6}\times 0 & \dfrac{0}{6}\times 1 + \dfrac{0\times 6}{6\times 6}\times 0 \end{bmatrix} = \begin{bmatrix} 0 & 0 & 1 \\ 0 & 1 & 0 \\ -1 & 0 & 0 \end{bmatrix}$$

$$r^1 = \begin{bmatrix} 0 & 0 & 1 & 0 & 0 & 0 & 0 & 0 & 0 & 0 & 0 & 0 \\ 0 & 1 & 0 & 0 & 0 & 0 & 0 & 0 & 0 & 0 & 0 & 0 \\ -1 & 0 & 0 & 0 & 0 & 0 & 0 & 0 & 0 & 0 & 0 & 0 \\ 0 & 0 & 0 & 0 & 0 & 1 & 0 & 0 & 0 & 0 & 0 & 0 \\ 0 & 0 & 0 & 0 & 1 & 0 & 0 & 0 & 0 & 0 & 0 & 0 \\ 0 & 0 & 0 & -1 & 0 & 0 & 0 & 0 & 0 & 0 & 0 & 0 \\ 0 & 0 & 0 & 0 & 0 & 0 & 0 & 0 & 1 & 0 & 0 & 0 \\ 0 & 0 & 0 & 0 & 0 & 0 & 0 & 1 & 0 & 0 & 0 & 0 \\ 0 & 0 & 0 & 0 & 0 & 0 & -1 & 0 & 0 & 0 & 0 & 0 \\ 0 & 0 & 0 & 0 & 0 & 0 & 0 & 0 & 0 & 0 & 0 & 1 \\ 0 & 0 & 0 & 0 & 0 & 0 & 0 & 0 & 0 & 0 & 1 & 0 \\ 0 & 0 & 0 & 0 & 0 & 0 & 0 & 0 & 0 & -1 & 0 & 0 \end{bmatrix} \tag{10.14}$$

From (10.11)

$$k^1 = (r^1)^T \bar{k}^1 r^1$$

$k^1 =$ (10.15)

13041	0	0	0	39122	0	−13041	0	0	0	39122	0
0	526	0	−1577	0	0	0	−526	0	−1577	0	0
0	0	623000	0	0	0	0	0	−623000	0	0	0
0	−1577	0	6307	0	0	0	1577	0	3154	0	0
39122	0	0	0	156488	0	−39122	0	0	0	78244	0
0	0	0	0	0	29	0	0	0	0	0	−29
−13041	0	0	0	−39122	0	13041	0	0	0	−39122	0
0	−526	0	1577	0	0	0	526	0	1577	0	0
0	0	−623000	0	0	0	0	0	623000	0	0	0
0	−1577	0	3154	0	0	0	1577	0	6307	0	0
39122	0	0	0	78244	0	−39122	0	0	0	156488	0
0	0	0	0	0	−29	0	0	0	0	0	29

Member 2, L = 4

From (10.5)

$\bar{k}^2 =$

934500	0	0	0	0	0	−934500	0	0	0	0	0
0	1774	0	0	0	3548	0	−1774	0	0	0	3548
0	0	44012	0	−88024	0	0	0	−44012	0	−88024	0
0	0	0	44	0	0	0	0	0	−44	0	0
0	0	−88024	0	234732	0	0	0	88024	0	117366	0
0	3548	0	0	0	9461	0	−3548	0	0	0	4730
−934500	0	0	0	0	0	934500	0	0	0	0	0
0	−1774	0	0	0	−3548	0	1774	0	0	0	−3548
0	0	−44012	0	88024	0	0	0	44012	0	88024	0
0	0	0	−44	0	0	0	0	0	44	0	0
0	0	−88024	0	117366	0	0	0	88024	0	234732	0
0	3548	0	0	0	4730	0	−3548	0	0	0	9461

(10.16)

The transformation matrix for this member represents a special case that arises when the local $\bar{x}$-axis of the member is coincident with the global y-axis, where $x_{ij} = 0$ and $z_{ij} = 0$. And since, $s = \sqrt{x_{ij}^2 + z_{ij}^2}$ then $x_{ij}/s = 0/0$ and $z_{ij}/s = 0/0$ are indeterminate quantities. To overcome this situation, we revert to the original transformation matrix (10.10) which is in terms of trigonometric functions and substitute $\varphi_{\bar{y}} = 0$, $\varphi_{\bar{z}} = +90^0$, and $\varphi_{\bar{x}} = 0$. The resulting transformation matrix will then be:

$$\rho^2 = \begin{bmatrix} \cos 90 \cos 0 & \sin 90 & -\cos 90 \sin 0 \\ \sin 0 \sin 0 - \sin 90 \cos 0 \cos 0 & \cos 90 \cos 0 & \sin 90 \cos 0 \sin 0 + \sin 0 \cos 0 \\ \sin 90 \sin 0 \cos 0 + \cos 0 \sin 0 & -\cos 90 \sin 0 & \cos 0 \cos 0 - \sin 90 \sin 0 \sin 0 \end{bmatrix}$$

$$\rho^2 = \begin{bmatrix} 0 & 1 & 0 \\ -1 & 0 & 0 \\ 0 & 0 & 1 \end{bmatrix}$$

$$r^2 = \begin{bmatrix} 0 & \mathbf{1} & 0 & 0 & 0 & 0 & 0 & 0 & 0 & 0 & 0 & 0 \\ \mathbf{-1} & 0 & 0 & 0 & 0 & 0 & 0 & 0 & 0 & 0 & 0 & 0 \\ 0 & 0 & \mathbf{1} & 0 & 0 & 0 & 0 & 0 & 0 & 0 & 0 & 0 \\ 0 & 0 & 0 & 0 & \mathbf{1} & 0 & 0 & 0 & 0 & 0 & 0 & 0 \\ 0 & 0 & 0 & \mathbf{-1} & 0 & 0 & 0 & 0 & 0 & 0 & 0 & 0 \\ 0 & 0 & 0 & 0 & 0 & \mathbf{1} & 0 & 0 & 0 & 0 & 0 & 0 \\ 0 & 0 & 0 & 0 & 0 & 0 & 0 & \mathbf{1} & 0 & 0 & 0 & 0 \\ 0 & 0 & 0 & 0 & 0 & 0 & \mathbf{-1} & 0 & 0 & 0 & 0 & 0 \\ 0 & 0 & 0 & 0 & 0 & 0 & 0 & 0 & \mathbf{1} & 0 & 0 & 0 \\ 0 & 0 & 0 & 0 & 0 & 0 & 0 & 0 & 0 & 0 & \mathbf{1} & 0 \\ 0 & 0 & 0 & 0 & 0 & 0 & 0 & 0 & 0 & \mathbf{-1} & 0 & 0 \\ 0 & 0 & 0 & 0 & 0 & 0 & 0 & 0 & 0 & 0 & 0 & \mathbf{1} \end{bmatrix} \quad (10.17)$$

From (10.11)

$$k^2 = (r^2)^T \bar{k}^2 r^2$$

$k^2 =$

1774	0	0	0	0	−3548	−1774	0	0	0	0	−3548
0	934500	0	0	0	0	0	−934500	0	0	0	0
0	0	44012	88024	0	0	0	0	−44012	88024	0	0
0	0	88024	234732	0	0	0	0	−88024	117366	0	0
0	0	0	0	44	0	0	0	0	0	−44	0
−3548	0	0	0	0	9461	3548	0	0	0	0	4730
−1774	0	0	0	0	3548	1774	0	0	0	0	3548
0	−934500	0	0	0	0	0	934500	0	0	0	0
0	0	−44012	−88024	0	0	0	0	44012	−88024	0	0
0	0	88024	117366	0	0	0	0	−88024	234732	0	0
0	0	0	0	−44	0	0	0	0	0	44	0
−3548	0	0	0	0	4730	3548	0	0	0	0	9461

(10.18)

Member 3, L = 5 m

From (10.5)

$$\overline{k}^3 = \left[\begin{array}{cccccc|cccccc}
747600 & 0 & 0 & 0 & 0 & 0 & -747600 & 0 & 0 & 0 & 0 & 0 \\
0 & 908 & 0 & 0 & 0 & 2271 & 0 & -908 & 0 & 0 & 0 & 2271 \\
0 & 0 & 22534 & 0 & -56336 & 0 & 0 & 0 & -22534 & 0 & -56336 & 0 \\
0 & 0 & 0 & 35 & 0 & 0 & 0 & 0 & 0 & -35 & 0 & 0 \\
0 & 0 & -56336 & 0 & 187785 & 0 & 0 & 0 & 56336 & 0 & 93893 & 0 \\
0 & 2271 & 0 & 0 & 0 & 7568 & 0 & -2271 & 0 & 0 & 0 & 3784 \\
\hline
-747600 & 0 & 0 & 0 & 0 & 0 & 747600 & 0 & 0 & 0 & 0 & 0 \\
0 & -908 & 0 & 0 & 0 & -2271 & 0 & 908 & 0 & 0 & 0 & -2271 \\
0 & 0 & -22534 & 0 & 56336 & 0 & 0 & 0 & 22534 & 0 & 56336 & 0 \\
0 & 0 & 0 & -35 & 0 & 0 & 0 & 0 & 0 & 35 & 0 & 0 \\
0 & 0 & -56336 & 0 & 93893 & 0 & 0 & 0 & 56336 & 0 & 187785 & 0 \\
0 & 2271 & 0 & 0 & 0 & 3784 & 0 & -2271 & 0 & 0 & 0 & 7568
\end{array}\right] \quad (10.19)$$

$x_i = 0, x_j = 5$ m, $x_{ij} = x_j - x_i = 5 - 0 = 5$ m

$y_i = 4$ m, $y_j = 4$ m, $y_{ij} = y_j - y_i = 4 - 4 = 0$

$z_i = 6$ m, $z_j = 6$ m, $z_{ij} = z_j - z_i = 6 - 6 = 0$

$$s = \sqrt{x_{ij}^2 + z_{ij}^2} = \sqrt{5^2 + 0^2} = 5\,\text{m}$$

$$L = \sqrt{x_{ij}^2 + y_{ij}^2 + z_{ij}^2} = \sqrt{5^2 + 0^2 + 0^2} = 5\,\text{m}$$

$$\varphi_{\bar{x}} = +90^0,\ \sin\varphi_{\bar{x}} = 1,\ \cos\varphi_{\bar{x}} = 0$$

From (10.12)

$$\rho^3 = \begin{bmatrix} \frac{5}{5} & \frac{0}{5} & \frac{0}{5} \\ -\frac{0}{5}\times 1 - \frac{0\times 5}{5\times 5}\times 0 & \frac{5}{5}\times 0 & -\frac{0\times 0}{5\times 5}\times 0 + \frac{5}{5}\times 1 \\ \frac{0\times 5}{5\times 5}\times 1 - \frac{0}{5}\times 0 & -\frac{5}{5}\times 1 & \frac{5}{5}\times 0 + \frac{0\times 0}{5\times 5}\times 1 \end{bmatrix} = \begin{bmatrix} 1 & 0 & 0 \\ 0 & 0 & 1 \\ 0 & -1 & 0 \end{bmatrix}$$

$r^3 =$ (10.20)

1	0	0	0	0	0	0	0	0	0	0	0
0	0	**1**	0	0	0	0	0	0	0	0	0
0	**–1**	0	0	0	0	0	0	0	0	0	0
0	0	0	**1**	0	0	0	0	0	0	0	0
0	0	0	0	0	**1**	0	0	0	0	0	0
0	0	0	0	**–1**	0	0	0	0	0	0	0
0	0	0	0	0	0	**1**	0	0	0	0	0
0	0	0	0	0	0	0	0	**1**	0	0	0
0	0	0	0	0	0	0	**–1**	0	0	0	0
0	0	0	0	0	0	0	0	0	**1**	0	0
0	0	0	0	0	0	0	0	0	0	0	**1**
0	0	0	0	0	0	0	0	0	0	**–1**	0

From (10.11)

$$k^3 = (r^3)^T \bar{k}^3 r^3$$

$k^3 =$

747600	0	0	0	0	0	−747600	0	0	0	0	0
0	22534	0	0	0	56336	0	−22534	0	0	0	56336
0	0	908	0	−2271	0	0	0	−908	0	−2271	0
0	0	0	35	0	0	0	0	0	−35	0	0
0	0	−2271	0	7568	0	0	0	2271	0	3784	0
0	56336	0	0	0	187785	0	−56336	0	0	0	93893
−747600	0	0	0	0	0	747600	0	0	0	0	0
0	−22534	0	0	0	−56336	0	22534	0	0	0	−56336
0	0	−908	0	2271	0	0	0	908	0	2271	0
0	0	0	−35	0	0	0	0	0	35	0	0
0	0	−2271	0	3784	0	0	0	2271	0	7568	0
0	56336	0	0	0	93893	0	−56336	0	0	0	187785

(10.21)

The overall structure stiffness matrix is assembled by inspection as:

K =

	δ_1	δ_2	δ_3	δ_4
δ_1	$K_{11} = k^1_{ii}$	$K_{12} = k^1_{ij}$	0	0
δ_2	$K_{21} = k^1_{ji}$	$K_{22} = k^1_{jj} + k^2_{ii}$	$K_{23} = k^2_{ij}$	0
δ_3	0	$K_{32} = k^2_{ji}$	$K_{33} = k^2_{jj} + k^3_{ii}$	$K_{34} = k^3_{ij}$
δ_4	0	0	$K_{43} = k^3_{ji}$	$K_{44} = k^3_{jj}$

where

$$\delta_1 = \begin{bmatrix} u_1 \\ v_1 \\ w_1 \\ \Phi_1 \\ \theta_1 \\ \Psi_1 \end{bmatrix},\ \delta_2 = \begin{bmatrix} u_2 \\ v_2 \\ w_2 \\ \Phi_2 \\ \theta_2 \\ \Psi_2 \end{bmatrix},\ \delta_3 = \begin{bmatrix} u_3 \\ v_3 \\ w_3 \\ \Phi_3 \\ \theta_3 \\ \Psi_3 \end{bmatrix},\ \text{and}\ \delta_4 = \begin{bmatrix} u_4 \\ v_4 \\ w_4 \\ \Phi_4 \\ \theta_4 \\ \Psi_4 \end{bmatrix}.$$

At the fixed supports 1 and 4, $\delta_1 = 0$ and $\delta_4 = 0$, respectively, hence rows and columns 1 and 4 are deleted to give the reduced structure stiffness matrix as:

K =

	δ_2	δ_3
δ_2	$K_{22} = k^1_{jj} + k^2_{ii}$	$K_{23} = k^2_{ij}$
δ_3	$K_{32} = k^2_{ji}$	$K_{33} = k^2_{jj} + k^3_{ii}$

K = (10.22)

u_2	v_2	w_2	Φ_2	θ_2	Ψ_2	u_3	v_3	w_3	Φ_3	θ_3	Ψ_3
14815	0	0	0	−39122	−3548	−1774	0	0	0	0	−3548
0	935026	0	1577	0	0	0	−934500	0	0	0	0
0	0	667012	88024	0	0	0	0	−44012	88024	0	0
0	1577	88024	241039	0	0	0	0	−88024	117366	0	0
−39122	0	0	0	156532	0	0	0	0	0	−44	0
−3548	0	0	0	0	9490	3548	0	0	0	0	4730
−1774	0	0	0	0	3548	749374	0	0	0	0	3548
0	−934500	0	0	0	0	0	957034	0	0	0	56336
0	0	−44012	−88024	0	0	0	0	44920	−88024	−2271	0
0	0	88024	117366	0	0	0	0	−88024	234767	0	0
0	0	0	0	−44	0	0	0	−2271	0	7612	0
−3548	0	0	0	0	4730	3548	56336	0	0	0	197246

Calculation of load vector

Actions on Member 1: n = –10 kN/m and L = 6 m

$$(\bar{Z}_1^1)_f = -\frac{nL}{2} = -\frac{(-10)\times 6}{2} = +30\ \text{kN}$$

$$(\bar{Z}_2^1)_f = -\frac{nL}{2} = -\frac{(-10)\times 6}{2} = +30\ \text{kN}$$

$$(\bar{M}_1^1)_f = +\frac{nL^2}{12} = +\frac{(-10)\times 6^2}{12} = -30\ \text{kNm}$$

$$(\bar{M}_2^1)_f = -\frac{nL^2}{12} = -\frac{(-10)\times 6^2}{12} = +30\ \text{kNm}$$

$$\bar{F}_f^1 = \begin{bmatrix} (\bar{X}_1^1)_f \\ (\bar{Y}_1^1)_f \\ (\bar{Z}_1^1)_f \\ (\bar{T}_1^1)_f \\ (\bar{M}_1^1)_f \\ (\bar{N}_1^1)_f \\ (\bar{X}_2^1)_f \\ (\bar{Y}_2^1)_f \\ (\bar{Z}_2^1)_f \\ (\bar{T}_2^1)_f \\ (\bar{M}_2^1)_f \\ (\bar{N}_2^1)_f \end{bmatrix} = \begin{bmatrix} 0 \\ 0 \\ +30 \\ 0 \\ -30 \\ 0 \\ 0 \\ 0 \\ +30 \\ 0 \\ +30 \\ 0 \end{bmatrix} \tag{10.23}$$

Loads on joints 1 and 2

$$(X_1^1)_S = \left(\bar{Z}_1^1\right)_f = +30 \text{ kN}$$

$$(X_2^1)_S = \left(\bar{Z}_2^1\right)_f = +30 \text{ kN}$$

$$\left(M_1^1\right)_S = -\left(\bar{M}_1^1\right)_f = +30 \text{ kN}$$

$$\left(M_2^1\right)_S = -\left(\bar{M}_2^1\right)_f = -30 \text{ kN}$$

$$F_S^1 = \begin{bmatrix} (X_1^1)_S \\ (Y_1^1)_S \\ (Z_1^1)_S \\ (T_1^1)_S \\ (M_1^1)_S \\ (N_1^1)_S \\ (X_2^1)_S \\ (Y_2^1)_S \\ (Z_2^1)_S \\ (T_2^1)_S \\ (M_2^1)_S \\ (N_2^1)_S \end{bmatrix} = \begin{bmatrix} +30 \\ 0 \\ 0 \\ 0 \\ +30 \\ 0 \\ +30 \\ 0 \\ 0 \\ 0 \\ -30 \\ 0 \end{bmatrix} \tag{10.24}$$

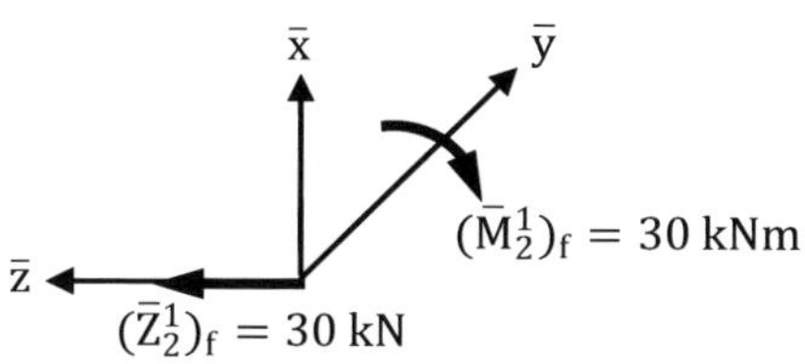

(b)

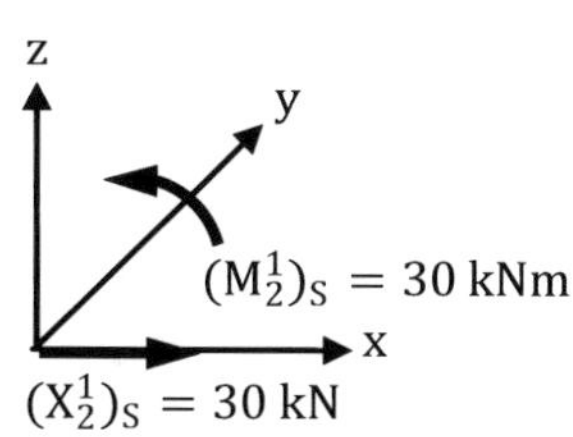

(d)

(*Continued*)

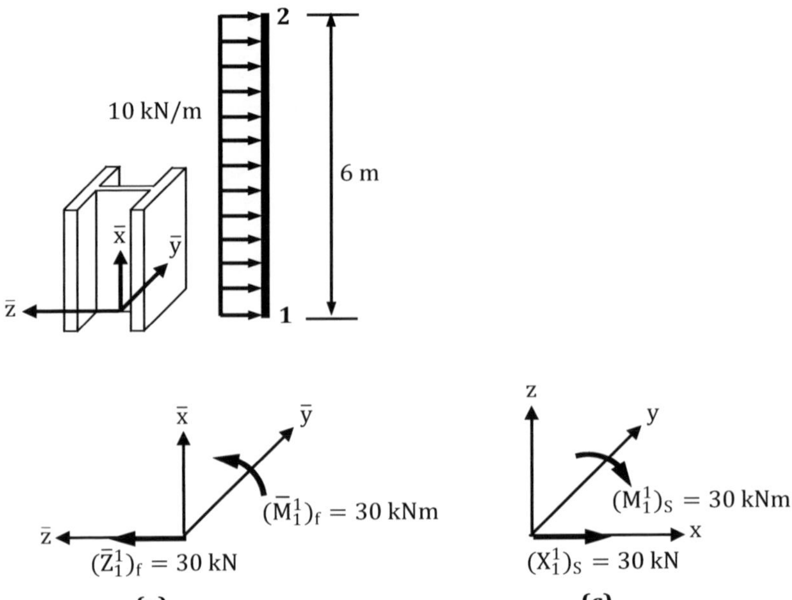

Figure 10.4 (a) and (b) actions on member 1 (relative to local coordinates), (c) and (d) loads on joints 1 and 2 (relative to global coordinates).

Alternatively, and to make the computations more systematic, the load vector on the joints of the structure, F_S, which is relative to global coordinates can be calculated from the actions at the ends of the member which are relative to the local coordinates of the member by using the transformation matrix as follows:

Consider the equilibrium of a section cut at the junction of the member and the joint $F_S + F_f = 0$ or $F_S = -F_f$, where F_f is the action vector on end of the member relative to global coordinates. The action vector relative to local coordinates is $\bar{F}_f$, therefore, $\bar{F}_f = rF_f$ or $F_f = r^{-1}\bar{F}_f = r^T\bar{F}_f$ (since $r^{-1} = r^T$) and hence $F_S = -r^T\bar{F}_f$.

$F_S^1 = -(r^1)^T\bar{F}_f^1$ and with r^1 from (10.14) and $\bar{F}_f^1$ from (10.23) we get

$$\mathrm{F}_S^1 = -(\mathrm{r}^1)^T \overline{\mathrm{F}}_f^1 = -\left[\begin{array}{cccccc|cccccc}
0 & 0 & -1 & 0 & 0 & 0 & 0 & 0 & 0 & 0 & 0 & 0 \\
0 & 1 & 0 & 0 & 0 & 0 & 0 & 0 & 0 & 0 & 0 & 0 \\
1 & 0 & 0 & 0 & 0 & 0 & 0 & 0 & 0 & 0 & 0 & 0 \\
0 & 0 & 0 & 0 & 0 & -1 & 0 & 0 & 0 & 0 & 0 & 0 \\
0 & 0 & 0 & 0 & 1 & 0 & 0 & 0 & 0 & 0 & 0 & 0 \\
0 & 0 & 0 & 1 & 0 & 0 & 0 & 0 & 0 & 0 & 0 & 0 \\
\hline
0 & 0 & 0 & 0 & 0 & 0 & 0 & 0 & -1 & 0 & 0 & 0 \\
0 & 0 & 0 & 0 & 0 & 0 & 0 & 1 & 0 & 0 & 0 & 0 \\
0 & 0 & 0 & 0 & 0 & 0 & 1 & 0 & 0 & 0 & 0 & 0 \\
0 & 0 & 0 & 0 & 0 & 0 & 0 & 0 & 0 & 0 & 0 & -1 \\
0 & 0 & 0 & 0 & 0 & 0 & 0 & 0 & 0 & 0 & 1 & 0 \\
0 & 0 & 0 & 0 & 0 & 0 & 0 & 0 & 0 & 1 & 0 & 0
\end{array}\right]
\left[\begin{array}{c}
0 \\ 0 \\ +30 \\ 0 \\ -30 \\ 0 \\ \hline 0 \\ 0 \\ +30 \\ 0 \\ +30 \\ 0
\end{array}\right]
=
\left[\begin{array}{c}
+30 \\ 0 \\ 0 \\ 0 \\ +30 \\ 0 \\ \hline +30 \\ 0 \\ 0 \\ 0 \\ -30 \\ 0
\end{array}\right]$$

which is the same result obtained in (10.24).

Actions on Member 2: P = –20 kN and L = 4 m

$$(\bar{Z}_2^2)_f = -\frac{P}{2} = -\frac{(-20)}{2} = +10\text{ kN}$$

$$(\bar{Z}_3^2)_f = -\frac{P}{2} = -\frac{(-20)}{2} = +10\text{ kN}$$

$$(\bar{M}_2^2)_f = +\frac{PL}{8} = +\frac{(-20)\times 4}{8} = -10\text{ kNm}$$

$$(\bar{M}_3^2)_f = -\frac{PL}{8} = -\frac{(-20)\times 4}{8} = +10\text{ kNm}$$

$$\bar{F}_f^2 = \begin{bmatrix} (\bar{X}_2^2)_f \\ (\bar{Y}_2^2)_f \\ (\bar{Z}_2^2)_f \\ (\bar{T}_2^2)_f \\ (\bar{M}_2^2)_f \\ (\bar{N}_2^2)_f \\ (\bar{X}_3^2)_f \\ (\bar{Y}_3^2)_f \\ (\bar{Z}_3^2)_f \\ (\bar{T}_3^2)_f \\ (\bar{M}_3^2)_f \\ (\bar{N}_3^2)_f \end{bmatrix} = \begin{bmatrix} 0 \\ 0 \\ +10 \\ 0 \\ -10 \\ 0 \\ 0 \\ 0 \\ +10 \\ 0 \\ +10 \\ 0 \end{bmatrix} \tag{10.25}$$

Loads on joints 2 and 3

$F_S^2 = -(r^2)^T \bar{F}_f^2$ and with r^2 from (10.17) and $\bar{F}_f^2$ from (10.25) we get

$$F_S^2 = -\left[\begin{array}{cccccc|cccccc} 0 & \mathbf{-1} & 0 & 0 & 0 & 0 & 0 & 0 & 0 & 0 & 0 & 0 \\ \mathbf{1} & 0 & 0 & 0 & 0 & 0 & 0 & 0 & 0 & 0 & 0 & 0 \\ 0 & 0 & \mathbf{1} & 0 & 0 & 0 & 0 & 0 & 0 & 0 & 0 & 0 \\ 0 & 0 & 0 & 0 & \mathbf{-1} & 0 & 0 & 0 & 0 & 0 & 0 & 0 \\ 0 & 0 & 0 & \mathbf{1} & 0 & 0 & 0 & 0 & 0 & 0 & 0 & 0 \\ 0 & 0 & 0 & 0 & 0 & \mathbf{1} & 0 & 0 & 0 & 0 & 0 & 0 \\ \hline 0 & 0 & 0 & 0 & 0 & 0 & 0 & \mathbf{-1} & 0 & 0 & 0 & 0 \\ 0 & 0 & 0 & 0 & 0 & 0 & \mathbf{1} & 0 & 0 & 0 & 0 & 0 \\ 0 & 0 & 0 & 0 & 0 & 0 & 0 & 0 & \mathbf{1} & 0 & 0 & 0 \\ 0 & 0 & 0 & 0 & 0 & 0 & 0 & 0 & 0 & 0 & \mathbf{-1} & 0 \\ 0 & 0 & 0 & 0 & 0 & 0 & 0 & 0 & 0 & \mathbf{1} & 0 & 0 \\ 0 & 0 & 0 & 0 & 0 & 0 & 0 & 0 & 0 & 0 & 0 & \mathbf{1} \end{array}\right] \left[\begin{array}{c} 0 \\ 0 \\ +10 \\ 0 \\ -10 \\ 0 \\ \hline 0 \\ 0 \\ +10 \\ 0 \\ +10 \\ 0 \end{array}\right] = \left[\begin{array}{c} 0 \\ 0 \\ -10 \\ -10 \\ 0 \\ 0 \\ \hline 0 \\ 0 \\ -10 \\ +10 \\ 0 \\ 0 \end{array}\right]$$

$$F_S^2 = \begin{bmatrix} (X_2^2)_S \\ (Y_2^2)_S \\ (Z_2^2)_S \\ (T_2^2)_S \\ (M_2^2)_S \\ (N_2^2)_S \\ (X_3^2)_S \\ (Y_3^2)_S \\ (Z_3^2)_S \\ (T_3^2)_S \\ (M_3^2)_S \\ (N_3^2)_S \end{bmatrix} = \begin{bmatrix} 0 \\ 0 \\ -10 \\ -10 \\ 0 \\ 0 \\ 0 \\ 0 \\ -10 \\ +10 \\ 0 \\ 0 \end{bmatrix} \tag{10.26}$$

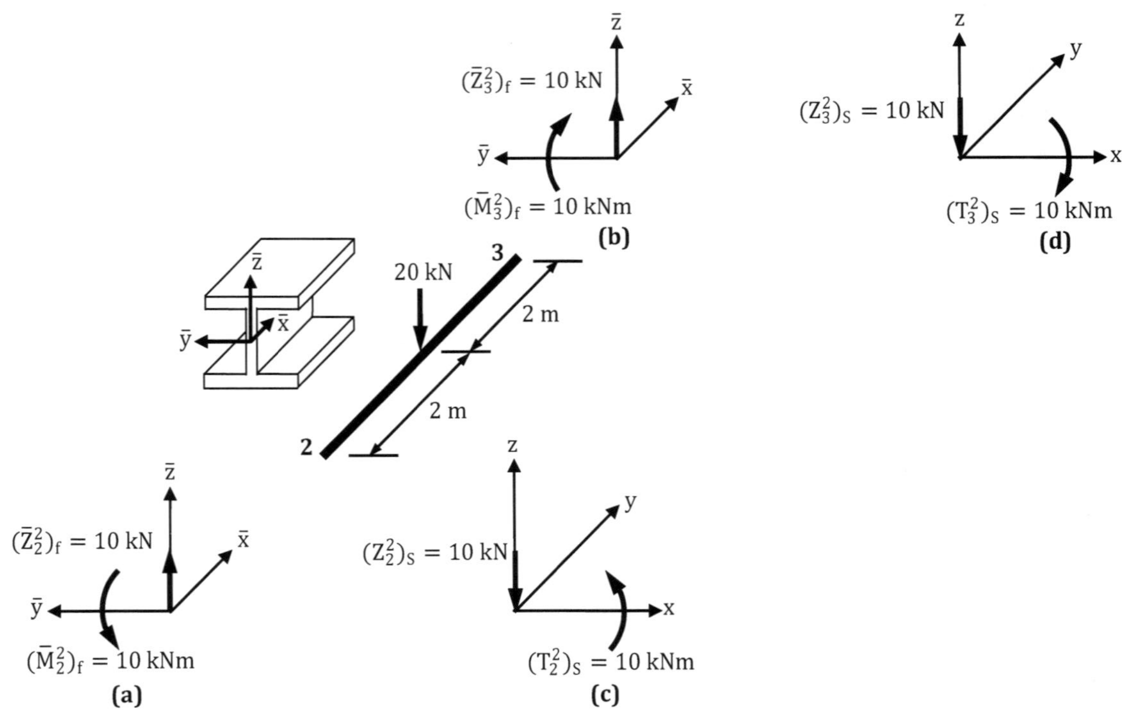

Figure 10.5 (a) and (b) actions on member 2 (relative to local coordinates), (c) and (d) loads on joints 2 and 3 (relative to global coordinates).

Actions on Member 3: P = –40 kN, a = 3 m and b = 2 m

$$(\bar{Z}_3^3)_f = -\frac{Pb\left(L^2+ab-a^2\right)}{L^3} = -\frac{(-40)\times 2\left(5^2+3\times 2-3^2\right)}{5^3} = +14.08\ \text{kN}$$

$$(\bar{Z}_4^3)_f = -\frac{Pa\left(L^2+ab-b^2\right)}{L^3} = -\frac{(-40)\times 3\left(5^2+3\times 2-2^2\right)}{5^3} = +25.92\ \text{kN}$$

$$(\bar{M}_3^3)_f = +\frac{Pab^2}{L^2} = +\frac{(-40)\times 3\times 2^2}{5^2} = -19.20\ \text{kNm}$$

$$(\bar{M}_4^3)_f = -\frac{Pa^2b}{L^2} = -\frac{(-40)\times 3^2\times 2}{5^2} = +28.80\ \text{kNm}$$

$$\bar{F}_f^3 = \begin{bmatrix} (\bar{X}_3^3)_f \\ (\bar{Y}_3^3)_f \\ (\bar{Z}_3^3)_f \\ (\bar{T}_3^3)_f \\ (\bar{M}_3^3)_f \\ (\bar{N}_3^3)_f \\ (\bar{X}_4^3)_f \\ (\bar{Y}_4^3)_f \\ (\bar{Z}_4^3)_f \\ (\bar{T}_4^3)_f \\ (\bar{M}_4^3)_f \\ (\bar{N}_4^3)_f \end{bmatrix} = \begin{bmatrix} 0 \\ 0 \\ +14.08 \\ \\ -19.20 \\ 0 \\ 0 \\ 0 \\ +25.92 \\ 0 \\ +28.80 \\ 0 \end{bmatrix} \tag{10.27}$$

Loads on joints 3 and 4

$F_S^3 = -(r^3)^T \bar{F}_f^3$ and with r^3 from (10.20) and $\bar{F}_f^3$ from (10.27) we get

$$F_S^3 = -\left(r^3\right)^T \overline{F}_f^3 = -\left[\begin{array}{cccccc|cccccc} 1 & 0 & 0 & 0 & 0 & 0 & 0 & 0 & 0 & 0 & 0 & 0 \\ 0 & 0 & -1 & 0 & 0 & 0 & 0 & 0 & 0 & 0 & 0 & 0 \\ 0 & 1 & 0 & 0 & 0 & 0 & 0 & 0 & 0 & 0 & 0 & 0 \\ 0 & 0 & 0 & 1 & 0 & 0 & 0 & 0 & 0 & 0 & 0 & 0 \\ 0 & 0 & 0 & 0 & 0 & -1 & 0 & 0 & 0 & 0 & 0 & 0 \\ 0 & 0 & 0 & 0 & 1 & 0 & 0 & 0 & 0 & 0 & 0 & 0 \\ \hline 0 & 0 & 0 & 0 & 0 & 0 & 1 & 0 & 0 & 0 & 0 & 0 \\ 0 & 0 & 0 & 0 & 0 & 0 & 0 & 0 & -1 & 0 & 0 & 0 \\ 0 & 0 & 0 & 0 & 0 & 0 & 0 & 1 & 0 & 0 & 0 & 0 \\ 0 & 0 & 0 & 0 & 0 & 0 & 0 & 0 & 0 & 1 & 0 & 0 \\ 0 & 0 & 0 & 0 & 0 & 0 & 0 & 0 & 0 & 0 & 0 & -1 \\ 0 & 0 & 0 & 0 & 0 & 0 & 0 & 0 & 0 & 0 & 1 & 0 \end{array}\right] \left[\begin{array}{c} 0 \\ 0 \\ +14.08 \\ 0 \\ -19.20 \\ 0 \\ \hline 0 \\ 0 \\ +25.92 \\ 0 \\ +28.80 \\ 0 \end{array}\right] = \left[\begin{array}{c} 0 \\ +14.08 \\ 0 \\ 0 \\ 0 \\ +19.20 \\ \hline 0 \\ +25.92 \\ 0 \\ 0 \\ 0 \\ -28.80 \end{array}\right]$$

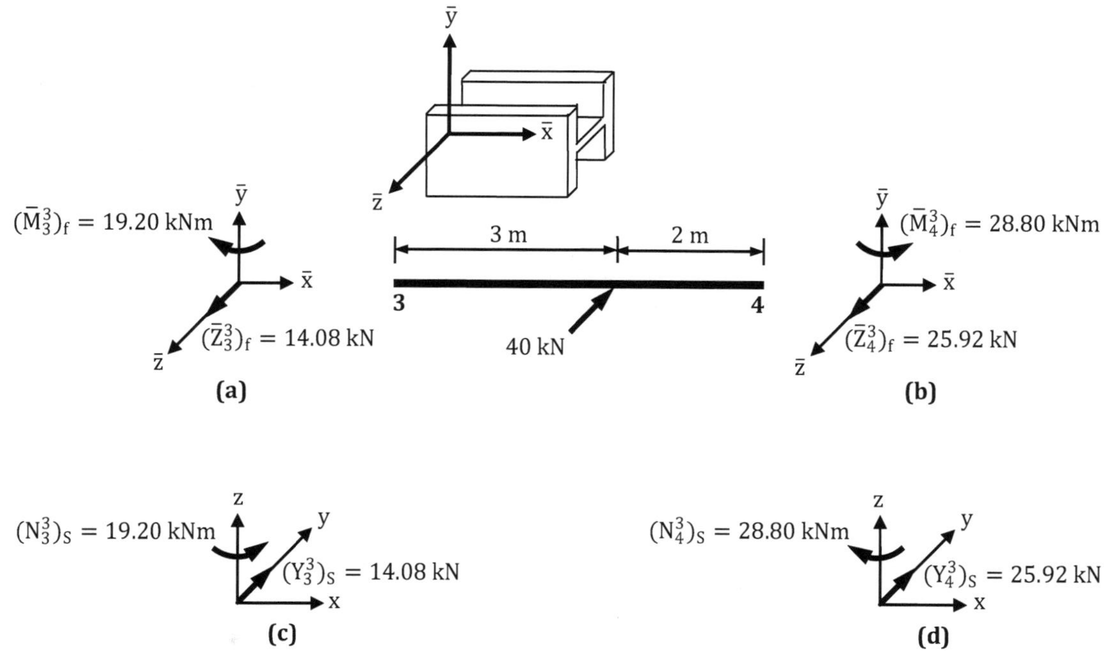

Figure 10.6 (a) and (b) actions on member 3 (relative to local coordinates), (c) and (d) loads on joints 3 and 4 (relative to global coordinates).

$$F_S^3 = \begin{bmatrix} (X_3^3)_S \\ (Y_3^3)_S \\ (Z_3^3)_S \\ (T_3^3)_S \\ (M_3^3)_S \\ (N_3^3)_S \\ (X_4^3)_S \\ (Y_4^3)_S \\ (Z_4^3)_S \\ (T_4^3)_S \\ (M_4^3)_S \\ (N_4^3)_S \end{bmatrix} = \begin{bmatrix} 0 \\ +14.08 \\ 0 \\ 0 \\ 0 \\ +19.20 \\ 0 \\ +25.92 \\ 0 \\ 0 \\ 0 \\ -28.80 \end{bmatrix} \tag{10.28}$$

The load vector on the joints of the structure due to the applied loads on the members is given by

$$F_S = F_S^1 + F_S^2 + F_S^3$$

where F_S^1, F_S^2, and F_S^3 are given in (10.24), (10.26), and (10.28), respectively, thus

$$F_S = \begin{bmatrix} +30.00 \\ 0 \\ 0 \\ 0 \\ +30.00 \\ 0 \\ +30.00 \\ 0 \\ 0 \\ 0 \\ -30.00 \\ 0 \\ 0 \\ 0 \\ 0 \\ 0 \\ 0 \\ 0 \\ 0 \\ 0 \\ 0 \\ 0 \\ 0 \\ 0 \end{bmatrix} + \begin{bmatrix} 0 \\ 0 \\ 0 \\ 0 \\ 0 \\ 0 \\ 0 \\ 0 \\ -10.00 \\ -10.00 \\ 0 \\ 0 \\ 0 \\ 0 \\ -10.00 \\ +10.00 \\ 0 \\ 0 \\ 0 \\ 0 \\ 0 \\ 0 \\ 0 \\ 0 \end{bmatrix} + \begin{bmatrix} 0 \\ 0 \\ 0 \\ 0 \\ 0 \\ 0 \\ 0 \\ 0 \\ 0 \\ 0 \\ 0 \\ 0 \\ 0 \\ +14.08 \\ 0 \\ 0 \\ 0 \\ +19.20 \\ 0 \\ +25.92 \\ 0 \\ 0 \\ 0 \\ -28.80 \end{bmatrix} = \begin{bmatrix} +30.00 \\ 0 \\ 0 \\ 0 \\ +30.00 \\ 0 \\ +30.00 \\ 0 \\ -10.00 \\ -10.00 \\ -30.00 \\ 0 \\ 0 \\ +14.08 \\ -10.00 \\ +10.00 \\ 0 \\ +19.20 \\ 0 \\ +25.92 \\ 0 \\ 0 \\ 0 \\ -28.80 \end{bmatrix} \tag{10.29}$$

The load vector due to loads directly applied at the nodes is zero since there are no direct loads at the joints, thus

$$F_N = \begin{bmatrix} 0 \\ 0 \\ 0 \\ 0 \\ 0 \\ 0 \\ 0 \\ 0 \\ 0 \\ 0 \\ 0 \\ 0 \\ 0 \\ 0 \\ 0 \\ 0 \\ 0 \\ 0 \\ 0 \\ 0 \\ 0 \\ 0 \\ 0 \\ 0 \end{bmatrix} \tag{10.30}$$

The load vector due to the reactions at the supports 1 and 4 is:

$$F_C = \begin{bmatrix} R_{X1} \\ R_{Y1} \\ R_{Z1} \\ R_{T1} \\ R_{M1} \\ R_{N1} \\ 0 \\ 0 \\ 0 \\ 0 \\ 0 \\ 0 \\ 0 \\ 0 \\ 0 \\ 0 \\ 0 \\ 0 \\ R_{X4} \\ R_{Y4} \\ R_{Z4} \\ R_{T4} \\ R_{M4} \\ R_{N4} \end{bmatrix} \tag{10.31}$$

where R_{X1}, R_{Y1}, and R_{Z1} are the reaction forces at support 1 and R_{T1}, R_{M1}, and R_{N1} are the reaction moments at support 1. Similar reaction forces and moments exist at support 4 and are given the subscript 4.

The total load vector for the whole structure is

$F = F_S + F_N + F_C$, and from (10.29), (10.30), and (10.31) we get

$$F=\begin{bmatrix} X_1\\ Y_1\\ Z_1\\ T_1\\ M_1\\ N_1\\ X_2\\ Y_2\\ Z_2\\ T_2\\ M_2\\ N_2\\ X_3\\ Y_3\\ Z_3\\ T_3\\ M_3\\ N_3\\ X_4\\ Y_4\\ Z_4\\ T_4\\ M_4\\ N_4 \end{bmatrix}=\begin{bmatrix} +30.00\\ 0\\ 0\\ 0\\ +30.00\\ 0\\ +30.00\\ 0\\ -10.00\\ -10.00\\ -30.00\\ 0\\ 0\\ +14.08\\ -10.00\\ +10.00\\ 0\\ +19.20\\ 0\\ +25.92\\ 0\\ 0\\ 0\\ -28.80 \end{bmatrix}+\begin{bmatrix} 0\\ 0\\ 0\\ 0\\ 0\\ 0\\ 0\\ 0\\ 0\\ 0\\ 0\\ 0\\ 0\\ 0\\ 0\\ 0\\ 0\\ 0\\ 0\\ 0\\ 0\\ 0\\ 0\\ 0 \end{bmatrix}+\begin{bmatrix} \mathbf{R}_{X1}\\ \mathbf{R}_{Y1}\\ \mathbf{R}_{Z1}\\ \mathbf{R}_{T1}\\ \mathbf{R}_{M1}\\ \mathbf{R}_{N1}\\ 0\\ 0\\ 0\\ 0\\ 0\\ 0\\ 0\\ 0\\ 0\\ 0\\ 0\\ 0\\ \mathbf{R}_{X4}\\ \mathbf{R}_{Y4}\\ \mathbf{R}_{Z4}\\ \mathbf{R}_{T4}\\ \mathbf{R}_{M4}\\ \mathbf{R}_{N4} \end{bmatrix}=\begin{bmatrix} +30.00+\mathbf{R}_{X1}\\ \mathbf{R}_{Y1}\\ \mathbf{R}_{Z1}\\ \mathbf{R}_{T1}\\ +30.00+\mathbf{R}_{M1}\\ \mathbf{R}_{N1}\\ +30.00\\ 0\\ -10.00\\ -10.00\\ -30.00\\ 0\\ 0\\ +14.08\\ -10.00\\ +10.00\\ 0\\ +19.20\\ \mathbf{R}_{X4}\\ +25.92+\mathbf{R}_{Y4}\\ \mathbf{R}_{Z4}\\ \mathbf{R}_{T4}\\ \mathbf{R}_{M4}\\ -28.80+\mathbf{R}_{N4} \end{bmatrix} \tag{10.32}$$

At the fixed support 1: $u_1 = 0$, $v_1 = 0$, $w_1 = 0$, $\Phi_1 = 0$, $\theta_1 = 0$, and $\Psi_1 = 0$. So, delete rows 1, 2, 3, 4, 5, and 6.

At the fixed support 4: $u_4 = 0$, $v_4 = 0$, $w_4 = 0$, $\Phi_4 = 0$, $\theta_4 = 0$, and $\Psi_4 = 0$. So, delete rows 19, 20, 21, 22, 23, and 24. Thus the resulting load vector becomes:

$$F=\begin{bmatrix} +30.00\\ 0\\ -10.00\\ -10.00\\ -30.00\\ 0\\ 0\\ +14.08\\ -10.00\\ +10.00\\ 0\\ +19.20 \end{bmatrix} \tag{10.33}$$

$K\delta = F$ and with K from (10.22) and F from (10.33) we get:

14815	0	0	0	−39122	−3548	−1774	0	0	0	0	−3548	u_2		+30.00
0	935026	0	1577	0	0	0	−934500	0	0	0	0	v_2		0
0	0	667012	88024	0	0	0	0	−44012	88024	0	0	w_2		−10.00
0	1577	88024	241039	0	0	0	0	−88024	117366	0	0	Φ_2		−10.00
−39122	0	0	0	156532	0	0	0	0	0	−44	0	θ_2		−30.00
−3548	0	0	0	0	9490	3548	0	0	0	0	4730	Ψ_2	=	0
−1774	0	0	0	0	3548	749374	0	0	0	0	3548	u_3		0
0	−934500	0	0	0	0	0	957034	0	0	0	56336	v_3		+14.08
0	0	−44012	−88024	0	0	0	0	44920	−88024	−2271	0	w_3		−10.00
0	0	88024	117366	0	0	0	0	−88024	234767	0	0	Φ_3		+10.00
0	0	0	0	−44	0	0	0	−2271	0	7612	0	θ_3		0
−3548	0	0	0	0	4730	3548	56336	0	0	0	197246	Ψ_3		+19.20

The solution of the above equations is:

$u_2 = +\ 0.005873$ m, $v_2 = +\ 0.001807$ m, $w_2 = -\ 0.000026$ m,

$\Phi_2 = -\ 0.004243$ rad, $\theta_2 = +\ 0.001275$ rad, $\Psi_2 = +\ 0.002378$ rad,

$u_3 = +\ 0.000004$ m, $v_3 = +\ 0.001801$ m, $w_3 = -\ 0.017199$ m,

$\Phi_3 = -\ 0.004275$ rad, $\theta_3 = -\ 0.005124$ rad, $\Psi_3 = -\ 0.000368$ rad.

Forces Developed in the Members

The resultant forces (and moments) developed in the members are given by $\overline{F}_r = \overline{F}_d + \overline{F}_f$, where the forces developed due to the displacements, $\overline{F}_d = \overline{k}\overline{\delta}$ ($\overline{\delta} = r\delta$) and $\overline{F}_f$ is the force vector due to the applied loads on the member.

Member 1:

$$\delta^1 = \begin{bmatrix} \delta_i^1 \\ \delta_j^1 \end{bmatrix} = \begin{bmatrix} u_1 \\ v_1 \\ w_1 \\ \Phi_1 \\ \theta_1 \\ \Psi_1 \\ u_2 \\ v_2 \\ w_2 \\ \Phi_2 \\ \theta_2 \\ \Psi_2 \end{bmatrix} = \begin{bmatrix} 0 \\ 0 \\ 0 \\ 0 \\ 0 \\ 0 \\ +0.005873 \\ +0.001807 \\ -0.000026 \\ -0.004243 \\ +0.001275 \\ +0.002378 \end{bmatrix}$$

and with r^1 from (10.14)

$$\bar{\delta}^1 = r^1\delta^1 = \left[\begin{array}{cccccc|cccccc} 0&0&1&0&0&0&0&0&0&0&0&0\\ 0&1&0&0&0&0&0&0&0&0&0&0\\ -1&0&0&0&0&0&0&0&0&0&0&0\\ 0&0&0&0&0&1&0&0&0&0&0&0\\ 0&0&0&0&1&0&0&0&0&0&0&0\\ 0&0&0&-1&0&0&0&0&0&0&0&0\\ \hline 0&0&0&0&0&0&0&0&1&0&0&0\\ 0&0&0&0&0&0&0&1&0&0&0&0\\ 0&0&0&0&0&0&-1&0&0&0&0&0\\ 0&0&0&0&0&0&0&0&0&0&0&1\\ 0&0&0&0&0&0&0&0&0&0&1&0\\ 0&0&0&0&0&0&0&0&0&-1&0&0 \end{array}\right] \left[\begin{array}{c} 0\\ 0\\ 0\\ 0\\ 0\\ 0\\ \hline 0.005873\\ 0.001807\\ -0.000026\\ -0.004243\\ 0.001275\\ 0.002378 \end{array}\right]$$

$$\bar{\delta}^1 = \begin{bmatrix} \bar{\delta}_i^1 \\ \bar{\delta}_j^1 \end{bmatrix} = \begin{bmatrix} \bar{u}_1\\ \bar{v}_1\\ \bar{w}_1\\ \bar{\Phi}_1\\ \bar{\theta}_1\\ \bar{\Psi}_1\\ \bar{u}_2\\ \bar{v}_2\\ \bar{w}_2\\ \bar{\Phi}_2\\ \bar{\theta}_2\\ \bar{\Psi}_2 \end{bmatrix} = \begin{bmatrix} 0\\ 0\\ 0\\ 0\\ 0\\ 0\\ -0.000026\\ +0.001807\\ -0.005873\\ +0.002378\\ +0.001275\\ +0.004243 \end{bmatrix}$$

and with $\bar{k}^1$ from (10.13)

$$\bar{F}_d^1 = \bar{k}^1 \bar{\delta}^1 = \left[\begin{array}{cccccc|cccccc} 623000 & 0 & 0 & 0 & 0 & 0 & -623000 & 0 & 0 & 0 & 0 & 0 \\ 0 & 526 & 0 & 0 & 0 & 1577 & 0 & -526 & 0 & 0 & 0 & 1577 \\ 0 & 0 & 13041 & 0 & -39122 & 0 & 0 & 0 & -13041 & 0 & -39122 & 0 \\ 0 & 0 & 0 & 29 & 0 & 0 & 0 & 0 & 0 & -29 & 0 & 0 \\ 0 & 0 & -39122 & 0 & 156488 & 0 & 0 & 0 & 39122 & 0 & 78244 & 0 \\ 0 & 1577 & 0 & 0 & 0 & 6307 & 0 & -1577 & 0 & 0 & 0 & 3154 \\ \hline -623000 & 0 & 0 & 0 & 0 & 0 & 623000 & 0 & 0 & 0 & 0 & 0 \\ 0 & -526 & 0 & 0 & 0 & -1577 & 0 & 526 & 0 & 0 & 0 & -1577 \\ 0 & 0 & -13041 & 0 & 39122 & 0 & 0 & 0 & 13041 & 0 & 39122 & 0 \\ 0 & 0 & 0 & -29 & 0 & 0 & 0 & 0 & 0 & 29 & 0 & 0 \\ 0 & 0 & -39122 & 0 & 78244 & 0 & 0 & 0 & 39122 & 0 & 156488 & 0 \\ 0 & 1577 & 0 & 0 & 0 & 3154 & 0 & -1577 & 0 & 0 & 0 & 6307 \end{array}\right] \left[\begin{array}{c} 0 \\ 0 \\ 0 \\ 0 \\ 0 \\ 0 \\ \hline -0.000026 \\ 0.001807 \\ -0.005873 \\ 0.002378 \\ 0.001275 \\ 0.004243 \end{array}\right] = \left[\begin{array}{c} +16.01 \\ +5.74 \\ +26.72 \\ -0.07 \\ -130.01 \\ +10.53 \\ \hline -16.01 \\ -5.74 \\ -26.72 \\ +0.07 \\ -30.28 \\ +23.91 \end{array}\right]$$

$$\bar{F}_d^1 = \begin{bmatrix} (\bar{F}_i^1)_d \\ (\bar{F}_j^1)_d \end{bmatrix} = \begin{bmatrix} (\bar{X}_1^1)_d \\ (\bar{Y}_1^1)_d \\ (\bar{Z}_1^1)_d \\ (\bar{T}_1^1)_d \\ (\bar{M}_1^1)_d \\ (\bar{N}_1^1)_d \\ (\bar{X}_2^1)_d \\ (\bar{Y}_2^1)_d \\ (\bar{Z}_2^1)_d \\ (\bar{T}_2^1)_d \\ (\bar{M}_2^1)_d \\ (\bar{N}_2^1)_d \end{bmatrix} = \begin{bmatrix} +16.01 \\ +5.74 \\ +26.72 \\ -0.07 \\ -130.01 \\ +10.53 \\ -16.01 \\ -5.74 \\ -26.72 \\ +0.07 \\ -30.28 \\ +23.91 \end{bmatrix}$$

and from (10.23), $$\bar{F}_f^1 = \begin{bmatrix} (\bar{F}_i^1)_f \\ (\bar{F}_j^1)_f \end{bmatrix} = \begin{bmatrix} (\bar{X}_1^1)_f \\ (\bar{Y}_1^1)_f \\ (\bar{Z}_1^1)_f \\ (\bar{T}_1^1)_f \\ (\bar{M}_1^1)_f \\ (\bar{N}_1^1)_f \\ (\bar{X}_2^1)_f \\ (\bar{Y}_2^1)_f \\ (\bar{Z}_2^1)_f \\ (\bar{T}_2^1)_f \\ (\bar{M}_2^1)_f \\ (\bar{N}_2^1)_f \end{bmatrix} = \begin{bmatrix} 0 \\ 0 \\ +30.00 \\ 0 \\ -30.00 \\ 0 \\ 0 \\ 0 \\ +30.00 \\ 0 \\ +30.00 \\ 0 \end{bmatrix}$$

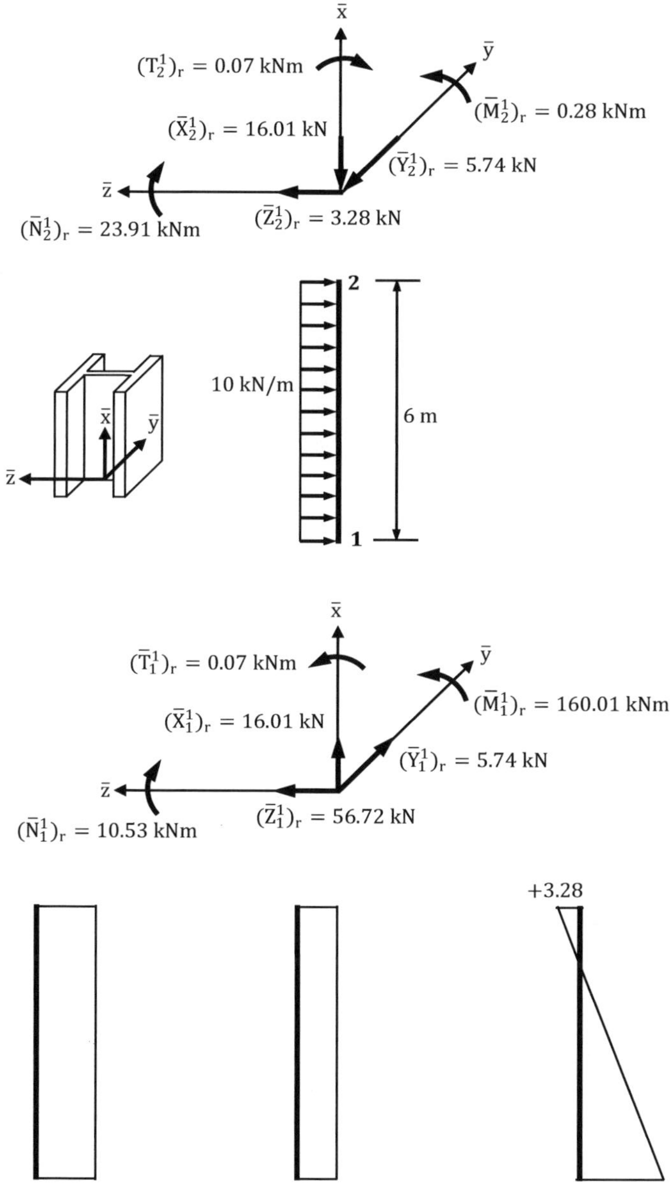

$\bar{P} = -16.01$ kN (compression) axial force in the $\bar{x}$ direction

$\bar{S} = -5.74$ kN shear force in the $\bar{y}$ direction

$\bar{V} = -56.72$ kN shear force in the $\bar{z}$ direction

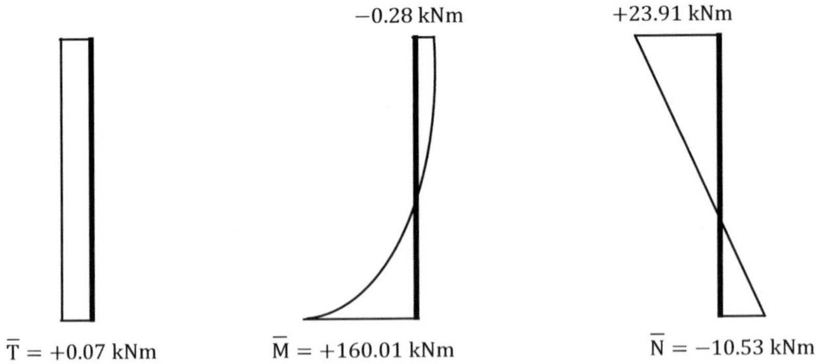

$$\bar{F}_r^1 = \begin{bmatrix} (\bar{F}_i^1)_r \\ (\bar{F}_j^1)_r \end{bmatrix} = \begin{bmatrix} (\bar{X}_1^1)_r \\ (\bar{Y}_1^1)_r \\ (\bar{Z}_1^1)_r \\ (\bar{T}_1^1)_r \\ (\bar{M}_1^1)_r \\ (\bar{N}_1^1)_r \\ (\bar{X}_2^1)_r \\ (\bar{Y}_2^1)_r \\ (\bar{Z}_2^1)_r \\ (\bar{T}_2^1)_r \\ (\bar{M}_2^1)_r \\ (\bar{N}_2^1)_r \end{bmatrix} = \bar{F}_d^1 + \bar{F}_f^1 = \begin{bmatrix} +16.01 \\ +5.74 \\ +26.72 \\ -0.07 \\ -130.01 \\ +10.53 \\ -16.01 \\ -5.74 \\ -26.72 \\ +0.07 \\ -30.28 \\ +23.91 \end{bmatrix} + \begin{bmatrix} 0 \\ 0 \\ +30.00 \\ 0 \\ -30.00 \\ 0 \\ 0 \\ 0 \\ +30.00 \\ 0 \\ +30.00 \\ 0 \end{bmatrix} = \begin{bmatrix} +16.01 \\ +5.74 \\ +56.72 \\ -0.07 \\ -160.01 \\ +10.53 \\ -16.01 \\ -5.74 \\ +3.28 \\ +0.07 \\ -0.28 \\ +23.91 \end{bmatrix}$$

Note that the sign in the above diagrams is for the forces and moments acting at a section at distance $\bar{x}$ from node i which is node 1 for member 1.

Member 2:

$$\delta^2 = \begin{bmatrix} \delta_i^2 \\ \delta_j^2 \end{bmatrix} = \begin{bmatrix} u_2 \\ v_2 \\ w_2 \\ \Phi_2 \\ \theta_2 \\ \Psi_2 \\ u_3 \\ v_3 \\ w_3 \\ \Phi_3 \\ \theta_3 \\ \Psi_3 \end{bmatrix} = \begin{bmatrix} +0.005873 \\ +0.001807 \\ -0.000026 \\ -0.004243 \\ +0.001275 \\ +0.002378 \\ +0.000004 \\ +0.001801 \\ -0.017199 \\ -0.004275 \\ -0.005124 \\ -0.000368 \end{bmatrix}$$

And r^2 from (10.17)

$$\bar{\delta}^2 = r^2\delta^2 = \begin{bmatrix} +0.001807 \\ -0.005873 \\ -0.000026 \\ +0.001275 \\ +0.004243 \\ +0.002378 \\ +0.001801 \\ -0.000004 \\ -0.017199 \\ -0.005124 \\ +0.004275 \\ -0.000368 \end{bmatrix}$$

and $\overline{k}^2$ from (10.16)

$$\overline{F}_d^2 = \overline{k}^2 \overline{\delta}^2 = \begin{bmatrix} +5.70 \\ -3.28 \\ +6.02 \\ +0.28 \\ -13.92 \\ -0.07 \\ -5.70 \\ +3.28 \\ -6.02 \\ -0.28 \\ -10.16 \\ -13.06 \end{bmatrix}$$

From (10.25)

$$\overline{F}_f^2 = \begin{bmatrix} 0 \\ 0 \\ +10 \\ 0 \\ -10 \\ 0 \\ 0 \\ 0 \\ +10 \\ 0 \\ +10 \\ 0 \end{bmatrix}$$

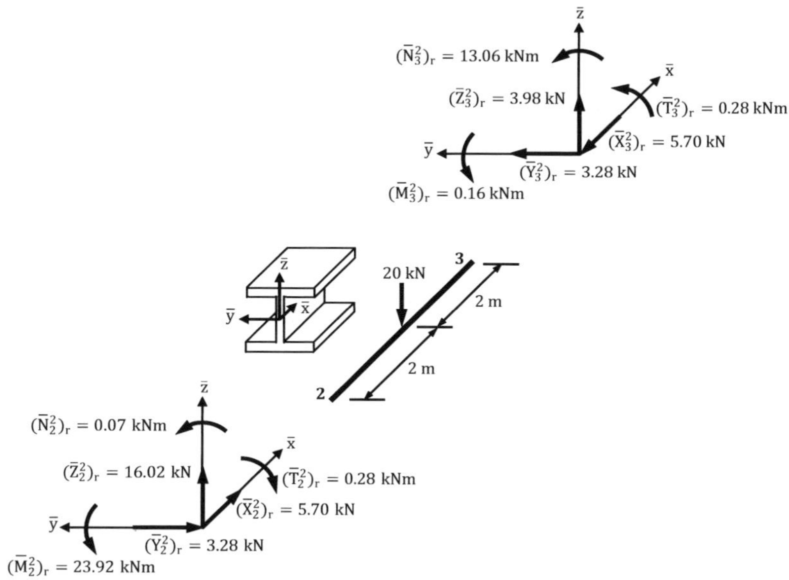
$\bar{z}$
$(\bar{N}_3^2)_r = 13.06$ kNm
$\bar{x}$
$(\bar{Z}_3^2)_r = 3.98$ kN
$(\bar{T}_3^2)_r = 0.28$ kNm
$(\bar{X}_3^2)_r = 5.70$ kN
$\bar{y}$
$(\bar{Y}_3^2)_r = 3.28$ kN
$(\bar{M}_3^2)_r = 0.16$ kNm
$\bar{z}$
$\bar{y}$
$\bar{x}$
20 kN
3
2 m
2 m
2
$\bar{z}$
$(\bar{N}_2^2)_r = 0.07$ kNm
$\bar{x}$
$(\bar{Z}_2^2)_r = 16.02$ kN
$(\bar{T}_2^2)_r = 0.28$ kNm
$(\bar{X}_2^2)_r = 5.70$ kN
$\bar{y}$
$(\bar{Y}_2^2)_r = 3.28$ kN
$(\bar{M}_2^2)_r = 23.92$ kNm

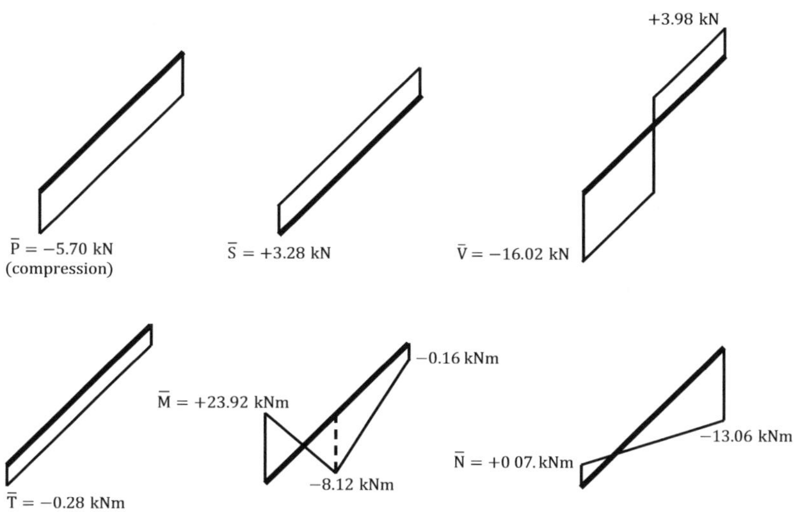
+3.98 kN
$\bar{P} = -5.70$ kN
(compression)
$\bar{S} = +3.28$ kN
$\bar{V} = -16.02$ kN
−0.16 kNm
$\bar{M} = +23.92$ kNm
−13.06 kNm
$\bar{N} = +0\ 07.$ kNm
−8.12 kNm
$\bar{T} = -0.28$ kNm

$$\overline{F}_r^2 = \begin{bmatrix} (\overline{F}_i^2)_r \\ (\overline{F}_j^2)_r \end{bmatrix} = \begin{bmatrix} (\overline{X}_2^2)_r \\ (\overline{Y}_2^2)_r \\ (\overline{Z}_2^2)_r \\ (\overline{T}_2^2)_r \\ (\overline{M}_2^2)_r \\ (\overline{N}_2^2)_r \\ (\overline{X}_3^2)_r \\ (\overline{Y}_3^2)_r \\ (\overline{Z}_3^2)_r \\ (\overline{T}_3^2)_r \\ (\overline{M}_3^2)_r \\ (\overline{N}_3^2)_r \end{bmatrix} = \overline{F}_d^2 + \overline{F}_f^2 = \begin{bmatrix} +5.70 \\ -3.28 \\ +6.02 \\ +0.28 \\ -13.92 \\ -0.07 \\ -5.70 \\ +3.28 \\ -6.02 \\ -0.28 \\ -10.16 \\ -13.06 \end{bmatrix} + \begin{bmatrix} 0 \\ 0 \\ +10.00 \\ 0 \\ -10.00 \\ 0 \\ 0 \\ 0 \\ +10.00 \\ 0 \\ +10.00 \\ 0 \end{bmatrix} = \begin{bmatrix} +5.70 \\ -3.28 \\ +16.02 \\ +0.28 \\ -23.92 \\ -0.07 \\ -5.70 \\ +3.28 \\ +3.98 \\ -0.28 \\ -0.16 \\ -13.06 \end{bmatrix}$$

Member 3:

$$\delta^3 = \begin{bmatrix} \delta_i^3 \\ \delta_j^3 \end{bmatrix} = \begin{bmatrix} u_3 \\ v_3 \\ w_3 \\ \Phi_3 \\ \theta_3 \\ \Psi_3 \\ u_4 \\ v_4 \\ w_4 \\ \Phi_4 \\ \theta_4 \\ \Psi_4 \end{bmatrix} = \begin{bmatrix} +0.000004 \\ +0.001801 \\ -0.017199 \\ -0.004275 \\ -0.005124 \\ -0.000368 \\ 0 \\ 0 \\ 0 \\ 0 \\ 0 \\ 0 \end{bmatrix}$$

and r^3 from (10.20)

$$\bar{\delta}^3 = r^3\delta^3 = \begin{bmatrix} +0.000004 \\ -0.017199 \\ -0.001801 \\ -0.004275 \\ -0.000368 \\ +0.005124 \\ 0 \\ 0 \\ 0 \\ 0 \\ 0 \\ 0 \end{bmatrix}$$

and $\bar{k}^3$ from (10.19)

$$\bar{F}_d^3 = \bar{k}^3\bar{\delta}^3 = \begin{bmatrix} +3.29 \\ -3.99 \\ -19.82 \\ -0.15 \\ +32.26 \\ -0.27 \\ -3.29 \\ +3.99 \\ +19.82 \\ +0.15 \\ +66.85 \\ -19.66 \end{bmatrix}$$

From (10.27)

$$\bar{F}_f^3 = \begin{bmatrix} 0 \\ 0 \\ +14.08 \\ 0 \\ -19.20 \\ 0 \\ 0 \\ 0 \\ +25.92 \\ 0 \\ +28.80 \\ 0 \end{bmatrix}$$

$$\bar{F}_r^3 = \begin{bmatrix} (\bar{X}_3^3)_r \\ (\bar{Y}_3^3)_r \\ (\bar{Z}_3^3)_r \\ (\bar{T}_3^3)_r \\ (\bar{M}_3^3)_r \\ (\bar{N}_3^3)_r \\ (\bar{X}_4^3)_r \\ (\bar{Y}_4^3)_r \\ (\bar{Z}_4^3)_r \\ (\bar{T}_4^3)_r \\ (\bar{M}_4^3)_r \\ (\bar{N}_4^3)_r \end{bmatrix} = \bar{F}_d^3 + \bar{F}_f^3 = \begin{bmatrix} +3.29 \\ -3.99 \\ -19.82 \\ -0.15 \\ +32.26 \\ -0.27 \\ -3.29 \\ +3.99 \\ +19.82 \\ +0.15 \\ +66.85 \\ -19.66 \end{bmatrix} + \begin{bmatrix} 0 \\ 0 \\ +14.08 \\ 0 \\ -19.20 \\ 0 \\ 0 \\ 0 \\ +25.92 \\ 0 \\ +28.80 \\ 0 \end{bmatrix} = \begin{bmatrix} +3.29 \\ -3.99 \\ -5.74 \\ -0.15 \\ +13.06 \\ -0.27 \\ -3.29 \\ +3.99 \\ +45.74 \\ +0.15 \\ +95.65 \\ -19.66 \end{bmatrix}$$

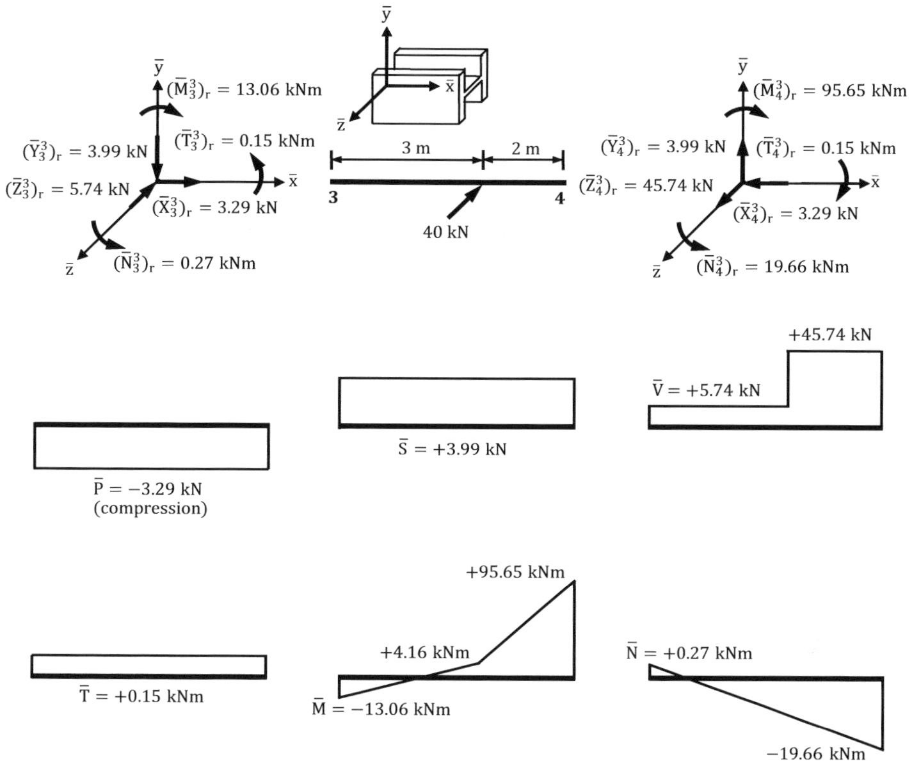

Problems

Analyse the rigidly connected space frames shown in Problems P10.1 and P10.2 for the given data.

P10.1 All members of the frame have the same rectangular cross section as shown in Fig. P10.1(b) with the dimensions b = 0.30 m and h = 0.54 m. The material of the frame has a modulus of elasticity, E = 29 × 10^6 kN/m^2 and modulus of rigidity, G = 12.6 × 10^6 kN/m^2.

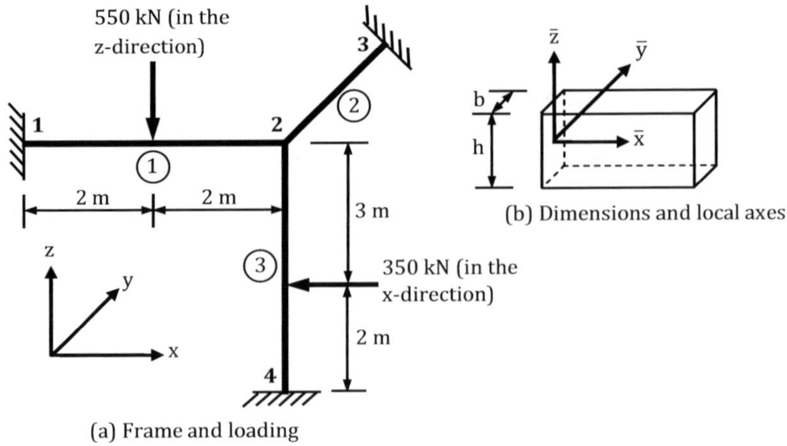

Figure P10.1

Node	x	y	z
1	0	0	0
2	4	0	0
3	4	3	2
4	4	0	-5

Member	Node i	Node j
1	1	2
2	2	3
3	2	4

$\varphi_{\bar{x}} = 0$ for all members

Answer:

$A = 0.162\ m^2$, $I_{\bar{y}} = 3937 \times 10^{-6}\, m^4$, $I_{\bar{z}} = 1215 \times 10^{-6}\, m^4$,

$J = 3076 \times 10^{-6}\ m^4$,

$u_2 = -0.000111$ m, $v_2 = +0.000166$ m, $w_2 = -0.000259$ m,

$\Phi_2 = +0.000090$ rad, $\theta_2 = -0.000439$ rad, $\Psi_2 = -0.000069$ rad.

$$\bar{F}_r^1 = \begin{bmatrix} (\bar{X}_1^1)_r \\ (\bar{Y}_1^1)_r \\ (\bar{Z}_1^1)_r \\ (\bar{T}_1^1)_r \\ (\bar{M}_1^1)_r \\ (\bar{N}_1^1)_r \\ (\bar{X}_2^1)_r \\ (\bar{Y}_2^1)_r \\ (\bar{Z}_2^1)_r \\ (\bar{T}_2^1)_r \\ (\bar{M}_2^1)_r \\ (\bar{N}_2^1)_r \end{bmatrix} = \begin{bmatrix} +130.00 \\ -2.01 \\ +299.36 \\ -0.87 \\ -311.17 \\ -3.41 \\ -130.00 \\ +2.01 \\ +250.65 \\ +0.87 \\ +213.75 \\ -4.62 \end{bmatrix}, \quad \bar{F}_r^2 = \begin{bmatrix} (\bar{X}_2^2)_r \\ (\bar{Y}_2^2)_r \\ (\bar{Z}_2^2)_r \\ (\bar{T}_2^2)_r \\ (\bar{M}_2^2)_r \\ (\bar{N}_2^2)_r \\ (\bar{X}_3^2)_r \\ (\bar{Y}_3^2)_r \\ (\bar{Z}_3^2)_r \\ (\bar{T}_3^2)_r \\ (\bar{M}_3^2)_r \\ (\bar{N}_3^2)_r \end{bmatrix} = \begin{bmatrix} -6.84 \\ +4.03 \\ -4.25 \\ -4.34 \\ +4.82 \\ +9.09 \\ +6.84 \\ -4.03 \\ +4.25 \\ +4.34 \\ +10.52 \\ +5.44 \end{bmatrix},$$

$$\bar{F}_r^3 = \begin{bmatrix} (\bar{X}_2^3)_r \\ (\bar{Y}_2^3)_r \\ (\bar{Z}_2^3)_r \\ (\bar{T}_2^3)_r \\ (\bar{M}_2^3)_r \\ (\bar{N}_2^3)_r \\ (\bar{X}_4^3)_r \\ (\bar{Y}_4^3)_r \\ (\bar{Z}_4^3)_r \\ (\bar{T}_4^3)_r \\ (\bar{M}_4^3)_r \\ (\bar{N}_4^3)_r \end{bmatrix} = \begin{bmatrix} +243.31 \\ +1.32 \\ +134.03 \\ +0.53 \\ -205.10 \\ +3.94 \\ -243.31 \\ -1.32 \\ +215.97 \\ -0.53 \\ +234.97 \\ +2.68 \end{bmatrix}$$

P10.2 All members have the same cross section with the properties:

$A = 0.0155\ m^2,\ I_{\bar{y}} = 760.40 \times 10^{-6}\, m^4,\ I_{\bar{z}} = 33.88 \times 10^{-6}\, m^4,$
$J = 1.78 \times 10^{-6}\ m^4,$
$E = 210 \times 10^6\ kN/m^2$ and $G = 81 \times 10^6\ kN/m^2$.

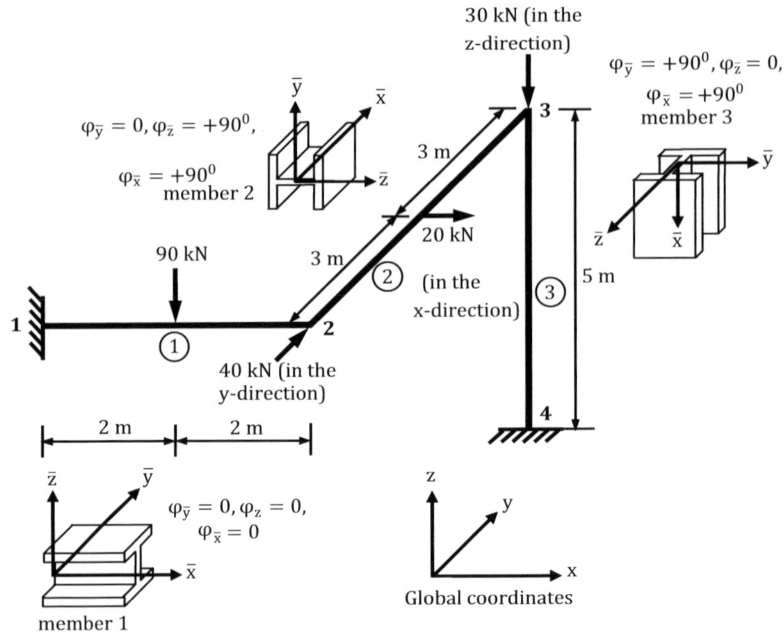

Figure P10.2

Node	x	y	z
1	0	0	0
2	4	0	0
3	4	6	0
4	4	6	–5

Member	Node i	Node j
1	1	2
2	2	3
3	3	4

Member 1: $\varphi_{\bar{y}} = 0,\ \varphi_{\bar{z}} = 0,\ \varphi_{\bar{x}} = 0$
Member 2: $\varphi_{\bar{y}} = 0,\ \varphi_{\bar{z}} = +90^0,\ \varphi_{\bar{x}} = +90^0$
Member 3: $\varphi_{\bar{y}} = +90^0,\ \varphi_{\bar{z}} = 0,\ \varphi_{\bar{x}} = +90^0$

Answer:

$u_2 = +0.000020$ m, $v_2 = +0.005627$ m, $w_2 = -0.003587$ m,

$\Phi_2 = +0.001679$ rad, $\theta_2 = +0.001064$ rad, $\Psi_2 = -0.003275$ rad,

$u_3 = +0.020854$ m, $v_3 = +0.005583$ m, $w_3 = -0.000048$ m,

$\Phi_3 = -0.001613$ rad, $\theta_3 = +0.006234$ rad, $\Psi_3 = -0.003429$ rad.

$$\bar{F}_r^1 = \begin{bmatrix} (\bar{X}_1^1)_r \\ (\bar{Y}_1^1)_r \\ (\bar{Z}_1^1)_r \\ (\bar{T}_1^1)_r \\ (\bar{M}_1^1)_r \\ (\bar{N}_1^1)_r \\ (\bar{X}_2^1)_r \\ (\bar{Y}_2^1)_r \\ (\bar{Z}_2^1)_r \\ (\bar{T}_2^1)_r \\ (\bar{M}_2^1)_r \\ (\bar{N}_2^1)_r \end{bmatrix} = \begin{bmatrix} -16.40 \\ -16.24 \\ +88.68 \\ -0.06 \\ -174.84 \\ -26.66 \\ +16.40 \\ +16.24 \\ +1.32 \\ +0.06 \\ +0.12 \\ -38.31 \end{bmatrix}, \quad \bar{F}_r^2 = \begin{bmatrix} (\bar{X}_2^2)_r \\ (\bar{Y}_2^2)_r \\ (\bar{Z}_2^2)_r \\ (\bar{T}_2^2)_r \\ (\bar{M}_2^2)_r \\ (\bar{N}_2^2)_r \\ (\bar{X}_3^2)_r \\ (\bar{Y}_3^2)_r \\ (\bar{Z}_3^2)_r \\ (\bar{T}_3^2)_r \\ (\bar{M}_3^2)_r \\ (\bar{N}_3^2)_r \end{bmatrix} = \begin{bmatrix} +23.76 \\ -1.32 \\ -16.40 \\ -0.12 \\ +38.31 \\ -0.06 \\ -23.76 \\ +1.32 \\ -3.60 \\ +0.12 \\ +0.10 \\ -7.87 \end{bmatrix},$$

$$\bar{F}_r^3 = \begin{bmatrix} (\bar{X}_3^3)_r \\ (\bar{Y}_3^3)_r \\ (\bar{Z}_3^3)_r \\ (\bar{T}_3^3)_r \\ (\bar{M}_3^3)_r \\ (\bar{N}_3^3)_r \\ (\bar{X}_4^3)_r \\ (\bar{Y}_4^3)_r \\ (\bar{Z}_4^3)_r \\ (\bar{T}_4^3)_r \\ (\bar{M}_4^3)_r \\ (\bar{N}_4^3)_r \end{bmatrix} = \begin{bmatrix} +31.32 \\ +3.60 \\ -23.76 \\ +0.10 \\ +7.87 \\ +0.12 \\ -31.32 \\ -3.60 \\ +23.76 \\ -0.10 \\ +110.91 \\ +17.87 \end{bmatrix}$$

Chapter 11

Stability of Struts and Frames

One of the considerations in the design of framed structures is to ensure that the structure is stable under the applied design loads. It is also important to consider the effect of change in the geometry of the structure as the loads are applied which results in a modification of the stiffness of the structure. This is particularly so when the deformation of the structure (or part of it) is large. For example, consider the simple case of a column that is subjected to an axial compressive force applied at an eccentricity at its top end. The resulting bending moment is the product of the force times its eccentricity relative to the centroidal axis of the column. This bending moment is assumed constant along the whole length of the column if it remains straight (or nearly so). But because the column deflects, its axis is no longer vertical and the eccentricity will increase by the amount of lateral deflection at the section considered. This is called second order effect which can be significant when the deflection is large. Obviously, when the deflection is small the second order effect is neglected. Another situation where bending moments develop even when the strut (or column) is axially loaded and that is due to imperfections (out of straightness) in the manufacture of the column which result in an unavoidable eccentricity.

The second order analysis is nonlinear due to the change in geometry of the structure as the applied loads are increased and

Analysis of Structures by Matrix Methods
Fathi Al-Shawi

ISBN 978-981-4968-19-5 (Hardcover), 978-1-003-29130-5 (eBook)
www.jennystanford.com

for a member that is subjected to a compressive force its stiffness is reduced as can be seen in the next section.

In this chapter, two aspects of stability are presented; the first consideration is to determine the magnitude of the axial compressive force required to produce buckling assuming an ideal strut that is perfectly straight before applying the load. The second consideration is to investigate the nonlinear behaviour of a strut under progressively increasing load up to the stage when deformations become excessively large and the strut approaches instability.

11.1 Derivation of Strut Buckling Matrix

Consider the strut shown in Fig. 11.1 which is acted upon by shear forces $\bar{Z}_i$ and $\bar{Z}_j$ and bending moments $\bar{M}_i$ and $\bar{M}_j$ at its ends. In addition, the strut is subjected to axial compressive forces, $\bar{P}$, at nodes i and j.

Summation of moments about node j

$$\bar{M}_i + \bar{M}_j + \bar{Z}_i L - \bar{P}\left(\bar{w}_j - \bar{w}_i\right) = 0$$

$$\bar{Z}_i = -\frac{\bar{M}_i + \bar{M}_j}{L} + \frac{\bar{P}\left(\bar{w}_j - \bar{w}_i\right)}{L} \tag{11.1}$$

Consider a section at a distance $\bar{x}$ from node i and the equilibrium of the left part of the beam, and take moments about point O. The bending moment $\bar{M}$ is given by:

$$\bar{M} + \bar{M}_i + \bar{Z}_i \bar{x} - \bar{P}\left(\bar{w} - \bar{w}_i\right) = 0$$

Substitute for $\bar{Z}_i$ from (11.1) to get

$$\bar{M} = -\bar{M}_i + \left[\frac{\bar{M}_i + \bar{M}_j}{L} - \frac{\bar{P}\left(\bar{w}_j - \bar{w}_i\right)}{L}\right]\bar{x} + \bar{P}(\bar{w} - \bar{w}_i) \tag{11.2}$$

The governing differential equation for the deflection of beams as derived in appendix 2 is:

$EI\dfrac{d^2\bar{w}}{d\bar{x}^2} = -\bar{M}$. Substitute for $\bar{M}$ from (11.2) to get

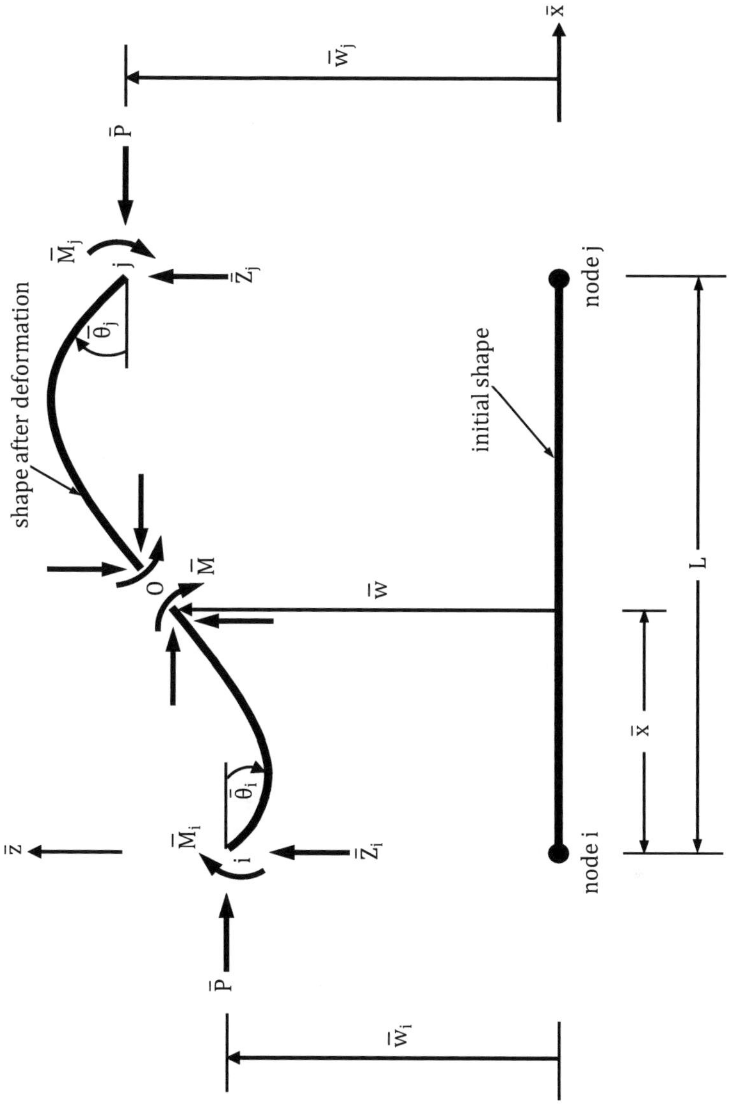

Figure 11.1 Beam-column element.

$$EI\frac{d^2\bar{w}}{d\bar{x}^2} = \bar{M}_i - \left[\frac{\bar{M}_i + \bar{M}_j}{L} - \frac{\bar{P}(\bar{w}_j - \bar{w}_i)}{L}\right]\bar{x} - \bar{P}(\bar{w} - \bar{w}_i)$$

$$\frac{d^2\bar{w}}{d\bar{x}^2} + \beta^2\bar{w} = \frac{\beta^2(\bar{M}_i + \bar{P}\bar{w}_i)}{\bar{P}} - \frac{\beta^2\left[\bar{M}_i + \bar{M}_j - \bar{P}(\bar{w}_j - \bar{w}_i)\right]}{\bar{P}L}\bar{x} \quad (11.3)$$

where $\beta^2 = \frac{\bar{P}}{EI}$.

The general solution to (11.3) is:

$$\bar{w} = C_1\sin\beta\bar{x} + C_2\cos\beta\bar{x} - \frac{\left[\bar{M}_i + \bar{M}_j - \bar{P}(\bar{w}_j - \bar{w}_i)\right]}{\bar{P}L}\bar{x} + \frac{(\bar{M}_i + \bar{P}\bar{w}_i)}{\bar{P}}. \quad (11.4)$$

Now apply the boundary conditions to find the constants C_1 and C_2 as follows:

At $\bar{x} = 0$, $\bar{w} = \bar{w}_i$, gives $C_2 = -\frac{\bar{M}_i}{\bar{P}}$

At $\bar{x} = L$, $\bar{w} = \bar{w}_j$, leads to $C_1 = \frac{\bar{M}_i\cos\beta L + \bar{M}_j}{\bar{P}\sin\beta L}$.

Substitute C_1 and C_2 in (11.4) to get

$$\bar{w} = \frac{(\bar{M}_i\cos\beta L + \bar{M}_j)}{\bar{P}\sin\beta L}\sin\beta\bar{x} - \frac{\bar{M}_i}{\bar{P}}\cos\beta\bar{x} - \frac{\left[\bar{M}_i + \bar{M}_j - \bar{P}(\bar{w}_j - \bar{w}_i)\right]}{\bar{P}L}\bar{x} + \frac{(\bar{M}_i + \bar{P}\bar{w}_i)}{\bar{P}}$$

$$\bar{\theta} = -\frac{d\bar{w}}{d\bar{x}}$$

$$\bar{\theta} = -\frac{\beta(\bar{M}_i\cos\beta L + \bar{M}_j)}{\bar{P}\sin\beta L}\cos\beta\bar{x} - \frac{\beta\bar{M}_i}{\bar{P}}\sin\beta\bar{x} + \frac{\left[\bar{M}_i + \bar{M}_j - \bar{P}(\bar{w}_j - \bar{w}_i)\right]}{\bar{P}L}$$

At $\bar{x} = 0$, $\bar{\theta} = \bar{\theta}_i$

$$\bar{\theta}_i = -\frac{\beta(\bar{M}_i\cos\beta L + \bar{M}_j)}{\bar{P}\sin\beta L} + \frac{\left[\bar{M}_i + \bar{M}_j - \bar{P}(\bar{w}_j - \bar{w}_i)\right]}{\bar{P}L} \quad (11.5)$$

$$\text{At } \bar{x} = L, \quad \bar{\theta} = \bar{\theta}_j$$

$$\bar{\theta}_j = -\frac{\beta(\bar{M}_i \cos\beta L + \bar{M}_j)}{\bar{P}\sin\beta L}\cos\beta L - \frac{\beta\bar{M}_i}{\bar{P}}\sin\beta L + \frac{\left[\bar{M}_i + \bar{M}_j - \bar{P}\left(\bar{w}_j - \bar{w}_i\right)\right]}{\bar{P}L} \tag{11.6}$$

Solving the simultaneous equations (11.5) and (11.6) for $\bar{M}_i$ and $\bar{M}_j$ we get:

$$\bar{M}_i = \frac{EI}{2\cos\beta L + \beta L\sin\beta L - 2}\Big[\bar{w}_i\beta^2(1-\cos\beta L) + \bar{\theta}_i\beta(\beta L\cos\beta L - \sin\beta L) + \bar{w}_j\beta^2(\cos\beta L - 1) + \bar{\theta}_j\beta(\sin\beta L - \beta L)\Big] \tag{11.7}$$

$$\bar{M}_j = \frac{EI}{2\cos\beta L + \beta L\sin\beta L - 2}\Big[\bar{w}_i\beta^2(1-\cos\beta L) + \bar{\theta}_i\beta(\sin\beta L - \beta L) + \bar{w}_j\beta^2(\cos\beta L - 1) + \bar{\theta}_j\beta(\beta L\cos\beta L - \sin\beta L)\Big] \tag{11.8}$$

In order that the analysis of stability is simplified, the process of calculations is linearized as explained below.

Equation (11.7) is written in an approximate form by using the infinite series for $\sin\beta L$ and $\cos\beta L$, i.e.

$$\sin\beta L = \beta L - \frac{(\beta L)^3}{3!} + \frac{(\beta L)^5}{5!} - \frac{(\beta L)^7}{7!} \ldots\ldots\ldots\ldots$$

$$\cos\beta L = 1 - \frac{(\beta L)^2}{2!} + \frac{(\beta L)^4}{4!} - \frac{(\beta L)^6}{6!} \ldots\ldots\ldots\ldots$$

Neglecting powers higher than six in the above two series and substituting in (11.7) and simplifying to get:

$$\bar{M}_i = \frac{EI\left\{\bar{w}_i\left[\frac{1}{2} - \frac{(\beta L)^2}{24}\right] - \bar{\theta}_i L\left[\frac{1}{3} - \frac{(\beta L)^2}{30}\right] - \bar{w}_j\left[\frac{1}{2} - \frac{(\beta L)^2}{24}\right] - \bar{\theta}_j L\left[\frac{1}{6} - \frac{(\beta L)^2}{120}\right]\right\}}{L^2\left[-\frac{1}{12} + \frac{(\beta L)^2}{180}\right]} \tag{11.7a}$$

The next step is to express the above equation in the form of a polynomial by using Taylor-Maclaurin infinite series as:

$\bar{M}_i = a_0 + a_1\beta + a_2\beta^2 + a_3\beta^3 + a_4\beta^4 + \ldots$, where the constants a_0, a_1, a_2,.... are found from (11.7a) as

$$a_0 = (\bar{M}_i)_{\text{at }\beta=0},\ a_1 = \left(\frac{\partial \bar{M}_i}{\partial \beta}\right)_{\text{at }\beta=0},\ a_2 = \frac{1}{2!}\left(\frac{\partial^2 \bar{M}_i}{\partial \beta^2}\right)_{\text{at }\beta=0},\ a_3 = \frac{1}{3!}\left(\frac{\partial^3 \bar{M}_i}{\partial \beta^3}\right)_{\text{at }\beta=0} \ldots.$$

As an approximation, consider only the first three terms of the above series. It is found that the first term gives the standard beam elastic stiffness matrix coefficients, the second term vanishes, and the third term represents the effect of the axial force, $\bar{P}$. The above equation is simplified and with the substitution of $\beta^2 EI = \bar{P}$, equation (11.7) becomes

$$\bar{M}_i = \frac{EI}{L^2}\left(-6\bar{w}_i + 4L\bar{\theta}_i + 6\bar{w}_j + 2L\bar{\theta}_j\right) - \bar{P}\left(-\frac{1}{10}\bar{w}_i + \frac{2}{15}L\bar{\theta}_i + \frac{1}{10}\bar{w}_j - \frac{1}{30}L\bar{\theta}_j\right) \tag{11.9}$$

Similarly, equation (11.8) can be written as:

$$\bar{M}_j = \frac{EI}{L^2}\left(-6\bar{w}_i + 2L\bar{\theta}_i + 6\bar{w}_j + 4L\bar{\theta}_j\right) - \bar{P}\left(-\frac{1}{10}\bar{w}_i - \frac{1}{30}L\bar{\theta}_i + \frac{1}{10}\bar{w}_j + \frac{2}{15}L\bar{\theta}_j\right) \tag{11.10}$$

Substitute (11.9) and (11.10) in (11.1) to get

$$\bar{Z}_i = \frac{EI}{L^3}\left(12\bar{w}_i - 6L\bar{\theta}_i - 12\bar{w}_j - 6L\bar{\theta}_j\right) - \bar{P}\left(\frac{6}{5L}\bar{w}_i - \frac{1}{10}\bar{\theta}_i - \frac{6}{5L}\bar{w}_j - \frac{1}{10}\bar{\theta}_j\right) \tag{11.11}$$

Also, the summation of the forces in the $\bar{z}$ direction is zero, i.e. $\bar{Z}_i + \bar{Z}_j = 0$, hence, $\bar{Z}_j = -\bar{Z}_i$ and from (11.11) we get

$$\bar{Z}_j = \frac{EI}{L^3}\left(-12\bar{w}_i + 6L\bar{\theta}_i + 12\bar{w}_j + 6L\bar{\theta}_j\right) - \bar{P}\left(-\frac{6}{5L}\bar{w}_i + \frac{1}{10}\bar{\theta}_i + \frac{6}{5L}\bar{w}_j + \frac{1}{10}\bar{\theta}_j\right) \tag{11.12}$$

Equations (11.9) to (11.12) are written in matrix form as:

$$\left[\begin{bmatrix} \frac{12EI}{L^3} & -\frac{6EI}{L^2} & -\frac{12EI}{L^3} & -\frac{6EI}{L^2} \\ -\frac{6EI}{L^2} & \frac{4EI}{L} & \frac{6EI}{L^2} & \frac{2EI}{L} \\ -\frac{12EI}{L^3} & \frac{6EI}{L^2} & \frac{12EI}{L^3} & \frac{6EI}{L^2} \\ -\frac{6EI}{L^2} & \frac{2EI}{L} & \frac{6EI}{L^2} & \frac{4EI}{L} \end{bmatrix} - \bar{P}\begin{bmatrix} \frac{6}{5L} & -\frac{1}{10} & -\frac{6}{5L} & -\frac{1}{10} \\ -\frac{1}{10} & \frac{2L}{15} & \frac{1}{10} & -\frac{L}{30} \\ -\frac{6}{5L} & \frac{1}{10} & \frac{6}{5L} & \frac{1}{10} \\ -\frac{1}{10} & -\frac{L}{30} & \frac{1}{10} & \frac{2L}{15} \end{bmatrix}\right]\begin{bmatrix} \bar{w}_i \\ \bar{\delta}_i \\ \bar{w}_j \\ \bar{\theta}_j \end{bmatrix} = \begin{bmatrix} \bar{Z}_i \\ \bar{M}_i \\ \bar{Z}_j \\ \bar{M}_j \end{bmatrix} \tag{11.13}$$

The above equation is of the general form $\bar{k}\bar{\delta} = \bar{F}$ and can be written as:

$$(\bar{k}_E - \bar{k}_G)\bar{\delta} = \bar{F} \tag{11.14}$$

where

$$\bar{k}_E = \begin{bmatrix} \frac{12EI}{L^3} & -\frac{6EI}{L^2} & -\frac{12EI}{L^3} & -\frac{6EI}{L^2} \\ -\frac{6EI}{L^2} & \frac{4EI}{L} & \frac{6EI}{L^2} & \frac{2EI}{L} \\ -\frac{12EI}{L^3} & \frac{6EI}{L^2} & \frac{12EI}{L^3} & \frac{6EI}{L^2} \\ -\frac{6EI}{L^2} & \frac{2EI}{L} & \frac{6EI}{L^2} & \frac{4EI}{L} \end{bmatrix} \tag{11.15}$$

is the elastic stiffness matrix and

$$\bar{k}_G = \bar{P}\begin{bmatrix} \frac{6}{5L} & -\frac{1}{10} & -\frac{6}{5L} & -\frac{1}{10} \\ -\frac{1}{10} & \frac{2L}{15} & \frac{1}{10} & -\frac{L}{30} \\ -\frac{6}{5L} & \frac{1}{10} & \frac{6}{5L} & \frac{1}{10} \\ -\frac{1}{10} & -\frac{L}{30} & \frac{1}{10} & \frac{2L}{15} \end{bmatrix} = \bar{P}\bar{k}_P \tag{11.16}$$

is called the geometric stiffness matrix.

It can be seen that the effect of the compressive force is a reduction of the elastic stiffness matrix. When the force $\bar{P}$ is tensile the solution of the differential equation will be in terms of sinhβL and coshβL instead of sinβL and cosβL. However, with the approximation used in the above derivation it is found that the sign of $\overline{P}$ is reversed resulting in an increase in the elastic stiffness matrix.

The above relationships can alternatively be derived by a finite element approach using the so-called interpolation polynomial which defines the displacement along the element as explained in Appendix 4.

For a member whose local axis does not lie along the global x-axis, equation (11.14) is written relative to global coordinates as

$$(k_E - k_G)\delta = F \tag{11.17a}$$

where δ and F are the displacement and load vectors relative to global coordinates with $k_E = r^T\overline{k}_E r$, $k_G = r^T\overline{k}_G r$, and r is the transformation matrix.

11.2 Stability of Struts

For the overall structure equation (11.17) is be written as:

$$(K_E - K_G)\delta = F \tag{11.17b}$$

For a perfectly straight strut subjected to only a direct compressive force the strut will remain straight as long as the force is less than a critical value defined by the point where the strut buckles. Bending of the strut will occur as a consequence of buckling and since there are no lateral forces acting on the span of the strut then the load vector, F = 0 and (11.17b) becomes

$$(K_E - K_G)\delta = 0 \tag{11.18}$$

The above relation represents a set of simultaneous equations whose trivial solution is δ = 0. The condition for a nontrivial solution of the displacement vector δ is that the determinant of the matrix $(K_E - K_G)$ is zero and this will lead to the values of the axial force that will cause buckling. Usually the smallest value is of most interest while higher values may apply when intermediate restraints at certain locations along the length of the strut are used.

This basically is an eigenvalue problem and the procedure for its solution is explained in Chapter 1.

Example 1:

Calculate the critical buckling load and draw the buckled shape of the strut shown in Fig. 11.2 which is pinned at point A and fixed at point B. The length (S) of the strut is 3.6 m, its second moment of area in the plane of buckling $I = 32 \times 10^{-6}\ m^4$ and its modulus of elasticity $E = 27 \times 10^6\ kN/m^2$. Assume that the strut is prevented against buckling out of plane.

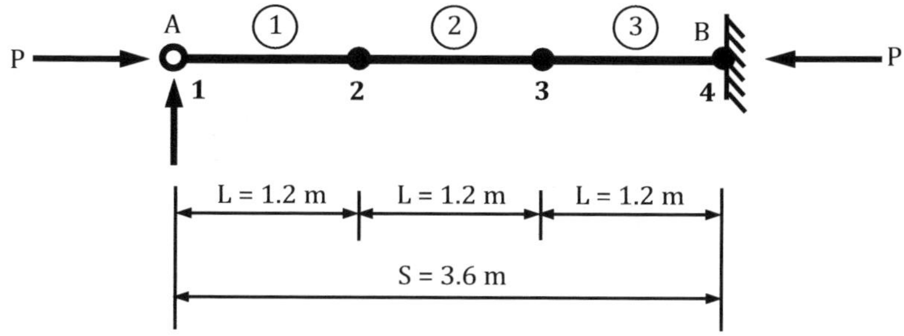

Figure 11.2

The strut is divided into three equal elements each of length L = 1.2 m.

For a member whose local axis is coincident with the global x-axis, the transformation matrix is the unit matrix and the stiffness matrix relative to the global coordinates is equal to the stiffness matrix relative to the local coordinates of the member. Thus, $k = \bar{k}$ and the elastic and geometric matrices for each element are given by (11.15) and (11.16), respectively.

$$k_E^1 = k_E^2 = k_E^3 = \begin{bmatrix} 6000 & -3600 & -6000 & -3600 \\ -3600 & 2880 & 3600 & 1440 \\ -6000 & 3600 & 6000 & 3600 \\ -3600 & 1440 & 3600 & 2880 \end{bmatrix}$$

$$k_G^1 = k_G^2 = k_G^3 = P\begin{bmatrix} 1.00 & -0.10 & -1.00 & -0.10 \\ -0.10 & 0.16 & 0.10 & -0.04 \\ -1.00 & 0.10 & 1.00 & 0.10 \\ -0.10 & -0.04 & 0.10 & 0.16 \end{bmatrix}$$

From (11.18) we have $k = k_E - k_G$, hence

$$k^1 = k^2 = k^3 = \begin{bmatrix} 6000-P & -3600+0.1P & -6000+P & -3600+0.1P \\ -3600+0.1P & 2880-0.16P & 3600-0.1P & 1440+0.04P \\ -6000+P & 3600-0.1P & 6000-P & 3600-0.1P \\ -3600+0.1P & 1440+0.04P & 3600-0.1P & 2880-0.16P \end{bmatrix}$$

For the overall structure

K =

	δ_1	δ_2	δ_3	δ_4
δ_1	k^1_{ii}	k^1_{ij}	0	0
δ_2	k^1_{ji}	$k^1_{jj}+k^2_{ii}$	k^2_{ij}	0
δ_3	0	k^2_{ji}	$k^2_{jj}+k^3_{ii}$	k^3_{ij}
δ_4	0	0	k^3_{ji}	k^3_{jj}

K =

	w_1	θ_1	w_2	θ_2	w_3	θ_3	w_4	θ_4
w_1	6000 −P	−3600 +0.1P	−6000 +P	−3600 +0.1P	0	0	0	0
θ_1	−3600 +0.1P	2880 −0.16P	3600 −0.1P	1440 +0.04P	0	0	0	0
w_2	−6000 +P	3600 −0.1P	12000− 2P	0	−6000 +P	−3600 +0.1P	0	0
θ_2	−3600 +0.1P	1440 +0.04P	0	5760− 0.32P	3600 −0.1P	1440 +0.04P	0	0
w_3	0	0	−6000 +P	3600 −0.1P	12000 −2P	0	−6000 +P	−3600 +0.1P
θ_3	0	0	−3600 +0.1P	1440 +0.04P	0	5760 −0.32P	3600 −0.1P	1440 +0.04P
w_4	0	0	0	0	−6000 +P	3600 −0.1P	6000 −P	3600 −0.1P
θ_4	0	0	0	0	−3600 +0.1P	1440 +0.04P	3600 −0.1P	2880 −0.16P

Equation (11.18) is represented by the above matrix with the right-hand side load vector F = 0. Apply the boundary conditions of $w_1 = 0$ (at the pinned end) and $w_4 = 0$ and $\theta_4 = 0$ (at the fixed end) by deleting rows and columns 1, 7, and 8 to get:

$$\begin{bmatrix} 2880-0.16P & 3600-0.10P & 1440+0.04P & 0 & 0 \\ 3600-0.10P & 12000-2.00P & 0 & -6000+1.00P & -3600+0.10P \\ 1440+0.04P & 0 & 5760-0.32P & 3600-0.10P & 1440+0.04P \\ 0 & -6000+1.00P & 3600-0.10P & 12000-2.00P & 0 \\ 0 & -3600+0.10P & 1440+0.04P & 0 & 5760-0.32P \end{bmatrix} \begin{bmatrix} \theta_1 \\ w_2 \\ \theta_2 \\ w_3 \\ \theta_3 \end{bmatrix} = \begin{bmatrix} 0 \\ 0 \\ 0 \\ 0 \\ 0 \end{bmatrix} \tag{11.19}$$

The trivial solution to the above set of simultaneous equations is $\delta = 0$. A non-trivial solution is obtained if the determinant of the above matrix is zero, i.e.

$$\begin{vmatrix} 2880-0.16P & 3600-0.10P & 1440+0.04P & 0 & 0 \\ 3600-0.10P & 12000-2.00P & 0 & -6000+1.00P & -3600+0.10P \\ 1440+0.04P & 0 & 5760-0.32P & 3600-0.10P & 1440+0.04P \\ 0 & -6000+1.00P & 3600-0.10P & 12000-2.00P & 0 \\ 0 & -3600+0.10P & 1440+0.04P & 0 & 5760-0.32P \end{vmatrix} = 0$$

Table 11.1

P (kN)	Det(K)/10^{16}
200	113.4
400	85.6
600	61.6
800	41.1
1000	23.7
1200	9.3
1400	-2.5
1600	-11.7

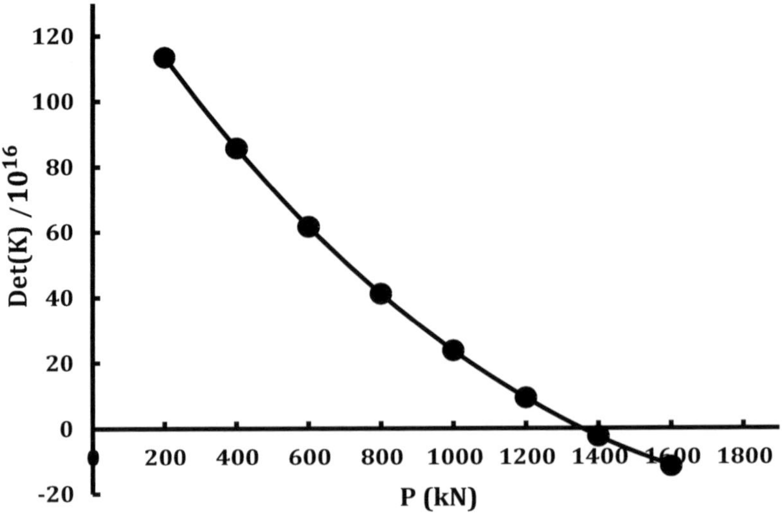

Figure 11.3 Determinant against applied axial force.

From Table 11.1 or from Fig. 11.3 the determinant changes sign and it is equal to zero when the value of P lies between 1200 kN and 1400 kN. By linear interpolation the critical value of P is

$$P_C = 1200 + \left(\frac{9.3}{9.3+2.5}\right) \times 200 = 1357.6 \text{ kN}$$

The exact value of the lowest critical load for a pinned-fixed strut is

$$P_C = \frac{\pi^2 EI}{(0.7S)^2} = \frac{\pi^2 \times 36 \times 10^6 \times 24 \times 10^{-6}}{(0.7 \times 3.6)^2} = 1342.8 \text{ kN}$$

The lowest critical load of 1357.6 kN is only 1.1% higher than the exact value. The number of elements in this example is three and the accuracy can be increased even further by increasing the number of elements.

We are often mostly interested in the lowest critical load so the above calculations will give the desired result.

Buckling Mode

For the buckling mode (shape) for the lowest critical load, substitute P = 1357.6 kN in (11.19) to get:

$$\left[\begin{bmatrix} 2880 & 3600 & 1440 & 0 & 0 \\ 3600 & 12000 & 0 & -6000 & -3600 \\ 1440 & 0 & 5760 & 3600 & 1440 \\ 0 & -6000 & 3600 & 12000 & 0 \\ 0 & -3600 & 1440 & 0 & 5760 \end{bmatrix} - 1357.6 \begin{bmatrix} 0.16 & 0.10 & -0.04 & 0 & 0 \\ 0.10 & 2.00 & 0 & -1.0 & -0.10 \\ -0.04 & 0 & 0.32 & 0.10 & -0.04 \\ 0 & -1.00 & 0.10 & 2.00 & 0 \\ 0 & -0.10 & -0.04 & 0 & 0.32 \end{bmatrix}\right] \begin{bmatrix} \theta_1 \\ w_2 \\ \theta_2 \\ w_3 \\ \theta_3 \end{bmatrix} = \begin{bmatrix} 0 \\ 0 \\ 0 \\ 0 \\ 0 \end{bmatrix}$$

Assume an arbitrary value for $\theta_1 = +1$, delete row 1 and substitute this value of θ_1 in the remaining four equations and rearrange the set of simultaneous equations to get:

$$9284.8w_2 + 0\theta_2 - 4642.4w_3 - 3464.2\theta_3 = -3464.2$$

$$0w_2 + 5325.6\theta_2 + 3464.2w_3 + 1494.3\theta_3 = -1494.3$$

$$-4642.4w_2 + 3464.2\theta_2 + 9284.8w_3 + 0\theta_3 = 0$$

$$-3464.2w_2 + 1494.3\theta_2 + 0w_3 + 5325.6\theta_3 = 0$$

The solution to the above set of simultaneous equations is:

$w_2 = -0.8732$ m, $\theta_2 = +0.2399$ rad, $w_3 = -0.5261$ m, $\theta_3 = -0.6354$ rad

And together with $w_1 = 0$, $\theta_1 = +1$, $w_4 = 0$, and $\theta_4 = 0$ the complete displacement vector is:

$$\delta = \begin{bmatrix} 0 \\ +1.0000 \\ -0.8732 \\ +0.2399 \\ -0.5261 \\ -0.6354 \\ 0 \\ 0 \end{bmatrix}$$ and the buckled shape is shown in Fig. 11.4.

Figure 11.4 Buckled shape of strut for the lowest critical load.

11.3 Nonlinear Analysis of Struts

In the previous chapters the analysis of structures was based on the assumption that the relationship between loads and displacements was linear. When the effect of axial forces developed in the members of a structure is taken into account then the stiffness of the member is modified because it is a function of the axial force as can be seen in (11.14). This will lead to nonlinear behaviour since the stiffness matrix is modified as the axial force is increased.

Example 2:

Consider the previous example with a point load of 30 kN in the z direction applied at node 2 as shown in Fig. 11.5. An axial load in the x direction is applied at node 1 which increases from P = 0 to P = 1200 kN in steps of 200 kN. Determine the relation between the deflection w_2 at node 2 and the load P.

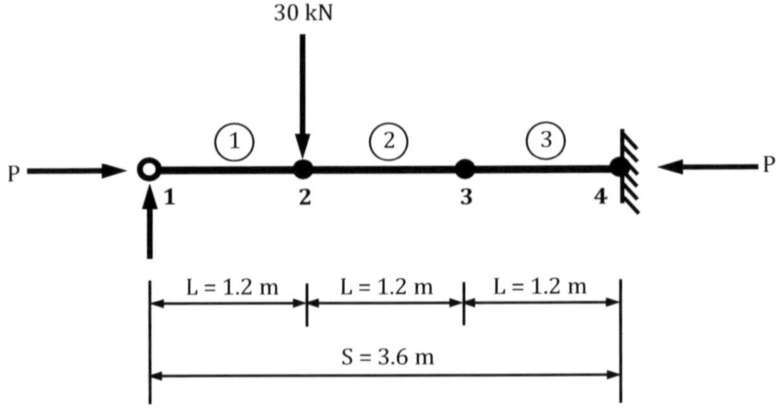

Figure 11.5

The load vector in this case is derived from the applied lateral loads on the strut as:

$$F = \begin{bmatrix} M_1 \\ Z_2 \\ M_2 \\ Z_3 \\ M_3 \end{bmatrix} = \begin{bmatrix} 0 \\ -30 \\ 0 \\ 0 \\ 0 \end{bmatrix}$$

The general relation (11.17b) becomes as:

$$\begin{bmatrix} 2880-0.16P & 3600-0.10P & 1440+0.04P & 0 & 0 \\ 3600-0.10P & 12000-2.00P & 0 & -6000+1.00P & -3600+0.10P \\ 1440+0.04P & 0 & 5760-0.32P & 3600-0.10P & 1440+0.04P \\ 0 & -6000+1.00P & 3600-0.10P & 12000-2.00P & 0 \\ 0 & -3600+0.10P & 1440+0.04P & 0 & 5760-0.32P \end{bmatrix}$$

$$\begin{bmatrix} \theta_1 \\ w_2 \\ \theta_2 \\ w_3 \\ \theta_3 \end{bmatrix} = \begin{bmatrix} 0 \\ -30 \\ 0 \\ 0 \\ 0 \end{bmatrix}$$

The displacement vector for a particular value of P is obtained from the solution of the above relation, for example let P = 200 kN and simplify to get the following set of linear simultaneous equations:

$$2848\theta_1 + 3580w_2 + 1448\theta_2 + 0w_3 + 0\theta_3 = 0$$

$$3580\theta_1 + 11600w_2 + 0\theta_2 - 5800w_3 - 3580\theta_3 = -30$$

$$1448\theta_1 + 0w_2 + 5696\theta_2 + 3580w_3 + 1448\theta_3 = 0$$

$$0\theta_1 - 5800w_2 + 3580\theta_2 + 11600w_3 + 0\theta_3 = 0$$

$$0\theta_1 - 3580w_2 + 1448\theta_2 + 0w_3 + 5696\theta_3 = 0$$

The solution of the above set of simultaneous equations is:

$$\theta_1 = +0.0196 \text{ rad}, w_2 = -0.0174 \text{ m}, \theta_2 = +0.0044 \text{ rad},$$

$$w_3 = -0.0100 \text{ m}, \theta_3 = -0.0120 \text{ rad}.$$

Similarly, the displacement vector is calculated for other values of P and the results are shown in Table 11.2 and a plot of the deflection at node 2 is shown in Fig. 11.6.

Table 11.2

P (kN)	Downward deflection at node 2, $-w_2$ (m)
0	0.0148
200	0.0174
400	0.0210
600	0.0265
800	0.0360
1000	0.0563
1200	0.1291
1300	0.3666
1350	4.6183

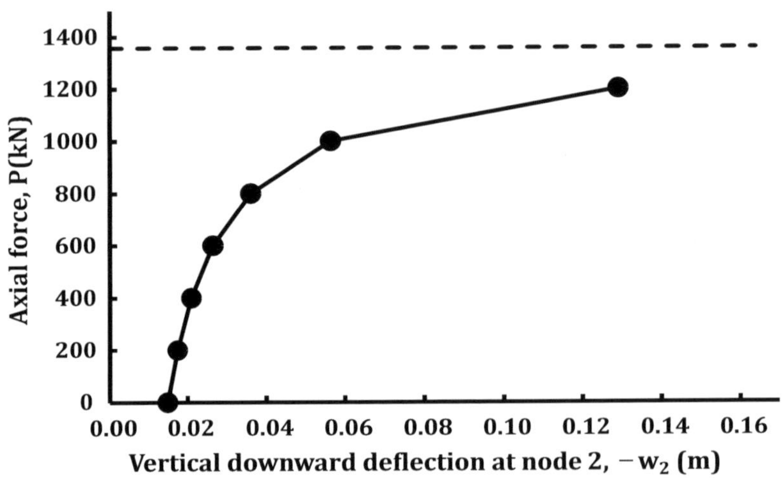

Figure 11.6 Axial compresssion against displacement.

For this case, the load is increased in steps and the displacements are calculated for each new value of the axial load, P. It should be noted that the rate of increase of displacements becomes larger as the load is increased due to the progressive decrease of the stiffness matrix. Instability is reached when the value of P is equal to the critical load of the strut. At this stage the stiffness matrix becomes singular and its determinant is equal to zero.

It will be seen later in the chapter that this is not the case when dealing with frames. The reason being that when increasing the loading on the structure from F_n (in the nth load level) to F_{n+1} the resulting displacements δ_{n+1} are not consistent with the axial forces developed in the members and the right-hand side of the equations which is a function of these axial forces. Therefore, an iterative approach is used for each load increment until convergence is reached and once that is achieved, the load is increased to a new level, i.e. F_{n+2} and so on as will be explained later.

11.4 Stability of Frames

If axial strains are considered as in the case of general analysis of frames, then the axial stiffness of the member (EA/L) is incorporated leading to the following relationship:

$$\left[\begin{bmatrix} \frac{EA}{L} & 0 & 0 & -\frac{EA}{L} & 0 & 0 \\ 0 & \frac{12EI}{L^3} & -\frac{6EI}{L^2} & 0 & -\frac{12EI}{L^3} & -\frac{6EI}{L^2} \\ 0 & -\frac{6EI}{L^2} & \frac{4EI}{L} & 0 & \frac{6EI}{L^2} & \frac{2EI}{L} \\ -\frac{EA}{L} & 0 & 0 & \frac{EA}{L} & 0 & 0 \\ 0 & -\frac{12EI}{L^3} & \frac{6EI}{L^2} & 0 & \frac{12EI}{L^3} & \frac{6EI}{L^2} \\ 0 & -\frac{6EI}{L^2} & \frac{2EI}{L} & 0 & \frac{6EI}{L^2} & \frac{4EI}{L} \end{bmatrix} - \bar{P}\begin{bmatrix} 0 & 0 & 0 & 0 & 0 & 0 \\ 0 & \frac{6}{5L} & -\frac{1}{10} & 0 & -\frac{6}{5L} & -\frac{1}{10} \\ 0 & -\frac{1}{10} & \frac{2L}{15} & 0 & \frac{1}{10} & -\frac{L}{30} \\ 0 & 0 & 0 & 0 & 0 & 0 \\ 0 & -\frac{6}{5L} & \frac{1}{10} & 0 & \frac{6}{5L} & \frac{1}{10} \\ 0 & -\frac{1}{10} & -\frac{L}{30} & 0 & \frac{1}{10} & \frac{2L}{15} \end{bmatrix}\right] \begin{bmatrix} \bar{u}_i \\ \bar{w}_i \\ \bar{\theta}_i \\ \bar{u}_j \\ \bar{w}_j \\ \bar{\theta}_j \end{bmatrix} = \begin{bmatrix} \bar{X}_i \\ \bar{Z}_i \\ \bar{M}_i \\ \bar{X}_j \\ \bar{Z}_j \\ \bar{M}_j \end{bmatrix} \tag{11.20}$$

or $\overline{k}\overline{\delta} = \overline{F}$.

For members that do not lie along the global x-axis then matrix transformation will be used to convert the member stiffness matrix from local to global coordinates and the resulting overall structure relationship is:

$$(K_E - K_G)\delta = F,$$

where K_E is the structure elastic stiffness matrix and K_G is the structure geometric stiffness matrix.

The above relationship is for the general second order analysis and for buckling (instability) analysis where the members are subjected to axial forces only then $F = 0$ and the resulting relationship is:

$$(K_E - K_G)\delta = 0.$$

The above relationship represents a set of homogeneous equations leading to the trivial solution vector of $\delta = 0$.

For a non-trivial solution, the determinant of the matrix $(K_E - K_G)$ must be equal to zero which is essentially an eigenvalue problem.

Example 3:

A rigidly connected frame is subjected to two loads each of magnitude Q and acting along the axis of the columns as shown in Fig. 11.7. All members of the frame have the same

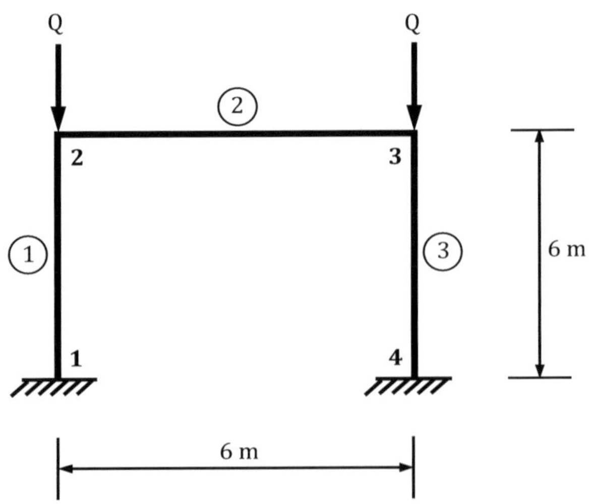

Figure 11.7

cross section with $I = 0.0001\ m^4$, $A = 0.005\ m^2$, and $E = 210 \times 10^6\ kN/m^2$. Determine the lowest value of Q which will cause the frame to buckle. Assume that buckling out of the plane is prevented.

From (11.20)

$$\bar{k} = \left[\begin{bmatrix} \frac{EA}{L} & 0 & 0 & -\frac{EA}{L} & 0 & 0 \\ 0 & \frac{12EI}{L^3} & -\frac{6EI}{L^2} & 0 & -\frac{12EI}{L^3} & -\frac{6EI}{L^2} \\ 0 & -\frac{6EI}{L^2} & \frac{4EI}{L} & 0 & \frac{6EI}{L^2} & \frac{2EI}{L} \\ -\frac{EA}{L} & 0 & 0 & \frac{EA}{L} & 0 & 0 \\ 0 & -\frac{12EI}{L^3} & \frac{6EI}{L^2} & 0 & \frac{12EI}{L^3} & \frac{6EI}{L^2} \\ 0 & -\frac{6EI}{L^2} & \frac{2EI}{L} & 0 & \frac{6EI}{L^2} & \frac{4EI}{L} \end{bmatrix} - \bar{P}\begin{bmatrix} 0 & 0 & 0 & 0 & 0 & 0 \\ 0 & \frac{6}{5L} & -\frac{1}{10} & 0 & -\frac{6}{5L} & -\frac{1}{10} \\ 0 & -\frac{1}{10} & \frac{2L}{15} & 0 & \frac{1}{10} & -\frac{L}{30} \\ 0 & 0 & 0 & 0 & 0 & 0 \\ 0 & -\frac{6}{5L} & \frac{1}{10} & 0 & \frac{6}{5L} & \frac{1}{10} \\ 0 & -\frac{1}{10} & -\frac{L}{30} & 0 & \frac{1}{10} & \frac{2L}{15} \end{bmatrix}\right]$$

Member 1:

The axial compressive force acting on member 1 is Q, thus $\bar{P} = Q$

$$\bar{k}^1 = \left[\begin{bmatrix} 175000 & 0 & 0 & -175000 & 0 & 0 \\ 0 & 1167 & -3500 & 0 & -1167 & -3500 \\ 0 & -3500 & 14000 & 0 & 3500 & 7000 \\ -175000 & 0 & 0 & 175000 & 0 & 0 \\ 0 & -1167 & 3500 & 0 & 1167 & 3500 \\ 0 & -3500 & 7000 & 0 & 3500 & 14000 \end{bmatrix}\right.$$

$$-Q\begin{bmatrix} 0 & 0 & 0 & 0 & 0 & 0 \\ 0 & 0.2 & -0.1 & 0 & -0.2 & -0.1 \\ 0 & -0.1 & 0.8 & 0 & 0.1 & -0.2 \\ 0 & 0 & 0 & 0 & 0 & 0 \\ 0 & -0.2 & 0.1 & 0 & 0.2 & 0.1 \\ 0 & -0.1 & -0.2 & 0 & 0.1 & 0.8 \end{bmatrix}\Bigg]$$

$$\bar{k}^1 = \begin{bmatrix} 175000 & 0 & 0 & -175000 & 0 & 0 \\ 0 & 1167-0.2Q & -3500+0.1Q & 0 & -1167+0.2Q & -3500+0.1Q \\ 0 & -3500+0.1Q & 14000-0.8Q & 0 & 3500-0.1Q & 7000+0.2Q \\ -175000 & 0 & 0 & 175000 & 0 & 0 \\ 0 & -1167+0.2Q & 3500-0.1Q & 0 & 1167-0.2Q & 3500-0.1Q \\ 0 & -3500+0.1Q & 7000+0.2Q & 0 & 3500-0.1Q & 14000-0.8Q \end{bmatrix}$$

$$x_i = 0,\ x_j = 0,\ x_{ij} = x_j - x_i = 0 - 0 = 0,$$

$$z_i = 0,\ z_j = 6\text{ m},\ z_{ij} = z_j - z_i = 6 - 0 = 6\text{ m},$$

$$L = \sqrt{x_{ij}^2 + z_{ij}^2} = \sqrt{0^2 + 6^2} = 6\text{ m}.$$

From Chapter 5, (5.6), the transformation matrix for rigidly connected plane frames is

$$r = \begin{bmatrix} x_{ij}/L & z_{ij}/L & 0 & 0 & 0 & 0 \\ -z_{ij}/L & x_{ij}/L & 0 & 0 & 0 & 0 \\ 0 & 0 & 1 & 0 & 0 & 0 \\ 0 & 0 & 0 & x_{ij}/L & z_{ij}/L & 0 \\ 0 & 0 & 0 & -z_{ij}/L & x_{ij}/L & 0 \\ 0 & 0 & 0 & 0 & 0 & 1 \end{bmatrix},$$

$$r^1 = \begin{bmatrix} 0 & 1 & 0 & 0 & 0 & 0 \\ -1 & 0 & 0 & 0 & 0 & 0 \\ 0 & 0 & 1 & 0 & 0 & 0 \\ 0 & 0 & 0 & 0 & 1 & 0 \\ 0 & 0 & 0 & -1 & 0 & 0 \\ 0 & 0 & 0 & 0 & 0 & 1 \end{bmatrix}$$

$$k^1 = (r^1)^T \bar{k}^1 r^1$$

$$k^1 = \begin{bmatrix} 1167-0.2Q & 0 & 3500-0.1Q & -1167+0.2Q & 0 & 3500-0.1Q \\ 0 & 175000 & 0 & 0 & -175000 & 0 \\ 3500-0.1Q & 0 & 14000-0.8Q & -3500+0.1Q & 0 & 7000+0.2Q \\ -1167+0.2Q & 0 & -3500+0.1Q & 1167-0.2Q & 0 & -3500+0.1Q \\ 0 & -175000 & 0 & 0 & 175000 & 0 \\ 3500-0.1Q & 0 & 7000+0.2Q & -3500+0.1Q & 0 & 14000-0.8Q \end{bmatrix}$$

Member 2:

Since the local $\bar{x}$-axis of member 2 coincides with the global x-axis then the transformation matrix r is equal to the unit matrix resulting in $k^2 = \bar{k}^2$. Also, it is not subjected to an axial force, therefore, $\bar{P} = 0$, thus

$$k^2 = \begin{bmatrix} 175000 & 0 & 0 & -175000 & 0 & 0 \\ 0 & 1167 & -3500 & 0 & -1167 & -3500 \\ 0 & -3500 & 14000 & 0 & 3500 & 7000 \\ -175000 & 0 & 0 & 175000 & 0 & 0 \\ 0 & -1167 & 3500 & 0 & 1167 & 3500 \\ 0 & -3500 & 7000 & 0 & 3500 & 14000 \end{bmatrix}$$

Member 3:

$$\bar{k}^3 = \begin{bmatrix} 175000 & 0 & 0 & -175000 & 0 & 0 \\ 0 & 1167-0.2Q & -3500+0.1Q & 0 & -1167+0.2Q & -3500+0.1Q \\ 0 & -3500+0.1Q & 14000-0.8Q & 0 & 3500-0.1Q & 7000+0.2Q \\ -175000 & 0 & 0 & 175000 & 0 & 0 \\ 0 & -1167+0.2Q & 3500-0.1Q & 0 & 1167-0.2Q & 3500-0.1Q \\ 0 & -3500+0.1Q & 7000+0.2Q & 0 & 3500-0.1Q & 14000-0.8Q \end{bmatrix}$$

$$x_i = 6 \text{ m},\ x_j = 6 \text{ m},\ x_{ij} = x_j - x_i = 6 - 6 = 0,$$

$$z_i = 6 \text{ m},\ z_j = 0,\ z_{ij} = z_j - z_i = 0 - 6 = -6 \text{ m},$$

$$L = \sqrt{x_{ij}^2 + z_{ij}^2} = \sqrt{0^2 + (-6)^2} = 6 \text{ m}$$

$$r = \begin{bmatrix} x_{ij}/L & z_{ij}/L & 0 & 0 & 0 & 0 \\ -z_{ij}/L & x_{ij}/L & 0 & 0 & 0 & 0 \\ 0 & 0 & 1 & 0 & 0 & 0 \\ 0 & 0 & 0 & x_{ij}/L & z_{ij}/L & 0 \\ 0 & 0 & 0 & -z_{ij}/L & x_{ij}/L & 0 \\ 0 & 0 & 0 & 0 & 0 & 1 \end{bmatrix},$$

$$r^3 = \begin{bmatrix} 0 & -1 & 0 & 0 & 0 & 0 \\ 1 & 0 & 0 & 0 & 0 & 0 \\ 0 & 0 & 1 & 0 & 0 & 0 \\ 0 & 0 & 0 & 0 & -1 & 0 \\ 0 & 0 & 0 & 1 & 0 & 0 \\ 0 & 0 & 0 & 0 & 0 & 1 \end{bmatrix}$$

$$k^3 = (r^3)^T \bar{k}^3 r^3$$

$$k^3 = \begin{bmatrix} 1167-0.2Q & 0 & -3500+0.1Q & -1167+0.2Q & 0 & -3500+0.1Q \\ 0 & 175000 & 0 & 0 & -175000 & 0 \\ -3500+0.1Q & 0 & 14000-0.8Q & 3500-0.1Q & 0 & 7000+0.2Q \\ -1167+0.2Q & 0 & 3500-0.1Q & 1167-0.2Q & 0 & 3500-0.1Q \\ 0 & -175000 & 0 & 0 & 175000 & 0 \\ -3500+0.1Q & 0 & 7000+0.2Q & 3500-0.1Q & 0 & 14000-0.8Q \end{bmatrix}$$

Structure Stiffness Matrix

$K =$

	δ_1	δ_2	δ_3	δ_4	
	k_{ii}^1	k_{ij}^1	0	0	δ_1
	k_{ji}^1	$k_{jj}^1 + k_{ii}^2$	k_{ij}^2	0	δ_2
	0	k_{ji}^2	$k_{jj}^2 + k_{ii}^3$	k_{ij}^3	δ_3
	0	0	k_{ji}^3	k_{ii}^1	δ_4

The boundary conditions are for the fixed supports 1 and 4 so, $\delta_1 = 0$ and $\delta_4 = 0$, therefore, delete rows 1 and 4 and columns 1 and 4.

$$K = \begin{array}{|c|c|c} \multicolumn{1}{c}{\delta_2} & \multicolumn{1}{c}{\delta_3} & \\ \hline k_{jj}^1 + k_{ii}^2 & k_{ij}^2 & \delta_2 \\ \hline k_{ji}^2 & k_{jj}^2 + k_{ii}^3 & \delta_3 \\ \hline \end{array}$$

$$K = \begin{bmatrix} 176167-0.2Q & 0 & -3500+0.1Q & -175000 & 0 & 0 \\ 0 & 176167 & -3500 & 0 & -1167 & -3500 \\ -3500+0.1Q & -3500 & 28000-0.8Q & 0 & 3500 & 7000 \\ -175000 & 0 & 0 & 176167-0.2Q & 0 & -3500+0.1Q \\ 0 & -1167 & 3500 & 0 & 176167 & 3500 \\ 0 & -3500 & 7000 & -3500+0.1Q & 3500 & 28000-0.8Q \end{bmatrix} \tag{11.21}$$

The values of the determinant of matrix K for different values of Q are as shown in Table 11.3.

Table 11.3

Q (kN)	**Det(K)/10²⁶**
1000	46.7
2000	30.6
3000	16.3
4000	3.8
5000	-7.1
6000	-16.5

By linear interpolation between Q = 4000 kN and Q = 5000 kN in Table 11.3 or from Fig. 11.8 the value of Det(K) = 0 occurs at Q = Q_C, where Q_C = 4000 + [3.8/(3.8 + 7.1)]1000 = 4349 kN, which is the smallest load that will cause instability of the frame.

The exact value of the lowest critical load given by Timoshenko is $Q_C = 0.75(\pi^2 EI/L^2) = 0.75(\pi^2 \times 210 \times 10^6 \times 0.0001/6^2) = 4318$ kN, and the difference between the calculated and the exact values is only +0.72%.

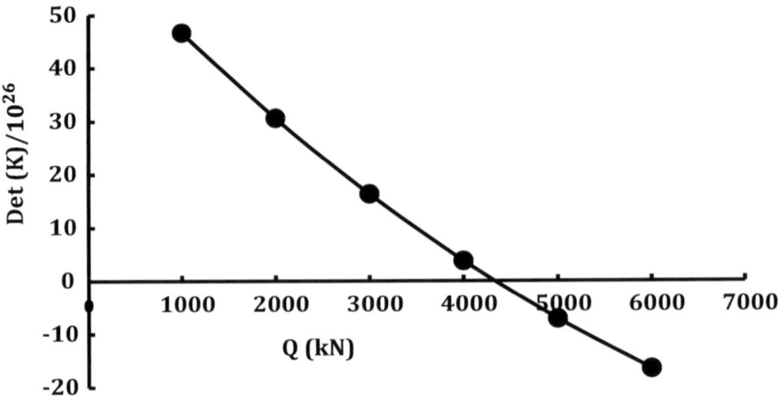

Figure 11.8 Valued of the determinant against load.

In order to determine the buckled shape (buckling mode) of the frame for the lowest critical load, the value of $Q_C = 4349$ kN is substituted in the K matrix as given by equation (11.21) and applying a zero load vector for the right-hand side to get:

$$\begin{bmatrix} 175297 & 0 & -3065 & -175000 & 0 & 0 \\ 0 & 176167 & -3500 & 0 & -1167 & -3500 \\ -3065 & -3500 & 24521 & 0 & 3500 & 7000 \\ -175000 & 0 & 0 & 175297 & 0 & -3065 \\ 0 & -1167 & 3500 & 0 & 176167 & 3500 \\ 0 & -3500 & 7000 & -3065 & 3500 & 24521 \end{bmatrix} \begin{bmatrix} u_2 \\ w_2 \\ \theta_2 \\ u_3 \\ w_3 \\ \theta_3 \end{bmatrix} = \begin{bmatrix} 0 \\ 0 \\ 0 \\ 0 \\ 0 \\ 0 \end{bmatrix}$$

Assume an arbitrary value for $u_2 = +1$, delete row 1 and substitute this value of u_2 in the remaining five equations, rearrange, and solve the resulting set of simultaneous equations to get:

$w_2 = +0.003872$ m, $\theta_2 = +0.098096$ rad, $u_3 = +1.000021$ m, $w_3 = -0.003872$ m, $\theta_3 = +0.098100$ rad.

With the boundary conditions $u_1 = 0$, $w_1 = 0$, $\theta_1 = 0$, $u_4 = 0$, $w_4 = 0$, $\theta_4 = 0$ together with $u_2 = +1$, the complete displacement vector is:

$$\delta = \begin{bmatrix} u_1 \\ w_1 \\ \theta_1 \\ u_2 \\ w_2 \\ \theta_2 \\ u_3 \\ w_3 \\ \theta_3 \\ u_4 \\ w_4 \\ \theta_4 \end{bmatrix} = \begin{bmatrix} 0 \\ 0 \\ 0 \\ +1.000000 \\ +0.003872 \\ +0.098096 \\ +1.000021 \\ -0.003872 \\ +0.098100 \\ 0 \\ 0 \\ 0 \end{bmatrix}$$

and the normalised displacement vector is obtained by dividing by the largest coefficient 1.000021 to get

$$\begin{bmatrix} u_1 \\ w_1 \\ \theta_1 \\ u_2 \\ w_2 \\ \theta_2 \\ u_3 \\ w_3 \\ \theta_3 \\ u_4 \\ w_4 \\ \theta_4 \end{bmatrix} = \begin{bmatrix} 0 \\ 0 \\ 0 \\ +0.999979 \\ +0.003872 \\ +0.098094 \\ +1.000000 \\ -0.003872 \\ +0.098098 \\ 0 \\ 0 \\ 0 \end{bmatrix}$$

and the buckled shape is shown in Fig. 11.9, which is a sway type of buckling.

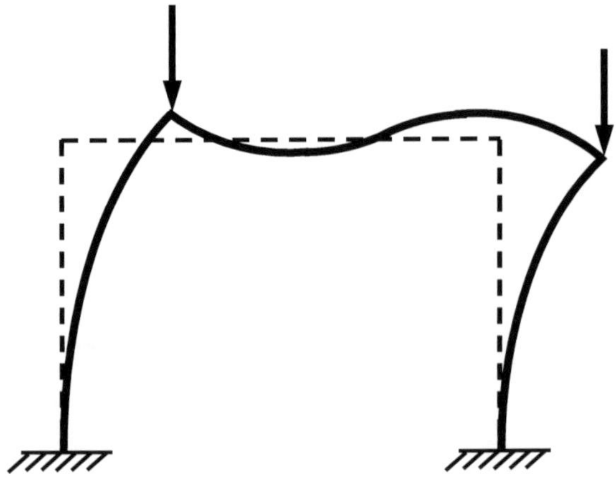

Figure 11.9 Buckled shape of portal frame.

11.5 Nonlinear Analysis of Frames

It was discussed earlier that changes to the geometry of a beam column lead to a nonlinear behaviour and magnified actions on the structure the so called second order effects. This principle applies to frames where some or all the members carry compressive forces resulting from the application of loads to the frame.

Consider the frame in the previous example but with applied loading as shown in Fig. 11.10.

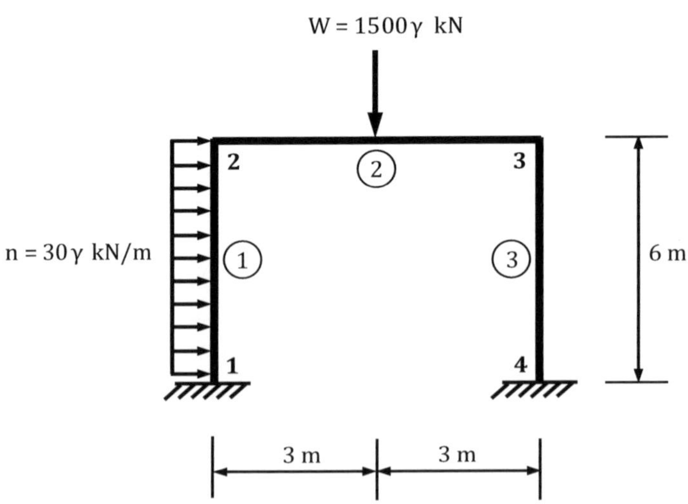

Figure 11.10 Portal frame and loading.

Calculation of End Moments

The fixed end moment for a beam carrying a uniformly distributed load, n per unit length, and subjected to an axial force, P as shown in Fig. 11.11 are given by:

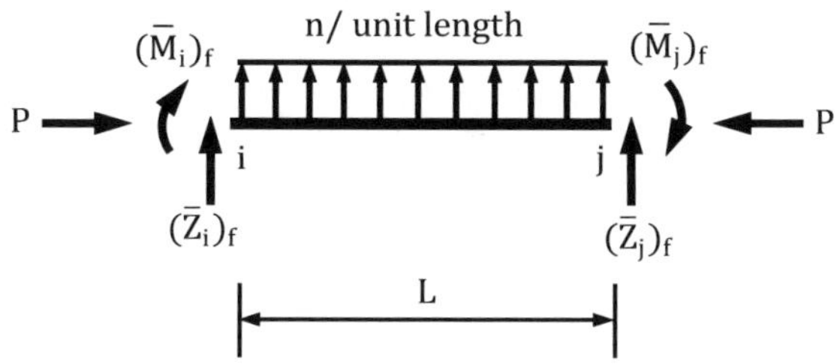

Figure 11.11

$$(\bar{M}_i)_f = +\left[\frac{3(\tan\alpha - \alpha)}{\alpha^2 \tan\alpha}\right]\frac{nL^2}{12}, \quad (\bar{M}_j)_f = -\left[\frac{3(\tan\alpha - \alpha)}{\alpha^2 \tan\alpha}\right]\frac{nL^2}{12}$$

$$(\bar{Z}_i)_f = -\frac{nL}{2}, \quad (\bar{Z}_j)_f = -\frac{nL}{2},$$

where $\alpha = \frac{\pi}{2}\sqrt{\rho}$, $\rho = \frac{P}{P_E}$, and the Euler load, $P_E = \frac{\pi^2 EI}{L^2}$.

When there is no axial load, i.e. P = 0, α will be equal to zero.

The limit of $\left[\frac{3(\tan\alpha - \alpha)}{\alpha^2 \tan\alpha}\right]$ as α approaches 0, is equal to 1.0 hence

$$(\bar{M}_i)_f = +\frac{nL^2}{12}, \quad (\bar{M}_j)_f = -\frac{nL^2}{12}$$

The fixed end moment for a beam carrying a concentrated load, W at mid-span, and subjected to an axial force, P as shown in Fig. 11.12 are given by:

$$(\bar{M}_i)_f = +\left[\frac{2(1 - \cos\alpha)}{\alpha \sin\alpha}\right]\frac{WL}{8}, \quad (\bar{M}_j)_f = -\left[\frac{2(1 - \cos\alpha)}{\alpha \sin\alpha}\right]\frac{WL}{8}$$

$$(\bar{Z}_i)_f = -\frac{W}{2}, \quad (\bar{Z}_j)_f = -\frac{W}{2}$$

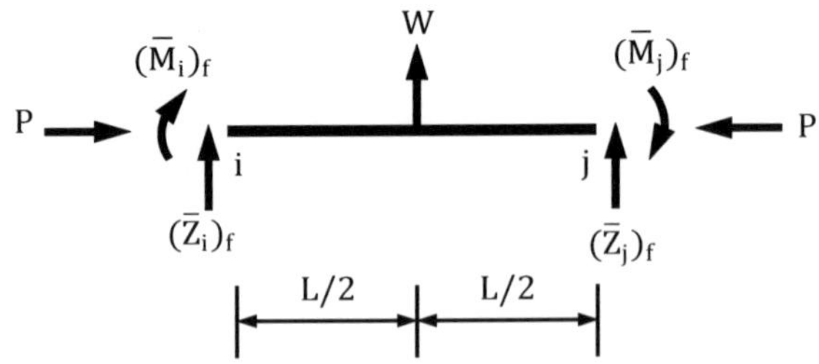

Figure 11.12

The limit of $\left[\dfrac{2(1-\cos\alpha)}{\alpha\sin\alpha}\right]$ as α approaches 0, is equal to 1.0 hence

$$(\bar{M}_i)_f = +\frac{WL}{8}, \quad (\bar{M}_j)_f = -\frac{WL}{8}$$

The moments and forces acting on the joints are in the opposite direction to the above moments.

Calculation of Axial Forces

The axial force in any member of the frame is found in the same way as in Chapter 2 as shown below.

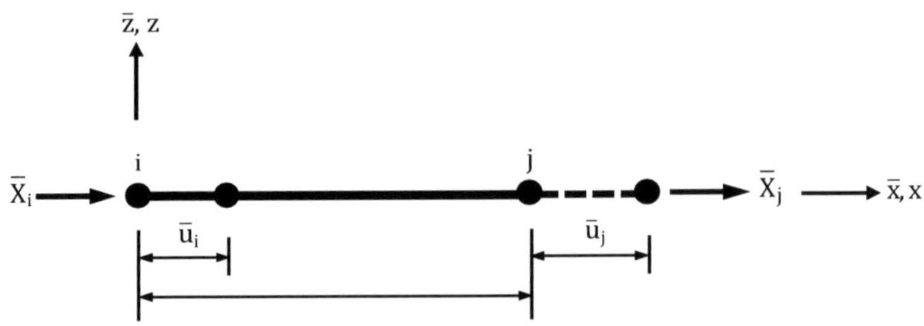

Figure 11.13 Bar element subjected to axial forces.

$$\begin{bmatrix} \overline{X}_i \\ \overline{X}_j \end{bmatrix} = \begin{bmatrix} \frac{EA}{L} & -\frac{EA}{L} \\ -\frac{EA}{L} & \frac{EA}{L} \end{bmatrix} \begin{bmatrix} \overline{u}_i \\ \overline{u}_j \end{bmatrix} \tag{11.13}$$

For a member that does not lie along the global x-axis, transformation is used as was followed in Chapter 3.

Start the incremental load process with $\gamma = 0$ and this means that all the displacements and the axial forces are equal to zero. Then increase γ from 0 to 1.0 and use the current values of the axial forces (i.e. zero) to determine new values for the displacements and calculate the new values of the axial forces from (11.22) as

$$\overline{P}_i = \overline{X}_i = \frac{EA}{L}\left(\overline{u}_i - \overline{u}_j\right) \tag{11.22a}$$

Notice that if $\overline{P}_i$ is positive the member is in compression.

Use these values of axial forces to calculate new values for the displacements which will in turn be used to calculate new values for the axial forces from the above equation. This iteration is carried out until the difference between the values of displacements from two successive iterations is within prescribed limits. At this cycle of iteration there will be consistency between the axial forces and the displacements used in their calculation for that value of γ. This will end the iteration for $\gamma = 1.0$.

Now increase the load by setting $\gamma = 2.0$ and use the current axial forces, i.e. the last values from previous iteration and start the iteration until consistency is reached between axial forces and displacements and this will end the iteration for this load increment.

At the end of iteration for each value of γ_{n+1} test if for this load factor the structure has become unstable by calculating the determinant of the structure matrix K. If the determinant of the matrix is positive then the structure is stable and the process of increasing the load is continued. If the determinant is negative (i.e. singularity has been passed) then the value of γ is reduced to a value between γ_{n+1} and γ_n and the iteration process is continued.

Denote the axial forces in members 1, 2, and 3 by P^1, P^2, and P^3, respectively and these are obtained from (11.22a). Then the stiffness matrices for the three members are:

$$
k^1 = \begin{bmatrix}
1167-0.2P^1 & 0 & 3500-0.1P^1 & -1167+0.2P^1 & 0 & 3500-0.1P^1 \\
0 & 175000 & 0 & 0 & -175000 & 0 \\
3500-0.1P^1 & 0 & 14000-0.8P^1 & -3500+0.1P^1 & 0 & 7000+0.2P^1 \\
-1167+0.2P^1 & 0 & -3500+0.1P^1 & 1167-0.2P^1 & 0 & -3500+0.1P^1 \\
0 & -175000 & 0 & 0 & 175000 & 0 \\
3500-0.1P^1 & 0 & 7000+0.2P^1 & -3500+0.1P^1 & 0 & 14000-0.8P^1
\end{bmatrix}
$$

$$
k^2 = \begin{bmatrix}
175000 & 0 & 0 & -175000 & 0 & 0 \\
0 & 1167-0.2P^2 & -3500+0.1P^2 & 0 & -1167+0.2P^2 & -3500+0.1P^2 \\
0 & -3500+0.1P^2 & 14000-0.8P^2 & 0 & 3500-0.1P^2 & 7000+0.2P^2 \\
-175000 & 0 & 0 & 175000 & 0 & 0 \\
0 & -1167+0.2P^2 & 3500-0.1P^2 & 0 & 1167-0.2P^2 & 3500-0.1P^2 \\
0 & -3500+0.1P^2 & 7000+0.2P^2 & 0 & 3500-0.1P^2 & 14000-0.8P^2
\end{bmatrix}
$$

$$k^3 = \begin{bmatrix} 1167-0.2P^3 & 0 & -3500+0.1P^3 & -1167+0.2P^3 & 0 & -3500+0.1P^3 \\ 0 & 175000 & 0 & 0 & -175000 & 0 \\ -3500+0.1P^3 & 0 & 14000-0.8P^3 & 3500-0.1P^3 & 0 & 7000+0.2P^3 \\ -1167+0.2P^3 & 0 & 3500-0.1P^3 & 1167-0.2P^3 & 0 & 3500-0.1P^3 \\ 0 & -175000 & 0 & 0 & 175000 & 0 \\ -3500+0.1P^3 & 0 & 7000+0.2P^3 & 3500-0.1P^3 & 0 & 14000-0.8P^3 \end{bmatrix}$$

The overall structure matrix is:

$$K = \begin{bmatrix} 176167-0.2P^1 & 0 & -3500+0.1P^1 & -17500 & 0 & 0 \\ 0 & 176167-0.2P^2 & -3500+0.1P^2 & 0 & -1167+0.2P^2 & -3500+0.1P^2 \\ -3500+0.1P^1 & -3500+0.1P^2 & 28000-0.8P^1-0.8P^2 & 0 & 3500-0.1P^2 & 7000+0.2P^2 \\ -175000 & 0 & 0 & 176167-0.2P^3 & 0 & -3500+0.1P^3 \\ 0 & -1167+0.2P^2 & 3500-0.1P^2 & 0 & 176167-0.2P^2 & 3500-0.1P^2 \\ 0 & -3500+0.1P^2 & 7000+0.2P^2 & -3500+0.1P^3 & 3500-0.1P^2 & 28000-0.8P^2-0.8P^3 \end{bmatrix} \quad (11.23)$$

$$P_E^1 = P_E^2 = P_E^3 = P_E = \frac{\pi^2 EI}{L^2} = \frac{\pi^2 \times 210 \times 10^6 \times 0.0001}{6^2} = 5757 \text{ kN}$$

$$\alpha_1 = \frac{\pi}{2}\sqrt{\rho_1} = \frac{\pi}{2}\sqrt{\frac{P^1}{P_E}} = 0.0207\sqrt{P^1}, \quad \alpha_2 = 0.0207\sqrt{P^2}, \text{ and } \alpha_3 = 0.0207\sqrt{P^3}.$$

Calculation of the load vector

Only joints 2 and 3 are considered since the reduced stiffness matrix includes displacements at these joints only.

Contribution of loads on member 1 to joint 2

$$(X_2^1)_s = -\frac{\gamma n L}{2}, \ \left(Z_2^1\right)_s = 0, \ \left(M_2^1\right)_s = +\left[\frac{3(\tan\alpha_1 - \alpha_1)}{\alpha_1^2 \tan\alpha_1}\right]\frac{\gamma n L^2}{12}$$

Contribution of loads on member 2 to joint 2

$$(X_2^2)_s = 0, (Z_2^2)_s = +\frac{\gamma W}{2}, \ \left(M_2^2\right)_s = -\left[\frac{2(1-\cos\alpha_2)}{\alpha_2 \sin\alpha_2}\right]\frac{\gamma W L}{8}$$

Contribution of loads on member 2 to joint 3

$$(X_3^2)_s = 0, (Z_3^2)_s = +\frac{\gamma W}{2}, \ \left(M_3^2\right)_s = +\left[\frac{2(1-\cos\alpha_2)}{\alpha_2 \sin\alpha_2}\right]\frac{\gamma W L}{8}$$

There are no loads on member 3 and hence no contribution from this member to joint 3.

The resultant load vector for the whole frame is

$$F = \begin{bmatrix} F_2 \\ F_3 \end{bmatrix} = \begin{bmatrix} X_2 \\ Z_2 \\ M_2 \\ X_3 \\ Z_3 \\ M_3 \end{bmatrix} = \begin{bmatrix} -\dfrac{\gamma n L}{2} \\ +\dfrac{\gamma W}{2} \\ +\left[\dfrac{3(\tan\alpha_1 - \alpha_1)}{\alpha_1^2 \tan\alpha_1}\right]\dfrac{\gamma n L^2}{12} - \left[\dfrac{2(1-\cos\alpha_2)}{\alpha_2 \sin\alpha_2}\right]\dfrac{\gamma W L}{8} \\ 0 \\ +\dfrac{\gamma W}{2} \\ +\left[\dfrac{2(1-\cos\alpha_2)}{\alpha_2 \sin\alpha_2}\right]\dfrac{\gamma W L}{8} \end{bmatrix} \tag{11.24}$$

where n = –30 kN/m and W = –1500 kN.

For the first load factor increment, let $\gamma = 1$.

Start by setting the axial forces in the members equal to zero, i.e. $P^1 = 0$, $P^2 = 0$, and $P^3 = 0$, thus, $\alpha_1 = 0$, $\alpha_2 = 0$, and $\alpha_3 = 0$. The load vector is calculated from (11.24) as:

$$F = \begin{bmatrix} +90 \\ -750 \\ +1035 \\ 0 \\ -750 \\ -1125 \end{bmatrix}$$, the stiffness matrix K from (11.23) and with $P^1 = 0$, $P^2 = 0$, and $P^3 = 0$, the relationship $K\delta = F$ becomes

$$\begin{bmatrix} 176167 & 0 & -3500 & -175000 & 0 & 0 \\ 0 & 176167 & -3500 & 0 & -1167 & -3500 \\ -3500 & -3500 & 28000 & 0 & 3500 & 7000 \\ -175000 & 0 & 0 & 176167 & 0 & -3500 \\ 0 & -1167 & 3500 & 0 & 176167 & 3500 \\ 0 & -3500 & 7000 & -3500 & 3500 & 28000 \end{bmatrix} \begin{bmatrix} u_2 \\ w_2 \\ \theta_2 \\ u_3 \\ w_3 \\ \theta_3 \end{bmatrix} = \begin{bmatrix} +90 \\ -750 \\ +1035 \\ 0 \\ -750 \\ -1125 \end{bmatrix}$$

The resulting displacement vector is

$$\delta = \begin{bmatrix} u_2 \\ w_2 \\ \theta_2 \\ u_3 \\ w_3 \\ \theta_3 \end{bmatrix} = \begin{bmatrix} +0.050338 \\ -0.004139 \\ +0.055248 \\ +0.049055 \\ -0.004432 \\ -0.047822 \end{bmatrix}$$

These displacements are used to calculate the axial forces in the members from (11.13) as:

$$\begin{bmatrix} \bar{X}_i \\ \bar{X}_j \end{bmatrix} = \begin{bmatrix} \frac{EA}{L} & -\frac{EA}{L} \\ -\frac{EA}{L} & \frac{EA}{L} \end{bmatrix} \begin{bmatrix} \bar{u}_i \\ \bar{u}_j \end{bmatrix}$$

Member 1: $\frac{EA}{L} = 175000$ kN/m, $\bar{u}_i = w_1 = 0$, and $\bar{u}_j = w_2 = -0.004139$ m.

$$\begin{bmatrix} \bar{X}_1 \\ \bar{X}_2 \end{bmatrix} = \begin{bmatrix} 175000 & -175000 \\ -175000 & 175000 \end{bmatrix} \begin{bmatrix} 0 \\ -0.004139 \end{bmatrix} = \begin{bmatrix} +724.3 \\ -724.3 \end{bmatrix},$$

i.e. member 1 is in compression with a force of $P^1 = +724.3$ kN.

Member 2: $\frac{EA}{L} = 175000$ kN/m, $\bar{u}_i = u_2 = +0.050338$ m, and $\bar{u}_j = u_3 = +0.049055$ m.

$$\begin{bmatrix} \bar{X}_2 \\ \bar{X}_3 \end{bmatrix} = \begin{bmatrix} 175000 & -175000 \\ -175000 & 175000 \end{bmatrix} \begin{bmatrix} +0.050338 \\ +0.049055 \end{bmatrix} = \begin{bmatrix} +224.5 \\ -224.5 \end{bmatrix},$$

i.e. member 2 is in compression with a force of $P^2 = +224.5$ kN.

Member 3: $\frac{EA}{L} = 175000$ kN/m, $\bar{u}_i = -w_3 = -(-004432)$ m, and $\bar{u}_j = -w_4 = 0$.

$$\begin{bmatrix} \bar{X}_3 \\ \bar{X}_4 \end{bmatrix} = \begin{bmatrix} 175000 & -175000 \\ -175000 & 175000 \end{bmatrix} \begin{bmatrix} +0.004432 \\ 0 \end{bmatrix} = \begin{bmatrix} +775.6 \\ -775.6 \end{bmatrix},$$

i.e. member 3 is in compression with a force of $P^3 = +775.6$ kN.

The above calculations complete the first cycle of iteration and it is seen that the values of the axial forces at the end of this cycle are different from those assumed at the beginning of the cycle. This means that the displacements are not consistent with the forces and the newly found axial forces are used in the second cycle of iteration.

Calculate new K from (11.23) with the substitution of $P^1 = +724.3$ kN, $P^2 = +224.5$ kN, and $P^3 = +775.6$ kN to get:

$$K = \begin{bmatrix} 176022 & 0 & -3428 & -175000 & 0 & 0 \\ 0 & 176122 & -3478 & 0 & -1122 & -3478 \\ -3428 & -3478 & 27241 & 0 & 3478 & 7045 \\ -175000 & 0 & 0 & 176011 & 0 & -3422 \\ 0 & -1122 & 3478 & 0 & 176122 & 3478 \\ 0 & -3478 & 7045 & -3422 & 3478 & 27200 \end{bmatrix}$$

$$\alpha_1 = 0.0207\sqrt{P^1} = 0.0207\sqrt{724.3} = 0.5571$$

$$\alpha_2 = 0.0207\sqrt{P^2} = 0.0207\sqrt{224.5} = 0.3102$$

The load vector from (11.24) is

$$F = \begin{bmatrix} +90 \\ -750 \\ +1042.2 \\ 0 \\ -750 \\ -1134.1 \end{bmatrix}$$

Kδ = F becomes

$$\begin{bmatrix} 176022 & 0 & -3428 & -175000 & 0 & 0 \\ 0 & 176122 & -3478 & 0 & -1122 & -3478 \\ -3428 & -3478 & 27241 & 0 & 3478 & 7045 \\ -175000 & 0 & 0 & 176011 & 0 & -3422 \\ 0 & -1122 & 3478 & 0 & 176122 & 3478 \\ 0 & -3478 & 7045 & -3422 & 3478 & 27200 \end{bmatrix} \begin{bmatrix} u_2 \\ w_2 \\ \theta_2 \\ u_3 \\ w_3 \\ \theta_3 \end{bmatrix} = \begin{bmatrix} +90 \\ -750 \\ +1042.2 \\ 0 \\ -750 \\ -1134.1 \end{bmatrix}$$

The resulting displacement vector is

$$\delta = \begin{bmatrix} u_2 \\ w_2 \\ \theta_2 \\ u_3 \\ w_3 \\ \theta_3 \end{bmatrix} = \begin{bmatrix} +0.060770 \\ -0.004102 \\ +0.058721 \\ +0.059460 \\ -0.004469 \\ -0.049376 \end{bmatrix}$$

From the above displacements a new set of axial forces in the members is calculated as: P^1 = +717.9 kN, P^2 = +229.3 kN, and P^3 = +782.1 kN and these values are substituted in (11.23) to give a new stiffness matrix

$$K=\begin{bmatrix} 176023 & 0 & -3428 & -175000 & 0 & 0 \\ 0 & 176121 & -3477 & 0 & -1121 & -3477 \\ -3428 & -3477 & 27242 & 0 & 3477 & 7046 \\ -175000 & 0 & 0 & 176011 & 0 & -3422 \\ 0 & -1121 & 3477 & 0 & 176121 & 3477 \\ 0 & -3477 & 7046 & -3422 & 3477 & 27191 \end{bmatrix}$$

$\alpha_1 = 0.0207\sqrt{717.9} = 0.5546$, $\alpha_2 = 0.0207\sqrt{229.3} = 0.3135$ and hence the load vector from (11.24) is

$$F=\begin{bmatrix} +90 \\ -750 \\ +1042.4 \\ 0 \\ -750 \\ -1134.3 \end{bmatrix}$$

$$\begin{bmatrix} 176023 & 0 & -3428 & -175000 & 0 & 0 \\ 0 & 176121 & -3477 & 0 & -1121 & -3477 \\ -3428 & -3477 & 27242 & 0 & 3477 & 7046 \\ -175000 & 0 & 0 & 176011 & 0 & -3422 \\ 0 & -1121 & 3477 & 0 & 176121 & 3477 \\ 0 & -3477 & 7046 & -3422 & 3477 & 27191 \end{bmatrix}\begin{bmatrix} u_2 \\ w_2 \\ \theta_2 \\ u_3 \\ w_3 \\ \theta_3 \end{bmatrix}=\begin{bmatrix} +90 \\ -750 \\ +1042.4 \\ 0 \\ -750 \\ -1134.3 \end{bmatrix}$$

The solution of the above set is

$$\begin{bmatrix} u_2 \\ w_2 \\ \theta_2 \\ u_3 \\ w_3 \\ \theta_3 \end{bmatrix}=\begin{bmatrix} +0.060784 \\ -0.004103 \\ +0.058738 \\ +0.059475 \\ -0.004469 \\ -0.049405 \end{bmatrix}$$

Since the difference between the displacements from two successive cycles (two and three in this case) is very small, the iteration for this load factor increment is stopped. Now increase the load factor γ to 2.0 and use the current values of P^1, P^2, and P^3.

The iteration is carried out as explained above to give the successive displacement vectors and the axial forces in the members as follows:

The current values of $P^1 = +717.9$ kN, $P^2 = +229.3$ kN, $P^3 = +782.1$ kN, $\alpha_1 = 0.5546$ and $\alpha_2 = 0.3135$, and with $\gamma = 2.0$ the load vector from (11.24) is

$$F = \begin{bmatrix} +180 \\ -1500 \\ +2084.8 \\ 0 \\ -1500 \\ -2268.6 \end{bmatrix}$$

$$\begin{bmatrix} 176023 & 0 & -3428 & -175000 & 0 & 0 \\ 0 & 176121 & -3477 & 0 & -1121 & -3477 \\ -3428 & -3477 & 27242 & 0 & 3477 & 7046 \\ -175000 & 0 & 0 & 176011 & 0 & -3422 \\ 0 & -1121 & 3477 & 0 & 176121 & 3477 \\ 0 & -3477 & 7046 & -3422 & 3477 & 27191 \end{bmatrix} \begin{bmatrix} u_2 \\ w_2 \\ \theta_2 \\ u_3 \\ w_3 \\ \theta_3 \end{bmatrix} = \begin{bmatrix} +180 \\ -1500 \\ +2084.8 \\ 0 \\ -1500 \\ -2268.6 \end{bmatrix}$$

$$\begin{bmatrix} u_2 \\ w_2 \\ \theta_2 \\ u_3 \\ w_3 \\ \theta_3 \end{bmatrix} = \begin{bmatrix} +0.121629 \\ -0.008205 \\ +0.117483 \\ +0.119010 \\ -0.008938 \\ -0.098805 \end{bmatrix}$$

From the above displacements calculate $P^1 = +1435.9$ kN, $P^2 = +458.3$ kN, $P^3 = +1564.2$ kN, $\alpha_1 = 0.5546$ and $\alpha_2 = 0.3135$ and these will lead to

$$\begin{bmatrix} 175880 & 0 & -3356 & -175000 & 0 & 0 \\ 0 & 176075 & -3454 & 0 & -1075 & -3454 \\ -3356 & -3454 & 26485 & 0 & 3454 & 7092 \\ -175000 & 0 & 0 & 175854 & 0 & -3344 \\ 0 & -1075 & 3454 & 0 & 176075 & 3454 \\ 0 & -3454 & 7092 & -3344 & 3454 & 26382 \end{bmatrix} \begin{bmatrix} u_2 \\ w_2 \\ \theta_2 \\ u_3 \\ w_3 \\ \theta_3 \end{bmatrix} = \begin{bmatrix} +180 \\ -1500 \\ +2099.7 \\ 0 \\ -1500 \\ -2287.6 \end{bmatrix}$$

$$\begin{bmatrix} u_2 \\ w_2 \\ \theta_2 \\ u_3 \\ w_3 \\ \theta_3 \end{bmatrix} = \begin{bmatrix} +0.153762 \\ -0.008090 \\ +0.126020 \\ +0.151089 \\ -0.009053 \\ -0.101310 \end{bmatrix}$$

Leading to $P^1 = +1415.8$ kN, $P^2 = +467.8$ kN, $P^3 = +1584.3$ kN, $\alpha_1 = 0.7789$ and $\alpha_2 = 0.4477$ and these will lead to

$$\begin{bmatrix} u_2 \\ w_2 \\ \theta_2 \\ u_3 \\ w_3 \\ \theta_3 \end{bmatrix} = \begin{bmatrix} +0.153860 \\ -0.008091 \\ +0.126084 \\ +0.151189 \\ -0.009052 \\ -0.101456 \end{bmatrix}$$

From the above displacements we get $P^1 = +1415.8$ kN, $P^2 = +467.8$ kN, and $P^3 = +1584.1$ kN.

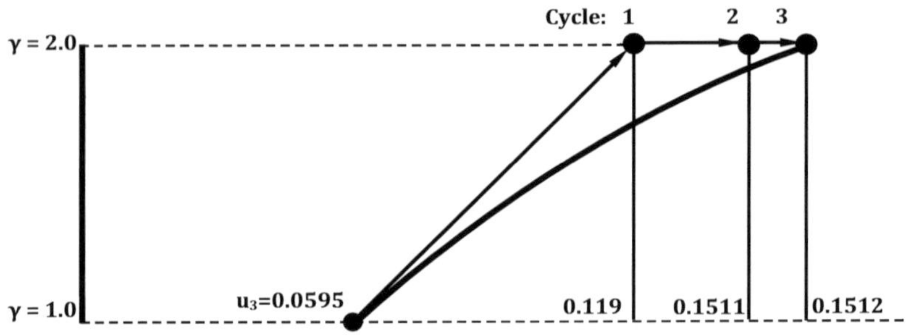

Figure 11.14 Cycles of iteration for convergence within one load increment from $\gamma = 1$ to $\gamma = 2$ (not to scale).

The difference between the last two cycles of iteration is small hence it can be assumed that convergence to the correct result is reached as shown in Fig. 11.14 and the iteration is stopped and a new load factor increment is applied, i.e. with $\gamma = 3$ and so on. The variation of u_3 with the load factor γ is shown in Fig. 11.15.

γ	**Horizontal deflection at node 3, u_3 (m)**
0	0
1	0.0595
2	0.1512
3	0.3110
4	0.6606
5	1.9930

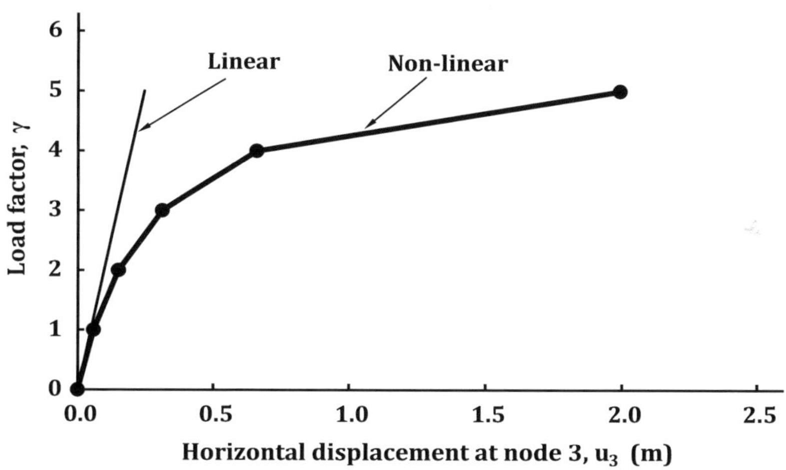

Figure 11.15 Load factor against displacement of frame.

Problems

P11.1 Find the smallest critical load that will cause the strut shown in Fig. P11.1 using the following data:

$E = 9 \times 10^6$ kN/m^2, $I_1 = 0.000019$ m^4, $I_2 = 0.000042$ m^4, and $I_3 = 0.000098$ m^4.

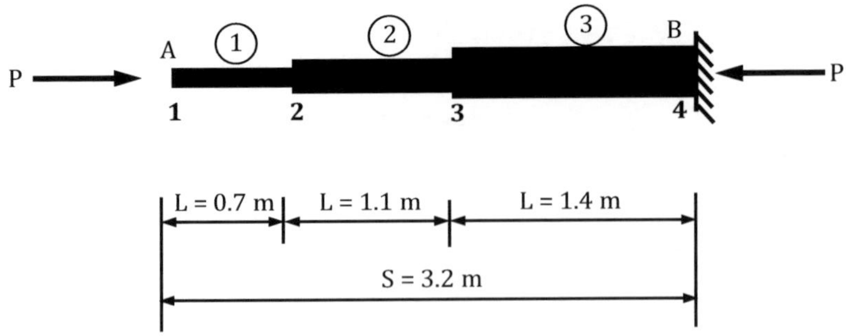

Figure P11.1

Answer:

P = 149.54 kN and the buckling mode is given by: $\begin{bmatrix} w_1 \\ \theta_1 \\ w_2 \\ \theta_2 \\ w_3 \\ \theta_3 \end{bmatrix} = \begin{bmatrix} +1.000 \\ +0.642 \\ +0.582 \\ +0.509 \\ +0.162 \\ +0.224 \end{bmatrix}$

P11.2 Find the smallest critical load that will cause the strut shown in Fig. P11.2 given that E = 210 × 10^6 kN/m^2 and I = 0.000006 m^4.

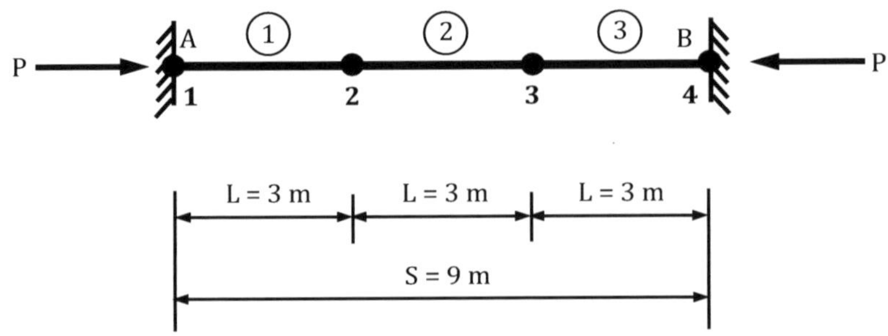

Figure P11.2

Answer:

$$P = 627.56\,\text{kN and the buckling mode is given by:} \begin{bmatrix} w_2 \\ \theta_2 \\ w_3 \\ \theta_3 \end{bmatrix} = \begin{bmatrix} +1.000 \\ -0.398 \\ +1.000 \\ +0.398 \end{bmatrix}$$

P11.3 Repeat Problem P11.2 and plot w_3 against P for the loading shown in Fig. 11.3.

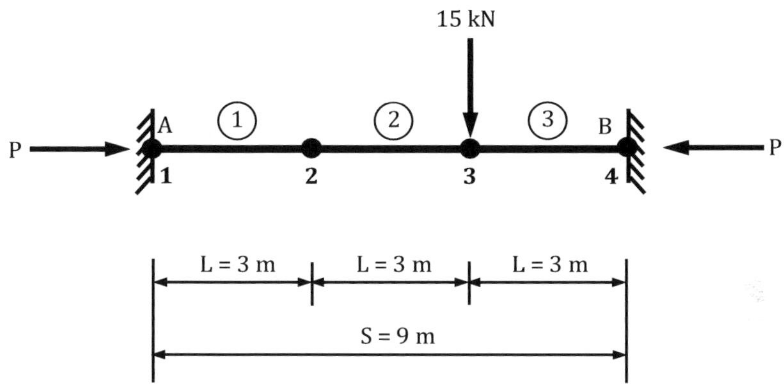

Figure P11.3

Answer:

P (kN)	**w_3 (m)**
0	-0.032
100	-0.037
200	-0.045
300	-0.057
400	-0.080
500	-0.136
600	-0.599

P11.4 Calculate the value of the load Q that will cause instability of the frame shown in Fig. 11.4 for the following data:

$E = 29 \times 10^6$ kN/m^2, $A_1 = 0.03$ m^2, $I_1 = 0.00012$ m^4, $A_2 = 0.09$ m^2, $I_2 = 0.00088$ m^4, $A_3 = 0.02$ m^2, and $I_3 = 0.00006$ m^4.

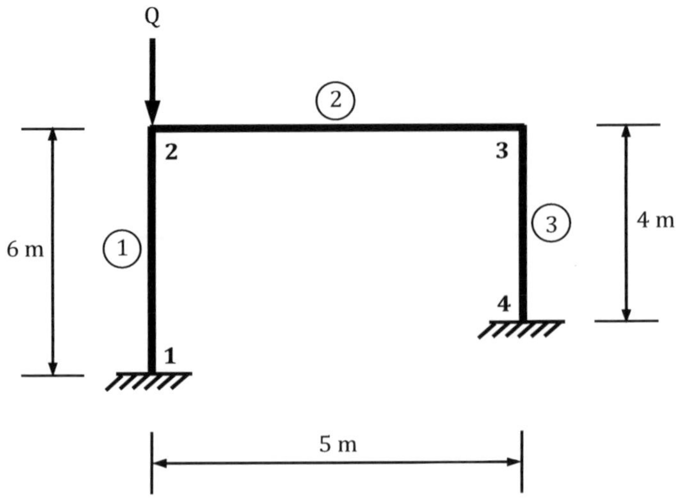

Figure P11.4

Answer:

Q = 2498.47 kN and the buckling mode is given by:

$$\begin{bmatrix} u_2 \\ w_2 \\ \theta_2 \\ u_3 \\ w_3 \\ \theta_3 \end{bmatrix} = \begin{bmatrix} +1.000000 \\ +0.001284 \\ +0.002381 \\ +0.999412 \\ -0.001284 \\ +0.029046 \end{bmatrix}$$

P11.5 Repeat Problem P11.4 and plot γ against u_2 for the loading shown in Fig. 11.5.

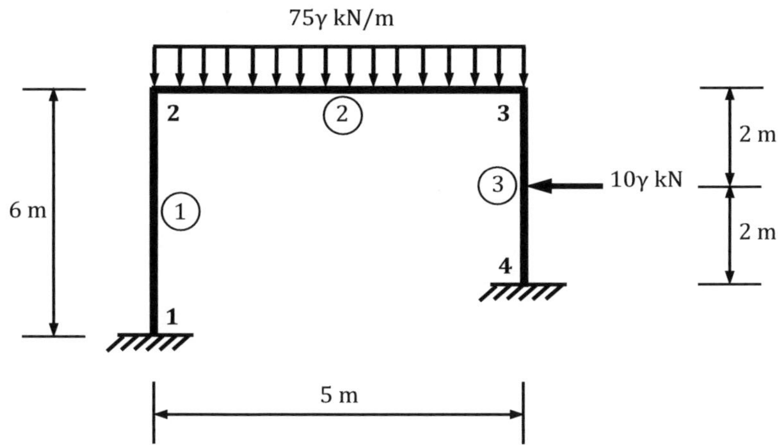

Figure P11.5

Answer:

γ	u_2 (m)
1	−0.014763
2	−0.038294
3	−0.081615
4	−0.186428
5	−0.678656

Chapter 12

Vibration of Beams and Frames

The preceding chapters dealt with the behaviour of structures when subjected to static loads. However, there are instances when the forces acting on the structure are dynamic which means that they are time varying. Examples of dynamic loading on structures include earthquakes, wind, human induced excitation, and dynamic disturbances from machinery. In such circumstances the response of the structure, i.e. its displacements and the developed internal actions in the members of the structure (shear forces, bending moments, etc.) will also be time varying. In order to assess the effect of dynamic loading on a structure, the free undamped vibration characteristics of the structure have to be determined first and this forms the main part of this chapter.

To illustrate the basic principles the case of free undamped vibration of a simple system with a single degree of freedom is considered first. The same principles are then applied in the treatment of vibration of structures with multi-degrees of freedom.

12.1 Systems with a Single Degree of Freedom

12.1.1 Free Undamped Vibration

Consider the single degree of freedom of spring/mass system shown in Fig. 12.1 which shows a horizontal spring of stiffness k with its

Analysis of Structures by Matrix Methods
Fathi Al-Shawi

ISBN 978-981-4968-19-5 (Hardcover), 978-1-003-29130-5 (eBook)
www.jennystanford.com

left end connected to a fixed support and has a mass m attached to its right end and can move on a smooth horizontal plane. The initial condition, i.e. at time t = 0 the system is disturbed by giving the mass a displacement u_o and velocity $\dot{u}_0$. In the absence of damping or an external force, i.e. natural vibration, the mass will oscillate freely indefinitely about the centre of vibration with a maximum displacement called the amplitude.

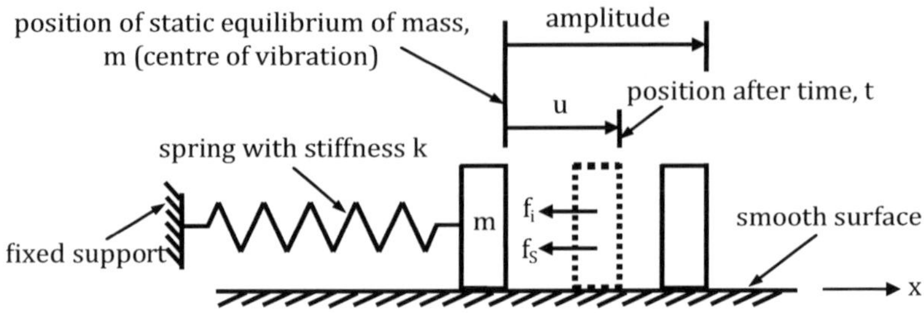

Figure 12.1 Mass with a horizontal spring.

This is a single degree of freedom system since there is only one translational displacement, u, along the x-axis.

At any instant in time t after the commencement of motion the displacement of the mass from the centre of vibration, defined by the position of static equilibrium, is u.

The inertia force = mass × acceleration, $f_i = m\ddot{u}$ which acts in the opposite direction of the acceleration and $\ddot{u}$ is the second derivative of displacement with respect to time, thus

$$f_i = m\frac{d^2u}{dt^2}$$

The tension developed in the spring is f_s = ku where k is the stiffness of the spring.

For dynamic equilibrium, the summation of the forces is zero, i.e. $-f_i - f_s = 0$

$$-m\frac{d^2u}{dt^2} - ku = 0 \tag{12.1}$$

$$\frac{d^2u}{dt^2} + \omega^2 u = 0 \tag{12.2}$$

where $\omega = \sqrt{\dfrac{k}{m}}$ and is called the circular frequency of natural vibration.

The general solution of the above differential equation for the displacement is:

$$u = A\sin\omega t + B\cos\omega t \tag{12.3a}$$

where A and B are constants determined from the initial conditions.

The velocity is the first derivative of the displacement with respect to time, i.e.

$$\dot{u} = \frac{du}{dt} = \omega A\cos\omega t - \omega B\sin\omega t \tag{12.3b}$$

At time $t = 0$, the initial displacement $u = u_o$ and the initial velocity $\dot{u} = \dot{u}_o$

From equation (12.3a), $u_o = A\sin 0 + B\cos 0$, thus $B = u_o$

From equation (12.3b), $u_o = \omega A\cos 0 - \omega B\sin 0$, so $A = \dfrac{\dot{u}_o}{\omega}$

Substitute A and B in equation (12.3a) to get

$$u = \frac{\dot{u}_o}{\omega}\sin\omega t + u_o\cos\omega t$$

The above equation can be written as

$$u = R\sin(\omega t + \eta) \tag{12.4}$$

where

$R = \sqrt{u_o^2 + \dfrac{\dot{u}_o^2}{\omega^2}}$ is the maximum displacement called the amplitude

and $\eta = \tan^{-1}\left(\dfrac{u_o}{\dot{u}_o / \omega}\right)$.

Equation (12.4) represents what is called simple harmonic motion since sine or cosine waves are called harmonic functions.

As an example let m = 2 kg, k = 450 N/m, u_o = 0.04 m, and $\dot{u}_o = 0.75\ \text{m/s}$.

$$\omega = \sqrt{\frac{k}{m}} = \sqrt{\frac{450}{2}} = 15\ \text{rad/s},\ R = \sqrt{u_o^2 + \frac{\dot{u}_o^2}{\omega^2}} = \sqrt{0.04^2 + \frac{0.75^2}{15^2}} = 0.064\ \text{m}$$

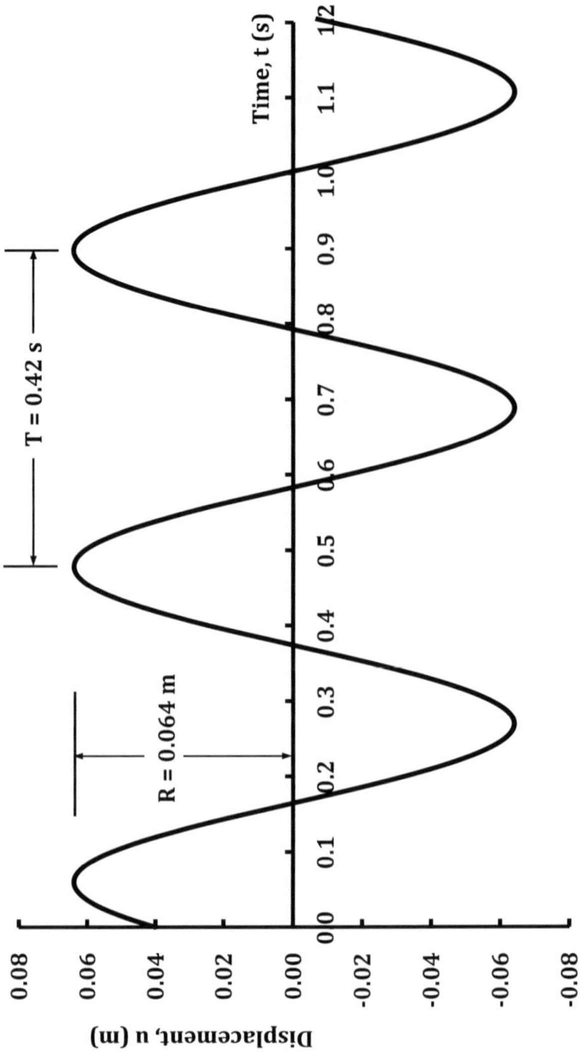

Figure 12.2 Displacement against time.

$$\eta = \tan^{-1}\left(\frac{u_o}{\dot{u}_o / \omega}\right) = \tan^{-1}\left(\frac{0.04}{0.75/15}\right) = 39^o = 0.68 \text{ rad}$$

Substituting the above values in (12.4) to get

$$u = 0.064\sin(15t + 0.68) \tag{12.4a}$$

Figure 12.2 shows a plot of the above equation.

The period of vibration is defined as the time taken for one complete cycle

$$T = \frac{2\pi}{\omega} = \frac{2\pi}{15} = 0.42 \text{ s}$$

The frequency of vibration is the number of cycles per second

$$f = \frac{1}{T} = \frac{\omega}{2\pi} = \frac{15}{2\pi} = 2.4 \text{ cycles/s}$$

If the spring is in the vertical direction the vibration of the mass will essentially be the same as that for the horizontal spring except that the centre of vibration will be the position of static equilibrium defined by an extension of Δ_{static} of the spring, where $\Delta_{static} = mg/k$.

12.1.2 Free Damped Vibration

Consider the single degree of freedom consisting of mass m, spring of stiffness k, and viscous damper with a damping coefficient c as shown in Fig. 12.3.

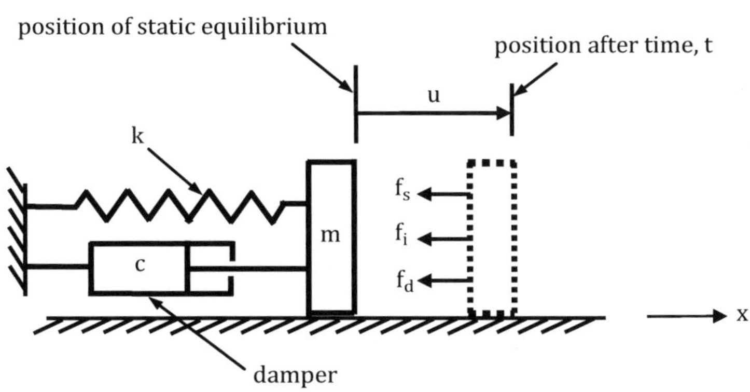

Figure 12.3 Mass with a horizontal spring and damper.

For dynamic equilibrium, the summation of the forces is zero, i.e.

$$-f_i - f_d - f_s = 0$$

where f_d is the force of resistance of the damper which is proportional to the velocity $\dot{u}$ of the piston inside the damper cylinder. Thus, $f_d = c\dot{u}$ where c is the damping coefficient. The equation of equilibrium becomes

$$-m\frac{d^2u}{dt^2} - c\dot{u} - ku = 0 \text{ which can be written as}$$

$$\frac{d^2u}{dt^2} + 2\zeta\omega\frac{du}{dt} + \omega^2u = 0 \qquad (12.5)$$

where $\zeta = \frac{c}{c_c}$ (is the damping factor), $\omega = \sqrt{\frac{k}{m}}$

and $c_c = 2m\omega$ (called the critical damping coefficient).

There are three possible cases of vibration depending on the value of the damping factor as follows:

Case one: when $\zeta < 1$ the vibration is under-damped and the solution of differential equation (12.5) is

$$u = e^{-\zeta\omega t}\left(A\sin\omega_d t + B\cos\omega_d t\right) \qquad (12.6a)$$

where $\omega_d = \omega\sqrt{1-\zeta^2}$.

Notice that for the case of no damping, i.e. $\zeta = 0$ the above equation is reduced to (12.3a).

Case two: when $\zeta = 1$, the vibration is critically damped and the motion will die out in the shortest time with the solution of (12.5) as

$$u = e^{-\omega t}(A + Bt). \qquad (12.6b)$$

Case three: when $\zeta > 1$, the vibration is over-damped and the solution is

$$u = e^{-\zeta\omega t}(Ae^{\omega_o t} + Be^{-\omega_o t}) \qquad (12.6c)$$

where $\omega_o = \omega\sqrt{\zeta^2 - 1}$.

As an example, consider a system consisting of a mass of 4 kg and a spring of stiffness 3600 N/m with an initial displacement $u_o = 0.05$ m and initial velocity $\dot{u}_o = 0$. Plot the equations of motion for the following cases of damping

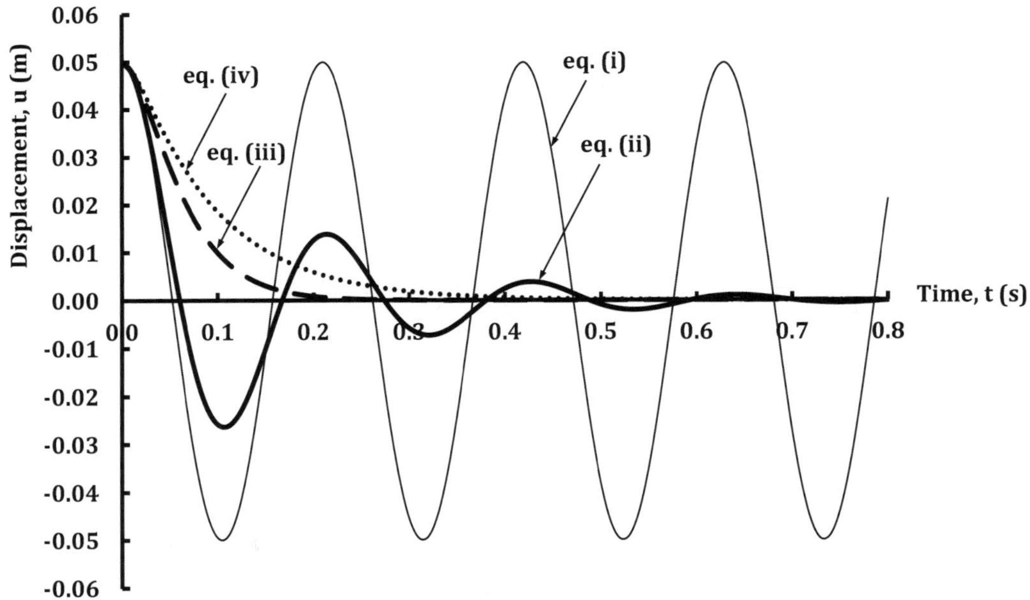

Figure 12.4 Displacement against time for different damping factors.

(i) c = 0, (ii) c = 48 Ns/m, (iii) c = 240 Ns/m, (iv) c = 360 Ns/m.

$$u = 0.05\cos 30t \quad \text{(i)}$$

$$u = e^{-6t}(0.01\sin 29.39t + 0.05\cos 29.39t) \quad \text{(ii)}$$

$$u = e^{-30t}(0.05 + 0.50t) \quad \text{(iii)}$$

$$u = e^{-45t}(0.05854e^{+33.54t} - 0.00854e^{-33.54t}) \quad \text{(iv)}$$

It can be seen from Fig. 12.4 that when there is no damping (Eq. (i)) the vibration continues indefinitely with constant amplitude while for 0.20 damping factor (under-damped) the vibration decays with decreasing amplitude until it dies out after about 0.8 seconds as shown by Eq. (ii). For the case of critical damping the vibration dies out without oscillations and in the shortest time which is about 0.3 seconds as indicated by Eq. (iii). When over-damping is applied as shown by Eq. (iv) the time taken for the vibration to die out is longer than that of the critically damped case and is about 0.5 seconds.

12.1.3 Forced Vibration Due to Harmonic Force Excitation

In practice it is often required to investigate the behaviour of a system when subjected to an external action that acts for a certain length of time or indefinitely and hence the name forced vibration. Let us consider first the case of a single degree of freedom system subjected to a harmonic exciting force $p_o \sin\Omega t$ as shown in Fig. 12.5. This will be used later as the bases for the treatment of multi-degrees of freedom systems.

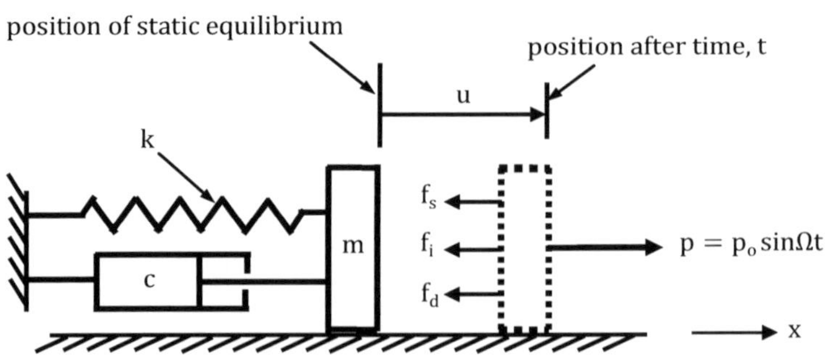

Figure 12.5 Forced vibration.

From dynamic equilibrium, $-f_i - f_d - f_s + p_o \sin\Omega t = 0$ and the differential equation is

$$-m\frac{d^2u}{dt^2} - c\frac{du}{dt} - ku + p_o \sin\Omega t = 0, \quad \text{or}$$

$$\frac{d^2u}{dt^2} + 2\xi\omega\frac{du}{dt} + \omega^2 u = \frac{p_o}{m}\sin\Omega t.$$

The solution of the above differential equation is

$$u = u_c + u_p$$

where u_c is the complementary function = $e^{-\omega t}(A\sin\omega_d t + B\cos\omega_d t)$

and u_p is the particular integral $= \dfrac{p_o[(1-\beta^2)\sin\Omega t - 2\xi\beta\cos\Omega t]}{m\omega^2[(1-\beta^2)^2 + (2\xi\beta)^2]}$,

therefore

$$u = e^{-\xi\omega t}(A\sin\omega_d t + B\cos\omega_d t) + \frac{\Delta_o}{\sqrt{[(1-\beta^2)^2 + (2\xi\beta)^2]}}\sin(\Omega t - \varphi),$$

where the frequency ration $\beta = \dfrac{\Omega}{\omega}$, the equivalent static displacement $\Delta_o = \dfrac{p_o}{k} = \dfrac{p_o}{m\omega^2}$, and the phase angle φ is given by $\tan\varphi = \dfrac{2\xi\beta}{1-\beta^2}$.

The constants A and B are determined from the initial conditions.

The first part of the solution which is given by the complementary function u_C will die out after a relatively short time being an exponential decay function. The second part represents the steady state response of the system due to the exciting force and has an amplitude, i.e. maximum displacement given by

$$u_{max} = \frac{\Delta_o}{\sqrt{[(1-\beta^2)^2 + (2\xi\beta)^2]}}$$

The dynamic amplification factor D is defined as the ratio of the maximum displacement u_{max} to the equivalent static displacement Δ_o, thus

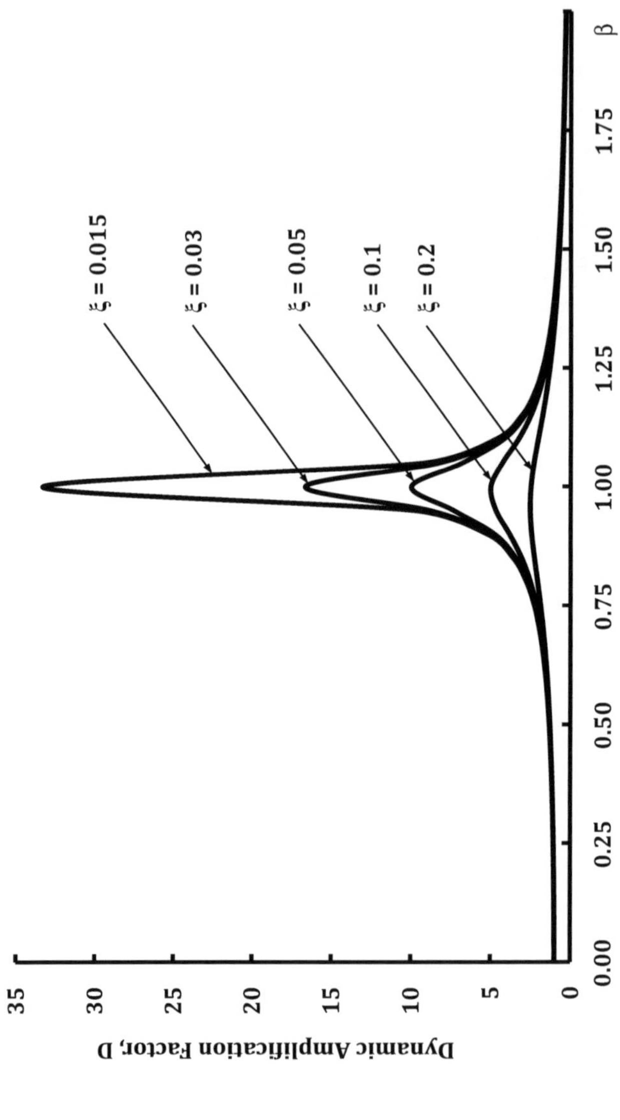

Figure 12.6 Dynamic amplification factor against frequency ratio.

$$D = \frac{u_{max}}{\Delta_o} = \frac{1}{\sqrt{[(1-\beta^2)^2 + (2\xi\beta)^2]}}$$

A plot of the above equation is shown in Fig. 12.6.

12.1.4 Forced Vibration Due to Base Motion Excitation

Consider the system of mass and spring shown in Fig. 12.7 excited by the motion of the support A. Let the displacement of the support after time t is u_B. The mass m moves by a distance u_R relative to B which is the new position of the support after time t. Therefore, the absolute displacement of the mass m relative to the original position A is $u = u_R + u_B$. Assume that the support motion is harmonic, i.e. $u_B = U\sin\sigma t$, where U and σ are the amplitude and circular frequency of the support motion, respectively.

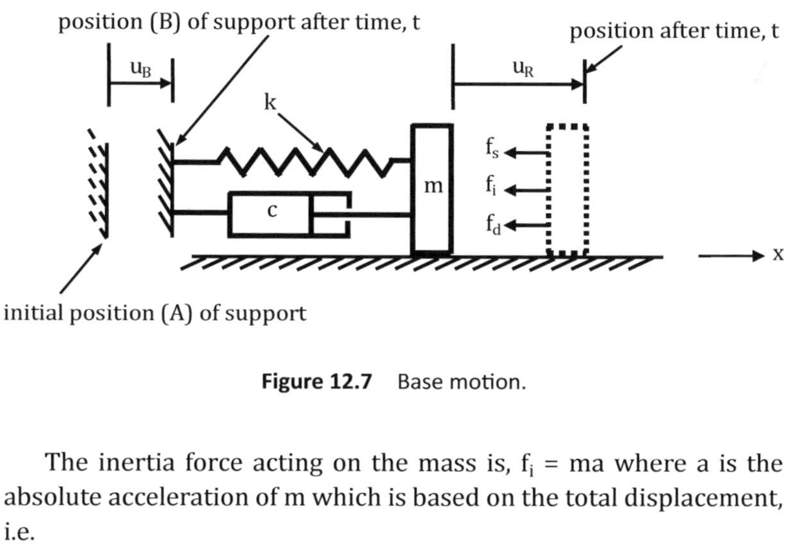

Figure 12.7 Base motion.

The inertia force acting on the mass is, f_i = ma where a is the absolute acceleration of m which is based on the total displacement, i.e.

$$a = \frac{d^2u}{dt^2} = \frac{d^2(u_B + u_R)}{dt^2} = \frac{d^2u_B}{dt^2} + \frac{d^2u_R}{dt^2} = -U\sigma^2\sin\sigma t + \frac{d^2u_R}{dt^2}$$

The spring force is based on the extension of the spring relative to the new position of the support at B, thus $f_s = ku_R$ and also the damping force $f_d = c\dot{u}_R$.

For dynamic equilibrium, the summation of the forces is zero, i.e. $-f_i - f_d - f_s = 0$

$$-m\left(-U\sigma^2\sin\sigma t+\frac{d^2u_R}{dt^2}\right)-c\frac{du_R}{dt}-ku_R=0$$

$$\frac{d^2u_R}{dt^2}+2\xi\omega\frac{du_R}{dt}+\omega^2u_R=U\sigma^2\sin\sigma t,\quad \text{where } \omega^2=\frac{k}{m}.$$

It can be seen that the above differential equation is similar to that of forced vibration, therefore, the rest of the analysis will follow in a similar manner. This procedure can be adopted for the analysis of multi-degrees of freedom such as the case of multi-storey buildings subjected to base motion excitation resulting from earthquakes. In this case the base motion is random and the treatment becomes more complex which is outside the scope of this book, but the above presentation gives the reader an introduction and a flavour of the subject matter.

12.2 Systems with Multi-degrees of Freedom

In the following sections of this chapter, damping is not considered and for damped systems the reader can refer to specialised textbooks on the subject of vibration of structures.

In using matrix methods for the vibration of structures where there are many degrees of freedom an alternative form of the governing equation may make the analysis more practical as shown below.

Equation 12.4, the displacement is given by

$$u=R\sin(\omega t+\eta), \text{ hence the second derivative } \frac{d^2u}{dt^2} \text{ is}$$

$$\ddot{u}=-\omega^2R\sin(\omega t+\varphi)=-\omega^2u$$

and this is substituted in equation 12.1 to give: $-\omega^2mu + ku = 0$, or

$$(k-\omega^2m)u = 0 \tag{12.5}$$

The above equation is for a system with a single degree of freedom and for a structure with multi-degrees of freedom it will take the following form:

$$(K - \omega^2 M)\delta = 0 \tag{12.6}$$

where K is the overall structure stiffness matrix as calculated in the previous chapters, M is the mass matrix of the structure, and δ is the column vector of the displacements at the nodes of the structure.

The first term in (12.6) represents the stiffness force, $F_{stiffness} = K\delta$, the second term is the inertia force, $F_{inertia} = \omega^2 M\delta$ and ω is the natural circular frequency of vibration of the structure. Equation (12.6) is a form of eigenvalue problem and its solution gives as many values for ω^2 as the number of degrees of freedom of the structure. For each eigenvalue there is an associated eigenvector δ called mode shape, which represents the relative amplitude of displacements at the various points in the structure.

The trivial solution of equation (12.6) is δ = 0, but the structure is vibrating which means that the displacement vector δ is not zero. The condition for δ to have a non-zero value is that the determinant of the quantity inside the brackets of equation (12.6) must be equal to zero and this will lead to the required eigenvalues, i.e.

$$|K - \omega^2 M| = 0. \tag{12.7}$$

The above equation gives the values of the circular frequency of vibration, ω, and equation (12.6) is used to determine the mode shapes of vibration as explained in the examples that follow.

When dealing with large matrices there are other techniques for determining the eigenvalues which are more efficient than the determinant method and these techniques are available in most relevant software packages.

12.3 Mass Matrix

The mass of a structure is continuously distributed throughout its members and this will lead to an infinite number of degrees of freedom. Therefore, the structure is divided into a number of elements (or members) with a finite number of degrees of freedom which will lead to a finite number of the values of ω^2. The next step is to translate the distributed mass of each element into 'equivalent' masses that are assumed to be concentrated at the end nodes of that element. To do this, one of the two possible approaches can be used, namely, the lumped mass and the consistent mass methods.

In the lumped mass matrix method, the element is assumed to have no rotational inertia and for elements with uniform cross section the mass is divided equally between the two end nodes as shown below with L = length of element (or member), A = cross-sectional area, and ρ = material density.

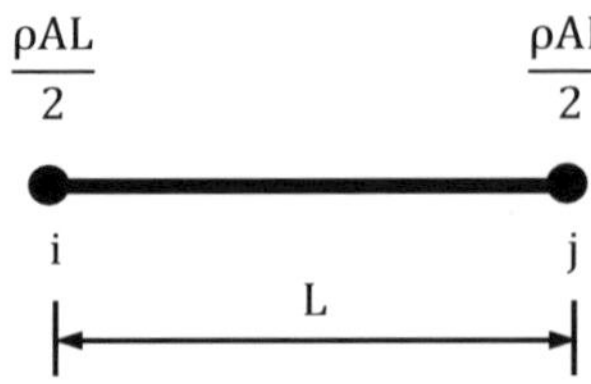

This is the simplest form of mass matrix that leads to a diagonal matrix which requires less storage space and, more importantly, less computer time compared with a more populated matrix.

The consistent mass matrix is derived from the same interpolation polynomial for the displacement as used in the derivation of the stiffness matrix hence the name consistent. This method leads to a more populated matrix than the lumped mass matrix, consequently, it requires more storage and computer time, but it is generally accepted that it gives better accuracy. In this chapter the lumped mass matrix is used for its simplicity, but the general procedure of analysis is the same in both methods.

12.4 Matrix Condensation

When dealing with large sets of simultaneous equations, economy can be achieved in obtaining a solution if the number of degrees of freedom is reduced. For example, consider the problem of bending of beams and using the lumped mass method where the rotational displacements are not included in the mass matrix, M, but they are present in the stiffness matrix, K. In this case the rotational displacements are regarded as unwanted and can be eliminated from the stiffness matrix by the so-called Guyan static condensation method as explained below.

The structure displacement vector δ in equation (12.6) is divided into two groups; the translational displacements vector δ_w and the second group of degrees of freedom is the vector of rotational displacements δ_θ to give

$$\delta = \begin{bmatrix} \delta_w \\ \delta_\theta \end{bmatrix}.$$

The next step is to write the matrices in equation (12.6) in a partitioned matrix form in terms of submatrices as

$$\begin{bmatrix} K_{ww} & K_{w\theta} \\ K_{\theta w} & K_{\theta\theta} \end{bmatrix} \begin{bmatrix} \delta_w \\ \delta_\theta \end{bmatrix} - \omega^2 \begin{bmatrix} M_{ww} & M_{w\theta} \\ M_{\theta w} & M_{\theta\theta} \end{bmatrix} \begin{bmatrix} \delta_w \\ \delta_\theta \end{bmatrix} = \begin{bmatrix} 0 \\ 0 \end{bmatrix}$$

For the simple case of lumped masses, the mass matrix is a diagonal matrix whose coefficients are associated with the translational displacements, δ_w only, therefore, $M_{w\theta} = M_{\theta w} = M_{\theta\theta} = 0$, thus

$$\begin{bmatrix} K_{ww} & K_{w\theta} \\ K_{\theta w} & K_{\theta\theta} \end{bmatrix} \begin{bmatrix} \delta_w \\ \delta_\theta \end{bmatrix} - \omega^2 \begin{bmatrix} M_{ww} & 0 \\ 0 & 0 \end{bmatrix} \begin{bmatrix} \delta_w \\ \delta_\theta \end{bmatrix} = \begin{bmatrix} 0 \\ 0 \end{bmatrix}.$$

The second equation in the above matrix is $K_{\theta w}\delta_w + K_{\theta\theta}\,\delta_\theta = 0$, pre-multiply by $K_{\theta\theta}^{-1}$ to give

$\delta_\theta = -K_{\theta\theta}^{-1}K_{\theta w}\delta_w$ and this is substituted in the first equation to give

$K_{ww}\delta_w + K_{w\theta}\left(-K_{\theta\theta}^{-1}K_{\theta w}\delta_w\right) - \omega^2 M_{ww}\delta_w = 0$, which can be written as

$$(K_c - \omega^2 M_c)\delta_a = 0$$

where the condensed stiffness matrix $K_c = K_{ww} - K_{w\theta}K_{\theta\theta}^{-1}K_{\theta w}$ and the condensed mass matrix $M_c = M_{ww}$.

It should be noted that in the case of lumped mass matrix the results obtained for the eigenvalues and eigenvectors from the condensed matrices are exactly the same as those obtained from the full uncondensed matrices.

12.5 Free Vibration of Pin-Connected Plane Frames

The lumped mass at each end of the element has components of acceleration in both the x and z directions and hence the use of the associated displacements in these directions and the mass matrix relative to local coordinates is

$$\bar{m} = \overbrace{\overbrace{\bar{u}_i \quad \bar{w}_i}^{\bar{\delta}_i} \quad \overbrace{\bar{u}_j \quad \bar{w}_j}^{\bar{\delta}_j}}^{\bar{\delta}} \begin{bmatrix} \rho AL/2 & 0 & 0 & 0 \\ 0 & \rho AL/2 & 0 & 0 \\ 0 & 0 & \rho AL/2 & 0 \\ 0 & 0 & 0 & \rho AL/2 \end{bmatrix} \begin{matrix} \bar{u}_i \\ \bar{w}_i \\ \bar{u}_j \\ \bar{w}_j \end{matrix} = \rho AL/2 \begin{bmatrix} 1 & 0 & 0 & 0 \\ 0 & 1 & 0 & 0 \\ 0 & 0 & 1 & 0 \\ 0 & 0 & 0 & 1 \end{bmatrix}$$

Since the unit matrix is unchanged when it is transformed from local to global coordinates it follows that the mass matrix relative to global coordinates, $m = \bar{m}$, hence

$$m = \frac{\rho AL}{2} \overbrace{\overbrace{u_i \quad w_i}^{\delta_i} \quad \overbrace{u_j \quad w_j}^{\delta_j}}^{\delta} \begin{bmatrix} 1 & 0 & 0 & 0 \\ 0 & 1 & 0 & 0 \\ 0 & 0 & 1 & 0 \\ 0 & 0 & 0 & 1 \end{bmatrix} \begin{matrix} u_i \\ w_i \\ u_j \\ w_j \end{matrix} \tag{12.18}$$

Example 1:

Calculate the natural frequencies and the corresponding modes of vibration of the pin-connected plane frame shown in Fig. 12.8 for the following data:

$E = 70 \times 10^9$ N/m^2, $\rho = 2500$ kg/m^3, $A_1 = 0.0024$ m^2, and $A_2 = 0.0018$ m^2.

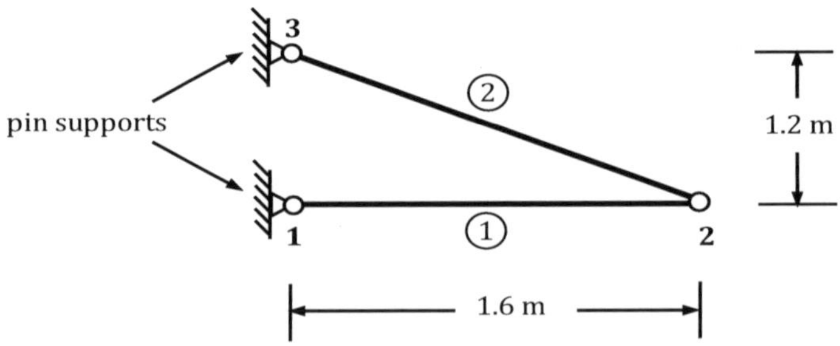

Figure 12.8

For Member 1

$$m^1 = \frac{2500 \times 0.0024 \times 1.6}{2} \begin{bmatrix} 1 & 0 & 0 & 0 \\ 0 & 1 & 0 & 0 \\ 0 & 0 & 1 & 0 \\ 0 & 0 & 0 & 1 \end{bmatrix} = \begin{bmatrix} 4.8 & 0 & 0 & 0 \\ 0 & 4.8 & 0 & 0 \\ 0 & 0 & 4.8 & 0 \\ 0 & 0 & 0 & 4.8 \end{bmatrix}$$

For Member 2

$$m^2 = \frac{2500 \times 0.0018 \times 2.0}{2} \begin{bmatrix} 1 & 0 & 0 & 0 \\ 0 & 1 & 0 & 0 \\ 0 & 0 & 1 & 0 \\ 0 & 0 & 0 & 1 \end{bmatrix} = \begin{bmatrix} 4.5 & 0 & 0 & 0 \\ 0 & 4.5 & 0 & 0 \\ 0 & 0 & 4.5 & 0 \\ 0 & 0 & 0 & 4.5 \end{bmatrix}$$

The structure mass matrix will be

$$M = \begin{array}{c} \begin{array}{cccccc} u_1 & w_1 & u_2 & w_2 & u_3 & w_3 \end{array} \\ \begin{bmatrix} 4.8 & 0 & 0 & 0 & 0 & 0 \\ 0 & 4.8 & 0 & 0 & 0 & 0 \\ 0 & 0 & 9.3 & 0 & 0 & 0 \\ 0 & 0 & 0 & 9.3 & 0 & 0 \\ 0 & 0 & 0 & 0 & 4.5 & 0 \\ 0 & 0 & 0 & 0 & 0 & 4.5 \end{bmatrix} \end{array} \begin{array}{c} u_1 \\ w_1 \\ u_2 \\ w_2 \\ u_3 \\ w_3 \end{array}$$

Apply the boundary conditions of $u_1 = 0$, $w_1 = 0$, $u_3 = 0$, and $w_3 = 0$ to get

$$M = \begin{array}{c} \begin{array}{cc} u_2 & w_2 \end{array} \\ \begin{bmatrix} 9.3 & 0 \\ 0 & 9.3 \end{bmatrix} \end{array} \begin{array}{c} u_2 \\ w_2 \end{array} \tag{12.19}$$

The stiffness matrix for a pin-connected frame member relative to global coordinates is given in Chapter 3 (3.10) as

$$k = \begin{bmatrix} \frac{EAx_{ij}^2}{L^3} & \frac{EAx_{ij}z_{ij}}{L^3} & -\frac{EAx_{ij}^2}{L^3} & -\frac{EAx_{ij}z_{ij}}{L^3} \\ \frac{EAx_{ij}z_{ij}}{L^3} & \frac{EAz_{ij}^2}{L^3} & -\frac{EAx_{ij}z_{ij}}{L^3} & -\frac{EAz_{ij}^2}{L^3} \\ -\frac{EAx_{ij}^2}{L^3} & -\frac{EAx_{ij}z_{ij}}{L^3} & \frac{EAx_{ij}^2}{L^3} & \frac{EAx_{ij}z_{ij}}{L^3} \\ -\frac{EAx_{ij}z_{ij}}{L^3} & -\frac{EAz_{ij}^2}{L^3} & \frac{EAx_{ij}z_{ij}}{L^3} & \frac{EAz_{ij}^2}{L^3} \end{bmatrix}$$

Member 1, (i,j:1,2)

$x_{ij} = x_j - x_i = 1.6 - 0 = 1.6$ m, $z_{ij} = z_j - z_i = 0 - 0 = 0$

$$L = \sqrt{x_{ij}^2 + z_{ij}^2} = \sqrt{1.6^2 + 0^2} = 1.6 \text{ m}$$

$$k^1 = 10^6 \begin{bmatrix} 105 & 0 & -105 & 0 \\ 0 & 0 & 0 & 0 \\ -105 & 0 & 105 & 0 \\ 0 & 0 & 0 & 0 \end{bmatrix}$$

Member 2, (i,j:2,3)

$x_{ij} = x_j - x_i = 0 - 1.6 = -1.6$ m, $z_{ij} = z_j - z_i = 1.2 - 0 = 1.2$ m

$$L = \sqrt{x_{ij}^2 + z_{ij}^2} = \sqrt{(-1.6)^2 + (1.2)^2} = 2.0 \text{ m}$$

$$k^2 = 10^6 \begin{bmatrix} +40.32 & -30.24 & -40.32 & +30.24 \\ -30.24 & +22.68 & +30.24 & -22.68 \\ -40.32 & +30.24 & +40.32 & -30.24 \\ +30.24 & -22.68 & -30.24 & +22.68 \end{bmatrix}$$

The overall structure stiffness matrix

$$K = 10^6 \begin{array}{c} \begin{array}{cccccc} u_1 & w_1 & u_2 & w_2 & u_3 & w_3 \end{array} \\ \begin{bmatrix} 105 & 0 & -105 & 0 & 0 & 0 \\ 0 & 0 & 0 & 0 & 0 & 0 \\ -105 & 0 & +145.32 & -30.24 & -40.32 & +30.24 \\ 0 & 0 & -30.24 & +22.68 & +30.24 & -22.68 \\ 0 & 0 & -40.32 & +30.24 & +40.32 & -30.24 \\ 0 & 0 & +30.24 & -22.68 & -30.24 & +22.68 \end{bmatrix} \end{array} \begin{array}{c} \\ u_1 \\ w_1 \\ u_2 \\ w_2 \\ u_3 \\ w_3 \end{array}$$

Apply the boundary conditions of $u_1 = 0$, $w_1 = 0$, $u_3 = 0$, and $w_3 = 0$ to get

$$K = 10^6 \begin{array}{c} \begin{array}{cc} u_2 & w_2 \end{array} \\ \begin{bmatrix} +145.32 & -30.24 \\ -30.24 & +22.68 \end{bmatrix} \end{array} \begin{array}{c} \\ u_2 \\ w_2 \end{array} \tag{12.20}$$

Substitute M and K from (12.19) and (12.20) respectively in equation (12.6) to get

$$\left[10^6 \begin{bmatrix} +145.32 & -30.24 \\ -30.24 & +22.68 \end{bmatrix} - \omega^2 \begin{bmatrix} 9.3 & 0 \\ 0 & 9.3 \end{bmatrix}\right] \begin{bmatrix} u_2 \\ w_2 \end{bmatrix} = \begin{bmatrix} 0 \\ 0 \end{bmatrix} \tag{12.21}$$

The trivial solution of the above set of simultaneous equations is $u_2 = 0$ and $w_2 = 0$, i.e. the displacement vector $\delta = 0$, i.e. no vibration. The condition that δ has a non-zero value is that the determinant of the quantity inside the brackets must be equal to zero and this will lead to the required eigenvalues. When dealing with large matrices there are other more efficient eigenvalue techniques that can be implemented on the computer and the reader is referred to specialist literature about the subject for details. However, the determinant method is used in this case because of its simplicity for hand calculations and the fact that the matrix is small, thus

$$\begin{vmatrix} +145.32 \times 10^6 - 9.3\lambda & -30.24 \times 10^6 \\ -30.24 \times 10^6 & +22.68 \times 10^6 - 9.3\lambda \end{vmatrix} = 0, \quad \text{where } \lambda = \omega^2$$

$$86.49\lambda^2 - 1562.40 \times 10^6\lambda + 2381.40 \times 10^{12} = 0$$

The above relationship is called the characteristic equation whose roots are the eigenvalues of the matrix, thus

$$\lambda_1 = 1.68 \times 10^6 \text{ and } \lambda_2 = 16.38 \times 10^6.$$

The circular frequency of vibration, $\omega = \sqrt{\lambda}$

$$\omega_1 = \sqrt{\lambda_1} = \sqrt{1.68 \times 10^6} = 1296.15 \text{ rad/s and}$$

$$\omega_2 = \sqrt{16.38 \times 10^6} = 4047.22 \text{ rad/s}$$

The frequency of vibration, $f = \dfrac{\omega}{2\pi}$

$$f_1 = \frac{\omega_1}{2\pi} = \frac{1296.15}{2\pi} = 206.29 \text{ cycles/s and } f_2 = \frac{4047.22}{2\pi} = 644.13 \text{ cycles/s}$$

Calculation of eigenvectors (mode shapes)

During vibration the displacement at each point along the member is described by the harmonic function u = Rsin(ωt + η) as given by equation (12.4) where R is the amplitude, i.e. the maximum value of the displacement. The mode shape represents the relative amplitudes at the various points and is given by the eigenvector of the matrix for a particular value of ω. Since the magnitude of the amplitude at any point is arbitrary the eigenvector is normalised by making the magnitude of the largest value equal to 1.0.

The shape of the truss for any mode of vibration is given by the eigenvector for that mode and is obtained from (12.21) as follows:

Let u_2 = +1.000 m arbitrarily to get

$$\left[10^6\begin{bmatrix}145.32 & -30.24\\ -30.24 & 22.68\end{bmatrix} - \omega^2\begin{bmatrix}9.3 & 0\\ 0 & 9.3\end{bmatrix}\right]\begin{bmatrix}+1.000\\ w_2\end{bmatrix} = \begin{bmatrix}0\\ 0\end{bmatrix}$$

For mode 1, ω = ω_1 = 1296.15 rad/s

$$\begin{bmatrix}129.70\times 10^6 & -30.24\times 10^6\\ -30.24\times 10^6 & 7.06\times 10^6\end{bmatrix}\begin{bmatrix}+1.000\\ w_2\end{bmatrix} = \begin{bmatrix}0\\ 0\end{bmatrix}$$

The second row of the above two equations give w_2 = +4.283 m. The mode shape is given by the eigenvector

$$\begin{bmatrix}u_2\\ w_2\end{bmatrix} = \begin{bmatrix}+1.000\\ +4.283\end{bmatrix}$$ which is normalised by dividing by the magnitude of the largest coefficient 4.283 to give mode 1 as

$$\psi_1 = \begin{bmatrix}u_2\\ w_2\end{bmatrix} = \begin{bmatrix}+0.233\\ +1.000\end{bmatrix}$$

Similarly, for mode 2 where, ω = ω_2 = 4047.22 rad/s the second normalised mode is calculated as

$$\psi_2 = \begin{bmatrix} u_2 \\ w_2 \end{bmatrix} = \begin{bmatrix} +1.000 \\ -0.233 \end{bmatrix}.$$

The mode shapes for the pin-connected frame are shown in Fig. 12.9.

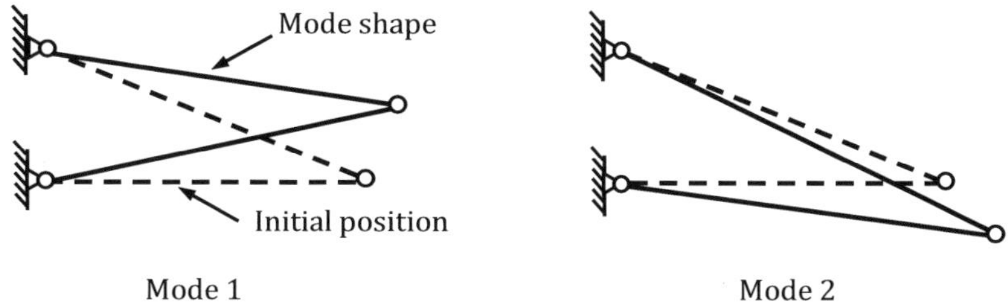

Figure 12.9 Mode shapes for the pin-connected frame.

12.6 Vibration of Beams

In the example below, the lumped mass matrix is used to determine the frequencies and modes of free undamped vibration of a fixed ended beam. These will be used later to investigate the behaviour of the beam under the action of an external force.

12.6.1 Free Vibration of Beams

Example 2:

Calculate the natural frequencies and the corresponding modes of vibration of the fixed ended beam shown in Fig. 12.10 using the following data: ρ = 2400 kg/m^3, A = 0.15 m^2, I = 0.0048 m^4, and E = 36 × 10^9 N/m^2.

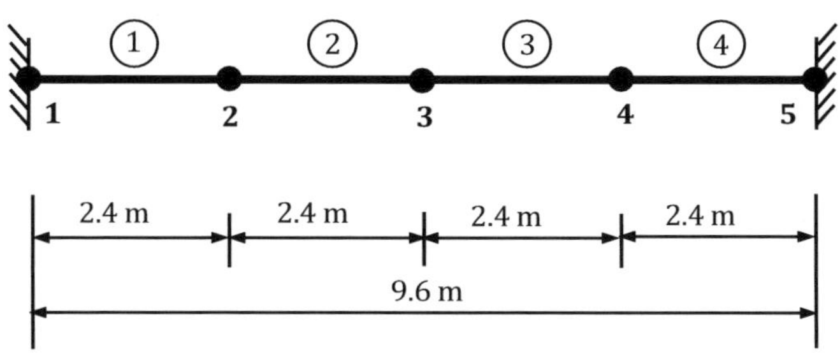

Figure 12.10 Fixed ended beam.

Calculation of the stiffness matrix

The stiffness matrix of a beam element whose local axis is coincident with the global x-axis was derived in Chapter 4 as given by (4.28) as:

$$k=\begin{bmatrix} \frac{12EI}{L^3} & -\frac{6EI}{L^2} & -\frac{12EI}{L^3} & -\frac{6EI}{L^2} \\ -\frac{6EI}{L^2} & \frac{4EI}{L} & \frac{6EI}{L^2} & \frac{2EI}{L} \\ -\frac{12EI}{L^3} & \frac{6EI}{L^2} & \frac{12EI}{L^3} & \frac{6EI}{L^2} \\ -\frac{6EI}{L^2} & \frac{2EI}{L} & \frac{6EI}{L^2} & \frac{4EI}{L} \end{bmatrix}$$

$$k^1=k^2=k^3=10^6\begin{bmatrix} 150 & -180 & -150 & -180 \\ -180 & 288 & 180 & 144 \\ -150 & 180 & 150 & 180 \\ -180 & 144 & 180 & 288 \end{bmatrix}$$

The stiffness matrix for the overall structure:

$K = 10^6$

w_1	θ_1	w_2	θ_2	w_3	θ_3	w_4	θ_4	w_5	θ_5	
150	-180	-150	-180	0	0	0	0	0	0	w_1
-180	288	180	144	0	0	0	0	0	0	θ_1
-150	180	300	0	-150	-180	0	0	0	0	w_2
-180	144	0	576	180	144	0	0	0	0	θ_2
0	0	-150	180	300	0	-150	-180	0	0	w_3
0	0	-180	144	0	576	180	144	0	0	θ_3
0	0	0	0	-150	180	300	0	-150	-180	w_4
0	0	0	0	-180	144	0	576	180	144	θ_4
0	0	0	0	0	0	-150	180	150	180	w_5
0	0	0	0	0	0	-180	144	180	288	θ_5

The boundary conditions of the fixed ends are $w_1 = \theta_1 = 0$ and $w_5 = \theta_5 = 0$, hence delete rows and columns 1, 2, 9, and 10 to get the reduced stiffness matrix as:

$K = 10^6$

w_2	θ_2	w_3	θ_3	w_4	θ_4	
300	0	−150	−180	0	0	w_2
0	576	180	144	0	0	θ_2
−150	180	300	0	−150	−180	w_3
−180	144	0	576	180	144	θ_3
0	0	−150	180	300	0	w_4
0	0	−180	144	0	576	θ_4

Condensation of the stiffness matrix K

The above matrix is condensed as explained earlier by first rearranging the coefficients in order to separate the w's from the θ's as shown below.

$K = 10^6$

w_2	w_3	w_4	θ_2	θ_3	θ_4	
300	−150	0	0	−180	0	w_2
−150	300	−150	180	0	−180	w_3
0	−150	300	0	180	0	w_4
0	180	0	576	144	0	θ_2
−180	0	180	144	576	144	θ_3
0	−180	0	0	144	576	θ_4

$$k_{ww} = 10^6 \begin{bmatrix} 300 & -150 & 0 \\ -150 & 300 & -150 \\ 0 & -150 & 300 \end{bmatrix}, k_{w\theta} = 10^6 \begin{bmatrix} 0 & -180 & 0 \\ 180 & 0 & -180 \\ 0 & 180 & 0 \end{bmatrix},$$

$$k_{\theta w} = 10^6 \begin{bmatrix} 0 & 180 & 0 \\ -180 & 0 & 180 \\ 0 & -180 & 0 \end{bmatrix}, \; k_{\theta\theta} = 10^6 \begin{bmatrix} 576 & 144 & 0 \\ 144 & 576 & 144 \\ 0 & 144 & 576 \end{bmatrix}$$

$$K_c = K_{ww} - K_{w\theta} K_{\theta\theta}^{-1} K_{\theta w} = 10^6 \begin{bmatrix} 235.714 & -150 & 64.286 \\ -150 & 187.5 & -150 \\ 64.286 & -150 & 235.714 \end{bmatrix} \begin{matrix} w_2 \\ w_3 \\ w_4 \end{matrix} \tag{12.22}$$

Calculation of the mass matrix

In the lumped mass method, it is assumed that the mass of the element is concentrated at the nodes and the rotational inertia of these masses is neglected. Also, if axial straining of the element is ignored then the mass matrix is

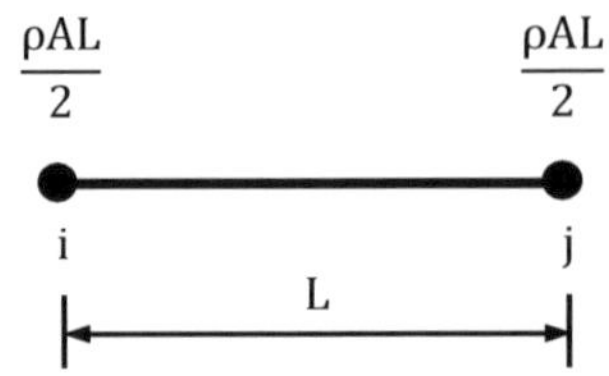

$$m = \frac{\rho AL}{2}\overbrace{\begin{matrix} \overbrace{w_i \quad \theta_i}^{\delta_i} & \overbrace{w_j \quad \theta_j}^{\delta_j} \end{matrix}}^{\delta} \begin{bmatrix} 1 & 0 & 0 & 0 \\ 0 & 0 & 0 & 0 \\ 0 & 0 & 1 & 0 \\ 0 & 0 & 0 & 0 \end{bmatrix} \begin{matrix} w_i \\ \theta_i \\ w_j \\ \theta_j \end{matrix} \quad \text{hence, } m^1 = m^2 = m^3 = \begin{bmatrix} 432 & 0 & 0 & 0 \\ 0 & 0 & 0 & 0 \\ 0 & 0 & 432 & 0 \\ 0 & 0 & 0 & 0 \end{bmatrix}$$

The mass matrix of the overall structure is:

M =

w_1	θ_1	w_2	θ_2	w_3	θ_3	w_4	θ_4	w_5	θ_5	
432	0	0	0	0	0	0	0	0	0	w_1
0	0	0	0	0	0	0	0	0	0	θ_1
0	0	864	0	0	0	0	0	0	0	w_2
0	0	0	0	0	0	0	0	0	0	θ_2
0	0	0	0	864	0	0	0	0	0	w_3
0	0	0	0	0	0	0	0	0	0	θ_3
0	0	0	0	0	0	864	0	0	0	w_4
0	0	0	0	0	0	0	0	0	0	θ_4
0	0	0	0	0	0	0	0	432	0	w_5
0	0	0	0	0	0	0	0	0	0	θ_5

Apply the boundary conditions of $w_1 = 0$ and $\theta_1 = 0$, hence delete rows and columns 1 and 2 to get the reduced mass matrix as:

The boundary conditions of the fixed ends are $w_1 = \theta_1 = 0$ and $w_5 = \theta_5 = 0$, hence delete rows and columns 1, 2, 9, and 10 to get the reduced mass matrix as:

M =

w_2	θ_2	w_3	θ_3	w_4	θ_4	
864	0	0	0	0	0	w_2
0	0	0	0	0	0	θ_2
0	0	864	0	0	0	w_3
0	0	0	0	0	0	θ_3
0	0	0	0	864	0	w_4
0	0	0	0	0	0	θ_4

Condensation of the mass matrix M

The coefficients of the above matrix are rearranged so as to separate the w's from the θ's as shown below.

M =

w_2	w_3	w_4	θ_2	θ_3	θ_4	
864	0	0	0	0	0	w_2
0	864	0	0	0	0	w_3
0	0	864	0	0	0	w_4
0	0	0	0	0	0	θ_2
0	0	0	0	0	0	θ_3
0	0	0	0	0	0	θ_4

The resulting condensed mass matrix is given by

$$M_c = M_{ww} = \begin{bmatrix} 864 & 0 & 0 \\ 0 & 864 & 0 \\ 0 & 0 & 864 \end{bmatrix} \begin{matrix} w_2 \\ w_3 \\ w_4 \end{matrix} \tag{12.23}$$

Substitute (12.22) and (12.23) in (12.6) to get

$$(K_c - \omega^2 M_c)\delta_w = 0$$

$$\left[10^6 \begin{bmatrix} 235.714 & -150 & 64.286 \\ -150 & 187.5 & -150 \\ 64.286 & -150 & 235.714 \end{bmatrix} - \omega^2 \begin{bmatrix} 864 & 0 & 0 \\ 0 & 864 & 0 \\ 0 & 0 & 864 \end{bmatrix} \right] \begin{bmatrix} w_2 \\ w_3 \\ w_4 \end{bmatrix} = 0 \tag{12.24}$$

For non-trivial solution the determinant of the above matrix is zero

$$\begin{vmatrix} 235.714\times10^6-864\lambda & -150\times10^6 & 64.286\times10^6 \\ -150\times10^6 & 187.5\times10^6-864\lambda & -150\times10^6 \\ 64.286\times10^6 & -150\times10^6 & 235.714\times10^6-864\lambda \end{vmatrix}=0$$

$$-644.973\times10^6(\lambda^3-762.648\times10^3\lambda^2+127.022\times10^9\lambda-2990.150\times10^{12})=0$$

$$\lambda_1=2.811\times10^4,\ \lambda_2=19.841\times10^4,\ \lambda_3=53.613\times10^4$$

The circular frequency of vibration, $\omega=\sqrt{\lambda}$,

$$\omega_1=\sqrt{2.811\times10^4}=167.660\text{ rad/s},$$

$$\omega_2=\sqrt{19.841\times10^4}=445.432\text{ rad/s},$$

$$\omega_3=\sqrt{53.613\times10^4}=732.209\text{ rad/s}.$$

The exact values of the circular frequency of vibration from appendix 7 are:

$$\omega_1=22.382\sqrt{\frac{EI}{\rho AL^4}}=22.382\sqrt{\frac{36\times10^9\times0.0048}{2400\times0.15\times9.6^4}}=168.259\text{ rad/s}$$

$$\omega_2=61.701\sqrt{\frac{EI}{\rho AL^4}}=61.701\sqrt{\frac{36\times10^9\times0.0048}{2400\times0.15\times9.6^4}}=463.842\text{ rad/s}$$

$$\omega_3=120.912\sqrt{\frac{EI}{\rho AL^4}}=120.912\sqrt{\frac{36\times10^9\times0.0048}{2400\times0.15\times9.6^4}}=908.966\text{ rad/s}$$

The error between the calculated and the exact values

Mode	1	2	3
Percentage error	-0.36%	-3.97%	-19.45%

It can be seen in the above table that excellent accuracy is obtained for the first mode, but the error increases rapidly for higher modes of vibration.

Calculation of eigenvectors (mode shapes)

The shape of the beam for any mode of vibration is given by the eigenvector for that mode from (12.24) and since the mode shapes are of relative magnitude assign one of the unknowns, say w_2, a value of +1.000, arbitrarily, then calculate the remaining unknows, w_3 and w_4.

For the first mode of vibration substitute $\omega_1 = 167.660$ rad/s in (12.24) to get

$$211.427w_2 - 150w_3 + 64.286w_4 = 0 \tag{12.25}$$

$$-150w_2 + 163.213w_3 - 150w_4 = 0 \tag{12.26}$$

$$64.286w_2 - 150w_3 + 211.427w_4 = 0 \tag{12.27}$$

Delete (12.25) and substitute $w_2 = 1.000$ m in (12.26) and (12.27) to get

$$+163.213w_3 - 150w_4 = +150$$

$$-150w_3 + 211.427w_4 = -64.286$$

The solution to the above simultaneous equations is

$w_3 = 1.838$ m and $w_4 = 1.000$ m and the full displacement vector is

$$\delta_w = \begin{bmatrix} w_2 \\ w_3 \\ w_4 \end{bmatrix} = \begin{bmatrix} +1.000 \\ +1.838 \\ +1.000 \end{bmatrix}$$, divide by the largest coefficient to get the normalised vector for the first mode as

$$\psi_1 = \begin{bmatrix} w_2 \\ w_3 \\ w_4 \end{bmatrix} = \begin{bmatrix} +0.544 \\ +1.000 \\ +0.544 \end{bmatrix} \tag{12.28}$$

Similarly, for $\omega_2 = 445.432$ rad/s, the normalised vector for the mode 2 is

$$\psi_2 = \begin{bmatrix} w_2 \\ w_3 \\ w_4 \end{bmatrix} = \begin{bmatrix} +1.000 \\ 0 \\ -1.000 \end{bmatrix} \tag{12.29}$$

and for $\omega_3 = 732.209$ rad/s, the normalised vector for mode 3 is

$$\psi_3 = \begin{bmatrix} w_2 \\ w_3 \\ w_4 \end{bmatrix} = \begin{bmatrix} +0.919 \\ -1.000 \\ +0.919 \end{bmatrix} \tag{12.30}$$

Equations (12.28), (12.29), and (12.30) are combined together to form the modal matrix, ψ, whose columns are the normalised mode shapes of vibration, thus

$$\psi = \begin{bmatrix} \psi_1 & \psi_2 & \psi_3 \end{bmatrix} = \begin{bmatrix} \psi_{11} & \psi_{12} & \psi_{13} \\ \psi_{21} & \psi_{22} & \psi_{23} \\ \psi_{31} & \psi_{32} & \psi_{33} \end{bmatrix} = \begin{bmatrix} +0.544 & +1.000 & +0.919 \\ +1.000 & 0 & -1.000 \\ +0.544 & -1.000 & +0.919 \end{bmatrix} \tag{12.31}$$

The three modes of natural vibration of the fixed-fixed beam are shown in Fig. 12.11.

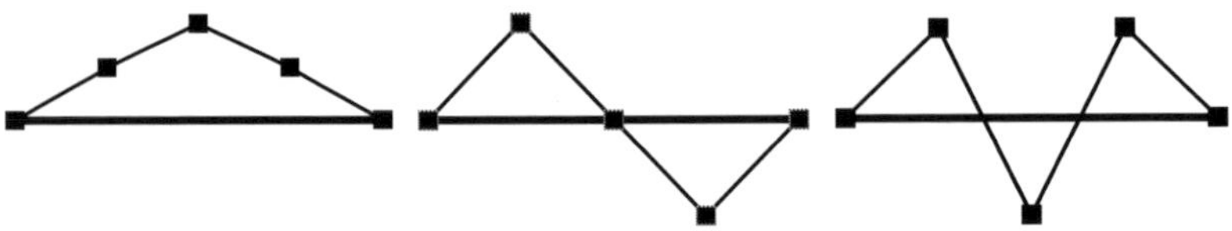

Figure 12.11 Mode shapes for the fixed ended beam.

12.6.2 Vibration of Beams Due to Harmonic Force Excitation

In practice, beams are subjected to some form of dynamic loading during their life in service and their behaviour in such circumstances must be investigated.

It will be assumed that there is no damping present in the system making the analysis less involved but in practice there is always some form of damping that is taken into account.

When an exciting force p(t) is acting on the system the equation of dynamic equilibrium will be

$$-M\ddot{w} - Kw + p(t) = 0 \tag{12.31}$$

where the acceleration, $\ddot{w} = \dfrac{d^2w}{dt^2}$.

In order to save on space, we will use the results obtained from the previous example on free vibration of beams but with the addition of an exciting force.

Assume that a harmonic exciting force $p = p_o \sin\Omega t$ is applied in the z-direction at node 2 so the column vector of force excitation is

$$p(t)=\begin{bmatrix} p_o \sin\Omega t \\ 0 \\ 0 \end{bmatrix}$$, then the equation of motion is

$$\begin{bmatrix} 864 & 0 & 0 \\ 0 & 864 & 0 \\ 0 & 0 & 864 \end{bmatrix}\begin{bmatrix} \ddot{w}_2 \\ \ddot{w}_3 \\ \ddot{w}_4 \end{bmatrix}+10^6\begin{bmatrix} 235.714 & -150 & 64.286 \\ -150 & 187.5 & -150 \\ 64.286 & -150 & 235.714 \end{bmatrix}\begin{bmatrix} w_2 \\ w_3 \\ w_4 \end{bmatrix}=\begin{bmatrix} p_o \sin\Omega t \\ 0 \\ 0 \end{bmatrix} \tag{12.32}$$

Each of the above three simultaneous differential equations contains the three variables w_2, w_3, and w_4, i.e. they are coupled and in order to find their solutions they require to be decoupled. The decoupling is particularly useful when dealing with a large number of simultaneous differential equations as explained below.

Transform the differential equations (12.32) to modal coordinates by introducing a new variable ξ such that $w = \psi\xi$, thus

$$w=\begin{bmatrix} w_2 \\ w_3 \\ w_4 \end{bmatrix},\ \xi=\begin{bmatrix} \xi_2 \\ \xi_3 \\ \xi_4 \end{bmatrix},\ \ddot{w}=\psi\ddot{\xi},$$

and the modal matrix, $\psi = [\psi_1\ \psi_2\ \psi_3]$.

Substitute in (12.31) and pre-multiply by ψ^T to get,

$$\psi^T M\psi\ddot{\xi}+\psi^T K\psi\xi=\psi^T p(t) \tag{12.33}$$

where $\psi^T p(t)$ is called the modal force vector.

The vibration modes ψ are orthogonal with respect to the mass and stiffness matrices, i.e. $\psi_i^T M\psi_j=(m_m)_{ij}=0$ and $\psi_i^T K\psi_j=(k_m)_{ij}=0$ for $i\neq j$ and for $i = j$, $\psi_i^T M\psi_i=(m_m)_{ii}$ and $\psi_i^T K\psi_i=(k_m)_{ii}$. Consequently, $\psi^T M\Psi = M_m$ and $\psi^T K\psi = K_m$ where M_m and K_m are called modal mass matrix and modal stiffness matrix respectively and they are both diagonal matrices with the values of $(m_m)_{ii}$ and $(k_m)_{ii}$ on their diagonals respectively.

It is more convenient to transform the modal mass matrix, M_m into a unit matrix by modifying the modal vectors using scaling factors which will not alter the relative values of the mode shapes.

Let the modified modal matrix, $\phi = \psi\mu$ where $\mu=\begin{bmatrix} \mu_1 \\ \mu_2 \\ \mu_3 \end{bmatrix}$ and $\psi = [\psi_1\ \psi_2\ \psi_3]$ such that

$$\phi = \begin{bmatrix} \phi_1 & \phi_2 & \phi_3 \end{bmatrix} = \begin{bmatrix} \psi_1 & \psi_2 & \psi_3 \end{bmatrix} \begin{bmatrix} \mu_1 \\ \mu_2 \\ \mu_3 \end{bmatrix} = \begin{bmatrix} \mu_1\psi_1 & \mu_2\psi_2 & \mu_3\psi_3 \end{bmatrix}.$$

We have, $\psi^T M \psi = M_m$ and to modify M_m and make it equal to the unit matrix, I, we should have $\phi^T M \phi = I$ from which the values of μ_1, μ_2, and μ_3 can be calculated and hence the modified modal matrix from the relation $\phi = \psi\mu$ with ψ as given by (12.31)

Modified

$$M_m = \phi^T M \phi = \begin{bmatrix} +0.544\mu_1 & +1.000\mu_1 & +0.544\mu_1 \\ +1000\mu_2 & 0\mu_2 & -1.000\mu_2 \\ +0.919\mu_3 & -1.000\mu_3 & +0.919\mu_3 \end{bmatrix} \begin{bmatrix} 864 & 0 & 0 \\ 0 & 864 & 0 \\ 0 & 0 & 864 \end{bmatrix}$$

$$\begin{bmatrix} +0.544\mu_1 & +1.000\mu_2 & +0.919\mu_3 \\ +1.000\mu_1 & 0\mu_2 & -1.000\mu_3 \\ +0.544\mu_1 & -1.000\mu_2 & +0.919\mu_3 \end{bmatrix} = \begin{bmatrix} 1 & 0 & 0 \\ 0 & 1 & 0 \\ 0 & 0 & 1 \end{bmatrix}$$

The above relation is simplified to

$$\begin{bmatrix} 1375.377\mu_1^2 & 0 & 0 \\ 0 & 1728\mu_2^2 & 0 \\ 0 & 0 & 2323.401\mu_3^2 \end{bmatrix} = \begin{bmatrix} 1 & 0 & 0 \\ 0 & 1 & 0 \\ 0 & 0 & 1 \end{bmatrix}$$

The quantities off the diagonal should be zeros and their small values that resulted from the multiplication operation are due to rounding of the numbers and they are ignored and shown equal to zero, hence

$\mu_1 = \sqrt{1/1375.377} = 0.026964$, $\mu_2 = \sqrt{1/1728} = 0.024056$, and $\mu_3 = \sqrt{1/2323.401} = 0.020746$.

The modified modal matrix is

$$\phi = \begin{bmatrix} +0.544 \times 0.026964 & +1.000 \times 0.024056 & +0.919 \times 0.020746 \\ +1.000 \times 0.026964 & 0 & -1.000 \times 0.020746 \\ +0.544 \times 0.026964 & -1.000 \times 0.024056 & +0.919 \times 0.020746 \end{bmatrix}$$

$$= \begin{bmatrix} 0.014668 & 0.024056 & 0.019066 \\ 0.026964 & 0 & -0.020746 \\ 0.014668 & -0.024056 & 0.019066 \end{bmatrix}$$

The stiffness matrix K_c was found from previous calculations, and dropping the subscript c for simplicity, as

$$K = 10^6 \begin{bmatrix} 235.714 & -150 & 64.286 \\ -150 & 187.5 & -150 \\ 64.286 & -150 & 235.714 \end{bmatrix}$$

Modified

$$K_m = \phi^T K \phi = \begin{bmatrix} 0.014668 & 0.026964 & 0.014668 \\ 0.024056 & 0 & -0.024056 \\ 0.019066 & -0.020746 & 0.019066 \end{bmatrix} 10^6 \begin{bmatrix} 235.714 & -150 & 64.286 \\ -150 & 187.5 & -150 \\ 64.286 & -150 & 235.714 \end{bmatrix}$$

$$\begin{bmatrix} 0.014668 & 0.024056 & 0.019066 \\ 0.026964 & 0 & -0.020746 \\ 0.014668 & -0.024056 & 0.019066 \end{bmatrix} = 10^4 \begin{bmatrix} 2.811 & 0 & 0 \\ 0 & 19.841 & 0 \\ 0 & 0 & 53.613 \end{bmatrix}$$

The small off-diagonal coefficients in K_m and M_m are not exactly zero, as they should be, due to rounding of the numbers and hence they are ignored.

Notice that the values of the coefficients in the diagonal of the above stiffness matrix are equal to those values of ω^2 calculated previously from the free vibration of the beam from the general relationship $(K - \omega^2 M)\delta = 0$ which gives $K_m - \omega^2 M_m = 0$ and with $M_m = I$ leads to

$$\omega^2 = K_m = 10^4 \begin{bmatrix} 2.8110 & 0 & 0 \\ 0 & 19.841 & 0 \\ 0 & 0 & 53.613 \end{bmatrix} = \begin{bmatrix} \omega_1^2 & 0 & 0 \\ 0 & \omega_2^2 & 0 \\ 0 & 0 & \omega_3^2 \end{bmatrix}.$$

This also serves as a check on the values of ω obtained previously. Replace ψ by ϕ in (12.33) to get,

$$\phi^T M \phi \ddot{\xi} + \phi^T K \phi \xi = \phi^T p(t) \qquad (12.33a)$$

$M_m \ddot{\xi} + K_m \xi = \phi^T p(t)$ premultiply by M_m^{-1}

$$\ddot{\xi} + M_m^{-1} K_m \xi = M_m^{-1} \phi^T p(t) \qquad (12.34)$$

where $M_m^{-1} = I$, $M_m^{-1} K_m = K_m = \omega^2$ and the right-hand side of the equation is simplified as

$$M_m^{-1}\phi^T p(t) = I\begin{bmatrix} 0.014668 & 0.026964 & 0.014668 \\ 0.024056 & 0 & -0.024056 \\ 0.019066 & -0.020746 & 0.019066 \end{bmatrix}\begin{bmatrix} p_o \sin\Omega t \\ 0 \\ 0 \end{bmatrix}$$

$$= 10^{-3}\begin{bmatrix} 14.668 p_o \sin\Omega t \\ 24.056 p_o \sin\Omega t \\ 19.066 p_o \sin\Omega t \end{bmatrix}$$

Substitute in (12.34) and write in matrix form to get

$$\begin{bmatrix} \ddot{\xi}_2 \\ \ddot{\xi}_3 \\ \ddot{\xi}_4 \end{bmatrix} + \begin{bmatrix} \omega_1^2 & 0 & 0 \\ 0 & \omega_2^2 & 0 \\ 0 & 0 & \omega_3^2 \end{bmatrix}\begin{bmatrix} \xi_2 \\ \xi_3 \\ \xi_4 \end{bmatrix} = 10^{-3}\begin{bmatrix} 14.668 p_o \sin\Omega t \\ 24.056 p_o \sin\Omega t \\ 19.066 p_o \sin\Omega t \end{bmatrix} \quad (12.35)$$

$$\ddot{\xi}_2 + \omega_1^2 \xi_2 = 14.668 \times 10^{-3} p_o \sin\Omega t \quad (12.35a)$$

$$\ddot{\xi}_3 + \omega_2^2 \xi_3 = 24.056 \times 10^{-3} p_o \sin\Omega t \quad (12.35b)$$

$$\ddot{\xi}_4 + \omega_3^2 \xi_4 = 19.066 \times 10^{-3} p_o \sin\Omega t \quad (12.35c)$$

The general solutions of the above set of differential equation are,

$$\xi_2 = C_1 \sin\omega_1 t + C_2 \cos\omega_1 t + \frac{14.668 \times 10^{-3} p_o}{\omega_1^2 - \Omega^2} \sin\Omega t \quad (12.36a)$$

$$\xi_3 = C_3 \sin\omega_2 t + C_4 \cos\omega_2 t + \frac{24.056 \times 10^{-3} p_o}{\omega_2^2 - \Omega^2} \sin\Omega t \quad (12.36b)$$

$$\xi_4 = C_5 \sin\omega_3 t + C_6 \cos\omega_3 t + \frac{19.066 \times 10^{3} p_o}{\omega_3^2 - \Omega^2} \sin\Omega t \quad (12.36c)$$

The constants C_1 to C_6 are found from the initial conditions as follows:

We have $w = \phi\xi$, pre-multiply by ϕ^{-1} to get $\xi = \phi^{-1}w$ and $\dot{\xi} = \phi^{-1}\dot{w}$. The calculations for determining the inverse of ϕ can be avoided by using the previous relation $\phi^T M\phi = M_m$ and $M_m = I$, hence $\phi^T M\phi = I$, post-multiply both sides by ϕ^{-1} to get $\phi^T M = \phi^{-1}$ hence $\xi = \phi^T Mw$, thus

$$\xi_o = \phi^T M w_o \text{ and } \dot{\xi}_o = \phi^T M \dot{w}_o \text{ with}$$

$$\phi^{-1} = \phi^{T}M = \begin{bmatrix} 0.014668 & 0.026964 & 0.014668 \\ 0.024056 & 0 & -0.024056 \\ 0.019066 & -0.020746 & 0.019066 \end{bmatrix}\begin{bmatrix} 864 & 0 & 0 \\ 0 & 864 & 0 \\ 0 & 0 & 864 \end{bmatrix}$$

$$= \begin{bmatrix} 12.673 & 23.297 & 12.673 \\ 20.784 & 0 & -20.784 \\ 16.473 & -17.925 & 16.473 \end{bmatrix}$$

At time t = 0, the displacement, $w_o = \begin{bmatrix} w_{2,o} \\ w_{3,o} \\ w_{4,o} \end{bmatrix} = \begin{bmatrix} 0 \\ 0 \\ 0 \end{bmatrix}$ and the velocity, $\dot{w}_o = \begin{bmatrix} \dot{w}_{2,o} \\ \dot{w}_{3,o} \\ \dot{w}_{4,o} \end{bmatrix} = \begin{bmatrix} 0 \\ 0 \\ 0 \end{bmatrix}$

$$\xi_0 = \begin{bmatrix} \xi_{2,o} \\ \xi_{3,o} \\ \xi_{4,o} \end{bmatrix} = \phi^{T}Mw_o = \begin{bmatrix} 12.673 & 23.297 & 12.673 \\ 20.784 & 0 & -20.784 \\ 16.473 & -17.925 & 16.473 \end{bmatrix}\begin{bmatrix} 0 \\ 0 \\ 0 \end{bmatrix} = \begin{bmatrix} 0 \\ 0 \\ 0 \end{bmatrix} \text{ and}$$

$$\dot{\xi}_o = \begin{bmatrix} \dot{\xi}_{2,o} \\ \dot{\xi}_{3,o} \\ \dot{\xi}_{4,o} \end{bmatrix} = \phi^{T}M\dot{w}_o = \begin{bmatrix} 12.673 & 23.297 & 12.673 \\ 20.784 & 0 & -20.784 \\ 16.473 & -17.925 & 16.473 \end{bmatrix}\begin{bmatrix} 0 \\ 0 \\ 0 \end{bmatrix} = \begin{bmatrix} 0 \\ 0 \\ 0 \end{bmatrix}$$

The above result could have been written by inspection, but it is shown here for generality of the principles, particularly when the initial conditions differ from $w_o = 0$ and $\dot{w}_o = 0$. Consider (12.36a) $\xi_{2o} = 0$ therefore, $C_2 = 0$

$$\dot{\xi}_{2o} = 0 \text{ gives } C_1 = -\frac{14.668 \times 10^{-3}\Omega p_o}{\omega_1(\omega_1^2 - \Omega^2)}$$

Substitute C_1 and C_2 in (12.36a) to get,

$$\xi_2 = \frac{14.668 \times 10^{-3} p_o}{\omega_1^2 - \Omega^2}\left(-\frac{\Omega}{\omega_1}\sin\omega_1 + \sin\Omega t\right)$$

Similarly

$$\xi_3 = \frac{24.056\times10^{-3}p_o}{\omega_2^2-\Omega^2}\left(-\frac{\Omega}{\omega_2}\sin\omega_2 t+\sin\Omega t\right)$$

$$\xi_4 = \frac{19.066\times10^{-3}p_o}{\omega_3^2-\Omega^2}\left(-\frac{\Omega}{\omega_3}\sin\omega_3 t+\sin\Omega t\right)$$

$$w = \begin{bmatrix} w_2 \\ w_3 \\ w_4 \end{bmatrix} = \phi\xi = \begin{bmatrix} \phi_{11} & \phi_{12} & \phi_{13} \\ \phi_{21} & \phi_{22} & \phi_{23} \\ \phi_{31} & \phi_{32} & \phi_{33} \end{bmatrix}\begin{bmatrix} \xi_2 \\ \xi_3 \\ \xi_4 \end{bmatrix} = \begin{bmatrix} \phi_{11}\xi_2+\phi_{12}\xi_3+\phi_{13}\xi_4 \\ \phi_{21}\xi_2+\phi_{22}\xi_3+\phi_{23}\xi_4 \\ \phi_{31}\xi_2+\phi_{32}\xi_3+\phi_{33}\xi_4 \end{bmatrix}$$

$$\begin{bmatrix} w_2 \\ w_3 \\ w_4 \end{bmatrix} = \begin{bmatrix} 0.014668 & 0.024056 & 0.019066 \\ 0.026964 & 0 & -0.020746 \\ 0.014668 & -0.024056 & 0.019066 \end{bmatrix} \begin{bmatrix} \frac{14.668\times10^{-3}p_o}{\omega_1^2-\Omega^2}(-\frac{\Omega}{\omega_1}\sin\omega_1 t+\sin\Omega t) \\ \frac{24.056\times10^{-3}p_o}{\omega_2^2-\Omega^2}(-\frac{\Omega}{\omega_2}\sin\omega_2 t+\sin\Omega t) \\ \frac{19.066\times10^{-3}p_o}{\omega_3^2-\Omega^2}(-\frac{\Omega}{\omega_3}\sin\omega_3 t+\sin\Omega t) \end{bmatrix}$$

The values of ω were found earlier as: ω_1 = 167.660 rad/s, ω_2 = 445.432 rad/s, and ω_3 = 732.209 rad/s and assume that p_o = 90000 N and Ω=140 rad/s we get

$w_2 = 10^{-3}[-0.012\ \sin(732.209t) - 0.092\ \sin(445.432t) - 1.900\ \sin(167.660t) + 2.630\ \sin(140t)]$

$w_3 = 10^{-3}[0.013\ \sin(732.209t) - 3.493\ \sin(167.660t) + 4.114\ \sin(140t)]$

$w_4 = 10^{-3}[-0.012\ \sin(732.209t) + 0.092\ \sin(445.432t) - 1.900\ \sin(167.660t) + 2.048\ \sin(140t)]$

A plot of w_2 and w_3 against time is shown in Fig. 12.12.

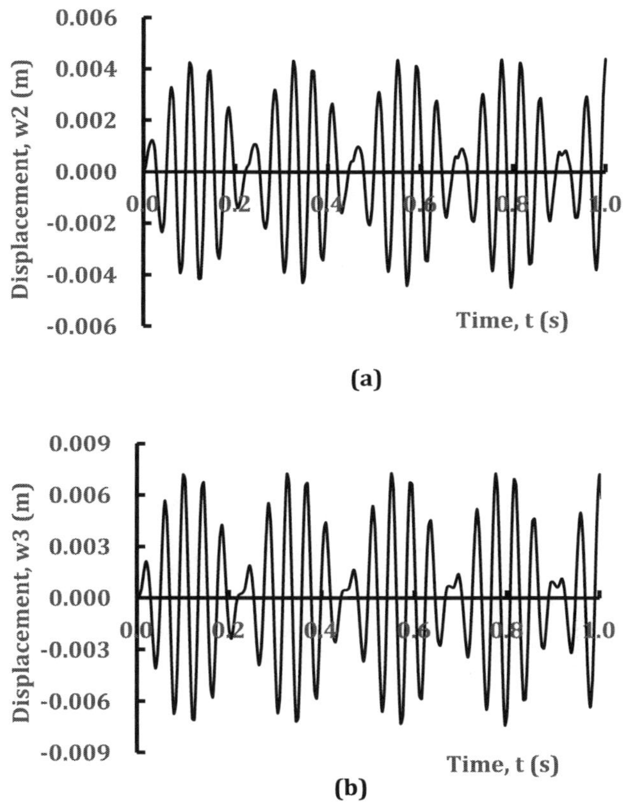

Figure 12.12 Vertical deflection against time, (a) at node 2 and (b) at node 3.

The inertia forces that act on the beam at the nodes is calculated from the product of the mass at the node times the acceleration $\ddot{w}$ which is obtained from the second derivative of the displacement w at that node.

12.7 Vibration of Rigidly Connected Plane Frames

Buildings are constructed of frames that consist of beams and columns where the beams support the floor slabs and are connected to the supporting columns. Shear frames are defined as frames that resist horizontal forces by the shear stiffness of their columns and the rigid connections between their beams and columns as opposed to frames with some type of cross bracing. In the vibration analysis of such frames, the horizontal floors together with their supporting beams are assumed to have infinite rigidity such that there is no rotation of the joints. The frame of the building is idealised as a vertical member with concentrated masses at the joints where the floor beams are connected to the columns. Each of the concentrated masses consists of the mass of the floor at that level plus the mass of half of the column length below and half above that level. The axial deformation of the columns is ignored and the joints are assumed to be fully restrained against rotation but they can displace in the x-direction. This idealisation is commonly used in the analysis of the response of building frames to earthquake excitation due to time varying base motion. The resulting model will have a reduced number of degrees of freedom leading to a significant computer time saving.

Consider a fixed ended vertical element with lateral displacements at the fixed ends as shown in Fig. 12.13.

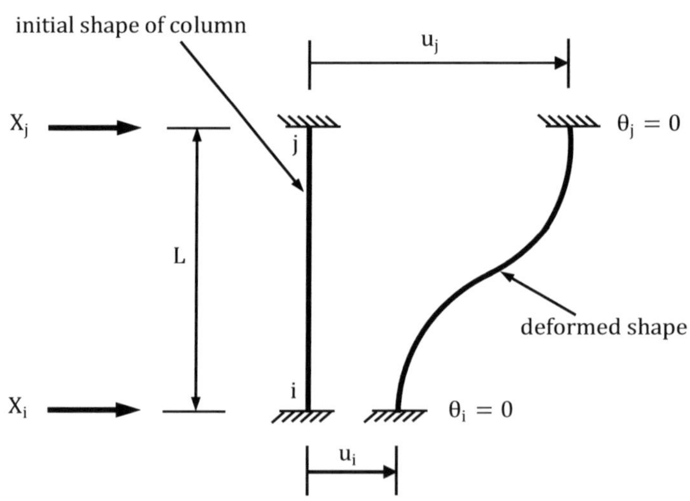

Figure 12.13 Vertical element (moments are not shown for clarity).

The stiffness matrix for a rigidly connected plane frame member derived in Chapter 5 as given by (5.7) can be modified for the column shown in Fig. 12.13 where $x_{ij} = 0$ and $z_{ij} = L$. Applying the end conditions of $\theta_i = 0$ and $\theta_j = 0$ and ignoring axial displacement, i.e. $v_i = 0$ and $v_j = 0$ will result in the following equations:

$$X_i = +\frac{12EI}{L^3}u_i - \frac{12EI}{L^3}u_j \tag{12.21}$$

$$M_i = +\frac{6EI}{L^2}u_i - \frac{6EI}{L^2}u_j \tag{12.22}$$

$$X_j = -\frac{12EI}{L^3}u_i + \frac{12EI}{L^3}u_j \tag{12.23}$$

$$M_j = +\frac{6EI}{L^2}u_i - \frac{6EI}{L^2}u_j \tag{12.24}$$

The moments at the ends of the member are not of interest in this case therefore, they are not considered. The resulting matrix, which represents the lateral stiffness of the column, is given by (12.21) and (12.23) as:

$$\begin{bmatrix} +\frac{12EI}{L^3} & -\frac{12EI}{L^3} \\ -\frac{12EI}{L^3} & +\frac{12EI}{L^3} \end{bmatrix} \begin{bmatrix} u_i \\ u_j \end{bmatrix} = \begin{bmatrix} X_i \\ X_j \end{bmatrix}, \text{ where}$$

$$k = \begin{bmatrix} +\frac{12EI}{L^3} & -\frac{12EI}{L^3} \\ -\frac{12EI}{L^3} & +\frac{12EI}{L^3} \end{bmatrix}, \ \delta = \begin{bmatrix} u_i \\ u_j \end{bmatrix}, \ \text{and } F = \begin{bmatrix} X_i \\ X_j \end{bmatrix}$$

Example 3:

Calculate the frequencies and mode shapes of natural vibration of the shear frame shown in Fig. 12.14 using the following data: The masses of floors including the contribution of columns are: m_2 = 9000 kg, m_3 = 8000 kg, and m_4 = 6000 kg. The properties of the columns are: $I_1 = 95 \times 10^{-6}$ m^4, $I_2 = 61 \times 10^{-6}$ m^4, $I_3 = 46 \times 10^{-6}$ m^4, L_1 = 5.0 m, L_2 = 4.5 m, L_3 = 4.0 m, ρ = 7850 kg/m^3, and $E = 210 \times 10^9$ N/m^2.

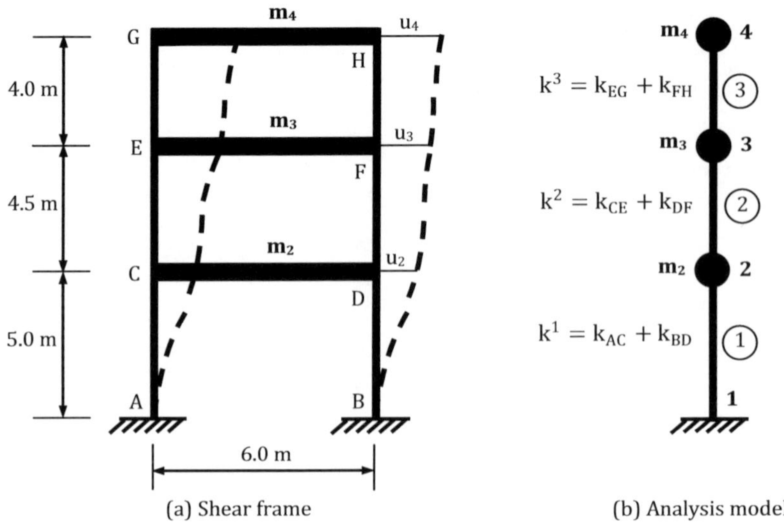

Figure 12.14 Modelling of a shear frames.

$$k_{AC} = \begin{bmatrix} \frac{12E_1I_1}{L_1^3} & -\frac{12E_1I_1}{L_1^3} \\ -\frac{12E_1I_1}{L_1^3} & \frac{12E_1I_1}{L_1^3} \end{bmatrix} = \begin{bmatrix} 1915\times10^3 & -1915\times10^3 \\ -1915\times10^3 & 1915\times10^3 \end{bmatrix} = k_{BD}$$

$$k^1 = k_{AC} + k_{BD} = 2\begin{bmatrix} 1915\times10^3 & -1915\times10^3 \\ -1915\times10^3 & 1915\times10^3 \end{bmatrix} = \begin{bmatrix} 3830\times10^3 & -3830\times10^3 \\ -3830\times10^3 & 3830\times10^3 \end{bmatrix}$$

$$k^2 = k_{CE} + k_{DF} = 2\begin{bmatrix} 1687\times10^3 & -1687\times10^3 \\ -1687\times10^3 & 1687\times10^3 \end{bmatrix} = \begin{bmatrix} 3374\times10^3 & -3374\times10^3 \\ -3374\times10^3 & 3374\times10^3 \end{bmatrix}$$

$$k^3 = k_{EG} + k_{FH} = 2\begin{bmatrix} 1811\times10^3 & -1811\times10^3 \\ -1811\times10^3 & 1811\times10^3 \end{bmatrix} = \begin{bmatrix} 3622\times10^3 & -3622\times10^3 \\ -3622\times10^3 & 3622\times10^3 \end{bmatrix}$$

K =

u_1	u_2	u_3	u_4	
k_{ii}^{1}	k_{ij}^{1}	0	0	u_1
k_{ji}^{1}	$k_{jj}^{1}+k_{ii}^{2}$	k_{ij}^{2}	0	u_2
0	k_{ji}^{2}	$k_{jj}^{2}+k_{ii}^{3}$	k_{ij}^{3}	u_3
0	0	k_{ji}^{3}	k_{jj}^{3}	u_4

Apply the boundary condition of $u_1 = 0$, i.e. delete row 1 and column 1 to get:

$$K = 10^3 \begin{array}{c} \begin{array}{ccc} u_2 & u_3 & u_4 \end{array} \\ \begin{bmatrix} 7204 & -3374 & 0 \\ -3374 & 6996 & -3622 \\ 0 & -3622 & 3622 \end{bmatrix} \end{array} \begin{array}{c} \\ u_2 \\ u_3 \\ u_4 \end{array} \quad \text{and}$$

$$M = \begin{bmatrix} m_1 & 0 & 0 \\ 0 & m_2 & 0 \\ 0 & 0 & m_3 \end{bmatrix} = \begin{bmatrix} 9000 & 0 & 0 \\ 0 & 8000 & 0 \\ 0 & 0 & 6000 \end{bmatrix}$$

Substitute in (12.8) to get

$$\left[10^3 \begin{bmatrix} 7204 & -3374 & 0 \\ -3374 & 6996 & -3622 \\ 0 & -3622 & 3622 \end{bmatrix} - \omega^2 \begin{bmatrix} 9000 & 0 & 0 \\ 0 & 8000 & 0 \\ 0 & 0 & 6000 \end{bmatrix} \right] \begin{bmatrix} u_2 \\ u_3 \\ u_4 \end{bmatrix} = \begin{bmatrix} 0 \\ 0 \\ 0 \end{bmatrix} \tag{12.25}$$

Let $\lambda = \omega^2$ and equate the determinant of the above matrix to zero to get

$$\begin{vmatrix} 10^3(7204-9\lambda) & -3374\times10^3 & 0 \\ -3374\times10^3 & 10^3(6996-8\lambda) & -3622\times10^3 \\ 0 & -3622\times10^3 & 10^3(3622-6\lambda) \end{vmatrix} = 0$$

$$432 \times 10^9(-\lambda^3 + 22.79 \times 10^2\lambda^2 - 127.97 \times 10^4\lambda + 108.35 \times 10^6) = 0$$

The roots of the above equation are: $\lambda_1 = 102.56, \lambda_2 = 730.82$, and $\lambda_3 = 1445.62$.

$$\omega = \sqrt{\lambda}; \quad \omega_1 = \sqrt{\lambda_1} = \sqrt{102.56} = 10.13 \text{ rad/s},$$

$$\omega_2 = \sqrt{\lambda_2} = \sqrt{730.82} = 27.03 \text{ rad/s, and}$$

$$\omega_3 = \sqrt{\lambda_3} = \sqrt{1445.62} = 38.02 \text{ rad/s}$$

The normalised vibration modes are calculated from (12.25) for the different values of ω in a similar way to the previous example as

$$\text{Mode 1, } \omega_1 = 10.13 \text{ rad/s, } \begin{bmatrix} u_2 \\ u_3 \\ u_4 \end{bmatrix} = \begin{bmatrix} +0.446 \\ +0.830 \\ +1.000 \end{bmatrix} = \psi_1$$

$$\text{Mode 2, } \omega_2 = 27.03 \text{ rad/s, } \begin{bmatrix} u_2 \\ u_3 \\ u_4 \end{bmatrix} = \begin{bmatrix} +1.000 \\ +0.184 \\ -0.873 \end{bmatrix} = \psi_2$$

$$\text{Mode 3, } \omega_3 = 38.02 \text{ rad/s, } \begin{bmatrix} u_2 \\ u_3 \\ u_4 \end{bmatrix} = \begin{bmatrix} +0.585 \\ -1.000 \\ +0.717 \end{bmatrix} = \psi_3$$

The three vibration modes of the frame are shown in Fig. 12.15.

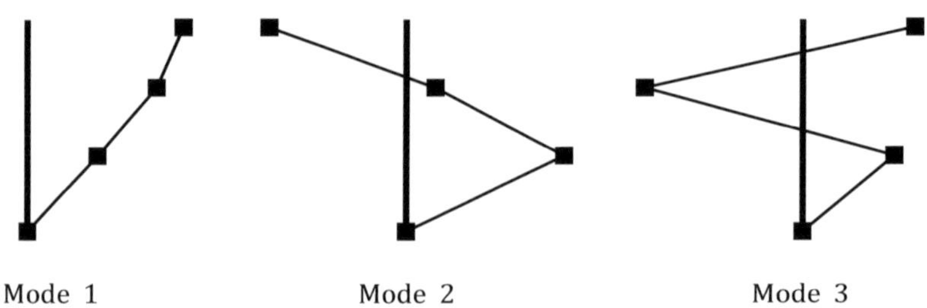

Figure 12.15 Vibration modes for shear frame.

If the axial compressive force in the columns is considered, the elastic stiffness matrix will decrease leading to lower values of ω. On the other hand, if the axial force is tensile, the elastic stiffness matrix will increase and the resulting natural frequencies of vibration will be higher.

Buildings located within a seismic zone are subjected to forces resulting from the ground motion in an earthquake. The vibration analysis follows the same principles as those for a single degree of freedom due to base motion as discussed in Section 12.2.2 with the inertia forces assumed to act at floor levels. A further complication arises from the fact that the ground motion in an earthquake is random and not as simple as, for example, a sine function that can be integrated analytically. This means that for the determination of earthquake forces acting on a building frame the differential equations governing the motion are integrated numerically step by step in small time increments to obtain the system response and hence the forces. The process is not suitable for hand calculations particularly for frames with large number of degrees of freedom for which specialised computer software or standard approximate procedures are usually used in practical design situations.

Problems

P12.1 Use the lumped mass method to calculate the natural frequencies and the corresponding modes of vibration of the pin-connected plane frame shown in Fig. P12.1 for the data shown below. The roller at support 4 can move in the x-direction but is restrained from movement in the z-direction during vibration. The properties of the members of the frame are: $A_1 = A_2 = A_3 = 0.003$ m^2, $E = 70 \times 10^9$ N/m^2, and $\rho = 2600$ kg/m^3.

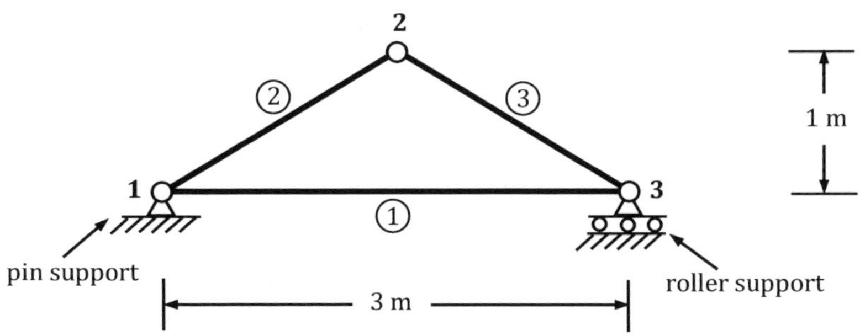

Figure P12.1

Answer:

$$\text{Mode 1,}\quad \omega_1 = 1398.28 \text{ rad/s,}\quad \begin{bmatrix} u_2 \\ w_2 \\ u_3 \end{bmatrix} = \begin{bmatrix} +0.495 \\ -1.000 \\ +0.822 \end{bmatrix}$$

$$\text{Mode 2,}\quad \omega_2 = 2694.15 \text{ rad/s,}\quad \begin{bmatrix} u_2 \\ w_2 \\ u_3 \end{bmatrix} = \begin{bmatrix} +0.770 \\ +1.000 \\ +0.565 \end{bmatrix}$$

$$\text{Mode 3,}\quad \omega_3 = 3923.87 \text{ rad/s,}\quad \begin{bmatrix} u_2 \\ w_2 \\ u_3 \end{bmatrix} = \begin{bmatrix} +1.000 \\ -0.254 \\ -0.685 \end{bmatrix}$$

P12.2 Calculate the natural frequencies and the corresponding modes of vibration of the beam shown in Fig. P12.2 which is fixed at node 1 and simply supported at node 4 using the lumped mass method. Determine the response of the beam if it starts from rest, i.e. w = 0 and $\dot{w} = 0$ and acted upon by a force $p(t) = p_o \cos\Omega t$ at node 3 in the z-direction. Use the following data: $A = 0.008 \text{ m}^2$, $I = 0.000162 \text{ m}^4$, $\rho = 7850 \text{ kg/m}^3$, $E = 210 \times 10^9 \text{ N/m}^2$, $p_o = 60000$ N and $\Omega = 500$ rad/s.

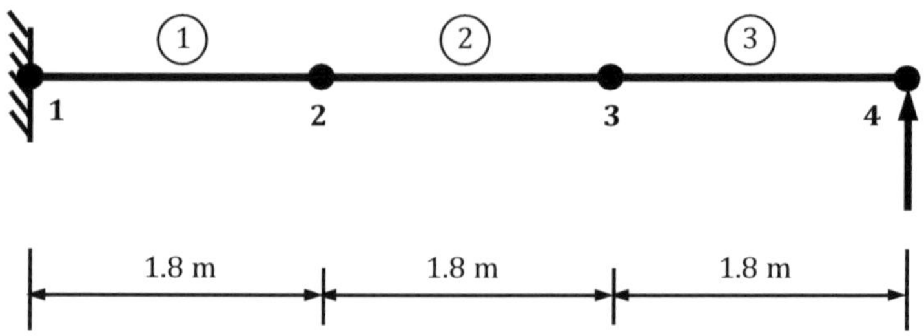

Figure P12.2 Fixed-pin beam.

Answer:

$K = 10^6$

	w_2	w_3	θ_2	θ_3	θ_4
w_2	140.0	-70.0	0	-63.0	0
w_3	-70.0	140.0	63.0	0	-63.0
θ_2	0	63.0	151.2	37.8	0
θ_3	-63.0	0	37.8	151.2	37.8
θ_4	0	-63.0	0	37.8	75.6

$$K_c = 10^6 \begin{bmatrix} 107.692 & -61.923 \\ -61.923 & 59.231 \end{bmatrix} \begin{matrix} w_2 \\ w_3 \end{matrix} \quad (\text{columns } w_2,\ w_3)$$

Mode 1: $\omega_1 = 387.418$ rad/s, $\psi_1 = \begin{bmatrix} w_1 \\ w_2 \end{bmatrix} = \begin{bmatrix} +0.6825 \\ +1.0000 \end{bmatrix}$

Mode 2: $\omega_2 = 1151.773$ rad/s, $\psi_1 = \begin{bmatrix} w_1 \\ w_2 \end{bmatrix} = \begin{bmatrix} +1.0000 \\ -0.6825 \end{bmatrix}$

$$\phi = \begin{bmatrix} +0.0530 & +0.0777 \\ +0.0777 & -0.0530 \end{bmatrix}, \quad M_m = \begin{bmatrix} 1 & 0 \\ 0 & 1 \end{bmatrix},$$

$$K_m = 10^6 \begin{bmatrix} 0.1501 & 0 \\ 0 & 1.3266 \end{bmatrix}$$

$w_2 = 10^{-3}$ [2.473 cos(387.418t) + 0.230 cos(1151.773t) −2.703 cos(500t)]
$w_3 = 10^{-3}$ [3.626 cos(387.418t) – 0.157 cos(1151.773t) −3.469 cos(500t)].

P12.3 Calculate the frequencies and mode shapes of natural vibration of the shear frame shown in Fig. P12.3 using the lumped mass method and the following data: The masses of floors including the contribution of the columns are: m_2 = 12000 kg, m_3 = 14000 kg, and m_4 = 8000 kg. The properties of the columns are: $I_1 = 0.0017$ m^4, $I_2 = 0.0015$ m^4, $I_3 = 0.0012$ m^4, $L_1 = 6$ m, $L_2 = 5$ m, $L_3 = 4$ m, and $E = 24 \times 10^9$ N/m^2.

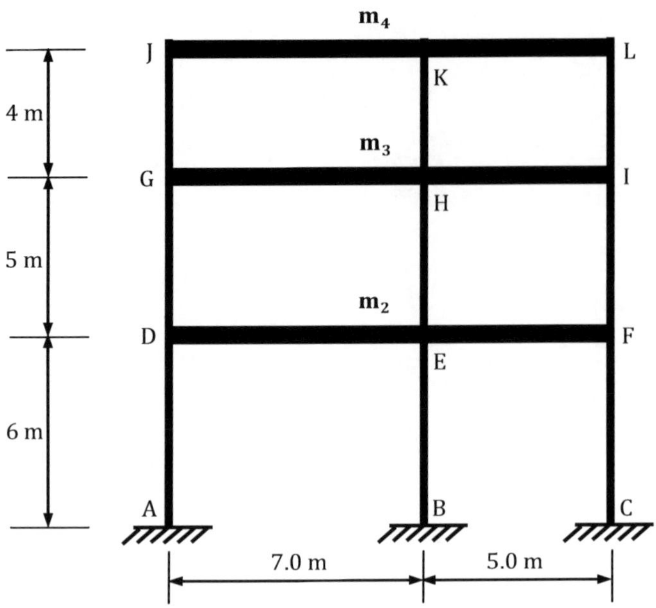

Figure P12.3

Answer:

Mode 1: $\omega_1 = 36.88$ rad/s, $\begin{bmatrix} u_2 \\ u_3 \\ u_4 \end{bmatrix} = \begin{bmatrix} +0.610 \\ +0.923 \\ +1.000 \end{bmatrix}$

Mode 2: $\omega_2 = 117.95$ rad/s, $\begin{bmatrix} u_2 \\ u_3 \\ u_4 \end{bmatrix} = \begin{bmatrix} +1.000 \\ -0.146 \\ -0.679 \end{bmatrix}$

Mode 3: $\omega_3 = 178.50$ rad/s, $\begin{bmatrix} u_2 \\ u_3 \\ u_4 \end{bmatrix} = \begin{bmatrix} +0.317 \\ -0.798 \\ +1.000 \end{bmatrix}$

Appendix 1

Bar Stiffness Matrix

In Chapter 1, the stiffness matrix was derived using direct relationship between the force acting on a bar of uniform cross section and the resulting displacement. A more general procedure may be employed based on assuming a polynomial to represent the variation of displacement along the bar. The degree of the polynomial is chosen so as to satisfy a certain state which in this case is a constant strain along the bar, i.e. $\partial\bar{u}/\partial\bar{x}$ = constant and this condition is satisfied by the polynomial

$$\bar{u} = a_0 + a_1\bar{x}. \tag{A1.1}$$

The above equation defines the displacement $\bar{u}$ at a distance $\bar{x}$ from node i as shown in Fig. A1.1. The constants a_0 and a_1 are found from the boundary conditions at the ends of the bar, i.e. in terms of the displacements at nodes i and j as follows:

At $\bar{x} = 0$, $\bar{u} = \bar{u}_i$, hence $a_o = \bar{u}_i$.

At $\bar{x} = L$, $\bar{u} = \bar{u}_j$, thus $\bar{u}_j = \bar{u}_i + a_1 L$, which gives

$$a_1 = \frac{\bar{u}_j - \bar{u}_i}{L}.$$

Therefore (A1.1) becomes, $\bar{u} = \bar{u}_i + \left(\dfrac{\bar{u}_j - \bar{u}_i}{L}\right)\bar{x}$, which can be written as

$$\bar{u} = \left(1 - \frac{\bar{x}}{L}\right)\bar{u}_i + \frac{\bar{x}}{L}\bar{u}_j. \tag{A1.2}$$

The above equation is called interpolation polynomial and the quantities $(1 - \bar{x}/L)$ and $\bar{x}/L$ are called shape functions.

The gain in strain energy in a bar subjected to an axial force $\bar{X}$ is given by:

$$E_S = \int_0^L \frac{\bar{X}^2}{2EA} d\bar{x}, \quad \text{where } \bar{X} = \sigma A = E\varepsilon A \text{ and the strain,}$$

$$\varepsilon = \frac{d\bar{u}}{d\bar{x}}$$

$$E_S = \int_0^L \frac{1}{2} EA \left(\frac{d\bar{u}}{d\bar{x}}\right)^2 d\bar{x}, \quad \text{and with } \frac{d\bar{u}}{d\bar{x}} = -\frac{\bar{u}_i}{L} + \frac{\bar{u}_j}{L}, \text{ from (A1.2) we get}$$

$$E_S = \frac{EA}{2L}\left(\bar{u}_i^2 - 2\bar{u}_i\bar{u}_j + \bar{u}_j^2\right).$$

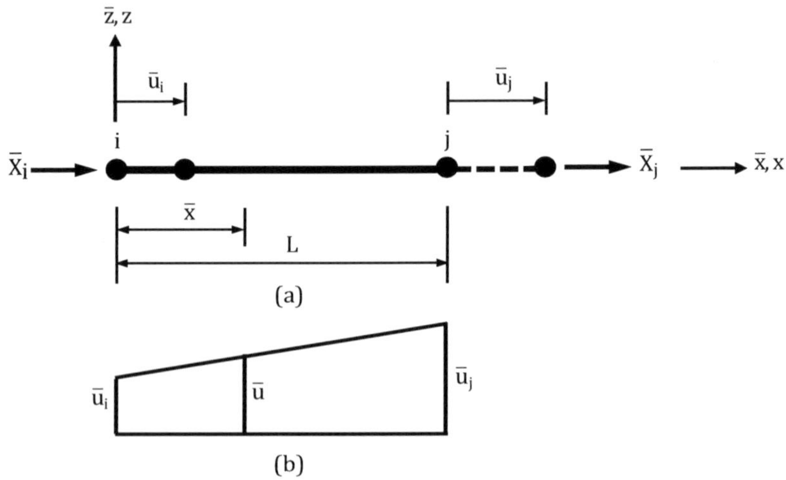

Figure A1.1 (a) Bar element and (b) variation of $\bar{u}$ with $\bar{x}$.

The work done by the actions at the ends of the element $E_W = -\left(\bar{X}_i\bar{u}_i + \bar{X}_j\bar{u}_j\right)$ (the minus sign because it is a loss in potential energy).

The total potential energy, $E_T = E_S + E_W$

$$E_T = \frac{EA}{2L}\left(\bar{u}_i^2 - 2\bar{u}_i\bar{u}_j + \bar{u}_j^2\right) - \left(\bar{X}_i\bar{u}_i + \bar{X}_j\bar{u}_j\right).$$

For the total potential energy to be minimum, its partial derivative with respect to the displacements is zero, i.e.

From $\dfrac{\partial E_T}{\partial \bar{u}_i} = 0$, we get $\dfrac{EA}{2L}\left(2\bar{u}_i - 2\bar{u}_j\right) - \bar{X}_i = 0$, therefore

$$\frac{EA}{L}\left(\bar{u}_i - \bar{u}_j\right) = \bar{X}_i. \tag{A1.3}$$

From $\dfrac{\partial E_T}{\partial \bar{u}_j} = 0$, we get $\dfrac{EA}{2L}\left(-2\bar{u}_i + 2\bar{u}_j\right) - \bar{X}_j = 0$, therefore

$$\frac{EA}{L}\left(-\bar{u}_i + \bar{u}_j\right) = \bar{X}_j. \tag{A1.4}$$

Equations (A1.3) and (A1.4) are written in matrix form as

$$\begin{bmatrix} \dfrac{EA}{L} & -\dfrac{EA}{L} \\ -\dfrac{EA}{L} & \dfrac{EA}{L} \end{bmatrix} \begin{bmatrix} \bar{u}_i \\ \bar{u}_j \end{bmatrix} = \begin{bmatrix} \bar{X}_i \\ \bar{X}_j \end{bmatrix}. \tag{A1.5}$$

Appendix 2

Beam Stiffness Matrix

A2.1 Bending about the $\bar{y}$-axis

In Chapter 4, the stiffness matrix for a beam element was derived using Castigilano's theorem to find a relationship between the actions at the ends of the beam and the resulting displacement. A more general procedure may be employed based on assuming a polynomial (interpolation function) to represent the variation of displacement along the beam.

We have from the theory of bending of beams

$$n = -\frac{d\bar{V}}{d\bar{x}}, \bar{V} = \frac{d\bar{M}}{d\bar{x}} \text{ and } \bar{M} = -EI_{\bar{y}}\frac{d^2\bar{w}}{d\bar{x}^2}, \text{ hence } EI_{\bar{y}}\frac{d^4\bar{w}}{d\bar{x}^4} = n$$

where n is the load intensity, $\bar{V}$ is the shear force, and $\bar{M}$ is the bending moment.

For any part of the beam where there is no load, n = 0, hence

$$\frac{d^4\bar{w}}{d\bar{x}^4} = 0.$$

A suitable choice for the interpolation polynomial which satisfies the above condition (i.e., its fourth derivative is equal to zero) may take the following form

$$\bar{w} = a_o + a_1\bar{x} + a_2\bar{x}^2 + a_3\bar{x}^3. \tag{A2.1}$$

The positive rotation $\bar{\theta}$ about the $\bar{y}$ -axis is equal to the negative slope of the deflection curve in the $\overline{xz}$ as shown in Fig. A2.1.

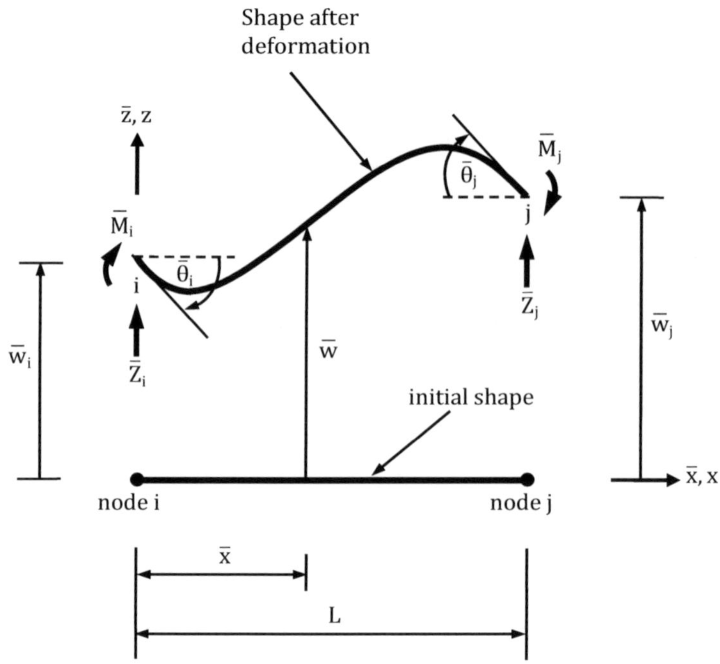

Figure A2.1 Beam element in the $\overline{xz}$ plane.

$$\bar{\theta} = -\frac{d\bar{w}}{d\bar{x}} = -a_1 - 2a_2\bar{x} - 3a_3\bar{x}^2. \tag{A2.2}$$

At $\bar{x} = 0$, $\bar{w} = \bar{w}_i$, and $\bar{\theta} = \bar{\theta}_i$.

From (A2.1) we get, $\bar{w}_i = a_o$ (A2.3)

From (A2.2) we get, $\bar{\theta}_i = -a_1$ (A2.4)

At $\bar{x} = L$, $\bar{w} = \bar{w}_j$, and $\bar{\theta} = \bar{\theta}_j$.

From (A2.1) we get, $\bar{w}_j = a_o + a_1 L + a_2 L^2 + a_3 L^3$. (A2.5)

From (A2.2) we get, $\bar{\theta}_j = -a_1 - 2a_2 L - 3a_3 L^2$. (A2.6)

Solving the simultaneous equations (A2.3) to (A2.6) for the unknowns a_0, a_1, a_2, and a_3 and substituting them in equation (A2.1) and rearranging to get

$$\bar{w} = \left(1 - \frac{3\bar{x}^2}{L^2} + \frac{2\bar{x}^3}{L^3}\right)\bar{w}_i + \left(-\bar{x} + \frac{2\bar{x}^2}{L} - \frac{\bar{x}^3}{L^2}\right)\bar{\theta}_i + \left(\frac{3\bar{x}^2}{L^2} - \frac{2\bar{x}^3}{L^3}\right)\bar{w}_j + \left(\frac{\bar{x}^2}{L} - \frac{\bar{x}^3}{L^2}\right)\bar{\theta}_j \tag{A2.7}$$

The quantities, $\left(1 - \frac{3\bar{x}^2}{L^2} + \frac{2\bar{x}^3}{L^3}\right)$, $\left(-\bar{x} + \frac{2\bar{x}^2}{L} - \frac{\bar{x}^3}{L^2}\right)$, $\left(\frac{3\bar{x}^2}{L^2} - \frac{2\bar{x}^3}{L^3}\right)$, and $\left(\frac{\bar{x}^2}{L} - \frac{\bar{x}^3}{L^2}\right)$ are called shape functions.

The gain in strain energy in bending is given by:

$$E_s = \int_0^L \frac{\bar{M}^2}{2EI_{\bar{y}}} d\bar{x}, \text{ but } \bar{M} = -EI_{\bar{y}} \frac{d^2\bar{w}}{d\bar{x}^2}$$

$$E_s = \int_0^L \frac{1}{2} EI_{\bar{y}} \left(\frac{d^2\bar{w}}{d\bar{x}^2}\right)^2 d\bar{x}$$

$$E_s = \frac{1}{2} EI_{\bar{y}} \int_0^L \left[\left(-\frac{6}{L^2} + \frac{12\bar{x}}{L^3}\right)\bar{w}_i + \left(\frac{4}{L} - \frac{6\bar{x}}{L^2}\right)\bar{\theta}_i + \left(\frac{6}{L^2} - \frac{12\bar{x}}{L^3}\right)\bar{w}_j + \left(\frac{2}{L} - \frac{6\bar{x}}{L^2}\right)\bar{\theta}_j\right]^2 d\bar{x}$$

$$E_s = \frac{EI_{\bar{y}}}{L^3}\left[6(\bar{w}_i - \bar{w}_j)^2 - 6L(\bar{w}_i - \bar{w}_j)(\bar{\theta}_i + \bar{\theta}_j) + 2L^2\left(\bar{\theta}_i^2 + \bar{\theta}_i\bar{\theta}_j + \bar{\theta}_j^2\right)\right]$$

The work done by the forces and moments at the end nodes of the element

$E_W = -(\bar{Z}_i\bar{w}_i + \bar{M}_i\bar{\theta}_i + \bar{Z}_j\bar{w}_j + \bar{M}_j\bar{\theta}_j)$ (the minus sign because it is a loss in potential energy).

Total potential energy, $E_T = E_S + E_W$

$$E_T = \frac{EI_{\bar{y}}}{L^3}\left[6(\bar{w}_i - \bar{w}_j)^2 - 6L(\bar{w}_i - \bar{w}_j)(\bar{\theta}_i + \bar{\theta}_j) + 2L^2\left(\bar{\theta}_i^2 + \bar{\theta}_i\bar{\theta}_j + \bar{\theta}_j^2\right)\right] - \left(\bar{Z}_i\bar{w}_i + \bar{M}_i\bar{\theta}_i + \bar{Z}_j\bar{w}_j + \bar{M}_j\bar{\theta}_j\right).$$

For minimum total potential energy we must have,

$$\frac{\partial E_T}{\partial \bar{w}_i} = 0, \ \frac{\partial E_T}{\partial \bar{\theta}_i} = 0, \ \frac{\partial E_T}{\partial \bar{w}_j} = 0, \text{ and } \frac{\partial E_T}{\partial \bar{\theta}_j} = 0$$

$$\frac{\partial E_T}{\partial \bar{w}_i} = \frac{EI_{\bar{y}}}{L^3}\left(12\bar{w}_i - 6L\bar{\theta}_i - 12\bar{w}_j - 6L\bar{\theta}_j\right) - \bar{Z}_i = 0$$

$$\bar{Z}_i = \frac{EI_{\bar{y}}}{L^3}\left(12\bar{w}_i - 6L\bar{\theta}_i - 12\bar{w}_j - 6L\bar{\theta}_j\right) \tag{A2.8}$$

$$\frac{\partial E_T}{\partial \bar{\theta}_i} = \frac{EI_{\bar{y}}}{L^3}\left(-6L\bar{w}_i + 4L^2\bar{\theta}_i + 6L\bar{w}_j + 2L^2\bar{\theta}_j\right) - \bar{M}_i = 0$$

$$\bar{M}_i = \frac{EI_{\bar{y}}}{L^3}\left(-6L\bar{w}_i + 4L^2\bar{\theta}_i + 6L\bar{w}_j + 2L^2\bar{\theta}_j\right) \tag{A2.9}$$

$$\frac{\partial E_T}{\partial \bar{w}_j} = \frac{EI_{\bar{y}}}{L^3}\left(-12\bar{w}_i + 6L\bar{\theta}_i + 12\bar{w}_j + 6L\bar{\theta}_j\right) - \bar{Z}_j = 0$$

$$\bar{Z}_j = \frac{EI_{\bar{y}}}{L^3}\left(-12\bar{w}_i + 6L\bar{\theta}_i + 12\bar{w}_j + 6L\bar{\theta}_j\right) \tag{A2.10}$$

$$\frac{\partial E_T}{\partial \bar{\theta}_j} = \frac{EI_{\bar{y}}}{L^3}\left(-6L\bar{w}_i + 2L^2\bar{\theta}_i + 6L\bar{w}_j + 4L^2\bar{\theta}_j\right) - \bar{M}_j = 0$$

$$\bar{M}_j = \frac{EI_{\bar{y}}}{L^3}\left(-6L\bar{w}_i + 2L^2\bar{\theta}_i + 6L\bar{w}_j + 4L^2\bar{\theta}_j\right) \tag{A2.11}$$

Equations (A2.8) to (A2.11) are written in matrix form as:

$$\begin{bmatrix} \bar{Z}_i \\ \bar{M}_i \\ \bar{Z}_j \\ \bar{M}_j \end{bmatrix} = \begin{bmatrix} \frac{12EI_{\bar{y}}}{L^3} & -\frac{6EI_{\bar{y}}}{L^2} & -\frac{12EI_{\bar{y}}}{L^3} & -\frac{6EI_{\bar{y}}}{L^2} \\ -\frac{6EI_{\bar{y}}}{L^2} & \frac{4EI_{\bar{y}}}{L} & \frac{6EI_{\bar{y}}}{L^2} & \frac{2EI_{\bar{y}}}{L} \\ -\frac{12EI_{\bar{y}}}{L^3} & \frac{6EI_{\bar{y}}}{L^2} & \frac{12EI_{\bar{y}}}{L^3} & \frac{6EI_{\bar{y}}}{L^2} \\ -\frac{6EI_{\bar{y}}}{L^2} & \frac{2EI_{\bar{y}}}{L} & \frac{6EI_{\bar{y}}}{L^2} & \frac{4EI_{\bar{y}}}{L} \end{bmatrix} \begin{bmatrix} \bar{w}_i \\ \bar{\theta}_i \\ \bar{w}_j \\ \bar{\theta}_j \end{bmatrix} \tag{A2.12}$$

The above relationship is the same as that derived by Castigliano's method as shown in Chapter 4.

A2.2 Bending about the $\bar{z}$-axis

For bending about the $\bar{z}$-axis, the same procedure is followed with the assumed interpolation function for the deflection polynomial as:

$$\bar{v} = a_o + a_1\bar{x} + a_2\bar{x}^2 + a_3\bar{x}^3 \tag{A2.13}$$

The positive rotation $\overline{\Psi}$ about the $\overline{z}$-axis shown in Fig. A2.2 is equal to the positive slope of the deflection curve in the $\overline{xy}$ plane, thus

$$\overline{\Psi} = +\frac{d\overline{v}}{d\overline{x}} = a_1 + 2a_2\overline{x} + 3a_3\overline{x}^2. \tag{A2.14}$$

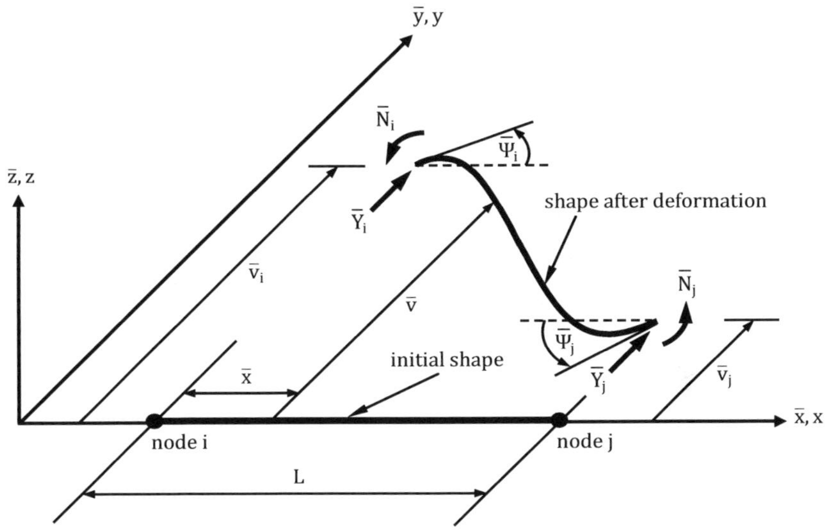

Figure A2.2 Beam element in the $\overline{xy}$ plane.

And the resulting stiffness matrix is:

$$\begin{bmatrix} \overline{Y}_i \\ \overline{N}_i \\ \overline{Y}_j \\ \overline{N}_j \end{bmatrix} = \begin{bmatrix} \frac{12EI_{\overline{z}}}{L^3} & \frac{6EI_{\overline{z}}}{L^2} & -\frac{12EI_{\overline{z}}}{L^3} & \frac{6EI_{\overline{z}}}{L^2} \\ \frac{6EI_{\overline{z}}}{L^2} & \frac{4EI_{\overline{z}}}{L} & -\frac{6EI_{\overline{z}}}{L^2} & \frac{2EI_{\overline{z}}}{L} \\ -\frac{12EI_{\overline{z}}}{L^3} & -\frac{6EI_{\overline{z}}}{L^2} & \frac{12EI_{\overline{z}}}{L^3} & -\frac{6EI_{\overline{z}}}{L^2} \\ \frac{6EI_{\overline{z}}}{L^2} & \frac{2EI_{\overline{z}}}{L} & -\frac{6EI_{\overline{z}}}{L^2} & \frac{4EI_{\overline{z}}}{L} \end{bmatrix} \begin{bmatrix} \overline{v}_i \\ \overline{\Psi}_i \\ \overline{v}_j \\ \overline{\Psi}_j \end{bmatrix} \tag{A2.15}$$

Notice the change in sign of some of the coefficients in the above matrix in comparison with the matrix for bending about the $\overline{y}$-axis.

Appendix 3

Bar Torsion Matrix

In Chapter 7, the stiffness matrix for a uniform bar subjected to torques acting at its ends was derived using the direct relationship between the torques and the resulting angles of twist. This appendix employs an alternative approach based on assuming a polynomial to represent the variation of angle of twist along the bar. It is assumed that the rate of change of angle twist is constant along the bar, i.e. $\partial\overline{\Phi} / \partial\overline{x}$ = constant. To satisfy this condition the linear interpolation polynomial given by (A3.1) is used to define the angle of twist $\overline{\Phi}$ at a distance $\overline{x}$ from node i as shown in Fig. A3.1.

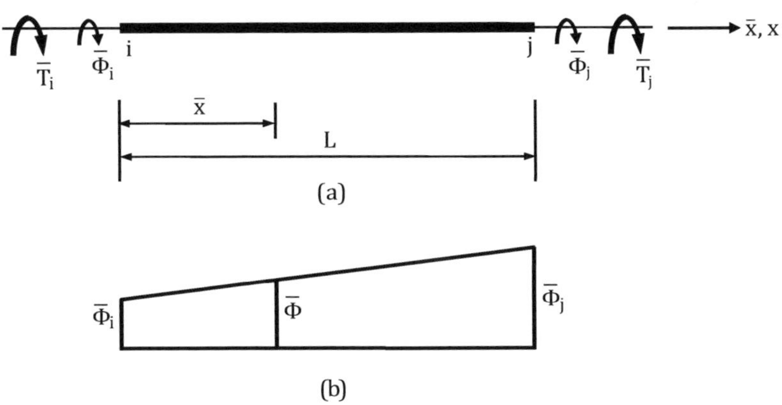

Figure A3.1 (a) Bar element and (b) variation of angle of twist $\overline{\Phi}$ with $\overline{x}$.

$$\overline{\Phi} = a_o + a_1\overline{x}. \tag{A3.1}$$

The constants a_0 and a_1 are found from the boundary conditions at the ends of the bar, i.e. in terms of the angles of twist at nodes i and j as follows:

$$\text{At } \bar{x}=0,\ \bar{\Phi}=\bar{\Phi}_i,\ \text{hence } a_o=\bar{\Phi}_i.$$

At $\bar{x}=L$, $\bar{\Phi}=\bar{\Phi}_j$, thus $\bar{\Phi}_j=\bar{\Phi}_i+a_1L$, which gives, $a_1=\dfrac{\bar{\Phi}_j-\bar{\Phi}_i}{L}$.

Therefore (A3.1) becomes $\bar{\Phi}=\bar{\Phi}_i+\left(\dfrac{\bar{\Phi}_j-\bar{\Phi}_i}{L}\right)\bar{x}$, which can be written as

$$\bar{\Phi}=\left(1-\frac{\bar{x}}{L}\right)\bar{\Phi}_i+\frac{\bar{x}}{L}\bar{\Phi}_j. \qquad \text{(A3.2)}$$

The above equation is called interpolation polynomial and the quantities $(1-\bar{x}/L)$ and $\bar{x}/L$ are called shape functions.

The gain in strain energy in a bar subjected to torque $\bar{T}$ is given by:

$$E_S=\int_0^L\frac{\bar{T}^2}{2GJ}d\bar{x},\quad \text{where } \bar{T}=GJ\frac{d\bar{\Phi}}{d\bar{x}}$$

$E_S=\displaystyle\int_0^L\frac{1}{2}GJ\left(\frac{d\bar{\Phi}}{d\bar{x}}\right)^2 d\bar{x}$, and with $\dfrac{d\bar{\Phi}}{d\bar{x}}=-\dfrac{\bar{\Phi}_i}{L}+\dfrac{\bar{\Phi}_j}{L}$ from (A4.2) we get

$$E_S=\frac{GJ}{2L}\left(\bar{\Phi}_i^2-2\bar{\Phi}_i\bar{\Phi}_j+\bar{\Phi}_j^2\right).$$

The work done by the actions at the ends of the element $E_W=-\left(\bar{T}_i\bar{\Phi}_i+\bar{T}_j\bar{\Phi}_j\right)$ (the minus sign because it is a loss in potential energy).

The total potential energy, $E_T = E_S + E_W$

$$E_T=\frac{GJ}{2L}\left(\bar{\Phi}_i^2-2\bar{\Phi}_i\bar{\Phi}_j+\bar{\Phi}_j^2\right)-\left(\bar{T}_i\bar{\Phi}_i+\bar{T}_j\bar{\Phi}_j\right).$$

For the total potential energy to be minimum, its partial derivative with respect to the angles of twist is zero, i.e.

From $\dfrac{\partial E_T}{\partial\bar{\Phi}_i}=0$, we get $\dfrac{GJ}{2L}\left(2\bar{\Phi}_i-2\bar{\Phi}_j\right)-\bar{T}_i=0$, therefore

$$\frac{GJ}{L}\left(\bar{\Phi}_i-\bar{\Phi}_j\right)=\bar{T}_i. \qquad \text{(A3.3)}$$

From $\dfrac{\partial E_T}{\partial \overline{\Phi}_j} = 0$, we get $\dfrac{GJ}{2L}\left(-2\overline{\Phi}_i + 2\overline{\Phi}_j\right) - \overline{T}_j = 0$, therefore

$$\frac{GJ}{L}\left(-\overline{\Phi}_i + \overline{\Phi}_j\right) = \overline{T}_j. \tag{A3.4}$$

Equations (A3.3) and (A3.4) are written in matrix form as

$$\begin{bmatrix} \dfrac{GJ}{L} & -\dfrac{GJ}{L} \\ -\dfrac{GJ}{L} & \dfrac{GJ}{L} \end{bmatrix} \begin{bmatrix} \overline{\Phi}_i \\ \overline{\Phi}_j \end{bmatrix} = \begin{bmatrix} \overline{T}_i \\ \overline{T}_j \end{bmatrix}. \tag{A3.5}$$

Appendix 4

Strut Stiffness Matrix

A4.1 Stability of Struts

Consider the strut shown in Fig. A4.1a which is pinned at its ends A and C and subjected to an axial compressive force $\bar{P}$. The strut is initially straight but when $\bar{P}$ is applied such that the strut buckles, bending will develop, and the strut will take the shape of a curve. The length of the neutral axis which is equal to the initial length of the straight strut is unchanged. Therefore, the pin at end A will have to move by an amount Δ and take the position B to accommodate the buckled curved shape.

The magnitude of Δ which is also the distance moved by $\bar{P}$ will be required in the formulation of the energy equation and is determined as follows:

Consider the infinitesimal length $d\bar{s}$ of the buckled curve shown in Fig. A4.1b. As $d\bar{x}$ gets smaller and smaller the length of the straight line (ac) which connects the ends of the curve will approach the curved length $d\bar{s}$, thus

$$(d\bar{s})^2 = (d\bar{x})^2 + (d\bar{w})^2$$

$$d\bar{s} = d\bar{x}\left[1 + \left(\frac{d\bar{w}}{d\bar{x}}\right)^2\right]^{\frac{1}{2}}$$

Using the Taylor-Maclaurin infinite series expansion of the quantity inside the square brackets we get

$$d\bar{s} = d\bar{x}\left[1 + \frac{1}{2}\left(\frac{d\bar{w}}{d\bar{x}}\right)^2 - \frac{1}{8}\left(\frac{d\bar{w}}{d\bar{x}}\right)^4 + \frac{1}{16}\left(\frac{d\bar{w}}{d\bar{x}}\right)^6 \cdots\cdots\cdots\right]$$

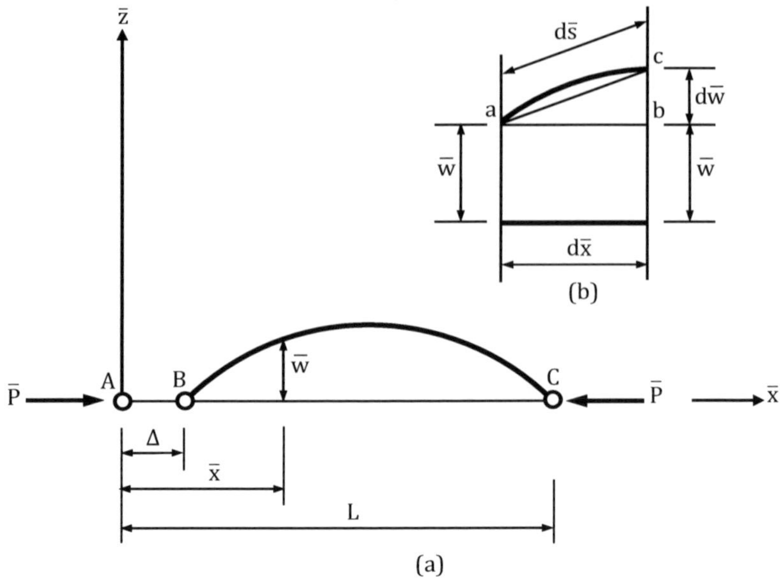

Figure A4.1 Pin ended strut subjected to an axial force $\bar{P}$.

The deflection $\bar{w}$ is small and so the slope $d\bar{w}/d\bar{x}$ of the deflection curve is small such that its high powers are neglected and only the first two terms of the above series are considered.

$$d\bar{s} = d\bar{x} + \frac{1}{2}\left(\frac{d\bar{w}}{d\bar{x}}\right)^2 d\bar{x}$$

The total length of the curve, S, can be found by integration as:

$$S = \int_{\Delta}^{L} d\bar{s} = \int_{\Delta}^{L}\left[d\bar{x} + \frac{1}{2}\left(\frac{d\bar{w}}{d\bar{x}}\right)^2 d\bar{x}\right] = \int_{\Delta}^{L} d\bar{x} + \int_{\Delta}^{L}\frac{1}{2}\left(\frac{d\bar{w}}{d\bar{x}}\right)^2 d\bar{x}$$

$$S = (L-\Delta) + \int_{\Delta}^{L}\frac{1}{2}\left(\frac{d\bar{w}}{d\bar{x}}\right)^2 d\bar{x}$$

Since the strain along the centroidal (neutral) axis is zero the length of the curved strut is the same as the original straight length, i.e. S = L, hence

$$L = (L-\Delta) + \int_{\Delta}^{L}\frac{1}{2}\left(\frac{d\bar{w}}{d\bar{x}}\right)^2 d\bar{x}$$

$$\Delta = \int_{\Delta}^{L} \frac{1}{2}\left(\frac{d\bar{w}}{d\bar{x}}\right)^2 d\bar{x} = \int_{0}^{L} \frac{1}{2}\left(\frac{d\bar{w}}{d\bar{x}}\right)^2 d\bar{x} - \int_{0}^{\Delta} \frac{1}{2}\left(\frac{d\bar{w}}{d\bar{x}}\right)^2 d\bar{x}$$

But $\bar{w}$ and $d\bar{w}/d\bar{x}$ are zero in the range of $\bar{x}=0$ to $\bar{x}=\Delta$, therefore

$$\int_{0}^{\Delta} \frac{1}{2}\left(\frac{d\bar{w}}{d\bar{x}}\right)^2 d\bar{x} = 0, \text{ hence}$$

$$\Delta = \int_{0}^{L} \frac{1}{2}\left(\frac{d\bar{w}}{d\bar{x}}\right)^2 d\bar{x}. \tag{A4.1}$$

A4.2 Stability of Beam-Columns

Now consider the general case of a strut where there are lateral forces and moments acting at the ends in addition to the axial compressive force $\bar{P}$, the socalled beam-column as shown in Fig. A4.2.

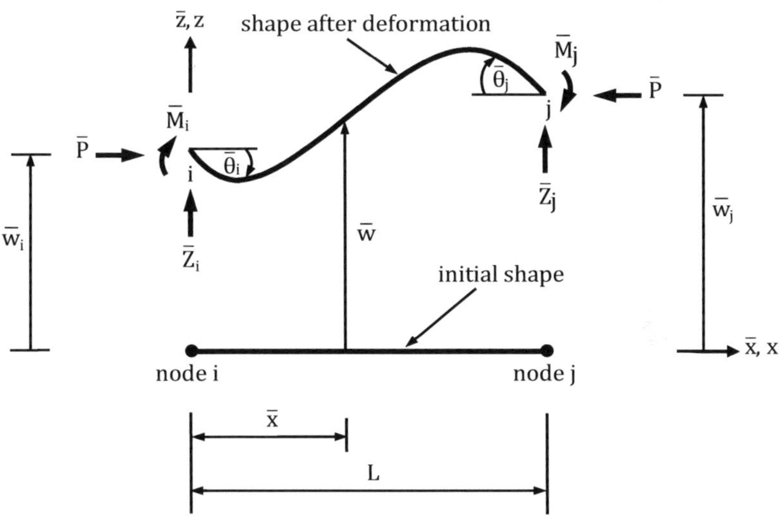

Figure A4.2 Beam-column element.

Assume that the beam-column ends i and j are subjected to forces $\bar{Z}_i$ and $\bar{Z}_j$ and moments $\bar{M}_i$ and $\bar{M}_j$, respectively. In addition, an axial compressive force of magnitude $\bar{P}$ has developed in the member. The presence of $\bar{P}$ will cause an increase in the displacements $\bar{w}$ and $\bar{\theta}$ reflecting a reduction in the bending stiffness of the member. For small values of the axial force this reduction is small and can be neglected. But as the load increases the reduction in bending stiffness becomes significant and has to be taken into account. Furthermore, at a critical value of the axial compressive force, the bending stiffness is reduced to zero causing the strut to become unstable. So, the behaviour of the member is essentially nonlinear since the bending stiffness is progressively reduced as the axial force increases.

The gain in strain energy due to bending is

$$E_S = \int_0^L \frac{\bar{M}^2}{2EI_{\bar{y}}} d\bar{x}, \text{ but } \bar{M} = -EI_{\bar{y}} \frac{d^2\bar{w}}{d\bar{x}^2}, \text{ hence}$$

$$E_S = \frac{EI_{\bar{y}}}{2} \int_0^L \left(\frac{d^2\bar{w}}{d\bar{x}^2} \right)^2 d\bar{x}.$$

The work done by the actions at the ends of the element is

$$E_w = -(\bar{P}\Delta + \bar{Z}_i\bar{w}_i + \bar{M}_i\bar{\theta}_i + \bar{Z}_j\bar{w}_j + \bar{M}_i\bar{\theta}_i), \text{ with } \Delta \text{ from (A4.1)}$$

(the minus sign because it is a loss in potential energy).

The total potential energy, $E_T = E_S + E_W$

$$E_T = +\frac{EI_{\bar{y}}}{2} \int_0^L \left(\frac{d^2\bar{w}}{d\bar{x}^2} \right)^2 d\bar{x} - \left[\bar{P} \int_0^L \frac{1}{2} \left(\frac{d\bar{w}}{d\bar{x}} \right)^2 d\bar{x} + \bar{Z}_i\bar{w}_i + \bar{M}_i\bar{\theta}_i + \bar{Z}_j\bar{w}_j + \bar{M}_i\bar{\theta}_i \right] \tag{A4.2}$$

The deflection equation is derived in Appendix 3 and given by the interpolation polynomial (A3.7) as

$$\bar{w} = \left(1 - \frac{3\bar{x}^2}{L^2} + \frac{2\bar{x}^3}{L^3} \right)\bar{w}_i + \left(-\bar{x} + \frac{2\bar{x}^2}{L} - \frac{\bar{x}^3}{L^2} \right)\bar{\theta}_i + \left(\frac{3\bar{x}^2}{L^2} - \frac{2\bar{x}^3}{L^3} \right)\bar{w}_j + \left(\frac{\bar{x}^2}{L} - \frac{\bar{x}^3}{L^2} \right)\bar{\theta}_j$$

Substituting $d\bar{w}/d\bar{x}$ and $d^2\bar{w}/d\bar{x}^2$, as found from the above equation, into (A4.2) and integrating to get an expression for the total potential energy, E_T.

For the total potential energy, E_T, to be minimum requires that

$$\frac{\partial E_T}{\partial \bar{w}_i}=0,\quad \frac{\partial E_T}{\partial \bar{\theta}_i}=0,\quad \frac{\partial E_T}{\partial \bar{w}_j}=0,\quad \text{and}\quad \frac{\partial E_T}{\partial \bar{\theta}_j}=0.$$

The above four conditions lead to the following relationships

$$\bar{Z}_i=\frac{EI_{\bar{y}}}{L^3}\left(12\bar{w}_i-6L\bar{\theta}_i-12\bar{w}_j-6L\bar{\theta}_j\right)-\bar{P}\left(\frac{6}{5L}\bar{w}_i-\frac{1}{10}\bar{\theta}_i-\frac{6}{5L}\bar{w}_j-\frac{1}{10}\bar{\theta}_j\right)$$

$$\bar{M}_i=\frac{EI_{\bar{y}}}{L^3}\left(-6L\bar{w}_i+4L^2\bar{\theta}_i+6L\bar{w}_j+2L^2\bar{\theta}_j\right)-\bar{P}\left(-\frac{1}{10}\bar{w}_i+\frac{2L}{15}\bar{\theta}_i+\frac{1}{10}\bar{w}_j-\frac{L}{30}\bar{\theta}_j\right)$$

$$\bar{Z}_j=\frac{EI_{\bar{y}}}{L^3}\left(-12\bar{w}_i+6L\bar{\theta}_i+12\bar{w}_j+6L\bar{\theta}_j\right)-\bar{P}\left(-\frac{6}{5L}\bar{w}_i+\frac{1}{10}\bar{\theta}_i+\frac{6}{5L}\bar{w}_j+\frac{1}{10}\bar{\theta}_j\right)$$

$$\bar{M}_j=\frac{EI_{\bar{y}}}{L^3}\left(-6L\bar{w}_i+2L^2\bar{\theta}_i+6L\bar{w}_j+4L^2\bar{\theta}_j\right)-\bar{P}\left(-\frac{1}{10}\bar{w}_i-\frac{L}{30}\bar{\theta}_i+\frac{1}{10}\bar{w}_j+\frac{2L}{15}\bar{\theta}_j\right)$$

The above four relationships are written in matrix form as

$$\left[EI_{\bar{y}}\begin{bmatrix}\frac{12}{L^3} & -\frac{6}{L^2} & -\frac{12}{L^3} & -\frac{6}{L^2}\\ -\frac{6}{L^2} & \frac{4}{L} & \frac{6}{L^2} & \frac{2}{L}\\ -\frac{12}{L^3} & \frac{6}{L^2} & \frac{12}{L^3} & \frac{6}{L^2}\\ -\frac{6}{L^2} & \frac{2}{L} & \frac{6}{L^2} & \frac{4}{L}\end{bmatrix}-\bar{P}\begin{bmatrix}\frac{6}{5L} & -\frac{1}{10} & -\frac{6}{5L} & -\frac{1}{10}\\ -\frac{1}{10} & \frac{2L}{15} & \frac{1}{10} & -\frac{L}{30}\\ -\frac{6}{5L} & \frac{1}{10} & \frac{6}{5L} & \frac{1}{10}\\ -\frac{1}{10} & -\frac{L}{30} & \frac{1}{10} & \frac{2L}{15}\end{bmatrix}\right]\begin{bmatrix}\bar{w}_i\\ \bar{\theta}_i\\ \bar{w}_j\\ \bar{\theta}_j\end{bmatrix}=\begin{bmatrix}\bar{Z}_i\\ \bar{M}_i\\ \bar{Z}_j\\ \bar{M}_j\end{bmatrix} \tag{A4.3}$$

A4.3 Stability of Frames

In the analysis of rigid frames when axial strains are considered, an additional relationship is included to take into account that effect as given by the following matrix which is given by Eq. (2.4) in Chapter 2

$$\begin{bmatrix}\frac{EA}{L} & -\frac{EA}{L}\\ -\frac{EA}{L} & \frac{EA}{L}\end{bmatrix}\begin{bmatrix}\bar{u}_i\\ \bar{u}_j\end{bmatrix}=\begin{bmatrix}\bar{X}_i\\ \bar{X}_j\end{bmatrix} \tag{A4.4}$$

The matrices given by (A4.3) and (A4.4) are combined to give the general stiffness matrix for the nonlinear (second order) analysis of rigid frames as

$$\left[\begin{bmatrix} \frac{EA}{L} & 0 & 0 & -\frac{EA}{L} & 0 & 0 \\ 0 & \frac{12EI_{\bar{y}}}{L^3} & -\frac{6EI_{\bar{y}}}{L^2} & 0 & -\frac{12EI_{\bar{y}}}{L^3} & -\frac{6EI_{\bar{y}}}{L^2} \\ 0 & -\frac{6EI_{\bar{y}}}{L^2} & \frac{4EI_{\bar{y}}}{L} & 0 & \frac{6EI_{\bar{y}}}{L^2} & \frac{2EI_{\bar{y}}}{L} \\ -\frac{EA}{L} & 0 & 0 & \frac{EA}{L} & 0 & 0 \\ 0 & -\frac{12EI_{\bar{y}}}{L^3} & \frac{6EI_{\bar{y}}}{L^2} & 0 & \frac{12EI_{\bar{y}}}{L^3} & \frac{6EI_{\bar{y}}}{L^2} \\ 0 & -\frac{6EI_{\bar{y}}}{L^2} & \frac{2EI_{\bar{y}}}{L} & 0 & \frac{6EI_{\bar{y}}}{L^2} & \frac{4EI_{\bar{y}}}{L} \end{bmatrix} - \bar{P}\begin{bmatrix} 0 & 0 & 0 & 0 & 0 & 0 \\ 0 & \frac{6}{5L} & -\frac{1}{10} & 0 & -\frac{6}{5L} & -\frac{1}{10} \\ 0 & -\frac{1}{10} & \frac{2L}{15} & 0 & \frac{1}{10} & -\frac{L}{30} \\ 0 & 0 & 0 & 0 & 0 & 0 \\ 0 & -\frac{6}{5L} & \frac{1}{10} & 0 & \frac{6}{5L} & \frac{1}{10} \\ 0 & -\frac{1}{10} & -\frac{L}{30} & 0 & \frac{1}{10} & \frac{2L}{15} \end{bmatrix}\right] \begin{bmatrix} \bar{u}_i \\ \bar{w}_i \\ \bar{\theta}_i \\ \bar{u}_j \\ \bar{w}_j \\ \bar{\theta}_j \end{bmatrix} = \begin{bmatrix} \bar{X}_i \\ \bar{Z}_i \\ \bar{M}_i \\ \bar{X}_j \\ \bar{Z}_j \\ \bar{M}_j \end{bmatrix} \qquad \text{(A4.5)}$$

For members whose local $\bar{x}$ -axis does not lie along the global x-axis then matrix transformation will be used to convert the stiffness matrix from local to global coordinates.

Appendix 5

Fixed End Moments and Forces

Consider a beam member (or element) which is fixed at its ends and subjected to an arbitrary loading as shown in Fig. A5.1.

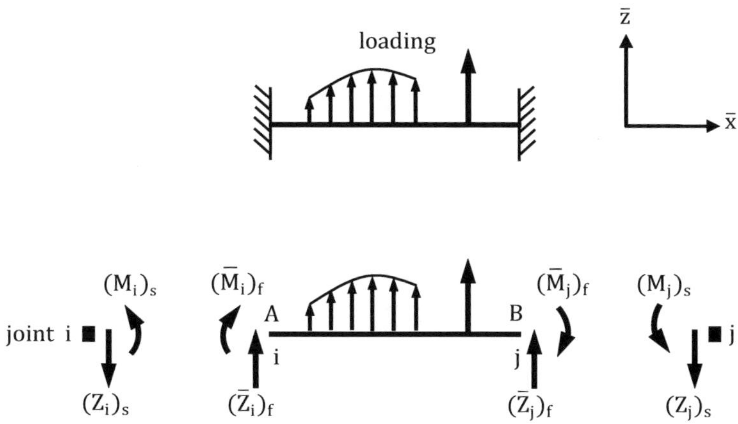

Figure A5.1

The forces and moments acting on the beam at its ends will be called actions on the beam with the subscript 'f' and are shown in the positive direction. From equilibrium at the joints the forces and moments acting on the joints are in the opposite direction to those acting at the ends of the beam and will be called loads on the joint with the subscript 's'. Four standard cases of loading are explained below. Other common forms of loading may be dealt with by superposition of the appropriate standard cases.

A5.1 Case 1: One Lateral Concentrated Load at Any Point along the Span of the Beam

(i) Both Ends of the Beam Are Fixed

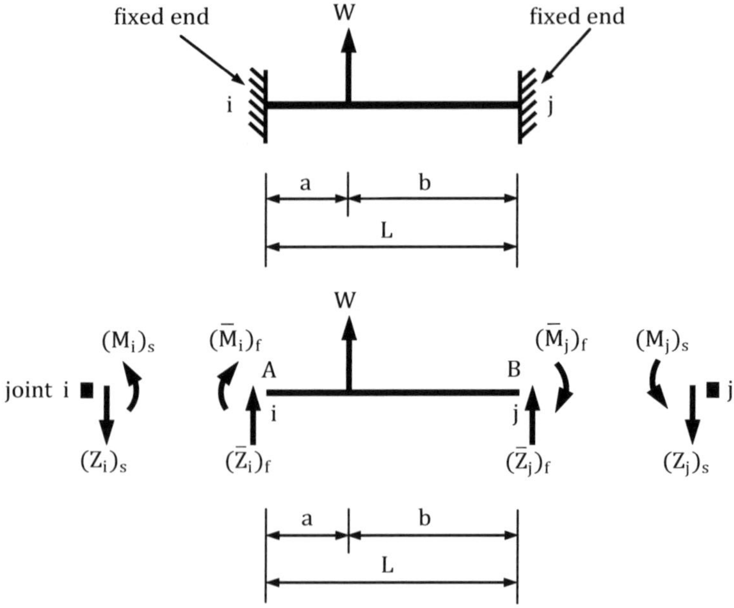

Figure A5.2

Actions on the beam:

$$(\bar{Z}_i)_f = -\frac{Wb}{L^3}(L^2 + ab - a^2), \quad (\bar{M}_i)_f = +\frac{Wab^2}{L^2}$$

$$(\bar{Z}_j)_f = -\frac{Wa}{L^3}(L^2 + ab - b^2), \quad (\bar{M}_j)_f = -\frac{Wa^2b}{L^2}$$

In matrix form, the action vector $\bar{F}_f$ on the beam is:

$$\bar{F}_f = \begin{bmatrix} (\bar{Z}_i)_f \\ (\bar{M}_i)_f \\ (\bar{Z}_j)_f \\ (\bar{M}_j)_f \end{bmatrix} = \begin{bmatrix} -\dfrac{Wb}{L^3}\left(L^2 + ab - a^2\right) \\ +\dfrac{Wab^2}{L^2} \\ -\dfrac{Wa}{L^3}\left(L^2 + ab - b^2\right) \\ -\dfrac{Wa^2b}{L^2} \end{bmatrix}$$

From equilibrium at the joints $\bar{F}_s = -\bar{F}_f$, where $\bar{F}_s$ is the load vector on the joints and is obtained simply by reversing the sign of the action vector. Thus:

$$\bar{F}_s = \begin{bmatrix} (\bar{Z}_i)_s \\ (\bar{M}_i)_s \\ (\bar{Z}_j)_s \\ (\bar{M}_j)_s \end{bmatrix} = \begin{bmatrix} +\dfrac{Wb}{L^3}\left(L^2 + ab - a^2\right) \\ -\dfrac{Wab^2}{L^2} \\ +\dfrac{Wa}{L^3}\left(L^2 + ab - b^2\right) \\ +\dfrac{Wa^2b}{L^2} \end{bmatrix}$$

For the special case of a beam with a concentrated load at mid-span, substitute a = b = L/2 to get

$$(\bar{Z}_i)_f = -\frac{W}{2},\ (\bar{M}_i)_f = +\frac{WL}{8},\ (\bar{Z}_j)_f = -\frac{W}{2},\ (\bar{M}_j)_f = -\frac{WL}{8}$$

(ii) One End of the Beam Is Fixed and the Other Is Pinned

Actions vector on the beam is

$$\bar{F}_f = \begin{bmatrix} (\bar{Z}_i)_f \\ (\bar{M}_i)_f \\ (\bar{Z}_j)_f \\ (\bar{M}_j)_f \end{bmatrix} = \begin{bmatrix} -\dfrac{Wb}{2L^3}(3L^2 - b^2) \\ +\dfrac{Wab}{2L^2}(L + b) \\ -\dfrac{Wa}{2L^3}(3La - a^2) \\ 0 \end{bmatrix}$$

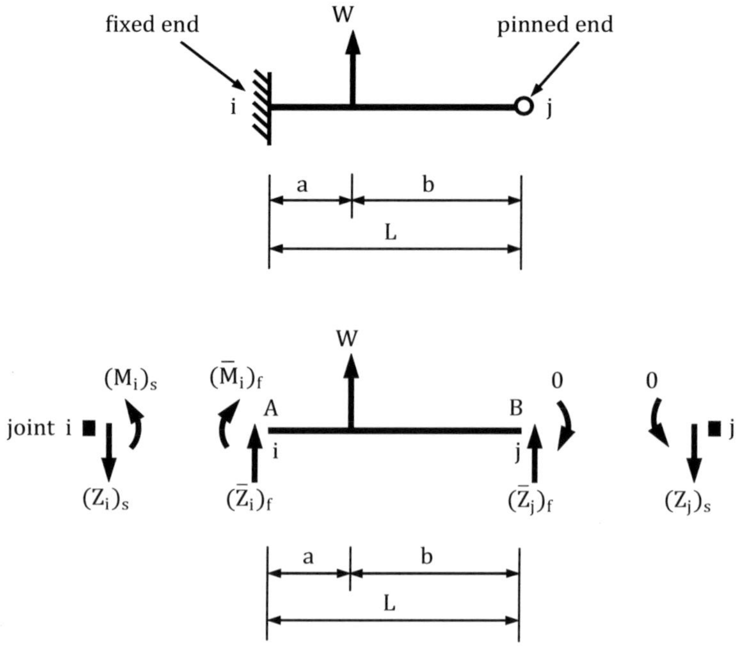

Figure A5.3

Load vector on the joint: $\bar{F}_s = -\bar{F}_f$

$$\bar{F}_s = \begin{bmatrix} (\bar{Z}_i)_s \\ (\bar{M}_i)_s \\ (\bar{Z}_j)_s \\ (\bar{M}_j)_s \end{bmatrix} = \begin{bmatrix} +\dfrac{Wb}{2L^3}(3L^2 - b^2) \\ -\dfrac{Wab}{2L^2}(L+b) \\ +\dfrac{Wa}{2L^3}(3La - a^2) \\ 0 \end{bmatrix}$$

Special case when the load is at mid-span, substitute a = b = L/2 to get

$$(\bar{Z}_i)_f = -\frac{11W}{16},\ (\bar{M}_i)_f = +\frac{3WL}{16},\ (\bar{Z}_j)_f = -\frac{5W}{16},\ (\bar{M}_j)_f = 0$$

A5.2 Case 2: Uniformly Distributed Lateral Load Covering Part of the Span of the Beam

(i) Both Ends of the Beam Are Fixed

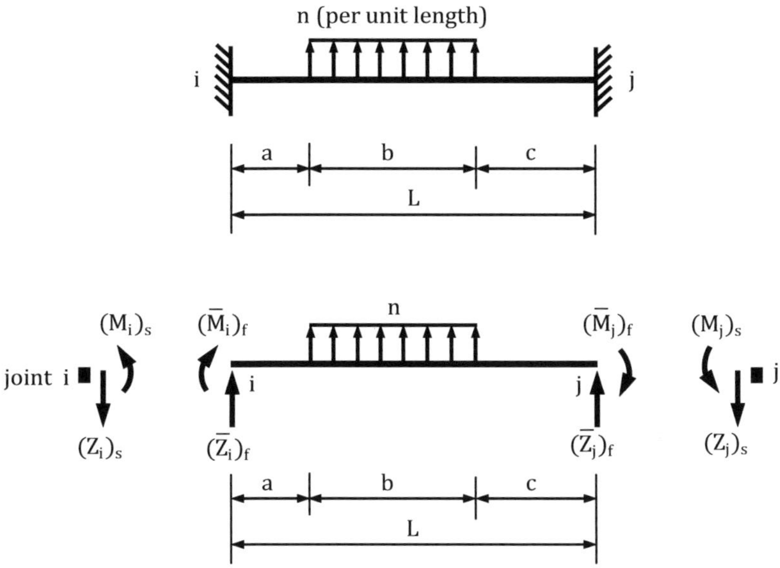

Figure A5.4

Action vector on the beam:

$$\bar{F}_f = \begin{bmatrix} (\bar{Z}_i)_f \\ (\bar{M}_i) \\ (\bar{Z}_j)_f \\ (\bar{M}_j)_f \end{bmatrix} = \begin{bmatrix} -\dfrac{n}{2L^3}\left[L^2(-a^2+b^2+c^2+2bc)+2L(a^3-c^3)-a^4+c^4\right] \\ +\dfrac{n}{12L^2}\left[(L-a)^3(L+3a)-c^3(4L-3c)\right] \\ -\dfrac{n}{2L^3}\left[L^2(a^2+b^2-c^2+2ab)-2L(a^3-c^3)+a^4-c^4\right] \\ -\dfrac{n}{12L^2}\left[(L-c)^3(L+3c)-a^3(4L-3a)\right] \end{bmatrix}$$

Load vector on the joint:

$$\bar{F}_s = \begin{bmatrix} (\bar{Z}_i)_s \\ (\bar{M}_i)_s \\ (\bar{Z}_j)_s \\ (\bar{M}_j)_s \end{bmatrix} = -\bar{F}_s = \begin{bmatrix} +\dfrac{n}{2L^3}\left[L^2(-a^2+b^2+c^2+2bc)+2L(a^3-c^3)-a^4+c^4\right] \\ -\dfrac{n}{12L^2}\left[(L-a)^3(L+3a)-c^3(4L-3c)\right] \\ +\dfrac{n}{2L^3}\left[L^2(a^2+b^2-c^2+2ab)-2L(a^3-c^3)+a^4-c^4\right] \\ +\dfrac{n}{12L^2}\left[(L-c)^3(L+3c)-a^3(4L-3a)\right] \end{bmatrix}$$

For the special case of a uniformly distributed load covering the whole length of the beam substitute a = 0, c = 0, and b = L to get

$$(\bar{Z}_i)_f = -\frac{nL}{2},\ (\bar{M}_i)_f = +\frac{nL^2}{12},\ (\bar{Z}_j)_f = -\frac{nL}{2},\ (\bar{M}_j)_f = -\frac{nL^2}{12}$$

Any combination of concentrated load(s) and uniformly distributed load(s) can be dealt with by superposition of the standard cases discussed in the previous examples.

(ii) One End of the Beam Is Fixed and the Other Is Pinned

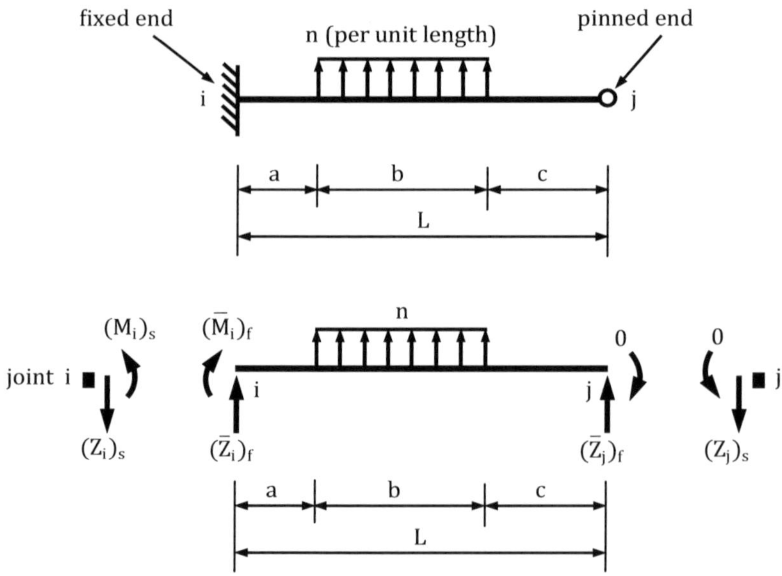

Figure A5.5

Action vector on the beam:

$$\bar{F}_f = \begin{bmatrix} (\bar{Z}_i)_f \\ (\bar{M}_i)_f \\ (\bar{Z}_j)_f \\ (\bar{M}_j)_f \end{bmatrix} = \begin{bmatrix} -\frac{nb}{8L^3}(b+2c)(6L^2-b^2-2c^2-2bc) \\ +\frac{n}{8L^2}(2L^2-b^2-2c^2-2bc)(b^2+2bc) \\ -\frac{nb}{8L^3}[L^2(8a+2b-4c)+bc(4b+6c)+(b^3+4c^3)] \\ 0 \end{bmatrix}$$

Load vector on the joint:

$$\bar{F}_s = \begin{bmatrix} (\bar{Z}_i)_s \\ (\bar{M}_i)_s \\ (\bar{Z}_j)_s \\ (\bar{M}_j)_s \end{bmatrix} = \begin{bmatrix} +\frac{nb}{8L^3}(b+2c)(6L^2-b^2-2c^2-2bc) \\ -\frac{n}{8L^2}(2L^2-b^2-2c^2-2bc)(b^2+2bc) \\ +\frac{nb}{8L^3}[L^2(8a+2b-4c)+bc(4b+6c)+(b^3+4c^3)] \\ 0 \end{bmatrix}$$

Special case when the uniformly distributed load covers the whole span substitute a = 0, c = 0, and b = L to get

$$(\bar{Z}_i)_f = -\frac{5nL}{8}, (\bar{M}_i)_f = +\frac{nL^2}{8}, (\bar{Z}_j)_f = -\frac{3nL}{8}, (\bar{M}_j)_f = 0$$

A5.3 Case 3: One Longitudinal Concentrated Load at Any Point Along the Span of the Beam

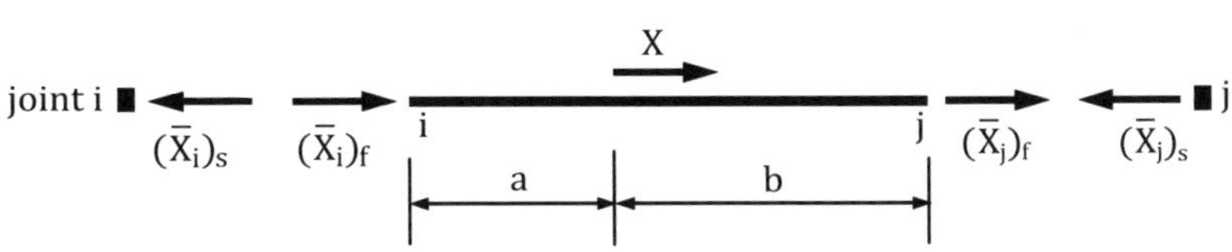

Figure A5.6

$$(\bar{X}_i)_f = -\frac{Xb}{L} \text{ and } (\bar{X}_j)_f = -\frac{Xa}{L}$$

$$\begin{bmatrix} (\bar{X}_i)_f \\ (\bar{X}_j)_f \end{bmatrix} = \begin{bmatrix} -\dfrac{Xb}{L} \\ -\dfrac{Xa}{L} \end{bmatrix}$$

From equilibrium at the joints, i.e. $(\bar{X}_i)_s = -(\bar{X}_i)_f$ and $(\bar{X}_j)_s = -(\bar{X}_j)_f$

$$\begin{bmatrix} (\bar{X}_i)_s \\ (\bar{X}_j)_s \end{bmatrix} = \begin{bmatrix} +\dfrac{Xb}{L} \\ +\dfrac{Xa}{L} \end{bmatrix}$$

For the special case of the load at mid-span, i.e. a = b = L/2

$$(\bar{X}_i)_f = -\frac{X}{2} \text{ and } (\bar{X}_j)_f = -\frac{X}{2}$$

A5.4 Case 4: Uniformly Distributed Longitudinal Load Covering Part of the Span of the Beam

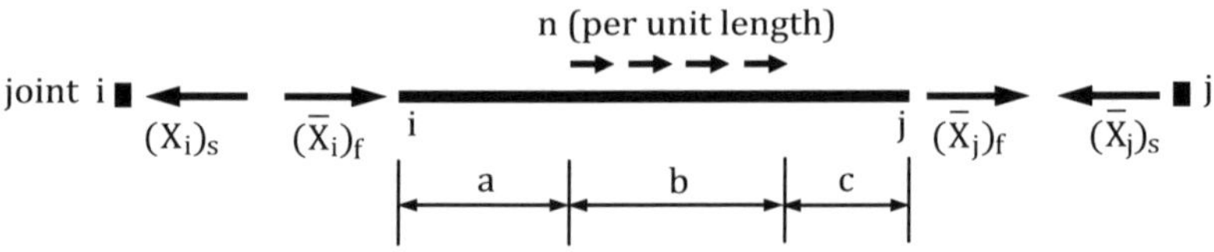

Figure A5.7

$$\begin{bmatrix} (\bar{X}_i)_f \\ (\bar{X}_j)_f \end{bmatrix} = \begin{bmatrix} -\dfrac{nb}{2L}(b+2c) \\ -\dfrac{nb}{2L}(b+2a) \end{bmatrix} \text{ and } \begin{bmatrix} (\bar{X}_i)_s \\ (\bar{X}_j)_s \end{bmatrix} = \begin{bmatrix} +\dfrac{nb}{2L}(b+2c) \\ +\dfrac{nb}{2L}(b+2a) \end{bmatrix}$$

For the special case when the distributed load covers the whole span, i.e. a = 0, c = 0, and b = L then

$$\begin{bmatrix} (\bar{X}_i)_f \\ (\bar{X}_j)_f \end{bmatrix} = \begin{bmatrix} -\dfrac{nL}{2} \\ -\dfrac{nL}{2} \end{bmatrix} \quad \text{and} \quad \begin{bmatrix} (\bar{X}_i)_s \\ (\bar{X}_j)_s \end{bmatrix} = \begin{bmatrix} +\dfrac{nL}{2} \\ +\dfrac{nL}{2} \end{bmatrix}$$

When the beam is subjected to more than one type of loading then superposition of the actions due to the various loads can be made.

A5.5 Transformation of Member End Actions to Joint Loads

For a member lying along the global x-axis the load vector on the joints, which is written relative to global coordinates, is obtained from the action vector on the member simply by reversing its sign as explained earlier. But for inclined members, for example in rigidly connected frames, the load vector on the joint is obtained by transformation of the action vector as explained below. Consider the inclined member shown in Fig. A6.8 with the actions $(\bar{X}_i)_f$, $(\bar{Z}_i)_f$, and $(\bar{M}_i)_f$ at its end i and $(\bar{X}_j)_f$, $(\bar{Z}_j)_f$, and $(\bar{M}_j)_f$ at end j as calculated relative to local coordinates for a given loading from the cases derived above.

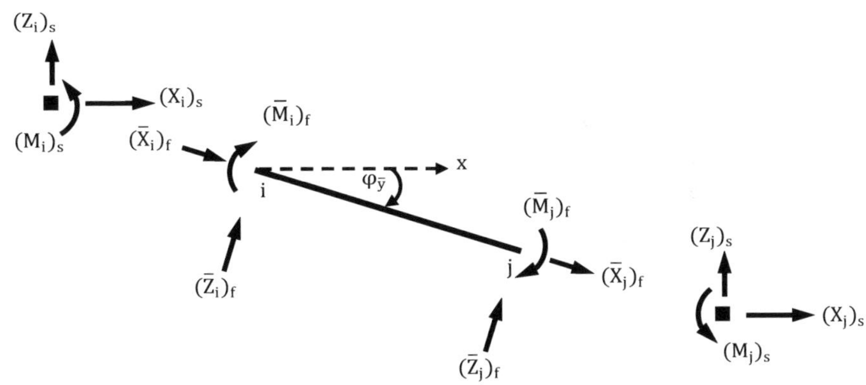

Figure A5.8

The loads on the joints are equal in magnitude and opposite in direction to the actions on the member and are written relative to global coordinates. Therefore, the actions on the member, which are relative to the local coordinates, are first written relative to the global coordinates as $(X_i)_f$, $(Z_i)_f$, $(M_i)_f$, $(X_j)_f$, $(Z_j)_f$, and $(M_j)_f$ and then their sign is reversed to get the loads on the joints as shown below.

$$(X_i)_s = -(X_i)_f = -(+(\bar{X}_i)_f \cos\varphi_{\bar{y}} + (\bar{Z}_i)_f \sin\varphi_{\bar{y}})$$

$$(Z_i)_s = -(Z_i)_f = -(-(\bar{X}_i)_f \sin\varphi_{\bar{y}} + (\bar{Z}_i)_f \cos\varphi_{\bar{y}})$$

$$(M_i)_s = -(M_i)_f = -(\bar{M}_i)_f$$

$$(X_j)_s = -(X_j)_f = -(+(\bar{X}_j)_f \cos\varphi_{\bar{y}} + (\bar{Z}_j)_f \sin\varphi_{\bar{y}})$$

$$(Z_j)_s = -(Z_j)_f = -(-(\bar{X}_j)_f \sin\varphi_{\bar{y}} + (\bar{Z}_j)_f \cos\varphi_{\bar{y}})$$

$$(M_j)_s = -(M_j)_f = -(\bar{M}_j)_f$$

The above equations may be written in matrix form as

$$\begin{bmatrix} (X_i)_s \\ (Z_i)_s \\ (M_i)_s \\ (X_j)_s \\ (Z_j)_s \\ (M_j)_s \end{bmatrix} = \begin{bmatrix} -\cos\varphi_{\bar{y}} & -\sin\varphi_{\bar{y}} & 0 & 0 & 0 & 0 \\ +\sin\varphi_{\bar{y}} & -\cos\varphi_{\bar{y}} & 0 & 0 & 0 & 0 \\ 0 & 0 & -1 & 0 & 0 & 0 \\ 0 & 0 & 0 & -\cos\varphi_{\bar{y}} & -\sin\varphi_{\bar{y}} & 0 \\ 0 & 0 & 0 & +\sin\varphi_{\bar{y}} & -\cos\varphi_{\bar{y}} & 0 \\ 0 & 0 & 0 & 0 & 0 & -1 \end{bmatrix} \begin{bmatrix} (\bar{X}_i)_f \\ (\bar{Z}_i)_f \\ (\bar{M}_i)_f \\ (\bar{X}_j)_f \\ (\bar{Z}_j)_f \\ (\bar{M}_j)_f \end{bmatrix}$$

or $$F_s = r_s \bar{F}_f$$

where r_s is a transformation matrix to transform the local actions on the member to global loads on the joints. Notice that $r_s = -r^T$, where r is the coordinates transformation matrix derived in Chapter 5. This relationship could have been derived directly from $\bar{F}_f = rF$, i.e. $F = r^{-1}\bar{F}_f$ and since $r^{-1} = r^T$ and $F_s = -F$, therefore, $F_s = -r^T \bar{F}_f = r_s \bar{F}_f$.

Bibliography

Bhatt, P. (1999) *Structures*. Addison Wesley Longman.

Chen, W. F. and Lui, E. W. (1988) *Structural Stability*. Elsevier.

Clough, R. W. and Penzien, J. (1993) *Dynamics of Structures*. 2nd ed., McGraw-Hill.

Coates, R. C., Coutie, M. G., and Kong, F. K. (1988) *Structural Analysis*. 3rd ed., Van Nostrand Reinhold.

Craig, R. R. (1981) *Structural Dynamics*. John Wiley.

Dawe, D. J. (1984) *Matrix and Finite Element Displacement Analysis of Structures*. Oxford University Press.

Ghali, A. and Neville A. M. (1989) *Structural Analysis*. 3rd ed., Chapman and Hall.

Horn, M. R. and Merchant, W. (1965) *The Stability of Frames*. Pergamon Press.

Kassimali, A. (2012) *Matrix Analysis of Structures*. 2nd ed., Cengage Learning.

Kreyszig, E. (1993) *Advanced Engineering Mathematics*. 7th ed., John Wiley.

Livesley, R. K. (1975) *Matrix Methods of Structural Analysis*. 2nd ed., Pergamon Press.

McGuire, W., Gallagher, R. H., and Ziemian, R. D. (2014) *Matrix Structural Analysis*. 2nd ed., John Wiley.

Timoshenko, S. P. and Gere, J. M. (1961) *Theory of Elastic Stability*. McGraw-Hill.

Williams, M. (2016) *Structural Dynamics*. CRC Press.

Williams, M. S. and Todd, J. D. (2000) *Structures*. Palgrave Macmillan.

Young, W.C. (1989) *Roark's Formulas for Stress & Strain*. 6th ed., McGraw-Hill.

Index